Environment and Natural Resources: Challenges and Solutions

IOP Conference Series: Earth and Environmental Science
Volume 1090

Online
24-25 November 2021

Editors:

Bouachrine Mohammed
Abba El Hassan
Khaffou Mhamed
Nait M'Barek Hasna

ISBN: 978-1-7138-6436-3
ISSN: 1755-1307

Printed from e-media with permission by:

Curran Associates, Inc.
57 Morehouse Lane
Red Hook, NY 12571

Institute of Physics
Dirac House, Temple Back
Bristol BS1 6BE UK

Phone: 44 1 17 929 7481
Fax: 44 1 17 920 0979

techtracking@iop.org

Additional copies of this publication are available from:

Curran Associates, Inc.
57 Morehouse Lane
Red Hook, NY 12571 USA
Phone: 845-758-0400
Fax: 845-758-2633
Email: curran@proceedings.com
Web: www.proceedings.com

TABLE OF CONTENTS

IOP Conf. Series: Earth and Environmental Science 1090 (2022) 011001 doi:10.1088/1755-1315/1090/1/011001

The First Edition of the International Scopus Congress on:

Environment and Natural Resources: Challenges and Solutions (ENRSC2021)

24-25th, November 2021, Khenifra, Morocco

- Online -

About the ENRSC2021

Anthropogenic activities have progressed considerably over the last century, especially after the 1950s. The growth of the world population, industrialization, new technologies, the extension of agricultural activities at all levels due to the growing needs of the world population, the overexploitation of natural resources in different environments, etc., have been and remain the driving force behind an unbalanced model of development that has a negative impact on the environment and on natural resources. Indeed, the world is currently facing challenges at all scales, from micro- to macro-environmental issues. The serious challenges to be taken up are major concerning all physical environments - air, soil, and water. The degradation of habitats by various anthropogenic and/or natural activities has a direct impact on biodiversity and consequently on socio-economic development. The International Conference on Environment & Natural Resources: Challenges and Solutions allowed scientists from different universities and institutes to exchange and debate different themes on the environment and natural resources, their challenges, and probable solutions to reduce the environmental impact and suggest better natural resources management practices. Finally, the current global context proves to us, every day, that we must contribute each from one's position to the mitigation of negative impacts, and that the urgency is to set up concrete and sustainable-integrated actions.

IOP Conf. Series: Earth and Environmental Science 1090 (2022) 011001 doi:10.1088/1755-1315/1090/1/011001

Congress Chairs

Prof. Bouachrine Mohammed
Director of the Higher School of Technology Khenifra (ESTK), Sultan Moulay Slimane University, Morocco
Member of the « Biotechnology, Bioressources and Bioinformatics » laboratory, Sultan Moulay Slimane University, Morocco
Full Professor and member of the « Molecular Chemistry and Natural Substances » laboratory, Faculty of Sciences of Meknes, Moulay Ismail University, Morocco

Prof. Abba El Hassan
Deputy Director of the Higher School of Technology Khenifra (ESTK), Sultan Moulay Slimane University, Morocco
Member of the « Biotechnology, Bioressources and Bioinformatics » laboratory, Sultan Moulay Slimane University, Morocco
Head of the « Biotechnology, Analytical Sciences and Natural Resources Management » research team

Keynote speakers

Prof. Riedacker Arthur
Co-winner of the Nobel peace prize, 2007
Research director at the National Research Institute for Agriculture, Food and the Environment (INRAE), France
Head of the « Food Security » chair, *Oikos* Scientific and technical NGO for globally sustainable national developments

IOP Conf. Series: Earth and Environmental Science 1090 (2022) 011001 doi:10.1088/1755-1315/1090/1/011001

Prof. Lakhlifi Tahar
Emeritus Professor of Chemistry, Faculty of Sciences of Meknes, Moulay Ismail University, Morocco
Head of the « Molecular Chemistry and Natural Substances » laboratory, Faculty of Sciences of Meknes, Moulay Ismail University, Morocco

Invited speakers

Prof. Al-Aquarbeh Marwa
National Agricultural Research Center NARC, Amman, Jordan

Prof. Mohamed Ahmed
Institute of Medicinal Research, Center for Study and Research of Djibouti, Djibouti

Prof. Bhim Bahadur Ghaley
Department of Plant and Environmental Sciences, University of Copenhagen, Danmark

Prof. Hasnaoui Mohamed
Sultane Moulay Slimane University, Morocco

Prof. Soudani Wafa.
Faculty de Medicine Badji Mokhrar, Annaba University, Algeria

Editorial Committee

Prof. Bouachrine Mohammed
USMS, Morocco

Prof. Abba El Hassan
USMS, Morocco

Prof. Khaffou Mhamed
USMS, Morocco

Dr. Nait M'Barek Hasna
CentraleSupélec, France

Organizing Committee

- **Pr. Abba El Hassan**, ESTK, Sultan Moulay Slimane University, Morocco.
- **Pr. Bouachrine Mohammed**, ESTK, Sultan Moulay Slimane University, Morocco.
- **Pr. Boutsougame Abdelaziz,** ESTK, Sultan Moulay Slimane University, Morocco
- **Pr. Ainane Tarik**, ESTK, Sultan Moulay Slimane University, Morocco.
- **Pr. El Yaacoubi Adnane,** ESTK, Sultan Moulay Slimane University, Morocco.
- **Pr. Hamdali Hanane,** FST. Sultan Moulay Slimane University, Morocco.
- **Pr. Chahlaoui Abdelkader**, FS, Moulay Ismail University of Meknes, Morocco.
- **Pr. Chillasse Lahcen,** FS, Moulay Ismail University of Meknes, Morocco.
- **Pr. Merzouki Mohamed**, FST, Sultan Moulay Slimane University. Morocco.
- **Pr. Khaffou Mhamed**, ESTK, Sultan Moulay Slimane University, Morocco.
- **Pr. Aboulkacem Asmae,** ESTK, Sultan Moulay Slimane University, Morocco.
- **Pr. Ben Aziz Mohamed**, ESTK, Sultan Moulay Slimane University, Morocco.
- **Pr. Cherroud Sanaa**, ESTK, Sultan Moulay Slimane University, Morocco.
- **Pr. Zaki Hanane,** ESTK, Sultan Moulay Slimane University, Morocco.
- **Pr. Abbas Younes**, FP, Sultan Moulay Slimane University, Morocco.
- **Pr. Jedouali Mohammed**, ESTK, Sultan Moulay Slimane University, Morocco.
- **Pr. Driowya Mohsine**, ESTK, Sultan Moulay Slimane University, Morocco.
- **Pr. Sadiki Yassine**, ESTK, Sultan Moulay Slimane University, Morocco.
- **Pr. Amechrouq Ali**, FS, Moulay Ismail University, Morocco.
- **Pr. Chtita Samir**, FS, University Hassan II, Casablanca, Morocco.
- **Pr. Bennani Mohammed**, FS, Moulay Ismail University, Morocco.
- **Pr. Marzouk Abdelkrim**, Al Akhawayn University. Morocco.
- **Dr. Nait M'Barek Hasna**, CentraleSupélec, Paris-Saclay University, France.
- **Dr. Ouizgane Anouar**, FST, Sultane Moulay Slimane University, Morocco.
- **Dr. Hachi Touria**, Faculty of science, Ibn Tofail University Morocco.
- **Dr. Mehanned Smahane**, FS, Moulay Ismail University of Meknes, Morocco.
- **Dr. Belhassan Assia**, FS, Moulay Ismail University of Meknes, Morocco.
- **Dr. El Moustaine Radouane,** FS, Moulay Ismail University of Meknes, Morocco.
- **Phd Student. Boumalkha Oumaima**, FST. Sultane Moulay Slimane University, Morocco.
- **Phd Student. El Hammioui Youssef**, FST. Sultan Moulay Slimane University, Morocco.
- **Phd Student. Essabiri Hamza**, FST. Sultan Moulay Slimane University, Morocco.
- **Phd Student. Idrissi Imame**, FST. Sultane Moulay Slimane University, Morocco,
- **Phd Student. Laiboud Chaimae**, FST, Sultane Moulay Slimane University, Morocco.

- **Phd Student. El Khatabi Khalil**, FS, Moulay Ismail University of Meknes, Morocco.
- **Phd Student. Kacimi Rachid**, FS, Moulay Ismail University of Meknes, Morocco.
- **Phd Student. Abdessaddak Oumaima**, FS, Moulay Ismail University, Morocco.
- **Dr. Doubi Mostapha.** Faculty of science, Ibn Tofail University Morocco.
- **Phd Student. Hajji Halima**, FS, Moulay Ismail University of Meknes, Morocco.
- **Phd Student. Elmchichi Larbi**, FS, Moulay Ismail University of Meknes, Morocco.
- **Phd Student. Sabiri Meriem**, FS, Moulay Ismail University of Meknes, Morocco.
- **Phd Student. Aguennouz Rachid,** FST. Sultan Moulay Slimane University, Morocco.

Scientific Committee

- **Pr. Abba. E.H.,** Sultan Moulay SlimaneUniversity Morocco.
- **Pr. Adey S.A.,** University of Ndjamena. Tchad.
- **Pr. Alaqarbeh M.,** National Agricultural Research Center. Amman Jordan.
- **Pr. Hammouti B..** Mohamed First University. Oujda. Morocco.
- **Pr. Benkadour R.,** Mohamed First University. Oujda. Morocco.
- **Pr. Alimi K.,** University of Monastir, Tunisia.
- **Pr. Bouachrine M.,** Sultan Moulay Sliman University. Morocco.
- **Pr. El Yaacoubi A.,** Sultan Moulay Sliman University. Morocco.
- **Pr. Ainane T**., Sultan Moulay Sliman University. Morocco.
- **Pr. Chahlaoui A**., Moulay Ismail University. Morocco.
- **Pr. Chillasse L.,** Moulay Ismail University. Morocco.
- **Pr. Fekhaoui M.,** Director of Scientific Institute of Rabat. Morocco.
- **Pr. Lakhlifi T**., Moulay Ismail University. Meknès. Morocco.
- **Pr. Jeffali F.,** Mohamed First University. Oujda. Morocco.
- **Pr. Ghaley B. B.,** University of Copenhagen. Danmark.
- **Pr. Hajji L.,** Moulay Ismail University. Morocco.
- **Pr. Hasnaoui M.,** Sultan Moulay Sliman University. Morocco.
- **Pr. Khaffou M.,** Sultan Moulay Slimane University. Morocco.
- **Pr. Laabou Z.,** Faculty of Pharmacy, Mansoura, Delta University, Egypt.
- **Pr. Lei G.,** School of Materials and Chemical Engineering, Tongren. China.
- **Pr. Marzouk A.,** Al Akhawayn University. Morocco.
- **Pr. Merimi I.,** Mohamed First University. Oujda. Morocco.
- **Pr. Mohamed A.,** Center for Study and Research, Djibouti.
- **Pr. Randazzo G.,** Campus Papardo. Università Degli Studi. Messina. Italia.
- **Pr. Serdaroglu G.,** Sivas Cumhuriyet University, Turkey.
- **Pr. Soudani W.,** Badji Mokhrar University. Annaba. Algérie.
- **Pr. Szumiata T.,** University of Technology and Human Sciences. Poland.
- **Pr. Zarrouk. A.** Mohamed V University. Morocco.
- **Pr. Zine N. E.,** Moulay Ismail University. Morocco.
- **Pr. Hamidi M.,** Moulay Ismail University, Morocco.fM
- **Pr. Khalil F.,** Sidi Mohamed Ben Abdellah University, Morocco.
- **Pr. Chokrad M.,** Moulay Ismail University, Morocco.
- **Pr. Messaoudi L.,** Moulay Ismail University, Morocco.
- **Pr. Bouzzine S.M.,** Regional Centre for Education and Training, Errachidia, Morocco
- **Dr. Nait M'Barek H.,** CentraleSupélec, Paris-Saclay University, France

IOP Conf. Series: Earth and Environmental Science 1090 (2022) 011001 doi:10.1088/1755-1315/1090/1/011001

ENRSC2021 photos

Opening ceremony in the presence of the president of Sultan Moulay Slimane University (USMS), the director of the Higher School of Technology of Khenifra (ESTK) and the congress chairmen, November 24th, 2021

IOP Conf. Series: Earth and Environmental Science 1090 (2022) 011001 doi:10.1088/1755-1315/1090/1/011001

« Thawing effect on the growth and histamine formation of Klebsiella aerogenes in broth and artificially contaminated mackerel »
Novalia Rachmawati et al.
Session 1 : Chemistry and Environment, November 24th, 2021

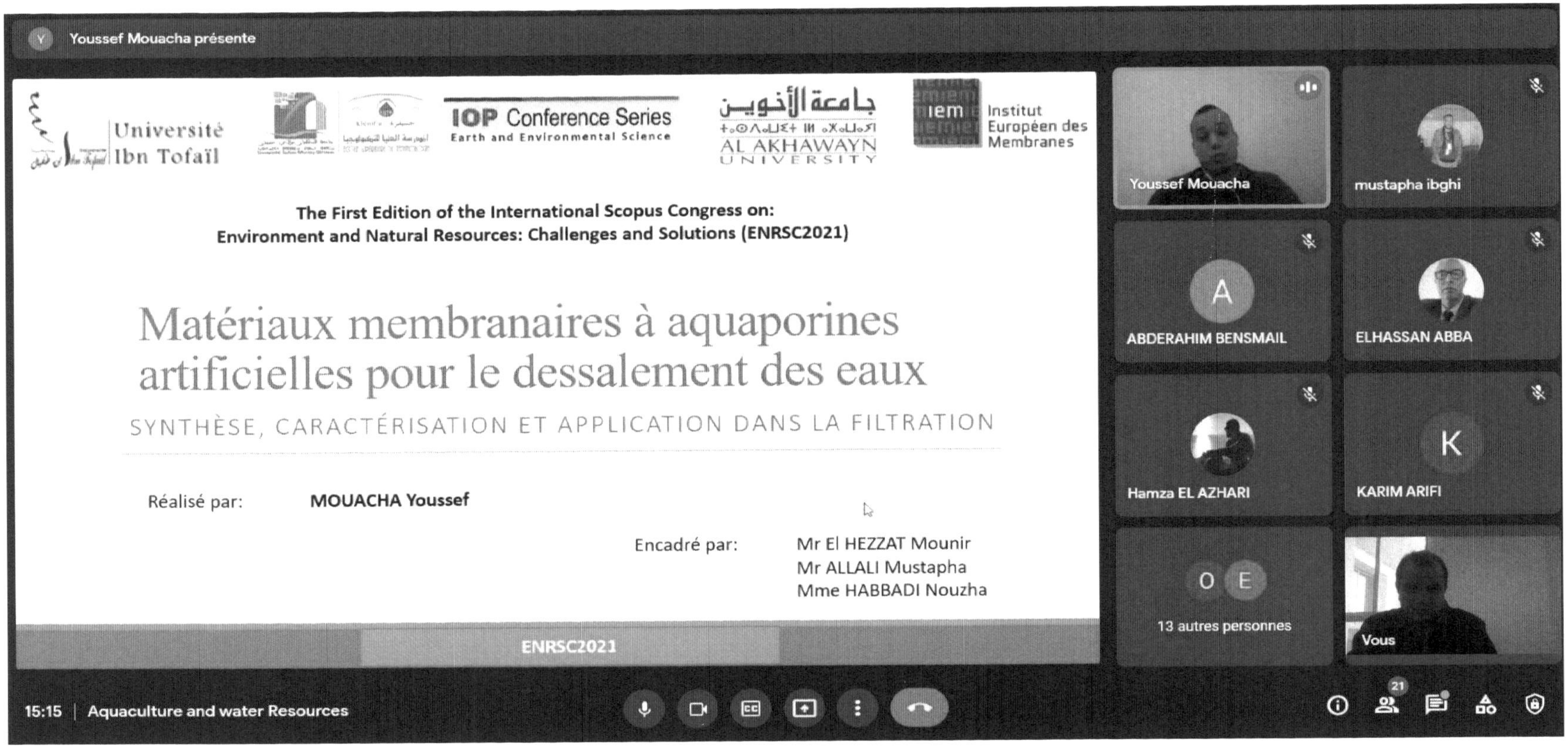

« Artificial water channels in water desalination membranes, exhibiting high water permeability » Mouacha Youssef et al.
Session 6 : Aquaculture and water Resources, November 25th, 2021

IOP Conf. Series: Earth and Environmental Science 1090 (2022) 011002 doi:10.1088/1755-1315/1090/1/011002

Peer Review Statement

All papers published in this volume have been reviewed through processes administered by the Editors. Reviews were conducted by expert referees to the professional and scientific standards expected of a proceedings journal published by IOP Publishing Publishing.

- **Type of peer review:** Double Anonymous
- **Conference submission management system:** Morressier
- **Number of submissions received:** 62
- **Number of submissions sent for review:** 51
- **Number of submissions accepted:** 40
- **Acceptance Rate (Submissions Accepted / Submissions Received × 100):** 64.5
- **Average number of reviews per paper:** 2.11
- **Total number of reviewers involved:** 38
- **Contact person for queries:**
  Name: Dr. Hasna Nait M'Barek
  Email: h.naitmbarek@edu.umi.ac.ma
  Affiliation: CentraleSupélec, Paris-Saclay University, France

Mapping the Site of Biological and Ecological Interest of Rganat-Bouchkal (Tsili) Argan forest (Moroccan Central Plateau) using remote sensing

Y Sahel[1], Y Dellahi[2]and D Chahhou[1]

[1] Faculty of literature and human Sciences, Laboratoiry of Lands Equilibriumand and Territories Planning, Physical Geographie Team, Mohammed V University Rabat,Morocco.
[2] Cadi Ayyad University, Laboratoiry of Microbial Biotechnologies, Agrosciences and Environment , Team of Agrosciences and Environment, Faculty of Sciences Semlalia, Marrakech, Morocco.

*Corresponding author yassir.sahel@gmail.com

Abstract. This work aims at developing a precise map of the distribution of Argan tree within Rganat-Bouchkal (Tsili) site which is considered as Sites of Biological and Ecological Interest (SBEI) with high biogeographic value. The methodological approach is based on the use of PLEIADES satellite image with a very high spatial resolution of 0.5m. A supervised classification was performed to develop an original map of the distribution of the Argan tree and the other accompanying species. The numerical method was also adopted to highlight the different factors influencing the distribution and the stat of the Rganat-Bouchkal (Tsili) SBEI stand. The results of this classification made it possible to map more than 770 feet of Argan trees located mainly on the slopes oriented South and South-East, on steepest slopes. The numerical method shows that the exposure factor and the edaphic factor explain mainly the distribution of SBEI species. Small stand either in the form of clumps or isolated feet, the Argan tree of Rganat-Bouchkal (Tsili) appears in a very degraded state, translating a massive exploitation of the tree cover and shrubby by the local population (example: abusive cutting and overgrazing).

Keywords: Argan tree, Rganat-Bouchkal (Tsili), Satellite imagery, Supervised classification, Numerical method, Ecological environments

1. Introduction

Due to its geographical situation, its climate and its relief, Morocco offers a great bioclimatic variety and important bioecological diversity. A wealth that has developed over millennia, making Morocco one of the regions with the highest endemic concentration of biological diversity around the Mediterranean basin [1]. The Argan tree (*Argania Spinosa L. Skeels*), endemic to the wild in Morocco, is in this context, the second forest species of the country after the green oak, (*Quercus rotundifolia)* and just before the Thuya of Barbarie. The Argan tree area is estimated at 800,000 hectares and is mainly located in the south-west of Morocco, along the oceanic coastline; from the mouth of the Tensift river in the north, to the mouth of the river Draa in the south, passing through the Argana corridor, the western and southern slopes of the High Atlas, the plain of Souss and the North-West of the Anti-Atlas slopes, up to altitudes of 1300-1500 m [2]. Apart from its natural biogeographic area, the SBEI [1] of Rganat-Bouchkal (Tsili) is one of the two most northerly areas in the moroccan Argan tree. Little studied and poorly known by the Moroccan and international scientific community, it remains without any exhaustive study of the ecological factors that conditioned its installation in this region. Some scientific works have mentioned the site in specialized articles through which this one is part of the morphological and dendrometric diversity analyzes of a multitude of Argan populations at the national level [3]. A historical and sociological approach was adopted by [4] to identify types of exploitation of the Argan tree by the SBEI users. A study on the aromatic and medicinal plants potentialities of the Moroccan Argan tree [5] has also tried to approach the site concerned and afterwards demonstrate a remarkable richness in AMP [2]. Taking into account the importance of these contributions, the present work is concerned with the site in its entirety trying to develop a detailed map of the Argan tree stand, accompanying the species and environment ecological factors.

2. Study area and data

Rganat-Bouchkal (Tsili) site is part of the valley of the Oued Grou which crosses the central Moroccan plateau. It is classified as Site of Biological and Ecological Interest (SBEI) priority 3, as the master plan protected areas in Morocco (Figure 1). It is a sylvo-pastoral vocation area spread over an area of 1200 ha. SIBE is located as a whole in the semi-arid bioclimate with temperate winter (m ranges from 3C° to -7C°). Its climate is of continental type with (M-m) [3] between 25 and 35 C°, and annual rainfall between 400 and 450 mm, providing favorable conditions for the development of this sapotacea of tropical origin. Almost all of the Rganat-Bouchkal SBEI (Tsili) is based on a geological substratum belonging to the Paleozoic which is reflected on the lithology (Schists, Quartzits, Sandstone and conglomerates) [20]. The soils and surface formations covering the SBEI of Rganat-Bouchkal (Tsili) are relatively recent. Crude mineral soils and poorly developed soils extend over more than 80% of the site surface. The steep slopes, the decrease in tree cover and the rainfall aggressiveness do not favor the soil formation process. The browned soils are very limited and develop on slopes or low slopes with considerable vegetation cover. The surface soil iron sesquioxydes does not exceed 22 ha, so it develops on bare surfaces. Regarding its topographic configuration, the SIBE is located in a very steep valley of the Oued Grou occupying a terrain that is very rugged and inaccessible (Figure 2). Slope classes greater than 25% cover more than 89% of the site surface. The altitude classes are generally moderate (ranging from 200 to 800m) dominated by the class between 200 and 600 m but rarely exceed the 900m. The slopes exposed to the South cover most of the site with more than 30%, followed by Northeast exposures with 13.44%. The other orientation classes are evenly distributed. The plant cover of the SBEI Rganat-Bouchkal (Tsili) is in the form of a matorral gathering a multitude of forest and secondary species (*Argania spinosa, Quercus rotundifulia, Pistacia atlantica, Tetraclinis articulata, Pistacia lentiscus, Olea europea, Rhus pentaphylla* ...).

3. Methodology

With the aim of elaborating a precise map of the vegetation of the SBEI of Rganat-Bouchkal (Tsili) and to know the potential of this site in Argan tree in the first place and to study the main factors contributing to the installation of this species. We used two different methods (The cartographic and numerical method).

[1] Site of Biological and Ecological Interest.
[2] Aromatic and Medicinal Plants.
[3] M= The average maximum of the hottest month, m= The average minimum of the coldest month

Figure 1. Oued Grou valley with some Argan tree feet (**Photos credit**: Sahel Yassir, April 2012)

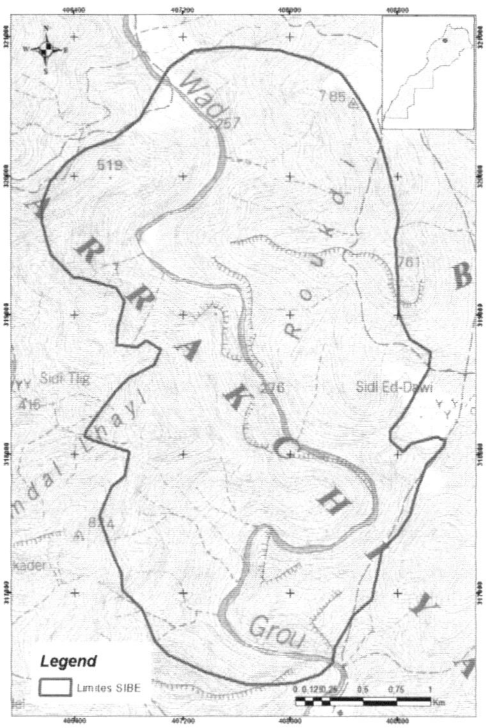

Figure 2. Rganat-Bouchkal SBEI location Map

3.1. Cartographic method

In order to develop a precise and detailed map of the vegetation distribution of the Rganat-Bouchka SIBE (Tsili), the approach adopted is the processing of a PLEIADES satellite image acquired in June 2015 at ORTHO level. It was chosen for its very high spectral and spatial resolution of 0.5m. The image processing process is summarized in figure 3.

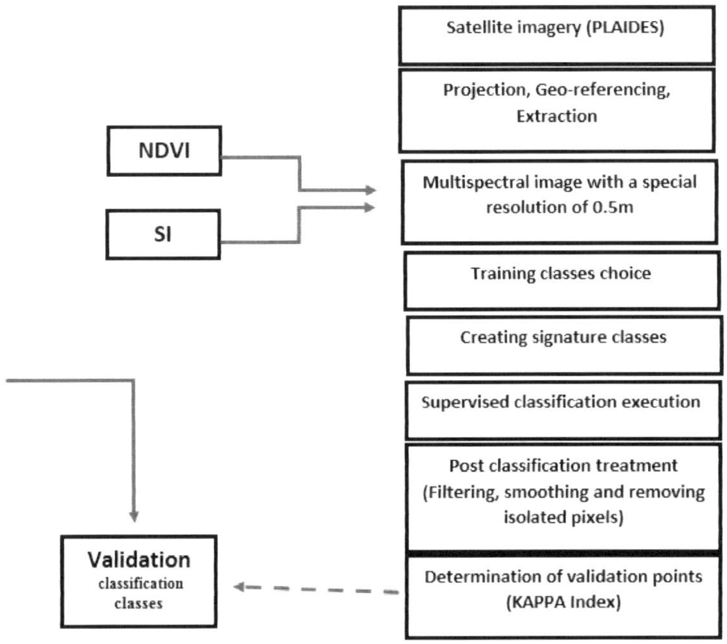

Figure 3. Diagrammatic description of Pleiades satellite imagery processing.

The different satellite image classification processes that have been adopted are as follows:

- **Geometric corrections:** this operation consists of georeferencing the satellite image using the Lambert Morocco Zone 1 projection system to make it superimposable to other reference cartographic material.
- **Radiometric corrections:** in order to improve the satellite image quality, and to discriminate the disturbing elements during the classification, the shaded areas on the image have been removed by applying a ratio that allows them to be recognized and eliminated. This is the Shadow Index (SI), which is a very useful treatment for better classification of forest vegetation and stand types [6].
- **The Normalized Difference Vegetation Index (NDVI)** [7], is also applied to this image, allowing us to spectrally characterize the differences between vegetation and other objects on the image (bare soil, water ...). It is an index often used for the vegetation monitoring [8]
- The fusion of the NDVI and SI with the image resulted in an improved and easily operable image.
- **Supervised classification:** was carried out on the basis of the control points collected by GPS on the ground (200 points recorded in two missions March and April 2015). These points have been subdivided into two groups; 60% of these points served to create the learning base of spectral signals, the rest was later used to validate the supervised classification quality, through the confusion matrix development and the Kappa index calculation. Thus, a classes of information list is created, not only for the Argan trees (*Argania spinosa*) but also, for the other species constituting the floristic procession of the Tsili Argan tree (*Tetraclinis articulata, Pistacia lentiscus, Olea europea, Rhus pentaphylla*), after which, a spectral signature specific to each of the created classes has been realized, making these classes a learning or training base (test data), to recognize the numerical properties of the classes with available field information . Subsequently, the classification supervised by the Maximum likelihood algorithm was launched. Pixels or

aggregations of pixels designating any feet of trees are attributed to the spectral classes having the greatest degrees of the radiometric values similarity.

- **Post-classification treatment:** The image obtained contains incorrect cells (random noise) requiring post-classification processing techniques, for cleaning isolated pixels that do not correspond to the information classes (filtering, smoothing and removing isolated pixels).
- **Validation of the classification:** For the validation of the results of the classification, we carried out the realization of a confusion matrix based on the collected control points, independently constituting the reality field. The objective is to compare the results obtained by the supervised classification with the validation samples. Then determining the Kappa index, which indicates an assessment of the quality classification. According to Landis & Koch [9], the Kappa index value (K) varies between 0.2 and 0.8. (Table 1):

Table 1. Table describing the KAPPA index.

Kappa index	**Quality classification**
K > 0.8	Excellent
K between 0.8 and 0.6	Good
K between 0.6 and 0.2	Moderate
K < 0.2	Bad

3.2. Study of the relationship between vegetation and environmental factors

In order to study the relationship between vegetation and environmental factors, we proceeded to superimpose the vegetation map, already developed, with various maps relating to the SBEI environment. These maps are:

- Map of slopes, altitudes and exposures made from a 30m resolution elevation model (DEM);
- Map of the SBEI geological substrates, realized by using the Maatla Chams geological map at 1/50000;
- Pedological map, extracted from the map of "Zaer and low Chaouia" (1981) at 1/200000 [10].

3.3. Numerical method

For the ecological factors analysis that influence the distribution of the Argan tree and other shrub species we also opted for numerical analysis, through the use of the Correspondence Factor Analysis (CFA). One of the most used methods by researchers in Morocco and elsewhere, for the study of vegetation; notably [11], [12], [13], [14]. Such an analysis serves both to reduce the information in the basic summary table, in the form of a simplified graph, and to explain how environmental conditions can limit the abundance of a living species, or the opposite contribute to its spread. For this, 200 surveys were conducted in the field; at the rate of 40 identified species (*Argania spinosa, Rhus pentaphylla, Tetraclinis articulata, Pistacia lentiscus, Olea europea var. oleaster*). At the level of each survey, we proceeded to the ecological descriptors collection; namely: altitude, slope, exposure, soil and substrate type (Table 2).

Species and ecological parameters have been classified in a dual-entry matrix; where species are arranged in rows and ecological descriptors in columns. The values 0 and 1 are respectively assigned to each survey to indicate the absence or the presence of an ecological character for each survey. The data processing using the Xlstat software allowed later to distinguish the ecological affinities of each species.

Table 2. Ecological descriptors modalities (topoctlimatic and edaphic)

Descriptors	Classes	Modalities
The substrate nature	Sub1	Schiste
	Sub2	Quartzit
Ground type	Gro 1	Raw minerals
	Gro2	Little evolved
	Gro3	Browned soil
Altitude	Alt1	200m-400m
	Alt2	400m-600m
	Alt3	>600
Slope	Slope1	<5%
	Slope 2	05%-25%
	Slope 3	25%-45%
	Slope 4	>45%
Exposure	Expo1	E-NE (East- North East)
	Expo2	S-SE (South - South East)
	Expo3	N-NO (North - North West)
	Expo4	O-SO (West- South West)

The results interpretation is based on reviewing the AFC main outputs; namely:

- The eigenvalues corresponding to the points cloud inertia taken along a factorial axis.
- The maximum of the cloud's inertia is absorbed by the first factorial axis, then the second, and so on.
- The inertia rate reflects the percentage of eigenvalues, compared to the overall inertia of the point cloud.

The point's distribution with respect to the axes indicates that, the more the factorial axes have large eigenvalues, the more easily the points become individualized. In the opposite case, the points gather at the center of the axes, making their delineation into homogeneous groupings difficult [15].

4. Results and discussions

4.1. Mapping method results

4.1.1. Distribution of the Argan tree mapping and other SBEI formations.

The supervised classification results of the satellite image (PLEIADES), and the validation process, show that the Kappa index indicates an excellent level of satisfaction (K= 0.86), which allows the validation of the classification results. This supervised classification shows that SBEI contains more than 775 feet of Argan tree, covering only 33.85 ha of its total area (Figure 4).

Excluding a dozen of hectares hosting the best-preserved stand, the rest of the distribution is widely dispersed with isolated, distant bunches and feet attesting the evidence of an eventual assisted regeneration. However, the Argan tree stand seems highly mutilated, especially on the steeper slopes, and those most subject to intensive course, coupled with an almost total absence of control or protection. These decreases reduce the spontaneous regeneration possibilities of this noble and endemic forest species.

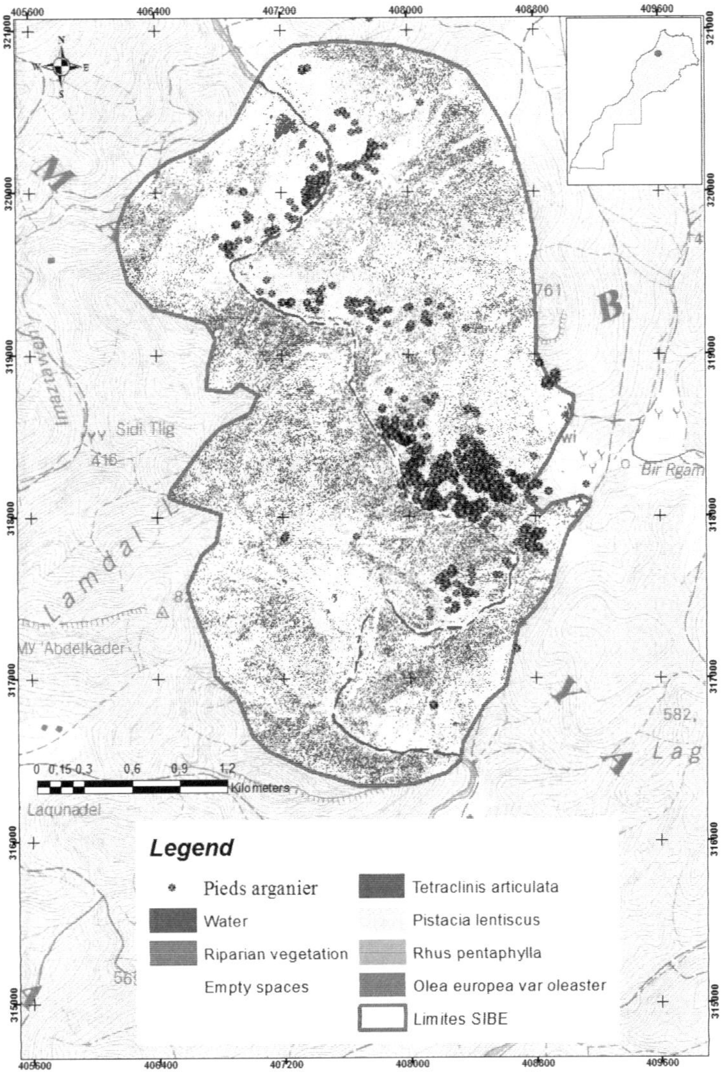

Figure 4. Stand distribution map of the SBEI Rganat-Bouchkal (Tsili).

4.1.2. Analysis of vegetation-environmental factors relations.

The superposition of the distribution map of the Argan tree and geological maps, soils and topographical factors (slopes, altitudes and exposures), indicates that the Argan tree develops, especially, on schistous substrates and occasionally on Quartzite. Indifferent to all soil types but with a tendency to slightly evolved soils, preferring south, east and south-east exposures, but occasionally be found on the slopes-oriented North, West and Southwest. Its density increases remarkably on the steeper slopes exceeding 30%; mixed with *Rhus pentaphylla* on the sunny sides, and occasionally with *Olea europea* and *Pistacia lentiscus* on wetter slopes. This proves that the very rugged nature of the site has played a key role in this species preservation. On the slopes North and West oriented, the Argan tree gives way to the Thuya of Barbarie (*Tetraclinis articulata*) thus undergoing a very marked degradation, and covering a large portion of SBEI, with nearly 159 ha. It grows better on relatively deep soils, and on the wettest slopes, often mixed with *Pistacia lentiscus*.

Secondary species such as: *Pistacia lentiscus, Olea europea, Rhus pentaphylla*, cover 208 ha, or about 21% of the area of Rganat-Bouchkal (Tsili). They are considered as stages of degradation of the climax forest, based on *Tetraclinis articulata*; mainly due to a strong anthropo-zoogenic pressure. But the most important manifestation is the surface area of the empty spaces covering 680 ha, which means 68% of the entire SBEI.

4.2. The numerical method Results

The eigenvalue of the F1 axis is 0.51 which is very close to the minimum threshold required for easy individualization of the groups [16]. The sum of the inertia of factorial plan F1-F2 absorbs 46% of the total variation; and explains the maximum amount of information. The review of the distribution of measured points and modalities of selected ecological descriptors, allowed highlighting two ecological stations (Table3).

Table 3 Eigenvalues table

	F1	F2	F3
Eigenvalue	0,509	0,490	0,342
Inertia (%)	23,533	22,664	15,807
Cumulated %	23,533	46,197	62,004

4.2.1. Factorial axes interpretation.

The F1-F2 factorial analysis shows that the F1 factorial axis opposes the environment types according to the altitude, the exposure, and the slopes classes; therefore, the axis F1 represents the topo-climatic parameters. While the factorial axis 2 reflects a gradient of soil conditions related to both substrates and soil types translating the edaphic parameters (Figure 5).

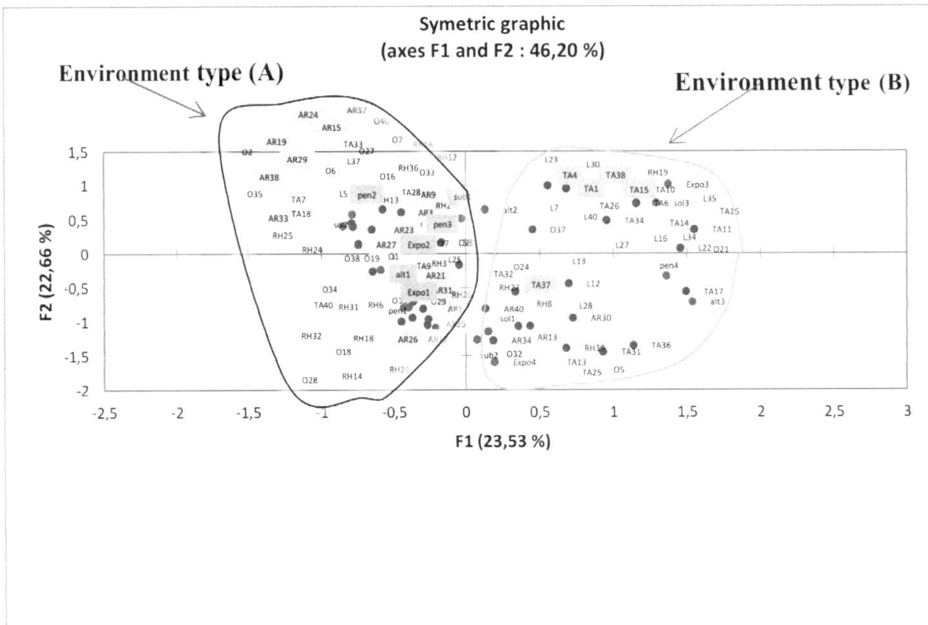

Figure 5. FCA graphical representation in the factorial plan (1x2) of species and ecological descriptors. TA : *Tetraclinis articulata* ; AR : *Argania spinosa* , RH : *Rhus pentaphylla*, L : *Pistacia lentiscus*, O : *Olea europea, sub1 : Schiste, sub2 : Quartzite, sol1 : Raw minerals, sol2 : Little evolved, sol3 : Browned soil, alt1 : 200m-400m, alt2 : 400m-600m, alt3 : >600m, slopes : <5%, slope2 : 05%-25%, slope 3 : 25%-45%, slope 4 : >45, expo1 : E-NE, expo2 : S-SE, expo3 : N-NO , expo4 : O-SO.*

4.2.2. Environments typology.

The examination of the figure 5 revealed only two distinct types of environments:

Type A environment: This type of environment groups the records located on the negative side of the F2 axis. It is characterized by *Argania spinosa* abundance, which is in mixture with *Rhus pentaphylla and Olea europea var, oleaster*; and by *Tetraclinis articulata* and *Pistacia lentiscus* scarcity. From the edaphic point of view, this environment is individualized by schistous substrates and poorly developed soils and raw minerals. It is also dominated by East, Northeast, South and Southeast exposition. This station is also characterized by moderate slopes, in addition to an altitudinal range from 200 to 600m.

Type B environment: This environment includes *Tetraclinis articulata* stands mixed with *Pistacia lentiscus*. The presence of *Argania spinosa* and *Rhus pentaphylla* is quite rare; on the other hand, *Olea europea var, oleaster* contains a species overlapping between the two types of ecological station. From an edaphic point of view, the B-type environment is characterized by schistose and quartzitic substrates, and browned soils in general, with raw mineral soils and little evolved in places. This station corresponds to North, North-West, West, South-West exposures. The slopes are steeper (over 45%), with an altitudinal range between 400 and 650m.

4.3. Discussion

This study results reveal a perfect match between the cartographic and numerical method. These methods show that the Argan tree is growing, mainly on the sunniest exposures, and on little evolved soils and raw minerals. This explains the negative balance of water reserves in these soils, and their condition very eroded by surface runoff [19]. These conditions also reflect the sparse and degraded nature of the vegetation. The presence of *Rhus pentaphylla* reveals the xerophilic and thermophilic aspect of the ecological conditions of this station. The *Tetraclinis articulata* with *Pistacia lentiscus* mixture replaces the Argan tree, and sumac on fresh slopes sheltered from the sun, and on deeper soils. This explains the

conserved state of the soil, and a very reduced manifestation of water erosion phenomena of browned soils, rich in clay-humic complexes enhancing the stability of their texture; due to a good capacity of water reserves retention [19]. It should be noted that *Olea europea var. oleaster* not lacking on both stations; but its strength is more remarkable on the second station. The ecological conditions of the first station remind us of the climax at *Argania spinosa* and *Rhus pentaphylla* described by Peltier on the eastern slopes of the south piedmont of the High Atlas. It develops on Schists, Quartzites and skeletal soils occupying an altitudinal range from 600 to 1000m. It abounds between Mnizla and Tafingoulte; but it is perfectly individualized on the southern slopes of the Ida-Ou-Tanan massif on ridges and rocky outcrops of more or less difficult access [17]. Regarding the second station it closely resembles the group *Tetraclinis articulata. Olea europea* and *Pistacia lentiscus* described by Benabid in the Amsittene region of Essaouira. This grouping develops on the North, North-East and North-West slopes and on steep slopes and low altitudes. Soils are more or less deep, but frequently eroded, evolving in a semi-arid temperate bioclimate. [18].

5. Conclusion

According to this study results, it can be deduced that the Argan tree (*Argania spinosa*) of the SBEI of Rganat-Bouchkal (Tsili) is in the form of a very degraded scrub forest (matorral). The supervised classification of the satellite image allowed us to elaborate the map distribution of Argan trees, as well as other species present at the level of SBEI. The results verification, in situ, largely confirm the classification results, including the discontinuous and sparse nature of this Argan tree stand. The numerical method reinforces the results of the mapping analysis. Two types of ecological environments were identified. The first environment distinguished by the presence and sometimes the abundance of *Argania spinosa* of *Rhus pentaphylla* and another marked by *Tetraclinis articulata*, *Olea europea var, oleaster* and *Pistacia lentiscus*. Both methods show that the "exposure" factor is the most determining factor in the vegetation distribution; Soil type factor is also important but rather reflects topo-climatic conditions. Field visits covering almost the entire central Moroccan plateau, have led to the conclusion that the Argan tree of the SBEI of Rganat-Bouchkal (Tsili), located in the middle Grou valley, is considered to be one of the most degraded forest formations in the central plateau. This is mainly due to the anthropogenic pressure exerted on the site. The SBEI of Rganat-Bouchkal (Tsili) is classified by the master plan of the protected areas in Morocco as a SBEI of priority 3 [21]. The measures taken to maintain the sustainability of this site are very derisory. It has never been the subject of any management, which has repercussions on its current state, marked by the detrimental effects of the unmaintained exploitation of the sole representative of *Argania spinosa* (L. Skeels) in the central Moroccan plateau. This work presents a new decision-making support, which can contribute to a better knowledge for the development, and programming interventions of this site.

6. References

[1] Fassi D 1999 *Le grand livre de la forêt marocaine*. Mhirit O & Blérot P Coordination. MARDAGA edition pp 10-12.

[2] Msanda F, El Aboudi A, Peltier JP 2005 *Biodiversite et biogeographie de l'Arganeraie marocaine*. Cahiers Agriculture vol 14 pp 357–364

[3] Bani-Aamer F 2006 *Description dendrométrique des populations naturelles de l'Arganier au Maroc*. Actes des 1éres assises de la recherche forestière ; L'Arganier : un rempart contre la désertification (Essaouira) 25 et 26 2006. Annales de la recherche forestière au Maroc pp 82-87.

[4] Sahel Y, Chahhou D, El Bahloul Y 2011 *Rganat-Bouchkal (Tsili), ultime refuge de l'Arganeraie d'Oued Grou dans le plateau central marocain*. Les actes du Congrès *int* de l'Arganier *2ème* édition. (Agadir) pp 388-390.

[5] Aafi A, Benabid A 2015 Les plantes aromatiques et médicinales : un levier de développement, conservation et valorisation des écosystèmes à arganier du Maroc (sud-ouest marocain, vallée de l'Oued Grou/ khémisset et Béni-Snassen/Berkane). Acte de congrès *int* d'arganier 5eme édition. (Agadir) p 25.

[6] Ono A, Kajiwara K, Honda Y *2010 Development of new vegetation indexes, shadow index (si) and water stress trend (wst)*, XXXVIII, Part 8, 710-714.

[7] Rouse JW., Jr, Haas R H, Schell J A and Deering DW 1974 Monitoring vegetation systems in the Great Plains with ERTS. *NASA special publication*, 351 309.

[8] Ossmane J 2015 *Connaissances expertes et modélisation pour l'exploitation d'images d'observation de la Terre à hautes résolutions spatiale, spectrale et temporelle*. Thése, Université de Toulouse pp 24-62.

[9] Landis J R, Koch G G 1977 The measurement of observer agreement for categorical data. *Biometrics* 33 159–174.

[10] Ghanem H 1981 *Contribution à la connaissance des sols du Maroc : genèse, classification et répartition des sols des régions des Zaers, de la Basse Chaouia et des Sehouls (Meseta Atlantique marocaine)*. Cahiers de la Recherche Agronomique, vol 37 et 38. INRA (Rabat)pp 18-21

[11] M'Hirit O 1982 *Etude écologique et forestière des cédraies du Rif marocain. Essai sur une approche multidimensionnelle de la phytoécologie et de la productivité du cèdre* Thèse Doct. Univ. Droit Eco. et Sci. Aix-Marseille III p 436 + annexes

[12] Nefaou, M 1996 Dendroécologie, productivité et dynamique *de la croissance radiale du pin maritime naturel du Maroc (Pinus pinaster Soland. subsp. Hamiltoni (Ten) H. Del villar var. maghrebiana H. Del villar)*. Tome 2. Thèse Doctorat en sciences,spécialité : Ecologie p 157.

[13] Chahhou, D 1992 *Etude phytogéographique des causses d'Azrou-Ifrane (moyen Atlas marocain)*. Thèse de doctorat en Géographie. Université de Bordeaux III pp 106-138.

[14] Aafi A 2007 *Etude de la diversité floristique de l'écosystème de chêne-liège de la forêt de la Mâamora*. Thèse de Doctorat d'Etat Es-Science Agronomiques, I.A.V Hassan II (Rabat) pp 59-93.

[15] Chkhichekh A 2015 *Cartographie et Etude des Groupements Végétaux du Plateau Central Marocain : Cas du Site d'Intérêt Biologique et Ecologique (SIBE) d'El Harcha*. Thèse de Doctorat en Biologie (Rabat) 37-72.

[16] Benzecri J P 1973 *L'analyse des données II. L'analyse des correspondances. Dunod*, Paris, p 619.

[17] Peltier J P 1982 *La Végétation du Bassin Versant de l'Oued Sous (Maroc)*. Thèse Doct. ès-Sc. Université Scientifique et Medicale (Grenoble) pp 54-136.

[18] Benabid A 1976 *Etude Phytoécologique, Phytosociologique et Sylvo- Pastorale de la Tétraclinaie de l'Amsittène*. Doct. 3ème cycle, Fac. Sci. St. Jerôme, Marseille III, France pp 84-86.

[19] Sahel Y 2011 *Essai d'évaluation de la sensibilité des sols à l'érosion selon leurs modes d'occupation dans le Grou moyen, cas du sous-bassin versant Rwaai*. Mémoire de Master Université Hassan II (Mohammedia) pp 55-71.

[20] Beaudet G 1969 *Le Plateau Central et ses Bordures, étude géomorphologique*. Ouvrage publié avec l'aide du Ministère de l'Enseignement Supérieur du Maroc et des Ministère francisent de l'Education Nationale et des Affaires Etrangères (Rabat) pp 28-47.

[21] AFM (Administration Forestière du Maroc) 1996. *Plan National des Aires Protégées*. HCEFLCD, Rabat, Maroc.

IOP Conf. Series: Earth and Environmental Science 1090 (2022) 012002 doi:10.1088/1755-1315/1090/1/012002

Structure and trophic organisation of the macrobenthic community of the Hassar stream after installation of the Mediouna wastewater treatment plant (Casablanca, Morocco)

A Nahli[1*], F E Sghiouer[1,2], S Oubraim[1] and M Chlaida[1]

[1]Laboratory of Ecology and Environment, Hassan II University of Casablanca, Faculty of Sciences Ben M'sik, PB 7955 - Sidi Othman, Casablanca, Morocco
[2]Laboratory of Natural Resources and Environment (LRNE), Sidi Mohamed Ben Abdallah University, Polydisciplinary Faculty of Taza, Morocco

* abdelmottalib.nahli@gmail.com

Abstract. The goal of this research is to assess changes in the structure of benthic macroinvertebrate Functional Feeding Groups (FFGs) in the Hassar Stream (north-east of Casablanca) after the Mediouna wastewater treatment plant (WWTP) was installed. Over the course of a year, from November 2013 to October 2014, data on water quality and macroinvertebrate fauna are gathered at seven sampling stations. To measure the resilience of FFGs, macroinvertebrates are used to analyse the influence of physicochemical water quality on their organisation. Based on FFGs abundance, Collector-gatherers make up 39.06 percent of the macroinvertebrates' assemblage, followed by shredders (28.41 percent) and collector-filters (28.41 percent) (18.76 percent). Except for the upstream stations (S1 and S2), the FFGs ratios indicated that the environment is largely heterotrophic, has a significant relationship with riparian inputs, and generally stable substrates. These findings emphasise the necessity of researching macroinvertebrate FFGs as a supplement to measuring the ecological integrity and resilience of rivers after anthropogenic impact reduction.

1. Introduction

Aquatic ecosystems are subjected to various anthropogenic pressures, including the discharge of various types of liquid waste, which have an impact on aquatic life [1-2]. These anthropogenic disruptions can overwhelm these habitats' ability to self-purify, reducing their resilience. However, the alteration degree of these hydrosystems' quality depends on their potential and their biological richness [3].

The biological monitoring of these natural environments, in addition to physical and chemical parameters, is crucial conservation of their integrity. Benthic macroinvertebrates can integrate the disturbances of rivers thanks to their sensitivity to the degradation of water quality [4]. As a result, the biotic characteristics of these benthic macroinvertebrates (composition, richness, abundance) change according to the nature, frequency, and amplitude of the disturbance of the watercourse.

Also, these benthic macroinvertebrates can be classified into FFGs according to their functional feeding mode and can therefore be used to assess the integrity of aquatic ecosystems [5]. Any changes in the composition or relative abundance of these trophic groups are often driven by changes in nutrients [6] and thus reveal the effect of anthropogenic activities on the aquatic environment [7]. This functional approach can therefore provide information on the chemical composition of the hydrosystem [8], its ecological functioning and its degree of resilience.

The suburban centres of Casablanca (i.e., Mediouna, Bouskoura and Nouaceur) have experienced, for several years, a significant and rapid urban development that has impacted several suburban streams, in particular the course of the Hassar Stream. Indeed, the latter has been receiving raw effluent from Mediouna for more than a decade, contributing to a significant change in its ecological status [9]. The mitigation of the impact of wastewater discharge by the installation of the Mediouna wastewater treatment plant (WWTP) has allowed the gradual recovery of its ecological equilibrium [10]. The functional structure assessment of Hassar Stream's macroinvertebrates community is essential to understand its ecological functioning and evaluate its resilience. The present work aims to assess the macroinvertebrates community's functional structure of the Hassar stream and estimate its attributes.

2. Material and methods

2.1. Study site and sampling stations

Hassar stream is a small river with a low flow rate (0.026 - 0.067 $m^3 s^{-1}$) located to the southeast of Casablanca (Figure 1). It is a tributary of the El Malleh river, which, together with the Nfifikh river, constitutes the principal rivers of the Mohammedia-Benslimane Sahel region [11]. This stream has long been associated with the partial drainage of the Berrechid aquifer, which gave rise to the mainspring located about 3 km north of Mediouna [12] in addition to several secondary, slightly saline springs that exist along its course (Figure 1).

Subsequently, with the succession of drought years experienced by the region, the mainstream springs and only a few secondary springs continued to flow. Regarding anthropogenic impacts and despite the construction of the Mediouna WWTP, which treats all the municipal wastewater of this locality, the Hassar stream still suffers from other effects of the activity of the riparian populations, notably through the infiltration of greywater and the discharge of solid waste [10, 13]. Table 1 lists the physicochemical characteristics of this stream waters in the different studied stations.

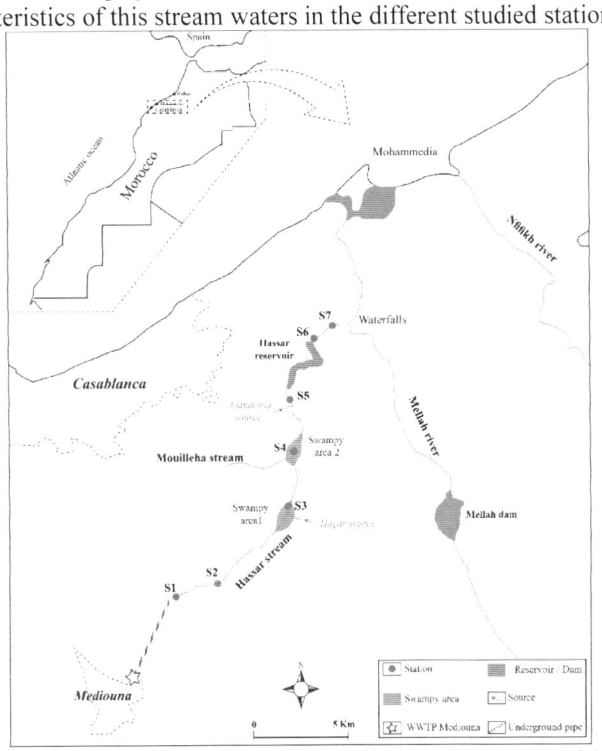

Figure 1. Location map of the study site and sampling stations (S1, S2, S3, S4, S5, S6, S7) [13]

Table 1. Means and standard deviations (SD) of the different physicochemical variables of the Hassar stream waters [13].

	S1	S2	S3	S4	S5	S6	S7
T (°c)	25.02±2.75	24.79±3.78	21.78±2.66	22.18±3.90	19.07±3.68	19.46±3.32	17.41±4.27
pH	8.23±0.70	8.21±0.91	8.21±0.73	8.31±0.64	7.91±0.78	8.60±0.85	8.21±0.87
EC (mS cm^{-1})	3.54±0.43	3.58±0.42	5.35±0.93	5.43±0.59	5.03±0.60	5.64±0.91	5.68±0.83
DO (mg O2 L^{-1})	4.91±1.15	3.86±0.82	6.45±1.58	6.25±1.01	4.88±1.24	6.83±1.56	5.91±1.14
TSS (mg L^{-1})	10.92±8.25	34.31±23.13	23.86±11.15	21.53±9.95	22.28±13.79	22.02±13.44	43.88±26.54
BOD_5 (mg L^{-1})	29.58±15.73	38.22±17.24	16.50±8.23	18.89±10.93	31.68±17.04	18.15±15.65	22.26±9.15
NH_4^+ (mg L^{-1})	8.99±15.52	9.24±5.32	6.79±5.32	4.44±3.21	7.66±6.81	4.12±3.55	3.93±3.40
PO_4^{3-} (mg L^{-1})	7.62±3.14	7.81±3.08	3.19±1.63	2.40±1.21	2.56±0.66	1.77±1.25	1.51±0.76
Cl^- (g L^{-1})	0.83±0.25	0.86±0.21	1.46±0.25	1.52±0.32	1.43±0.30	1.60±0.40	1.60±0.39

2.2. *Macroinvertebrate sampling*
A Surber type net with a mesh size of 500 m was used to sample macroinvertebrates, monthly from November 2013 to October 2014, over a sampling area of 0.05 m^2. The number of samples is fixed at six (06) per station [14-16], i.e. a total sampling surface of 0.3 m^2. The benthic macroinvertebrate samples were preserved in a 10% formalin solution. The identification of the gathered specimens is carried out using referenced works and determination keys [4, 17-19].

2.3. *Faunal data processing*
The identification of macrobenthic species allowed the calculation of taxonomic richness (S), the Shannon Diversity Index (H'), and the Pielou Equitability Index (E). The FFGs were determined based on information provided by [4, 7, 20]. The FFG categories used in this study are collector-gatherer (CG), collector-filter (CF), scraper (Sc), shredder (Sh) and predator (P). Following that, the percentage contribution of each FFG to the various communities was determined at the different studied stations.

Cummins [7] extensively discussed the effects of changes in river water quality on the nutritional resources of ecosystems and consequently on the abundances of their macrobenthic FFGs. FFGs rations give a range of information on how these hydrosystems work concerning changes in baseline nutritional supplies. The different FFGs rations were calculated and interpreted according to the definitions and threshold values proposed by Cummins [7].

3. Results
A total of 85 taxa were identified in the seven stations of Hassar Stream. The average taxonomic richness remained low and increased upstream (Table 2). Similarly, The H' index shows a relatively increasing spatial gradient upstream to downstream. The average Evenness shows slight spatial variations along the stream and varies between 0.5 (S2) and 0.64 (S3).

Concerning the relative abundance of FFGs, collector-gatherers are the most abundant in Hassar Stream, accounting for 39.06 per cent of the macroinvertebrate assemblages, followed by shredders (28.41%) and collector-filters (18.76 %). Spatially, the macrobenthic population of the upstream stations is dominated by collector-gatherers, followed by the filter-collectors and scrapers. The predators and shredders groups are very rarely represented (Figure 2). In the marshy areas (S3 and S4), the average relative abundance of collector-gatherers declines favouring scrapers, shredders and predators. In station S5, the average relative abundance of these scrapers declines sharply in favour of collector-gatherers and shredders.

Downstream of the Hassar dam (S6), the macrobenthic community is dominated by shredders (35%), followed by collector-filters (32%) and scrapers (22%), while the rest of the FFGs are poorly represented. Further downstream (S7), the abundance of the collector-gatherer group increases (55%) to the detriment of the collector-filters (14%) and scrapers (7%) groups.

Table 2. Average taxonomic richness (S), Shannon diversity (H') and Evenness (E) of the Hassar stream's macrobenthic community.

	S1	S2	S3	S4	S5	S6	S7
S	7.58±1.93	9.66±3.77	15.5±2.69	18.83±4.37	21.08±4.23	19.75±5.17	16.27±3.74
H'	1.68 ±0.38	1.49±0.25	2.52±0.42	2.38±0.49	2.09±0.55	2.62±0.30	2.08±0.70
E	0.58±	0.49±0.11	0.64±0.11	0.57±0.11	0.48±0.12	0.62±0.07	0.53±0.19

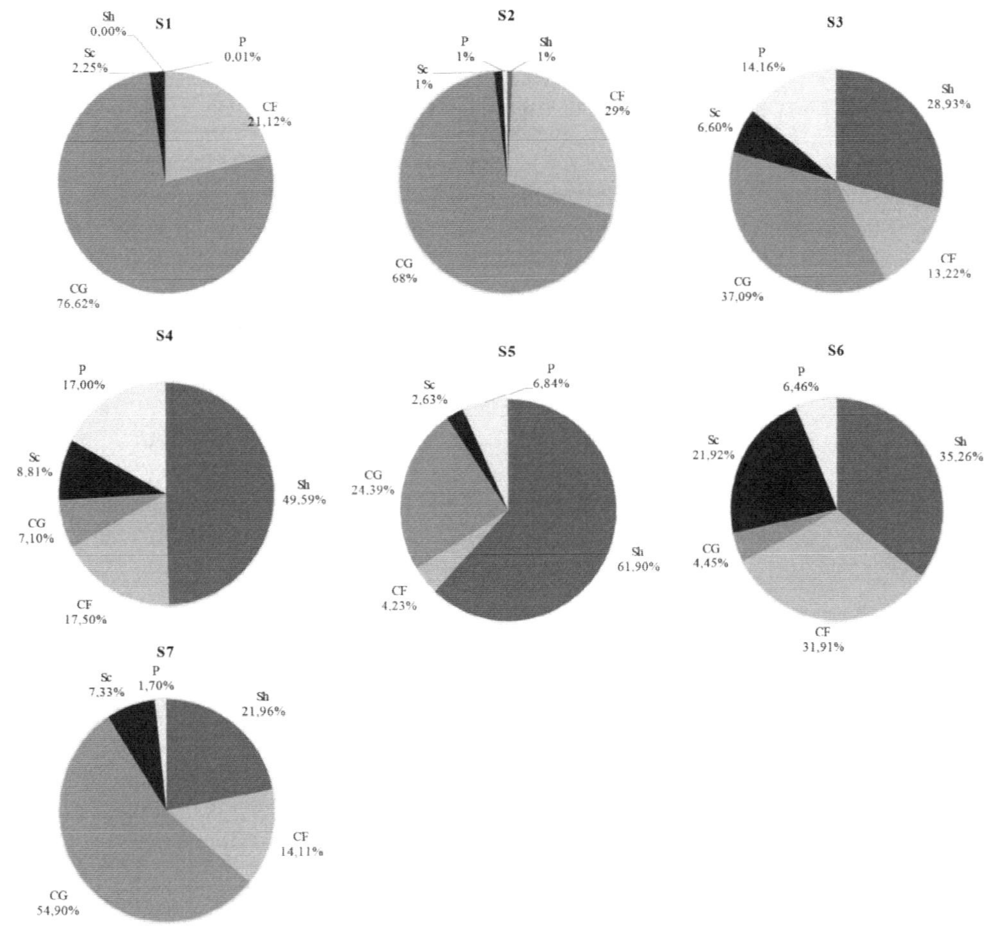

Figure 2. Average relative abundances (in %) of FFGs at the different studied stations (CG: collector-gatherers, CF: collector-filters, Sh: shredders, Sc: scrapers, P: predators).

The use of FFGs ratios as indicators of hydrosystem attributes (Table 3) demonstrates that all of the studied stations are significantly heterotrophic. At the same time, only stations S4 and S5 demonstrate a strong relationship to riparian inputs, and most of these stations, except stations S1 and S5, have relatively stable substrates. In all sites, the P/P ratio is less than 0.15, indicating a normal predator-prey interaction.

Table 3. Means and standard deviations (SD) of the different FFGs ratios (P/R: Autotrophy/heterotrophy index, CPOM/FPOM: Coarse particle organic matter (CPOM)/fine particle organic matter (FPOM), P/P: Predator/prey).

	S1	S2	S3	S4	S5	S6	S7
P/R	0.02±0.06	0.01±0.02	0.33±0.26	0.33±0.26	0.04±0.05	0.68±0.49	0.23±0.27
CPOM/FPOM	0.00±0.00	0.00±0.00	0.01±0.03	0.25±0.88	0.93±2.26	0.01±0.01	0.01±0.00
Substrate stability	0.49±0.56	1.37±2.86	0.71±0.56	1.21±2.22	0.09±0.08	2.1±0.79	0.67±0.65
P/P	0.00±0.00	0.00±0.00	0.08±0.07	0.04±0.04	0.03±0.03	0.02±0.02	0.02±0.02

4. Discussion

Analysis of the results shows that the punctual input of exogenous organic matter and nutrients would be at the origin of this observation since several faunal groups are eliminated from the upstream stations (Ephemeroptera, Trichoptera, Odonata). While the dominance of a few taxa (Diptera and Oligochaetes) favoured by the water's nutrients and organic matter richness reduces macroinvertebrates diversity [21]. The upstream-downstream increase of the taxonomic richness and diversity could be attributed to the recolonisation of the environment by tolerant taxa of moderate organic pollution, such as Baetis sp. and Hydropsyche maroccana [22, 23], which appear progressively from the first marshy zone (S3). At the same time, the appearance of these taxa at these stations would almost certainly be linked to the decrease in organic pollution as a result of the self-purification phenomenon, as well as the diversity of habitats between S3 and S5. However, the taxonomic richness and diversity decrease slightly at station S7 due to the disappearance of Trichoptera under the effect of the high organic and particulate loads.

The average relative abundances of the functional trophic groups undergo significant spatial variations in Hassar Stream. Thus, collectors (CG and CF), which feed on detritus [24], are the most dominant FFGs, particularly upstream. The abundance of collectors group, particularly Oligochaetes and Diptera considered generalist taxa and community pioneers [25], would be essentially linked to allochthonous inputs of fine organic matter that is deposited on the surface of the sediments as well as the low degree of oxygenation of these waters. The increased levels of this fine organic matter in the waters generally provide significant nutritional value in areas downstream of discharge points [26].

The shredders are almost absent from the upstream stations S1 and S2, respectively, due to agitation related to the WWTP treated water's continuous discharge and greywater infiltration from riparian settlements. These low shredders abundances upstream could indicate disturbed aquatic ecosystem functioning through a trophic cascade effect [27]. Further downstream, the increase of the shredders abundance in marshy zones 1 (S3 and S4) and at the entrance to the Hassar dam reservoir (S5) is thought to be linked to the presence of coarse organic matter from the riparian vegetation and the distance from the liquid discharge points (S1 and S2). Further downstream, the absence of riparian vegetation and the impacts of anthropogenic activities reduce the abundance of this group [28].

Scrapers are relatively more abundant in marshy areas and downstream of the Hassar dam (S6), probably because of the high algal production (periphyton), which serves as food for this macroinvertebrate [29, 30]. On the other hand, alterations of the water quality upstream (S1 and S2) and the Hassar dam upstream (S5) and downstream (S7) would be responsible for the reduction in the abundance of this specialised trophic group.

The low abundance of predators compared to other macroinvertebrates FFGs is common in most natural ecosystems. Nevertheless, these organisms are relatively more abundant in this stream's marshy areas, probably due to improved water quality, availability and diversity of prey, and low competition.

The FFGs ratios have indicated the overall functioning of the Hassar Stream and its attributes. Indeed, in this stream, the P/R ratio reveals a predominance of heterotrophy to the benefit of autotrophy with a low abundance of scrapers due to the scarcity of periphyton and the high abundance of shredders and collectors that use the detritus of riparian origin as a food resource [31]. This finding would be related to the state of the low-diversity riparian vegetation. In addition, the inflow of relatively warm and agitated treated wastewater, greywater infiltration, and various discharges from riparian populations along with this stream favour the development of collector-gatherers, decreasing the P/R ratio [28]. Furthermore, the low CPOM/FPOM ratio values recorded in the different disturbed stations indicate a low dominance of shredders against collectors and confirm that the riparian zone of the Hassar stream is non-functional. We believe that the loss of riparian zone integrity is associated with anthropogenic activities' impact [32]. Shredders disappear from points affected by anthropogenic disturbances such as wastewater discharges and washing of vegetable and textile products. Instead, the increase of the shredders abundance from the marshy areas would indicate an abundance of allochthonous coarse organic matter and thus a relative recovery of the riparian zone function. The low channel stability ratios recorded at stations S1 and S5 would be related to the high degree of heterotrophy. The low P/P ratios for the different stations would reflect an abundance of prey species with short life cycles [33]. This result entirely agrees with Fu [34], who showed that predators and shredders have a low abundance in disturbed streams. We believe that the high degree of mineralization of waters also influences the abundance of these predators [35]. Indeed, the water salinity acts on the trophic cascade by influencing the abundance of predators and thus modifies their top-down control over prey.

5. Conclusion

The mitigation of the impacts of the pollution of liquid discharges after the installation of the Mediouna WWTP allowed the improvement of the physicochemical quality of the water and the diversification of habitats which favoured their recolonisation by relatively rich and varied macrofauna of invertebrates. The various structural and functional changes in the macrobenthic population observed along this hydrosystem indicate a progressive recovery of its ecological integrity, despite previous losses in biocenosis diversity.

Our results reveal the importance of studying the functional structure of macroinvertebrates in bioassessing the integrity of disturbed lotic ecosystems in the Mediterranean region. This functional approach can provide consistent ecological information to understand the functioning of a resilient stream after the mitigation of anthropogenic disturbances.

6. References

[1] Grizzetti B Pistocchi A Liquete C Udias A Bouraoui F and van de Bund W 2017 *Sci Rep* **7** 205
[2] Bediako A A 2020 *Pol J Environ Stud* **29** 1–9
[3] Oliver T H Heard M S Isaac N J B Roy D B Procter D Eigenbrod F Freckleton R Hector A Orme C D L Petchey O L Proença V Raffaelli D Suttle K B Mace G M Martín-López B Woodcock B A and Bullock J M 2015 *Trends in Ecology & Evolution* **30** 673-684
[4] Tachet H Richoux P Bournaud M and Usseglio-Polatera P 2010 *Invertébrés d'eau douce : Systématique biologie écologie* (CNRS Editions Paris) p 601
[5] Bhawsar A Bhat M A Vyas V 2015 *International Journal of Scientific Research and Engineering* **3** 385–393
[6] Smith A J Bode R W and Kleppel G S 2007 *Ecol Indic* **7** 371–386
[7] Cummins K W Merritt R W Andrade P C N 2005 *Neotrop Fauna Environ* **40** 69–89
[8] Couceiro S Hamada N Forsberg B Pimentel T Luz S 2012 *Ecol Indic* **18** 118–125
[9] Fouad S 2013 PhD thesis, Hassan II University Casablanca Morocco p 140
[10] Nahli A Oubraim S and Chlaida M 2018 *Bulletin de l'Institut Scientifique Section Science de la vie* **40** 37-56
[11] Khatami A 1992 PhD thesis (Louis Pasteur University Strasbourg France p 205
[12] Oubraim S 2002 PhD Thesis Hassan II University of Mohammedia Morocco p 208
[13] Nahli A 2017 PhD Thesis Hassan II University of Casablanca Morocco p 252
[14] El Agbani M A 1984 PhD thesis Claude Bernard University Lyon France p 147
[15] Fekhaoui M 1990 PhD Thesis Mohammed V University Rabat Morocco p 165
[16] Berrahou A Chavanan G and Richoux P 2000 *Bull Mens Soc Limn Lyon* **69** 44-50

[17] Epler J H 2001 *Identification manual for the larval Chironomidae (Diptera) of North and South Carolina* (Department of Environment and Natural Resources Division of Water Quality North Carolina) p 526
[18] Martin M and Ait Boughrous A 2012 *Guide taxonomique des oligochètes dulçaquicoles du Maghreb* (Bruxelles: Abc Taxa Vol 12 : I-VIII) p 186
[19] Leung A Edward D and Pinder A 2013 *Photographic guide and keys to the larvae of Chironomidae (Diptera) of south-west Western Australia: Part II Orthocladiinae* (Crawley: University of western Australia Department of environment and conservation) p 56
[20] Merritt R W Cummins K W and Berg M B 2008 *An introduction to aquatic insects of North America* (Dubuque: Kendall/Hunt Publ Co) p 1158
[21] Prenda J and Gallardo-Mayenco A 1996 *Archiv fur Hydrobiologie* **136** 159–170
[22] Barbour M T Gerritsen J Snyder B D and Stribling J B 1999 *Rapid Bioassessment Protocols for Use in Streams and Wadeable Rivers: Periphyton Benthic Macroinvertebrates and Fish* (Washington DC: US Environmental Protection Agency Office of Water) p 337
[23] Pires A M Cowx I G and Coelho M 2000 *Hydrobiologia* **435** 167-175
[24] Allan J D and Castillo M M 2007 *Detrital energy sources* (Dordrecht: Springer) pp 135-163
[25] Dong-Jun C Dong-Jun R T J and Tae-ho R 2004 *Korean J Limnol* **37** 137-148
[26] Dalu T Wasserman R J Tonkin J D Alexander M E Dalu M T B Motitsoe S N Manungo K I Bepe O and Dube T 2017 *Sci Total Environ* **601–602** 1340–1348
[27] Baer S G Siler E R Eggert S L and Wallace B 2001 *Freshw Biol* **46** 347–365
[28] Mangadze T Wasserman R J Froneman P W Dalu T 2019 *Science of the Total Environment* **695** 133910
[29] Grubaugh J W Wallace J B and Houston E S 1996 *Canadian Journal of Fisheries and Aquatic Sciences* **53** 896-909
[30] Muñoz I 2003 *Limnetica* **22** 107-116
[31] Merritt R W and Cummins K W 1996 *Trophic relations of macroinvertebrates* Hauer F R and Lamberti G A (eds) Methods (San Diego: Academic Press) pp 453 – 474
[32] Masese F O Kitaka N Kipkemboi J Gettel G M Irvine K and McClain M E 2014 *Freshwater Science* **33** 435-450
[33] Merritt D M Manning M E Hough-Snee N 2017 *The National Riparian Core Protocol: A riparian vegetation monitoring protocol for wadeable streams of the conterminous* (United States Gen Tech Rep RMRS-GTR-367 (Fort Collins: Department of Agriculture Forest Service Rocky Mountain Research Station) p 37
[34] Fu L Jiang Y Ding J Liu Q Peng Q Z and Kang M Y 2015 *J of Freshw Ecol* **31** 21–35
[35] Canedo-Argüelles M Sala M Peixoto G Prat N Faria M Soares A M V M Barata C and Kefford B 2016 *Sci Total Environ* **540** 3–10

Hydrological response of the Assif Melloul basin in the Central High Atlas (Morocco): Monitoring and Creating a rating curve.

M Goumih[1], M El Ghachi [1] H Ouakhir [1], N Ennaji [2] and A Midaoui[3]

1 Laboratory Dynamic of Landscape, Risk and heritage, Sultan Moulay Slimane University, Béni Mellal, Morocco.
2 Laboratory of space, History, Dynamic and Development, Sidi Mohamed ben Abdellah university (Fez), Faculty of polydisciplinary, Taza.
3 Laboratory data science for sustainable earth, Sultan Moulay Slimane University, Beni Mellal, Morocco.
Corresponding author: m.goumih@usms.ma

Abstract: Dynamic of landscapes related to climate change and modification of land uses are the crucial elements that controlling hydrological responses within basins. Besides, in the Mediterranean areas, most of the authors indicate the role played by dynamic and change of land cover, and climate characteristics for determining and generating the hydrological processes and the resulting alterations of the surface at spatial and temporal scales and subsurface flow patterns. The purpose of this article is to compare the dynamic of both rating curves in order to extract the discharge for understanding the response and hydrological behaviour of the two selected headwaters (i.e., area = Kousser 311.09 km^2 and Annergui 69.81 km^2), which are situated within middle part of high Atlas Mountains (Morocco) at central part of Assif Melloul River basin. The dynamic of rainfall, and land cover dynamics were been considered in order to extract their observed changes in discharge at the studied headwaters. Besides, the estimation of the discharge in the outlet of both catchments is based on monitoring of the water level by limnimeter. The value of the obtained discharge is extracted by the rating curve method. The obtained results show a clear seasonality of the Assif Melloul basin, which is characterized by variability of the hydrological regime with high inter-annual in space and time. The maximum water level-discharge value at the two catchments is recorded during wet periods (19 m^3/s^{-1} at Annergui and 1m^3/s^{-1} at Kousser catchments), while the minimum level is registered during dry seasons (0.46 m^3/s^1 and 0.21 m$^{3/}$s$^{-1)}$. Consequently, this presented work will be a good investigation for managers to decrease the future impacts and risks related to flash flood events at catchment scale.

Keyword: Hydrological response, rating curve, water level-discharge, Assif Melloul basin, Kousser and Annergui catchments.

1. Introduction

In the Mediterranean regions, climate is marked with significant annual fluctuations of rainfall, and seasonal patterns of precipitation with long period of dry seasons [1]. Furthermore, irregular distribution of rainfall controls the huge differences in terms of water availability across these regions [2]. Besides, in the mountainous areas, most amount of precipitation (from 40% to 85%) are transformed and generated as runoff especially in semi-arid regions [3], consequently, the anthropogenic activities accelerate this response, and decrease sometimes the water availability in the downstream, where the urbanised parts are situated [4]. As a result, the hydrological behaviour of Mediterranean mountain areas is of interest because of the strong seasonality with historical impact of dynamic of land cover, as well as their strategic role in providing water resources to lowlands [5]. Situated in the south part of the

Mediterranean area, Morocco is located in the semi-arid regions where more than 80% of water resources are exploited in agriculture. Moreover, climate is Mediterranean where precipitation is irregular related to the influence of High Atlas Mountain range, is also concentrated, rains in a significant proportion, as contribution of snow to storage in winter followed by snowmelt that creates a baseflow in during summer [6]. Situated in the mountainous of central high atlas (Morocco), Assif Melloul basin is a mountainous part, where the mean annual average rainfall can reach 700 mm yr^{-1} in high summits [7]. According to these variations, the hydrological behaviour of this basin is influenced by this clear seasonality of rainfall [8]. Besides, this response can be linked to the characteristics of catchment landscape [9]. As result, this article tries to assess the response and the hydrological behaviour in the two tributaries located in Assif Melloul basin. This assessment, based on comparing two contrasted rating curves and their effects on hydrological functioning at both studied catchments. This study contributes also to analyse the hydrological response in bare areas within high semi-arid mountains. During very last centuries, hydrologists have attempted to assess and understand the relationships between streamflow and water level or discharge for different objectives such as, controlling flush flood, water supply, and generation of power [10]. In our case, for a good investigating of the behaviour and response of these studied areas, it is crucial to monitor and estimate water level in order to get discharge by realizing the rating curve, which extracted basing on by the field measurements of level so as to determine finally the discharge (Q) values on both headwaters.

2. Methodology and data creation

2.1 Presentation of the study area:

Assif Melloul basin (1773.53 km^2) is an area situated in Central High Atlas of Morocco and has a mountain range that culminates at 3243 m a.s.l (Figure 1). Slopes are also important in most parts of the basin, and their average, estimated from a digital elevation model (DEM), is about 34.7% [11]. The hydrometric stations; Tilouguit (600 m a.s.l) and Zaouit Ahenssal (700) serve as rainfall measurement stations, monitored by the Hydraulic Agency of Oum Err Bia (ABHOER). In terms of geology, the formation of central high Atlas Mountains was related to the complex inversion of the Mesozoic rift in the Northern Africa crust during collision of the Africa Cenozoic [7]. Besides, the Assif Melloul basin is situated in the limestone area of high atlas, which is characterized with a large depression developed by narrow valleys such Annergui and Imilchil valleys [12]. Thus, bare soils in this basin without vegetation cover are the most dominant and impacted by high rates of soil loss increasing the fragility of landscapes to high intensity of soil erosion [13]. Furthermore, the climate of the basin is Mediterranean with a clear influence of continentality in the northwest of the basin and sub-arid in the east. The average annual precipitation is 700 mm. Precipitation mainly occurs during winter, with a dry season taking place from May to September. The average of snowing estimated at 20 Day^{yr} at an altitude of more than 800 m [14]. Temperature registered 15.4 c° of mean annual, with a Max value of 27.3 c° and a Min of 4 c° [15]. Winter considered as the wettest season, then autumn, spring, and summer repectively. Evapotranspiration (ETP) can exceed sometimes 2450 mm^{-1} where the maximum at monthly scale is 250 mm registered during August or July [16]. Besides, rainfall intensities are greatly influenced by topographic gradients [17]. Furthermore, the principal streams and tributaries of the Assif Melloul are located in south-east part of basin, namely the Kousser and Anergui tributaries, which are the two investigated catchments in this study (Figure 1). Furthermore, the seasonal variability of discharge, reflects and relates to the distribution of precipitation. As a result, the maximum discharge is regetred during January while the minimum is recorded during August [15]. In January 2019, two rain stations are built at the downstream part of the two both studied catchments, (i.e., the Kousser catchment (311.09 km^2) and the Annergui catchment (69.81 km^2), for measuring water level and rainfall parameters. However, two other gauging stations are constructed in the southwest part of the Assif Melloul basin at an altitude of 1012 m. The aim of these hydrometric stations is to measure the water stage using limnimeter which help to create a rating curve for extracting the discharge in both studied headwaters. The aim also is to monitor the rainfall with two pluviometers installed in these gauging stations (**Figure 1**).

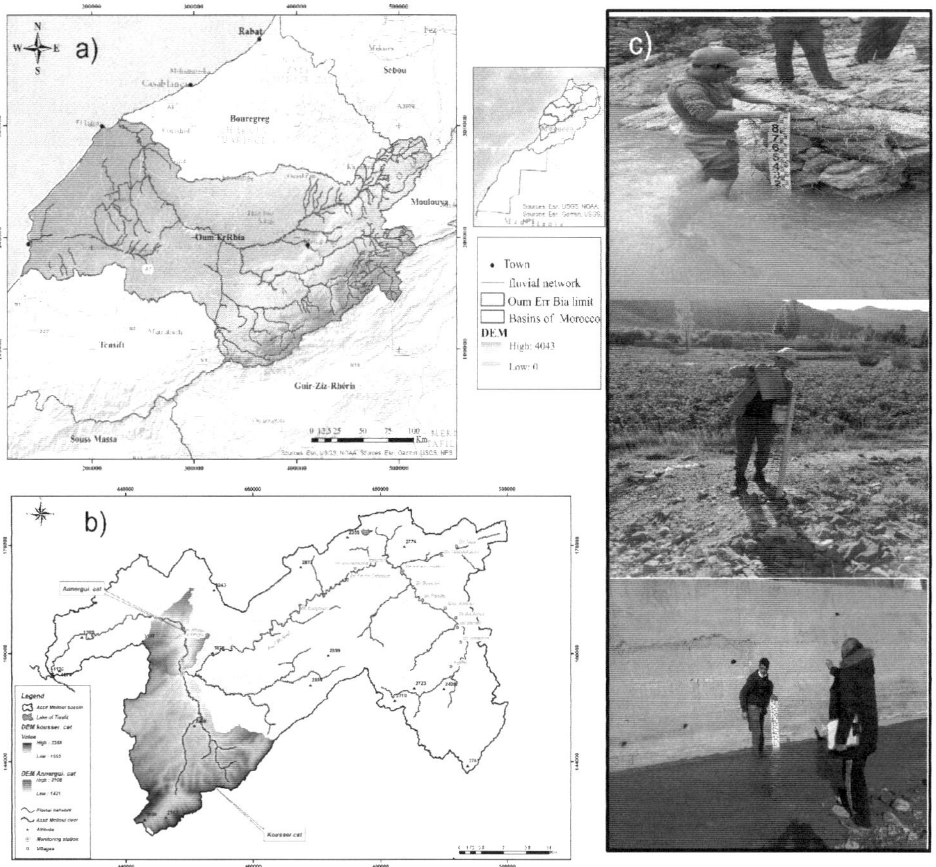

Figure 1: a) Localisation of basin Oum Err Bia River within Morocco, b) Assif Melloul basin and the studied catchments; Kousser and Anergui, and c) installed hydrometric gauging stations.

2.2 Instrumentation and data production:
Analyzing flow and rainfall through two installed hydrometric gauging stations (i.e., limnimeter and pluviometer) allows the understanding of hydrological response within two studied catchments in the Assif Melloul basin. In January 2019, the limnimeter scale was installed at the outlet of both studied catchments, with a cement rectangular broad-crested section at the bank of the river channel constructed by dry-stone in order to control the section. At each section, the water level was recorded using this scale controlled and monitored by a farmer located near to this station. This instrument was graduated 1 m in level. During the starting of measuring discharge by using limnimeter scale, different field works were done in order to maintain the instability of cross section for getting a credible rating curve. The installed data in the studied catchments is complemented with the obtained data from the Hydraulic Agency of Oum Err Rbia (ABHOER). This official data was used to calibrate and correlate the obtained results.

2.3 Methods and procedures: extracting and realizing rating curve:
The rating curve method is a common empirical method used in hydrology to estimate discharge in open natural and artificial channels [18][1]. Rating curve is a good element for hydrologist in order to estimate discharges basing on field investigations [10]. The purpose of this rating curve is measured Q in streams by monitoring water level [19]. The relationships between water stage cross-section parameters lead to get Q values at the monitored cross section. This correlation between the water level and (Q) is

constructed using a statistical analysis and is based on multiple investigations of water level (hi) and Q (Qi) in the field and is formulated as the followed mathematical equation:

$$Q_i = AH_i^b \quad (1)$$

Where A and B are the elements of the correlation, since the direct measurement of water stage by limnimeter is done, the final draw of the experimental curve, which represents the quantity of flood discharge, is often done by calibrating of water stage and discharge relationship. During the hydrological year of 2019, when the measurements are done for determining the stage-discharge rating curve, the river discharge is measured at the sections by the Moulinet tool (TOT) (figure 2).The relation between stage and discharge is calibrated by plotting measurements of discharge with corresponding daily and seasonal observations of the stage, taking into consideration whether the discharge is decreasing, increasing and steady, and also showing some change in stage (photos 1). However, it is standard practice to draw the stage on the ordinate and the rate of Q on the axis (x). Besides, in the abscissa, although when using the stage-discharge relationship to obtain the discharge from a measured value of the stage, then the stage parameter is treated as the independent variable.

Figure 2: a) Measuring the discharge by Moulinet tool in the section and b) measuring the section width in the outlet of Annergui catchment.

3. Results

3.1 Hydrological response of Assif Melloul: Rainfall-discharge relationships:

Rainfall and discharge are crucial parameters for estimating and understanding hydrological response and present the main important factors that controlling hydrological processes [20]. In the Assif Melloul basin, the regime is perennial and irregular, showed a variability and seasonality in terms of seasonal precipitation, intensity of rainfall in time and space; drought in summer; due to the different factors like relief and topographic gradients [1]. Besides, the irregular spatial distribution of precipitation, as figure 3 shows, present differences in water flow within the Assif Melloul basin. This figure presents the distribution of mean annual precipitations and discharges at Tilouguit rain gauge station managed by *ABHOER*. The station is situated in the outlet of the Assif Melloul basin. On an annual scale, the long hydrological years were recorded from 1980 to 2019 at the representative gauging station. Besides a strong variation in rainfall (mm) and discharge (Q m^3/s^{-1}) at each event, annual and seasonal scales were shown, where spring and winter seasons recording more than 85% of the water flow (**Figure 3**).

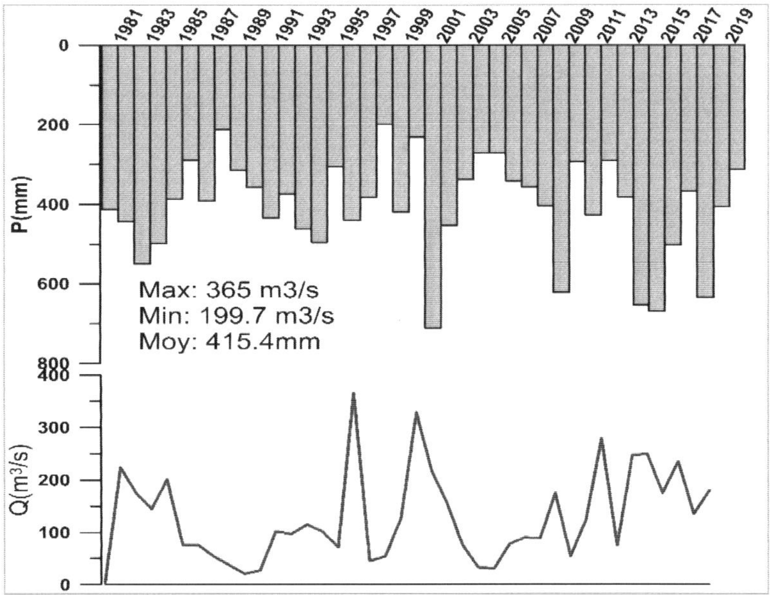

Figure 3: Response of annual discharge (Qm^{3}/$^{s-1}$) to mean annual rainfall (mm) for the long period (1981–2019) at Tilouguit rain gauge station *([11].;* ABHOER 2020).

3.2 Water level- discharge relationships: Creating a rating curve in both studied catchments:
The annual water level-discharge correlation of the Kousser catchment is expressed in the following equation (2). *y = 0.0596e0.055x* (2)
y is the Q (m^3/s^{-1}) and x is water level (cm). Moreover, the stage-discharge rating curve of Kousser headwater can be seen in figure 4.

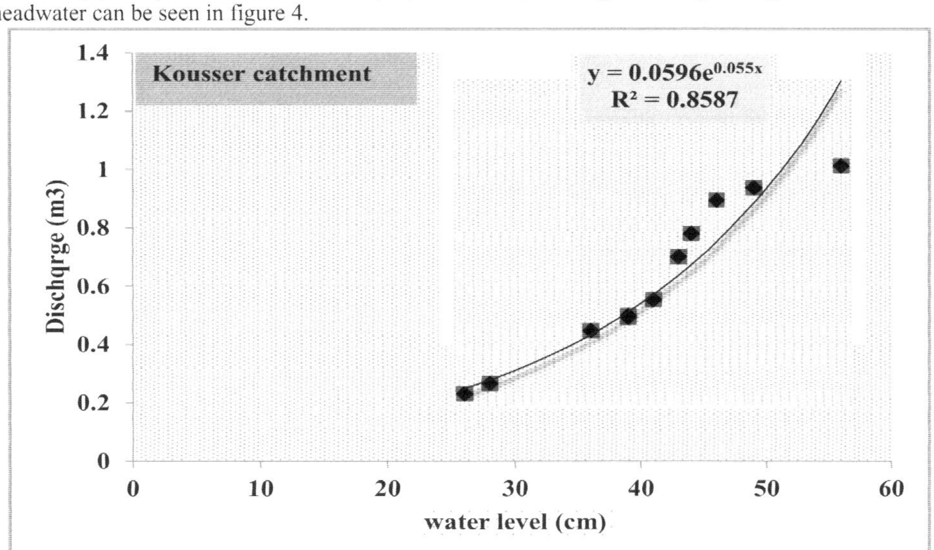

Figure 4: Stage-discharge rating curve of Kousser headwater during hydrological year of 2019-2020.

After determining the stage-discharge rating curve equation, the water level data collected is converted to discharge in both the studied catchments. Based on the rating curve results of the Kousser catchment, the maximum water level was 56 cm registered during 12/08/2019, presenting 1.01 m^3/s^{-1} of discharge.

While the minimum was 26 cm, corresponding to 0.21 m^3/s^{-1} of discharge, recorded during September 2019. However, intensive flush flood events started in July, which increase the discharge periodically during dry seasons. In inverse, located in the downstream part of the Kousser catchment, the Annergui river is different in terms of water level and amount of discharge due to the feeding of the Assif Mellol River. Figure 5 presents the rating curve of the Annergui catchment. The presented equation (2) shows the calibration water level-discharge relationships,

$$y = 0.002e^{0.1324x} \quad (3)$$

Based on these results, it is clear that the role played by Assif Mellol feeding in the increase of Annergui river discharge, especially during wet and dry seasons. The maximum water level-discharge value is recorded during summer (i.e., 68 cm = 19.60 m^3/s^{-1} during December 2019), while the minimum level is registered also during late dry seasons (i.e., 38 cm = 0.46 m^3/s^{-1} during September 2019) (**Figure 5**).

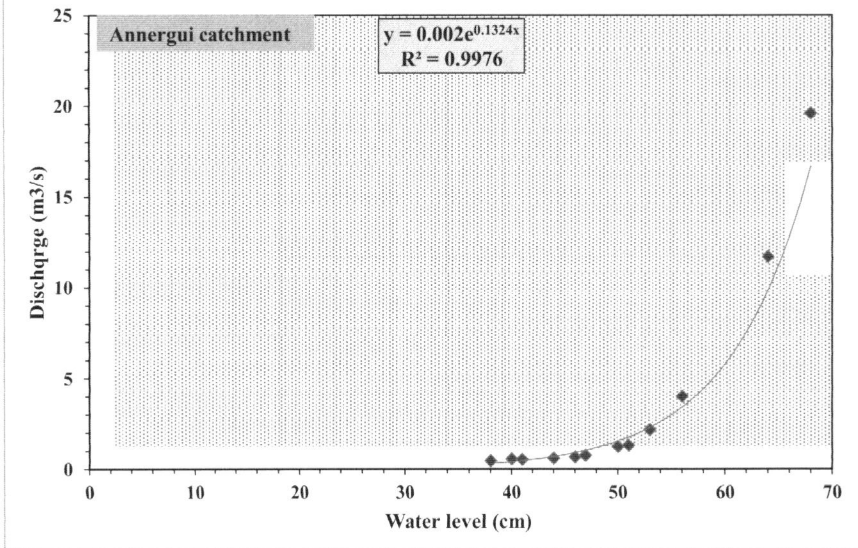

Figure 5: Stage-discharge rating curve of Annergui river during hydrological year of 2019-2020

As a result, the two extracted rating curves indicate the clear significant response between both studied catchments in the Assif Mellol basin. Besides, the aquifer of the Central High Atlas with high permeability plays a crucial role in feeding both streams, especially during rainy periods with a quick response to discharge generation [16].

3.3 Discharge response in the studied catchments:

Figure 6 represents some annual discharge time series in studied catchments that explains that discharge in the Annergui headwater can reach 19 m^3/s^{-1}, making it the most watered important headwater of the Assif Melloul River basin, while the discharge in the Kousser registered 1 m^3/s^{-1} as a maximum during the hydrological year of 2019. Furthermore, the magnitude, seasonality, and duration of peak discharge responses at the two experimental catchments are related to the Assif Melloul River basin's fundamental water balance and hydrological regime (figures 6a and b). The discharge response in these two catchments reveals a clear significance and contrast. The quick response of both studied catchments indicates that forest removal and forest roads often modify discharge response in the outlet of these headwaters. For these reasons, the relationship between forest cover and discharge response has been an important and controversial issue in many regions, especially in the Mediterranean basins [21].

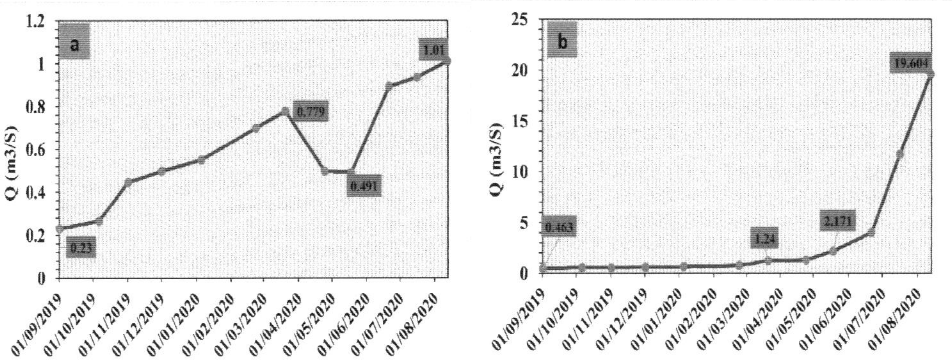

Figure 6: Measured discharge by moulinet tool at a) Kousser and b) Annergui catchments during hydrological year of 2019/2020.

4. Discussion

4.1 Discussion on the role of stage-discharge rating curve in hydrological response

It is known that the scarcity of data in most areas of North Africa is a common issue which impacts the study of hydrological response at different scales. Despite that, the produced rating curves are relatively good tools to obtain real findings at different time scales using calibrated elements to correlate the data. However, the monitored water level is important to understand the response and hydrological variability of Assif Melloul River basin in general and especially the behaviour of two studied headwaters (Annergui and Kousser) in the middle of high Atlas Mountains.

As a result, the obtained rating curves are extracted using the extrapolation method and updated when the water level registers a low or a high change that was observed and recorded on a limnimeter scale. Some errors could result from this extrapolation beyond the range of monitored and measured discharges. Furthermore, during high discharge seasons or flush flood events, the water level may pass or overflow the bank of the section, creating issues during the establishment of a credible relationship between stage and discharge [22]. Consequently, this problem will affect and impact the hydrological data calibration. But, in this case, a new limnimeter was installed near to bankfull in order to assess all the generated flood events during high discharge periods. Regardless the credibility of the obtained results in this work, the study of hydrological response during the past or future, requires a thorough monitoring of the current regime under current climatic and hydrological conditions. [23].

4.2 Factors affecting the hydrological response in the studied catchments:

The maximum discharge was registered during the dry season in both catchments (>19 m^3 s in Annergui and >1 m^3s in Kousser catchments) especially during August 2019. These flush events occurred during dry season, during the same day in both catchments, which confirm the quick response of these mountainous areas. As a result, the hydrological behaviour of the studied catchments illustrates that many components of this response are affected by multiple factors like; forest removal, land cover, lithology, slope, and evapotranspiration [24]. The analysis of discharge dynamics, precipitation, of Assif Melloul River has allowed the investigation of some flush flood events associated to the behaviour of both studied catchments. This behaviour can be compared with some Mediterranean catchments, like the results of [25; 26; 27; 28 and 30]. The findings of these studies, indicate that the quick response of the studied headwaters is related to different factors, which influence and control this hydrological response. Those factors are lithology and geology features, geomorphology, soil type, forest, land cover and climate, which interact and control the rainfall–discharge relationships in the two studied catchments. Besides, this response of the studied catchments can be observed also in multiple mountainous basins in the Maghreb regions and especially in Morocco, which are characterized by a clear seasonality and a quick response [31; 32; 33;34; and 16].

5. Conclusion

This work provides the first primary data of discharge measurement by creating two rating curves in the two mountainous studied catchments with high difficulty to monitoring the water level, especially during flush flood events. However, establishing reliable stream flow time series at different time scales is crucial for studying the hydrological response and water management. Besides, the obtained results of the stage-discharge rating curve are a good parameter to investigate the hydrological response of the Assif Melloul basin so as to have a better investigation of the two studied catchments behaviour.
The examined catchments in this study represent a portion of the range of geographic and climatic conditions of the Central High Atlas. Therefore, they illustrate a clear variability of hydrological response. Moreover, the high variability and intensity of rainfall events lead to the creation intense flash floods in the studied areas, especially during dry seasons.
Consequently, the studied headwaters areas are described with three main characteristics: clear variability in terms of hydrological response; intensity of precipitation; and dry summers that generate and produce a quick flash flood event.

Acknowledgement
The authors thank and appreciate the support and help of ABHOER which is offered a part of the hydrological data used in this work. Thanks to the farmers of Annergui and Kousser regions during the field work 2020 - 2021.

6. References

[1]. Merheb, M, Roger, M, Chadi, A, and Baghdadi, N. 2016. Hydrological Response Characteristics of Mediterranean Catchments at Different Time Scales : A Meta-Analysis. 6667 (January). https://doi.org/10.1080/02626667.2016.1140174.
[2]. Morán, T, Ceballos-Barbancho,A, and Manuel Llorente-Pinto, J. 2010. Hydrological Response of Mediterranean Headwaters to Climate Oscillations and Land-Cover Changes: The Mountains of Duero River Basin (Central Spain)." Global and Planetary Change 72 (1–2): 39–49. https://doi.org/10.1016/j.gloplacha.2010.03.003.
[3]. Viviroli, D., and Weingartner, R. 2004. The Hydrological Significance of Mountains: From Regional to Global Scale. Hydrology and Earth System Sciences 8 (6): 1017–30. https://doi.org/10.5194/hess-8-1017-2004.
[4]. Poyatos, R, Latron, J, and Llorens, P. 2003. Land Use and Land Cover Change After Agricultural Abandonment. Mountain Research and Development 23 (4): 362–68. https://doi.org/10.1659/0276-4741(2003)023[0362:lualcc]2.0.co;2.
[5]. Gallart, F, Latron, J, Llorens, P, and Rabadà, D. 1997. Hydrological Functioning of Mediterranean Mountain Basins in Vallcebre, Catalonia: Some Challenges for Hydrological Modelling.Hydrological Processes 11 (9): 1263–72. https://doi.org/10.1002/(sici)1099-1085(199707)11:9<1263::aid-hyp556>3.0.co;2-w.
[6]. Thomas, S. 2014. Hydrological Modelling in the High Atlas Mountains with the Help of Remote-Sensing Data : Milestones of the SudMed Project,no. 1.
[7]. Markhi, A., N. Laftouhi, Y. Grusson, and A. Soulaimani, S. 2019. Assessment of Potential Soil Erosion and Sediment Yield in the Semi-Arid N'fis Basin (High Atlas, Morocco) Using the SWAT Model. Acta Geophysica 67 (1):263–72.https://doi.org/10.1007/s11600-019-00251-z.
[8]. Teutschbein, C., Grabs, T., Karlsen, R-H., H. Laudon, and Bishop, K. 2015. Hydrological Response to Changing Climate Conditions: Spatial Streamflow Variability in the Boreal Region. Water Resources Research 51 (12): 9425–46. https://doi.org/10.1002/2015WR017337.
[9]. Nadal-Romero. E, Cammeraat, E, María Pilar S-M, Lana-Renault, N, and Regüés, D. 2016. "Hydrological Response of an Afforested Catchment in a Mediterranean Humid Mountain Area: A Comparative Study with a Natural Forest." Hydrological Processes 30 (15): 2717–33. https://doi.org/10.1002/hyp.10820.
[10]. Ajmera, T. K., & Goyal, M. K. 2012. Development of stage-discharge rating curve using model tree and neural networks: An application to Peachtree Creek in Atlanta. Expert Systems with Applications, 39(5), 5702–5710. https://doi.org/10.1016/j.eswa.2011.11.101.
[11]. Ouakhir. H, El Ghachi.M, and Goumih.M. 2019. Caractérisation de La Dynamique des lits fluviaux par la cartographie des morphologies fluviales : Amont du barrage Bin El Ouidane du

Haut Atlas Central Marocain" 7: 379–84. https://www.agromaroc.com/index.php/Actes_IAVH2/article/view/733

[12]. Ouakhir. H, El Ghachi. M, Goumih.M, and Lamti. H. 2020. Fluvial Dynamic in Oued El Abid Basin: Monitoring and Quantification at an Upstream River Section in Bin El Ouidane Dam - 2016 / 2017-(Central High Atlas / Morocco).American Journal of Mechanics and Applications 8 (4): 47. https://doi.org/10.11648/j.ajma.20200804.11.

[13]. Jazouli, Aafaf El, Ahmed Barakat, Rida Khellouk, Jamila Rais, and Mohamed El Baghdadi. 2019. "Remote Sensing and GIS Techniques for Prediction of Land Use Land Cover Change Effects on Soil Erosion in the High Basin of the Oum Er Rbia River (Morocco)." *Remote Sensing Applications: Society and Environment* 13: 361–74. https://doi.org/10.1016/j.rsase.2018.12.004.

[14]. Barakat. A, El Baghdadi.M, Rais. J, Aghezzaf. B, and Slassi. M. 2016. Assessment of Spatial and Seasonal Water Quality Variation of Oum Er Rbia River (Morocco) Using Multivariate Statistical Techniques. International Soil and Water Conservation Research 4 (4): 284–92. https://doi.org/10.1016/j.iswcr.2016.11.002.

[15]. Boudhar. A, Ouatiki. H, Bouamri. H, Lebrini. Y, Karaoui. I, Hssaisoune. M, Arioua. A, and Benabdelouahab.T.2020.Hydrological Response to Snow Cover Changes Using Remote Sensing over the Oum Er Rbia Upstream Basin, Morocco. Advances in Science, Technology and Innovation. Springer International Publishing.https://doi.org/10.1007/978-3-030-21166-.

[16]. El Ghachi. M, Ouakhir.H, Gomih.M, and Lamti.H. 2019.Dynamique fluviale dans l'oued El Abid : Suivi et quantification d'un tronçon fluvial à l'amont du barrage de Bin El Ouidane - 2016 / 2017-(Haut Atlas Central / Maroc)," vol. 3, pp. 45–63,. revues.imist.ma/index.php/Remageom/article/view/15404/0

[17]. Marchane. A, Tramblay.Y, Hanich. L, Ruelland. D, and Jarlan. l. 2017. Climate Change Impacts on Surface Water Resources in the Rheraya Catchment (High Atlas, Morocco).*Hydrological Sciences Journal* 62 (6): 979–95. https://doi.org/10.1080/02626667.2017.1283042.

[18]. Rashidinia. B, and Yekta, R. M. 1967. Evaluation of flow resistant equations by Einstein-Barbarossa and White et al's methods in determining the stage-discharge rating curve of river (Case study : Golgol river of Ilam)," pp. 27–29.

[19]. Morlot. T, Perret.P Anne Catherine. F, and Jalbert. J, 2014. Dynamic Rating Curve Assessment for Hydrometric Stations and Computation of the Associated Uncertainties: Quality and Station Management Indicators. *Journal of Hydrology* 517: 173–86. https://doi.org/10.1016/j.jhydrol.2014.05.007.

[20]. Cristiano. E, Marie Claire Ten. V, and Van De Giesen.N. 2017. Spatial and Temporal Variability of Rainfall and Their Effects on Hydrological Response in Urban Areas - A Review.*Hydrology and Earth System Sciences* 21 (7): 3859–78. https://doi.org/10.5194/hess-21-3859-2017.

[21]. Jones J. A., 2000. Hydrologic processes and peak discharge response to forest removal, regrowth, and roads in 10 small, experimental basins, western Cascades, Oregon. Water Resour. Res., vol. 36, no. 9, pp. 2621–2642, doi: 10.1029/2000WR900105.

[22]. Legesse. D, Vallet-Coulomb. C, and Gasse, F. 2003. Hydrological response of a catchment to climate and land use changes in Tropical Africa: Case study south central Ethiopia, J. Hydrol., vol. 275, no. 1–2, pp. 67–85, doi: 10.1016/S0022-1694(03)00019-2.

[23]. Mansanarez.V, Westerberg. I. K, Lam. N, and Lyon, S. W, 2019. Rapid Stage-Discharge Rating Curve Assessment Using Hydraulic Modeling in an Uncertainty Framework," Water Resour. Res., vol. 55, no. 11, pp. 9765–9787. doi: 10.1029/2018WR024176.

[24]. De Vente. J, Verduyn. R, Verstraeten. G, Vanmaercke. M, and Poesen. J. 2011.Factors controlling sediment yield at the catchment scale in NW Mediterranean geoecosystems. J. Soils Sediments, vol. 11, no. 4, pp. 690–707, doi: 10.1007/s11368-011-0346-3.

[25]. Ares M. G, Varni. M, and Chagas, C. 2020.Runoff response of a small agricultural basin in the argentine Pampas considering connectivity aspects. Hydrol. Process., vol. 34, no. 14, pp. 3102–3119, doi: 10.1002/hyp.13782.

[26]. Rodríguez-Blanco. M. L, Taboada-Castro. M. M, and Taboada-Castro. M. T. 2010. Factors controlling hydro-sedimentary response during runoff events in a rural catchment in the humid Spanish zone," Catena, vol. 82, no. 3, pp. 206–217, doi: 10.1016/j.catena.2010.06.007.

[27]. Oeurng. C, Sauvage. S., and Sánchez-Pérez. J. M. 2010.Temporal variability of nitrate transport through hydrological response during flood events within a large agricultural catchment in south-

west France," Sci. Total Environ., vol. 409, no. 1, pp. 140–149, doi: 10.1016/j.scitotenv.2010.09.006.

[28]. Morán-Tejeda. E, Ceballos-Barbancho. A., and Llorente-Pinto. J. M. 2010.Hydrological response of Mediterranean headwaters to climate oscillations and land-cover changes: The mountains of Duero River basin (Central Spain)," Glob. Planet. Change, vol. 72, no. 1–2, pp. 39–49, May, doi: 10.1016/j.gloplacha.2010.03.003.

[29]. Fortesa, J., Latron, J., García-Comendador, J., Company, J., & Estrany, J. 2020. Runoff and soil moisture as driving factors in suspended sediment transport of a small mid-mountain Mediterranean catchment. Geomorphology, 368, 107349. https://doi.org/10.1016/j.geomorph.2020.107349.

[30]. Estrany. J, Garcia. C, and Batalla. R. J 2009. .Hydrological response of a small mediterranean agricultural catchment," J. Hydrol., vol. 380, pp. 180–190,doi: 10.1016/j.jhydrol..10.035.

[31]. Fox H. R. and Moore H. M.1993.Fluvial suspended sediment transport and mechanical erosion in the maghreb (North africa)," Hydrol. Sci. J., vol. 38, no. 6, pp. 559–560, doi: 10.1080/02626669309492705.

[32]. Khanchoul. K, El Abidine Boukhrissa. Z, Acidi. A., and Altschul. R. 2012. Estimation of suspended sediment transport in the Kebir drainage basin, Algeria," Quat. Int., vol. 262, pp. 25–31, , doi: 10.1016/j.quaint.2010.08.016.

[33]. Vanmaercke. M, Poesen. J, Maetens. W, de Vente. J, and Verstraeten. G, 2011. Sediment yield as a desertification risk indicator," Sci. Total Environ., vol. 409, no. 9, pp. 1715–1725, , doi: 10.1016/j.scitotenv.2011.01.034.

[34]. Probst J. L. and Suchet P. A. 1992.Fluvial suspended sediment transport and mechanical erosionin the maghreb (north africa)," Hydrol. Sci. J., vol. 37, no. 6, pp. 621–637, doi: 10.1080/02626669209492628.

Assessment of soil erosion rate using the EPM model : case of Ouaoumana basin, Middle Atlas, Morocco.

N Ennaji [1*], H Ouakhir [2], S Halouan [1], and M Abahrour[1]

[1] Laboratory of Space, History, Dynamic and Development, Sidi Mohamed Ben Abdellah University, Faculty of polydisciplinary, Taza, Morocco.
[2] Laboratory of Dynamic of Landscapes, Risks, and Heritage, Sultan Moulay Slimane University, Faculty of Arts and Human sciences, Beni Mellal. Morocco.

* nadia.ennaji@usmba.ac.ma.

Abstract: Soil erosion and land degradation are considered serious problems, decrease soil fertility, flooding, and silting of dams, which makes soil conservation and control of erosion risk are basic issues in all countries. The crucial aim of this study is to evaluate the soil erosion rates and to spatialize its hazard, by applying an empirical model namely the erosion potential model (EPM), known also as the Gavrilovic method. Located in the Middle Atlas of Morocco, the chosen study area is known as the Ouaoumana basin, which is a small mountainous area. In terms of climate, it can be classified as the Mediterranean with mountainous influence. The studied area is characterized by an annual rainfall average range between 378 and 513 mm. Consequently, the obtained results were evaluated using GIS software and compared in terms of two different representative catchments. The criteria of selecting the two studied headwaters are; different localization within the Ouaoumana basin, land uses, slopes, and lithology. The Tiguert catchment (a) which is located in the downstream part, is moderately eroded with a rate of less than 39007 $m^3/km^2/yr^{-1}$. In inverse, the Wanmroud catchment (b) is situated in the upstream part of the studied basin, and is highly eroded due to lithology, slopes, and land use impacts. Besides, within this catchment (b) the produced soil erosion rate varies from 222 to more than 164420 $m^3/km^2/yr^1$.

Keywords: Soil erosion rate – Estimation- EPM - Gavrilovic - GIS software - Middle Atlas-Ouaoumana's basin.

1. Introduction

Soil erosion is a major issue in most Mediterranean regions, affecting soil quality and productivity because most arid and semi-arid soils are shallow in depth. At the national and regional levels, soil erosion is regarded as one of the primary mechanisms of desertification processes. [1]. Soil erosion has been identified as one of the most serious natural problems worldwide due to the catastrophic environmental and economic consequences of the various stages of soil loss and sedimentation. [2]. However, soil erosion is a process contributes to the shaping of landscapes and fertilization of soil in flood plains if it occurs naturally and slowly. Furthermore, it becomes a destructive process when it is exacerbated by a number of anthropogenic factors such as deforestation, overgrazing, incorrect methods of tillage, and unscientific agricultural practices [3].

In MoroccoSoil erosion by water is a major issue and one of the most common causes of decreased soil productivity and pollution of water resources; especially, in the Ouaoumana basin that located in the Middle Atlas. Consequently, the process of detachment, transport, and deposition of soil particles are involving by the modeling of soil erosion. Erosion rates are assessed based on the combination of several management and physical parameters. Several empirical models are proposed to predict the soil erosion by water and associated sediment yield. The problem with those models is that the most of them are not well tested and require several parameters, but they are used due to their simplicity [1].

The goal of this study is to evaluate the soil erosion mechanisms at studied area, and to monitor the dynamics of those processes in an integrative approach. In order to obtain input data for the mathematical models, GIS and satellite image analysis techniques are used. The Erosion Potential Model (EPM) was used to estimate erosion processes and to monitor the gullies [4].

2. Study area

Ouaoumana basin is located in the Middle Atlas of Morocco (figure 1), and covers an area of 172.4 km², characterized by a semi-arid climate with rainfall concentrated in time which creates flush flood events. Although, this rainfall is irregular in space and time, starting from October to April with predominance in January, and marked rainstorms from May to August. The annual rainfall in the studied catchment is 475 mm at the El Hanssali gauging station. The temperature varies from 5 to 45 °C. The lithology of the catchment is composed of dolomitic limestone, marls, and red clays. While quaternary deposits are covering the down parts of the basin. Furthermore, soils and their evolution follow essentially the lithological substrate, which under the influence of environmental factors including climate, vegetation, exhibition and topography gives birth to different soil types. However, the studied area is located in mountainous areas which are characterized by a great irregularity of the precipitation amounts and a regression of land cover due to human activities [5]. Two representative catchments were selected from Ouaoumana basin, in terms of working on small areas to have precise results. This choice was based on several criteria, that are considered in different land uses, which the Tiguert catchment (a) is characterized by the dominance of agricultural land, while the Wanmroud catchment (b) knows a spreading of damaged forest and dense cover. Furthermore, litholoy, different localization and slopes are token in consideration.

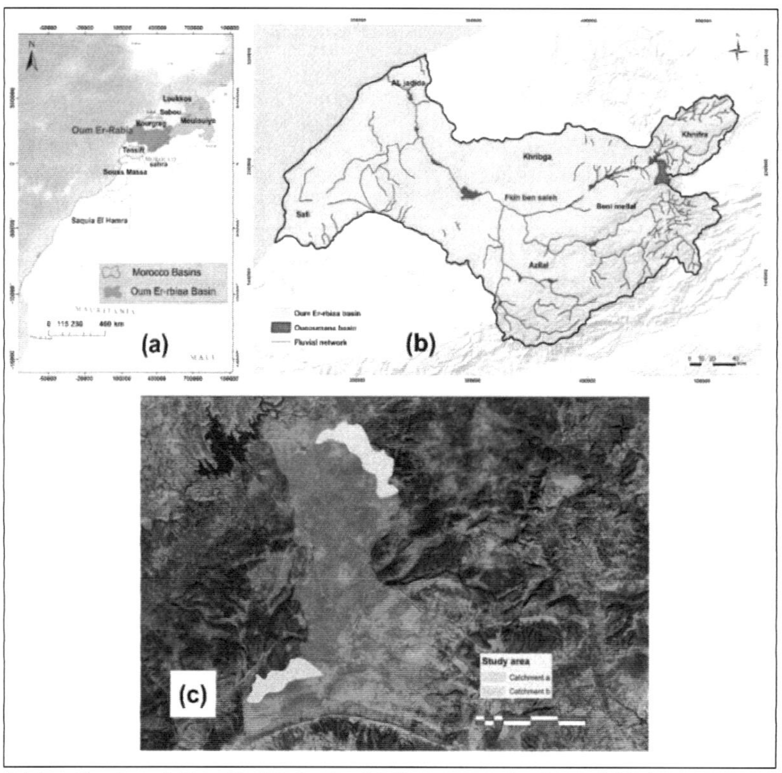

Figure 1. (a) localization of Oum Err Bia basin, (b) Ouaoumana basin, and (c) the representative catchments

3. Material and Methods

For the first time in 1988, Gavrilovic in Yugoslavia presented the EPM model, which is an abbreviation of Erosion Potential model (table 1). This method has been widely utilized for the prediction of soil erosion and sediment yield on a basin scale.[6] and has been developed for management practices in erosion protection, mainly in forest management and torrent control. This model is aimed to estimate the total annual sediment yield of a sub-catchment area and qualify the erosion severity [7], geology type (rock and soil), topography elements (elevation and slope), climatic conditions (mean annual rainfall and mean annual temperature), and land usage are considered as the basic inputs of this model. The erosion process is controlled by three naturally occurring elements (geology, topography, and climate), but the land-use aspect is totally human-controlled [8].

Table 1. Calculation of EPM index.

Index	Formulas
W	$W_{year} = T * H_{year} * \pi * Z^{3/2}$
T	$T = (t / 10 + 0.1)\ 0.5$
Z	$Z = Y * X_a * (\Psi + J_a{}^{0.5})$

The analytical equation for calculation of the average annual gross erosion W (m^3/year), according to Gavrilovic (1972)[9] is:

$$W_{year} = T * H_{year} * \pi * Z^{3/2} \quad (1)$$

Where W_{year} is the average annual gross erosion (m^3/year), **T** is the temperature coefficient [10] calculated according to two methods; from the mean annual air temperature (t, °C) or using the satellite images [11]. As it is shown below:

$$T = (t / 10 + 0.1)\ 0.5 \quad (2)$$

where ***T*** is the mean annual temperature (°C).

According to the EPM, the Z value or the erosion intensity coefficient is basing on the following equation which is the mean value for all EPM calculations [12] (Eq.3):

$$Z = Y * X_a * (\Psi + J_a{}^{0.5}) \quad (3)$$

where Y is the coefficient of the rock and soil resistance to the erosion (function of geology and soil type), range from 2 to 0.25, Xa is the land use coefficient, range from 1.0 to 0.05 (table 2 and 3), Ψ is the coefficient of the observed erosion process (taking into consideration clearly visible erosion processes), ranging from 1.0 to 0.1 based on the severity of erosion, and J_a is the average slope gradient of the watershed in % [13; 14; 15; 8].

Table 2. EPM erosion qualitative categorization and range of erosion coefficients (Z)

Erosion categories	Erosion intensity	Limit Values (Z)
I	Excessive erosion- deep erosion process (gullies, rills rockslides and similar)	$Z > 1$
II	Heavy or milder forms of excessive erosion	$0.71 < Z < 1$
III	Medium erosion	$0.41 < Z < 0.71$
IV	Slight erosion	$0.2 < Z < 0.4$
V	Very slight erosion	$Z < 0.19$

The applied methodology is based on determining the EPM model factors, to simulate the soil erosion rate over the last 20 years, using the monthly rainfall data that were recorded at the Taghzirt, Chacha Nmallah, Ahmed El Hanssali and Tarhat gauging stations. Therefore, the H, T factors and the Y factor were considered uniform in the study area, whereas the Ja, Ψ, and Xa factors were calculated considering a spatial variability in the study area (table 3) [16]. Besides, two satellite images processing, GIS and Grapher software's were used for the preparation of the required database and for analysis of the data [17].

Table 3. Descriptive factors used in the EPM model

Soil erodibility coefficient	Y
Hard rock, erosion resistant	0.1-0.3
Rock with moderate erosion resistance	0.3-0.5
Weak rock, schistose, stabilized	0.5-0.6
Sediments, moraines, clay and other rock with little resistance	0.6-0.8
Fine sediments and soils without erosion resistance	0.8-1.00
Erosion and stream network development coefficient	Ψ
Little erosion on watershed	0.1-0.2
Erosion in waterways on 20–50% of the catchment area	0.3-0.5
Erosion in rivers, gullies and alluvial deposits, karstic erosion	0.6-0.7
50–80% of catchment area affected by surface erosion and landslides	0.8-0.9
Whole watershed affected by erosion	1.00
Soil protection coefficient	Xa
Mixed and dense forest	0.05-0.20
Thin forest with grove	0.05-0.20
Coniferous forest with little grove, scarce bushes, bushy prairie	0.20-0.40
Damaged forest and bushes, pasture	0.40-0.60
Damaged pasture and cultivated land	0.60-0.80
Areas without vegetal cover	0.80-1.00

4. Results and discussion

4.1 Result of soil erosion rate estimation

The EPM model is used to estimate soil erosion rate based on distributed input data from GIS software. This model is including four essential factors [8]; where Ψ: Erosion and stream network development coefficient, Xa: Soil protection coefficient, Y: Soil erodibility coefficient, Ja: slope % (table 4). However, each one of those factors will have a proper value according to its contribution to the erosion process [18]. The data necessary for these factors are as follow:

Table 4. The basic factors in EPM model to estimate the soil erosion rate

Required data	Factor
Digital elevation model	Ja: slope %.
Land uses	Xa: Soil protection coefficient
Lithology map	Y: Soil erodibility coefficient
Satellite imagery	Ψ: Erosion and stream network development coefficient

To provide multisource data with a consistent scale and coordinate reference system, a data harmonization procedure was employed. GIS was used to process the supplied data. For the period 1990 to 2020, cartographic output (erosion intensity map and average yearly erosion map) was created [20].

4.1.1 Xa: *Soil protection coefficient*

Landsat 8 imagery of the year 2020 and fieldwork, were carried out to prepare the land use map of the two representative catchments. The land-uses classes include forest, agriculture land, and bare land in both catchments, the only difference is the damaged forest in the Wanmroud catchment (b) which gives the Xa index a different value. The spatial distribution of all the three land-uses classes is given in the table 5.

The harmonization of the soil protection index criteria (figure 2) range between 0.125 and 0.9 at Tiguert catchment (a), The value of 0.125 is covering the majority of the studied area, representing forest with 66.28 %, which matches forest and dense vegetation cover [21]. The value of Xa in Wanmroud catchment (b) 0.5 is reflecting the existence of damaged forest, this last can be explained by the need of the population to the heating which affects the wood resources negatively.

Table 5. Spatial distribution of land-uses at both representative catchments

Tiguert catchment (a)			
Xa	Land uses forms	Area km²	%
0.7	Agricultural land	3.01	28,31
0.9	Bare land	6.9	66.28
0.125	Forest	0.5	4.80
Wanmroud catchment (b)			
Xa	Land uses forms	Area km²	%
0.7	Agricultural land	0.4	6.05
0.9	Bare land	1.04	15.7
0.5	Damaged forest	5.17	78.21

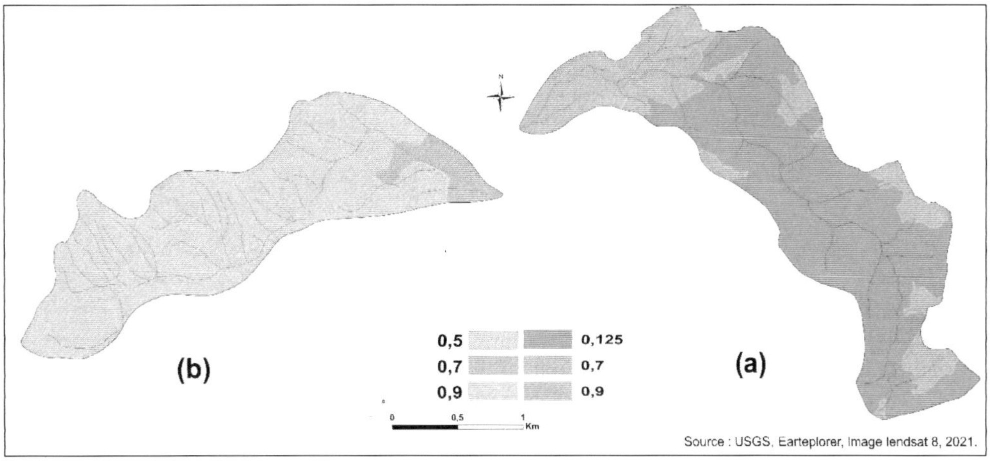

Figure 2. Xa index in (a) and (b) catchments

4.1.2. Ja: slope factor %

The slope is considering one of the most important determining elements in erosion development. It is frequently used in the calculation of soil erosion mapping [22]. The longer of the slope length is the amount of cumulative runoff. Also, the steeper of the area's slope, the higher is the velocity of the runoff that contributes to soil erosion [23]. The slope percent at the two representative catchments were assessed considering the DEM (30 m 2 grid cell size.).

The Ja values as observed in the map (figure 3), show that the slope values at Tiguert catchment (a) range from 0 to 11 %, while slope at Wanmroud catchment (b) varies between 0 to 7.05%, which is considered as moderately sloping area. The maximum percent of slope is observed in the middle part of Tiguert catchment (a) which is an area with dense forest. The topographic factor is essential in the implementation of the USLE since it describes the surface runoff speed and, therefore, works as an indicator of soil erosion risk in catchments [24].

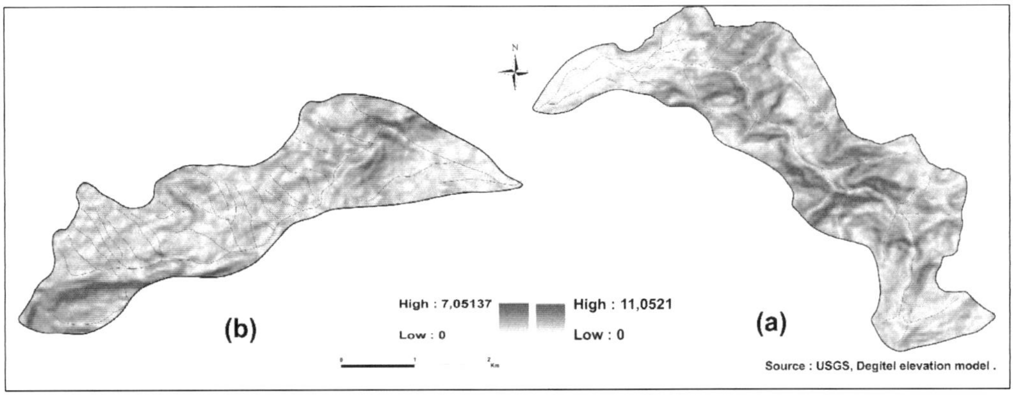

Figure 3. Ja index map of both a) Tiguert and b) Wanmroud catchments

4.1.3. Y: Soil erodibility coefficient

According to lithology structures in the basins, the geology factor plays an essential role in erosion formation, since these structures tend to contribute to differences in stone stability, strength, and soil texture [22].
In this study, the lithology type data was extracted from the geological map of Rabat at 1/200000 scale and field observation (2021). The presented map (figure 4) demonstrates that the two studied catchments are dominated by limestone, red clay, and marls. Tiguert catchment (a) is composed essentially of limestone, covering 74 %, and located in the upstream part of the catchment, the rest considering the downstream part is occupied by red clay and marl (26%). Whereas most area of Wanmroud catchment (b) is covered by laminated limestone presenting 69%, and white sandstone with 31%. The classification given to Y factor according to their lithology type is shown in figure 4.

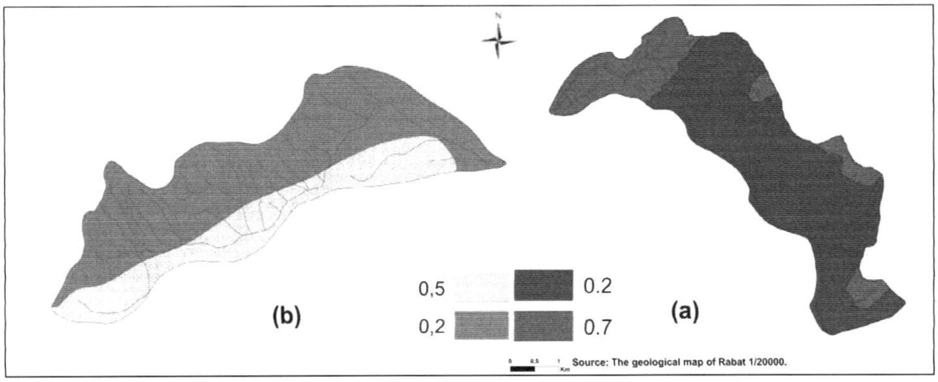

Figure 4. Y index map of both a) Tiguert and b) Wanmroud catchments

4.1.4 Ψ: Erosion and stream network development coefficient

The erosion and stream network development coefficient reflects the soil erosion type at the catchment [20]. Determines, and identifies the areas affected by the risk of erosion (rivers, streams, gullies, alluvial deposits, or the entire watershed) [21].

However, application of this factor is based on applying the following equation $\mathbf{\Psi} = \frac{\mathrm{R}}{Q\,MAX}$, using the red band of Landsat 8 imagery. The observed erosion process in figure 5, ranges from 0.3 to 0.6, based on the severity of erosion [7]. According to this index, a very high erosion risk is observed in Tiguert catchment (a) at the downstream part, which is considered an area of agricultural parcels. The Wanmroud catchment (b) knows a spreading of moderate to low severity of erosion but the upstream part of the catchment, which is an area knows a human intervention represented by using the wood trees in heating from cold winters.

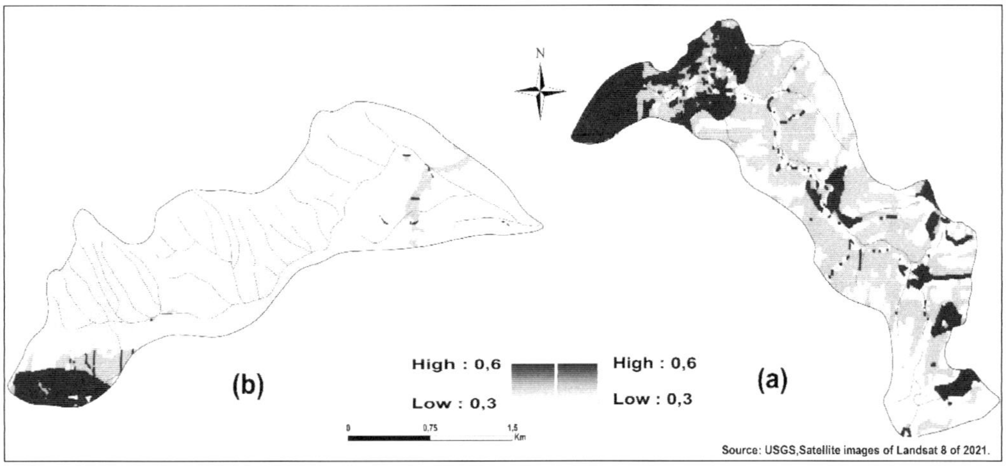

Figure 5. Ψ index map of both a) Tiguert and b) Wanmroud catchments

4.1.5 Z: erosion intensity coefficient

The erosion coefficient (Z) is a non-dimensional indicator, that shows how severe or intense erosion occurs in a catchment. According to the Gavrilovic table, its value (Z) specifies the erosion class (table 6) [21].

$$Z = Y * X_a * (\Psi + ja^{0.5}) \quad (4)$$

The combination of the previous four factors (Y-Xa -Ja- *Ψ),* leads to the calculation of the erosion intensity coefficient which illustrates in figure 6, the results show that most of the catchment area suffers from medium to very slight erosion, concerns more than 70 % of the area in both catchments. The heavy and aggressive erosion occupies a small area that reaches 15%, distributed over bare and agricultural lands [13].

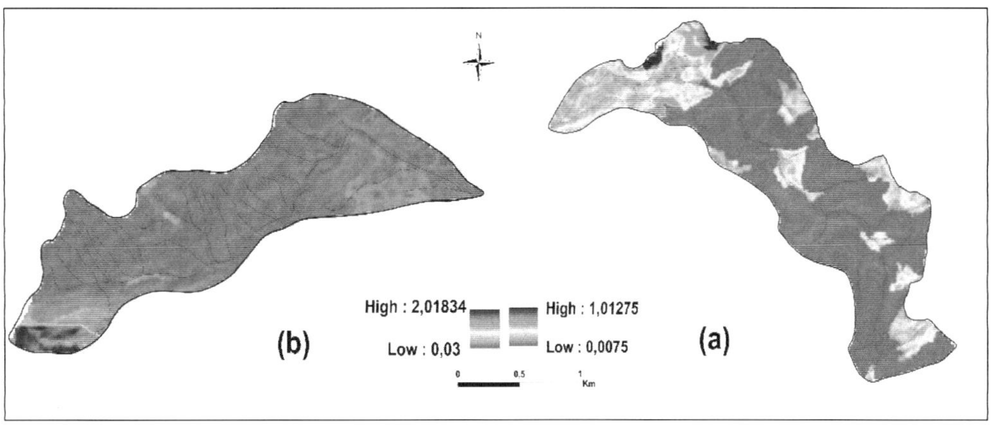

Figure 6. Spatial distribution of Z factor of both a) Tiguert and b) Wanmroud catchments

Table 6. The Z index classification

Erosion categories	Erosion intensity	Limit Values (Z)
I	Excessive erosion- deep erosion process (gullies, rills rockslides and similar)	Z > 1
II	Heavy or milder forms of excessive erosion	0.71 <Z< 1
III	Medium erosion	0.41 <Z< 0.71
IV	Slight erosion	0.2 <Z< 0.4
V	Very slight erosion	Z< 0.19

4.1.6. T: the temperature factor

Because thermal fluctuations allow for the dismantling of components and rock formations, Gavrilovic attempted to use temperature as an erosion factor in the EPM model [21]. Temperature data for this study (daily maximum, minimum, mean) are available from four gauging stations (Taghzirt, Chacha Nmallah, Ahmed El Hansali, and Tarhat). High temperatures affect biological activity in the soil and lead to high evaporation rates.

Two methods exist to calculate the temperature coefficient, the first method is based on the following equation T= (t / 10 + 0.1) 0.5 or can be calculated from the main annual temperature average. The obtained results (figure 7) show the maximum value at Tiguert (a) and Wanmroud (b) catchments are 35 C° and 28 C°, while the minimum values are 22 C° and 33 C° respectively (figure 7).

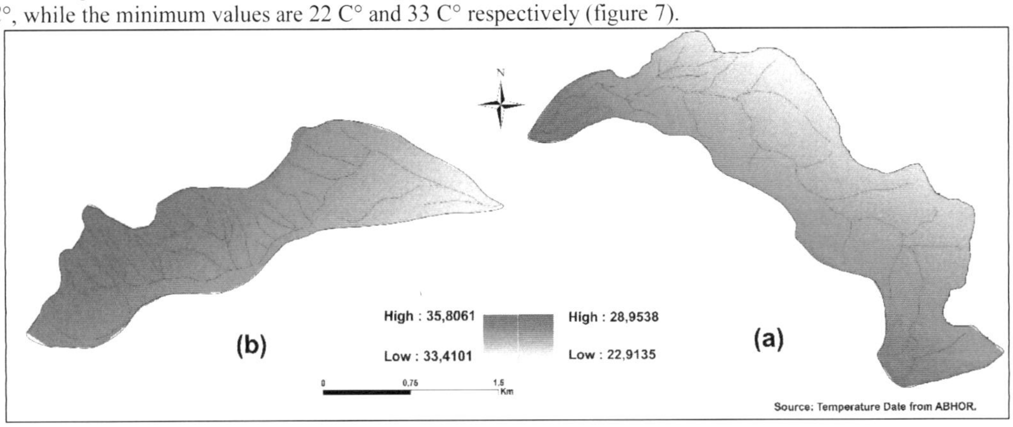

Figure 7. The temperature factor map at a) Tiguert and b) Wanmroud catchments

4.1.7 H: the precipitation factor

The rainfall erosivity has a positive linear connection with the precipitation volume, yearly precipitation from 1980 to 2020 (considering different climate scenarios) is provided for each gauging station used in the modeling of the H factor for the study [7].

The close geographic location of the two selected catchments (Tiguert (a), Wanmroud (b), gives a similar value of rainfall average, the maximum value reaches 500 mm at the two catchments covers the downstream part decreasing toward the upstream part (figure 8).

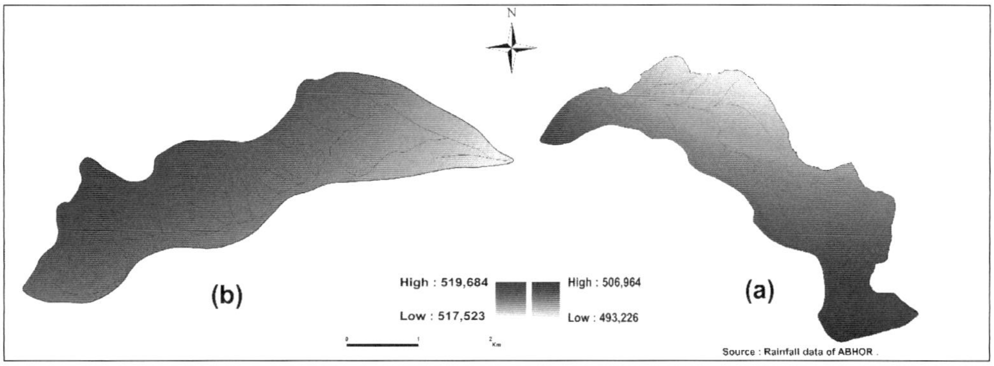

Figure 8. The spatial distribution of precipitation at a) Tiguert and b) Wanmroud catchments

4.1.8 W: the soil erosion loss map m^3 /km^2/year

The mean annual (compact) volume of detached soil due to surface erosion (W, m3/y) was calculated in relation to the inter-annual implementation of the EPM [26]. The annual total volume map (W) was created by applying the following equation after elaboration of all the maps representing the different factors of the EPM model:

$$W_{year} = T * H_{year} * \pi * Z^{3/2} \qquad (5)$$

With: **T:** temperature coefficient **H:** annual rainfall., ***Z*** : erosion coefficient.

According to [27] and [28] about the soil erosion rate classification, six categories were identified as shown in table 7.

Table 7. Soil erosion rate classification (W)

Soil erosion rate m^3 /km^2/year	Erosion intensity
<50	Very slight erosion
50 – 500	Slight erosion
500 – 1500	Moderate erosion
1500 -5000	Heavy erosion
5000-20000	Excessive erosion
> 20000	Catastrophic erosion

A map of soil losses caused by water erosion is created (figure 9), by combining the different elements that compose the Gavrilovic's EPM equation, as well as taking into consideration the numerical values of the seven factors. The maximum average reaching 39007 m^3 /km^2/yr^{-1} at Tiguert (a) catchment and 164420 m^3 /km^2/yr^{-1} at Wanmroud (b) catchment. While the minimum losses per catchment are 23,47 m^3 /km^2/yr^{-1} and 292 m^3 /km^2/yr^{-1} respectively [21].

The dominant erosion potential class is slight to very slight erosion, with 32.86% to 32.46% at the (a) catchment (table 8). Otherwise, Wanmroud (b) catchment knows a spreading of heavy erosion represents 68.63%. After applying the EPM model, the values of moderate and slight erosion potential categories ranged from 28.87 percent to 46.71 percent and 16.75 percent to 2.52 percent of the study area, respectively [8].

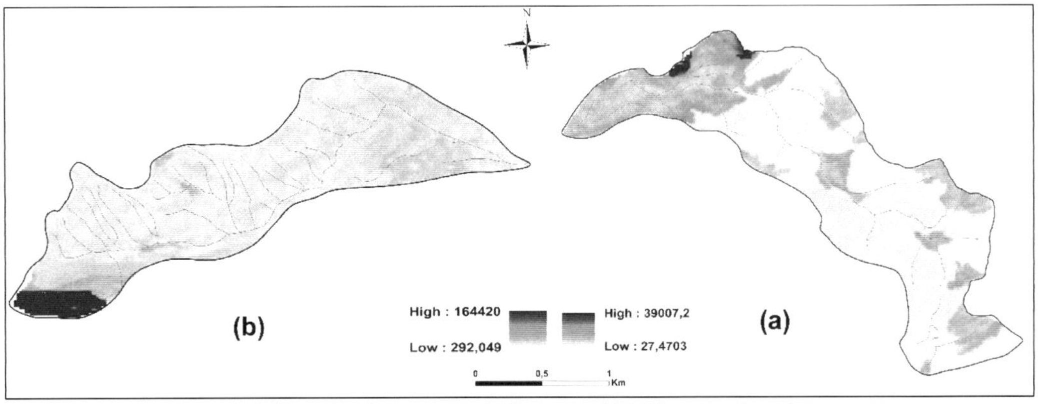

Figure 9. Soil erosion loss map at a) Tiguert and b) Wanmroud catchments

Table 8. Soil erosion rate classification

Classe m^3 /km^2/yr^{-1}	Erosion intensity	Area in Km^2 (a) catchment	%	Area in Km^2 (b) catchment	%
<50	Very slight erosion	3.27	32.86	------	----
50 – 500	Slight erosion	3.23	32.46	0.008	0.12
500 – 1500	Moderate erosion	1.82	18.29	0.06	0.93
1500 -5000	Heavy erosion	1.56	15.67	4.42	68.63
5000-20000	Excessive erosion	0.07	0.70	1.76	27.32
> 20000	Catastrophic erosion	0.0005	0.005	0.2	3.10

The obtained results of this study are compared with the findings of Irene Marzolff (2011), who studied the short-term versus medium-term monitoring for detecting gully-erosion variability in a Mediterranean environment, Deeper gullies show larger retreat rates than shallower gullies, and a positive exponential relationship exists between mean linear headcut retreat rate per year (Rlh a–1; in m/yr) and gully depth (D; in metres) at the headcut: where gully Casablanca erodes approximately 05 m3 of soil per year, gully Salada 1 grows by over 100 m^3 per year in general, deeper gullies show larger retreat rates R^2 = 053; n=9; Rlh a^{-1} = 00278 e 02338 D [29]. Furthermore, Martınez Casasnovas (2003) studied the use of spatial information technology to map and quantify gully erosion and discovered that the total area impacted by gully retreat between 1957 and 1993 was 76.5 hectares. During this time, farmers filled 38.8 ha of existing gullies, primarily to develop new grape plots [30].

4.2. Data validation

In the aim of the validation of the obtained results, three gullies are selected from the downstream part of the Tiguert catchment in order to monitor their evolution during the period 2021 to 2022.The measurement method is based on noting the changes in width.

Besides, table 9 and figure 10 present three monitoring gullies, that showed a high evolution of erosion dynamic in terms of widening especially at the Tiguert catchment. The results show a remarkable evolution of the gullies' width, increasing by a few centimetres, ranging from 1 to 10 cm over the two years. That's caused by several factors; the aggressiveness of rainfall and its concentration in time and space, the fragile lithology composed essentially of clay and marls, also the agricultural practices of the local population which make overexploitation of natural resources.

Table 9. The gullies evolution at Tiguert catchment

	2021				2022	
Gully 1 Sites	Date	X	Y	Width (cm)	Date	Width (cm)
S1	15/03/2021	32,72	5,77	230	24/1/2022	234
S2	15/03/2021	32,72	5,76	130	24/1/2022	137
S3	15/03/2021	32,73	5,78	150	24/1/2022	156
S4	15/03/2021	32,73	5,77	210	24/1/2022	215
Gully 2						
S1	15/03/2021	32,56	5,78	30	24/1/2022	37
S2	15/03/2021	32,73	5,79	20	24/1/2022	28
S3	15/03/2021	32,74	5,78	50	24/1/2022	59
S4	15/03/2021	32,74	5,76	95	24/1/2022	104
Gully 3						
S1	15/03/2021	32,74	5,78	120	24/1/2022	125
S2	15/03/2021	32,75	5,76	130	24/1/2022	136
S3	15/03/2021	32,71	5,74	110	24/1/2022	118
S4	15/03/2021	32,72	5,72	250	24/1/2022	255

Figure 10. Photos from Tiguert catchment of monitored gullies.

5. Conclusion

Soil erosion is assessed in the two representative catchments at the Ouaoumana basin, using the Gavrilovic's EPM model, as well as the geographic information system (GIS), to provide a quantitative assessment of soil losses. The study indicates that the Ououmana basin is susceptible to erosion, and it leading to the creation of mean factors that control soil erosion, such as rainfall, land use, and slope.

In terms of results, this study aimed to evaluate the EPM implementation in two different catchments selected from the Ouaoumana catchment, in order to determine the soil erosion loss taking into consideration different natural and human characteristics. The quantitative analysis of soil erosion risk obtained from the EPM model categorizes the study area into low, moderate, high, and very high soil erosion rates.

6. References

[1] M. Abdullah, R. Feagin, and L. Musawi, "The use of spatial empirical models to estimate soil erosion in arid ecosystems," *Environ. Monit. Assess.*, vol. 189, no. 2, 2017, doi: 10.1007/s10661-017-5784-y.

[2] D. T. Robinson, B. U. Meinen, and D. T. Robinson, "From hillslopes to watersheds : Variability in model outcomes with the USLE From hillslopes to watersheds : Variability in model outcomes with the USLE," *Environ. Model. Softw.*, vol. 146, no. October, p. 105229, 2021, doi: 10.1016/j.envsoft.2021.105229.

[3] A. Demirci and A. Karaburun, "Estimation of soil erosion using RUSLE in a GIS framework: A case study in the Buyukcekmece Lake watershed, northwest Turkey," *Environ. Earth Sci.*, vol. 66, no. 3, pp. 903–913, 2012, doi: 10.1007/s12665-011-1300-9.

[4] R. Zakerinejad and M. Maerker, "An integrated assessment of soil erosion dynamics with special emphasis on gully erosion in the Mazayjan," *Nat. Hazards*, 2015, doi: 10.1007/s11069-015-1700-3.

[5] A. El Jazouli, A. Barakat, A. Ghafiri, S. El Moutaki, and A. Ettaqy, "Soil erosion modeled with USLE , GIS , and remote sensing : a case study of Ikkour watershed in Middle Atlas (Morocco)," *Geosci. Lett.*, 2017, doi: 10.1186/s40562-017-0091-6.

[6] G. Petkovšek and J. R. Ć, "Applicability of the Gavrilovi ć method in erosion calculation using spatial data manipulation techniques," no. 279, pp. 224–233, 2003.

[7] M. H. Tangestani, "Comparison of EPM and PSIAC models in GIS for erosion and sediment yield assessment in a semi-arid environment: Afzar Catchment, Fars Province, Iran," *J. Asian Earth Sci.*, vol. 27, no. 5, pp. 585–597, 2006, doi: 10.1016/j.jseaes.2005.06.002.

[8] A. Bagherzadeh and M. R. M. Daneshvar, "Sediment yield assessment by EPM and PSIAC models using GIS data in semi-arid region," *Front. Earth Sci.*, vol. 5, no. 2, pp. 207–216, 2011, doi: 10.1007/s11707-011-0168-z.

[9] S. Kostadinov and S. Dragicevic, "Torrential flood prevention in the Kolubara river basin Torrential flood prevention in the Kolubara river basin," no. November, 2017, doi: 10.1007/s11629-017-4575-9.

[10] N. Efthimiou, E. Lykoudi, and C. Karavitis, "Comparative analysis of sediment yield estimations using different empirical soil erosion models," *Hydrol. Sci. J.*, vol. 00, no. 00, pp. 1–21, 2017, doi: 10.1080/02626667.2017.1404068.

[11] H. Noori, S. Mostafa, and B. Mojaradi, "International Soil and Water Conservation Research Assessment of sediment yield using RS and GIS at two sub-basins of Dez Watershed , Iran," *Int. Soil Water Conserv. Res.*, pp. 0–1, 2016, doi: 10.1016/j.iswcr.2016.06.001.

[12] P. Taylor and S. Maimouni, "Geo-spatial Information Science Water erosion risk mapping using derived parameters from digital elevation model and remotely sensed data," no. January 2015, pp. 37–41, doi: 10.1080/10095020.2012.715855.

[13] Z. Gavrilovic, M. Stefanovic, I. Milovanovic, and J. Cotric, "Torrent Classification – Base of Rational Management of Erosive Regions," vol. 4, pp. 1–8, 2008, doi: 10.1088/1755-1307/4/1/012039.

[14] R. Tošicá, S. Dragićević, and N. Lovrić, "Assessment of soil erosion and sediment yield changes using erosion potential model - case study: Republic of srpska (bih)," *Carpathian J. Earth Environ. Sci.*, vol. 7, no. 4, pp. 147–154, 2012.

[15] J. D. V. T and J. Poesen, "Predicting soil erosion and sediment yield at the basin scale : Scale issues and semi-quantitative models," vol. 71, pp. 95–125, 2005, doi: 10.1016/j.earscirev.2005.02.002.

[16] F. Berteni and G. Grossi, "Water soil erosion evaluation in a small alpine catchment located in northern italy: Potential effects of climate change," *Geosci.*, vol. 10, no. 10, pp. 1–24, 2020, doi: 10.3390/geosciences10100386.

[17] S. K. Jain, S. Kumar, and J. Varghese, "Estimation of Soil Erosion for a Himalayan Watershed Using GIS Technique," pp. 41–54, 2001.

[18] F. Amiri, "Air , Soil and Water Research Estimate of Erosion and Sedimentation in Semi-arid Basin using Empirical Models of Erosion Potential within a Geographic Information System," pp. 37–44.

[19] E. Tadi and N. Sad, "erosion intensity assessment using erosion potential method and geographic information systems : a case study of beočin municipality , Serbia," no. September, 2018, doi: 10.5937/ZbDght1801032T.

[20] E. Maliqi and S. K. Singh, "Quantitative Estimation of Soil Erosion Using Open-Access Earth Observation Data Sets and Erosion Potential Model," 2019.

[21] L. Marouane, B. Lahcen, and M. Valérie, "Assessment and mapping of water erosion by the integration of the Gavrilovic " EPM " model in the Inaouene watershed," vol. 03009, 2021.
[22] H. Shahabi, S. Khezri, B. Bin, and M. Hashim, "Catena Retraction notice to 'Landslide susceptibility mapping at central Zab basin, Iran : A comparison between analytical hierarchy process , frequency ratio and logistic regression models ' [CATENA 115 (2014) 55 – 70]," *Catena*, vol. 208, no. October 2021, p. 105784, 2022, doi: 10.1016/j.catena.2021.105784.
[23] B. Stefan, C. Horvath, and P. Cocean, "Implementation of the usle model using gis techniques. Case study the Some ş ean plateau implementation of the usle model using GIS techniques: case study the Somesean plateau, October 2015, 2009.
[24] C. P. Devatha and M. S. Renukaprasad, "Estimation of Soil loss using USLE model for Kulhan Watershed , Chattisgarh- A case study," vol. 4, no. Icwrcoe, pp. 1429–1436, 2015, doi: 10.1016/j.aqpro.2015.02.185.
[25] M. A. Nearing, V. G. Jetten, and C. Baffaut, "Modeling Response of Soil Erosion and Runoff to Changes in Modeling response of soil erosion and runoff to changes in precipitation and cover," no. June 2014, 2005, doi: 10.1016/j.catena.2005.03.007.
[26] N. E. & All, "Science of the Total Environment Soil erosion modelling: A global review and statistical analysis," vol. 780, 2021, doi: 10.1016/j.scitotenv.2021.146494.
[27] Zachar, "$102.25," pp. 140–141, 1982.
[28] I. Milevski, "Estimation of soil erosion risk in the upper part of Bregalnica watershed-republic of Macedonia , based on digital elevation estimation of soil erosion risk in the upper part of Bregalnica watershed-republic of Bacedonia , based on digital elevation model," no. May 2008, 2015.
[29] I. Marzolff, J. B. Ries, and J. Poesen, "Short-term versus medium-term monitoring for detecting gully-erosion variability in a Mediterranean environment," *Earth Surf. Process. Landforms*, vol. 36, no. 12, pp. 1604–1623, 2011, doi: 10.1002/esp.2172.
[30] J. A. Martínez-Casasnovas, "A spatial information technology approach for the mapping and quantification of gully erosion," *Catena*, vol. 50, no. 2–4, pp. 293–308, 2003, doi: 10.1016/S0341-8162(02)00134-0.

Pastoral pressure and degradation of forest and peri-forest rangelands in the central Middle Atlas: case of the Tigrigra watershed (Morocco)

H Achiban [(1) (2)] *, A Taous [(1)], H Achiban [(1)], I El-Khantoury [(1)], E. H Abba [(3)], R Addou [(1)]

[(1)] Sidi Mohamed Ben Abdallah University, Fez-Morocco
[(2)] Ifrane National Park, (Ministry of Agriculture, Maritime Fishing, Water and Forestry).
[(3)] Sultan Moulay Slimane University, Béni Mellal-Morocco

*Corresponding author: hassan.achiban@usmba.ac.ma

Abstract. Although the forest ecosystems of the western region of the central Middle Atlas have a significant fodder potential and a particular richness in terms of biodiversity. They are currently subject to a regressive dynamic under the influence of overgrazing. This study allows to draw up a qualitative and quantitative assessment of the pastoral pressure on the natural ecosystems of the Tigrigra watershed. This is a territory where climatic changes and social mutations (stability of herders and cultivation of land formerly reserved for grazing), combined with the increase in the number of herds, have led to the dependence of livestock feeding on forest fodder resources. Indeed, grazing practices in the study area are based on traditional concepts of extensive livestock rearing and misuse of pastoral resources, which leads to significant dysfunctions between pastoral supply and demand, and consequently to the degradation of natural ecosystems.

1. Introduction

In the Moroccan Middle Atlas, since the 1970s, the pastoral society has gradually abandoned transhumance to settle on grazing areas. The latter are increasingly limited, due to the rapid expansion of agriculture to the detriment of rangelands [1]. In addition, the Middle Atlas has experienced a population increase and an increase in the number of livestock, which has led to an increase in fodder needs, to the detriment of the balance of pastoral potential [2] and [3]. This situation is often accentuated by the Mediterranean climatic conditions marked by strong irregularities in rainfall and temperature.

This study addresses the issue of pastoral pressure and degradation of forest and peri-forest rangelands in the Tigrigra watershed. The analysis was based on the evaluation of pastoral pressure imposed by livestock and the estimation of forage production by vegetation type. The comparison of fodder demand and supply made it possible to determine the state of fodder deficit, the intensity of overgrazing imposed on the rangelands of the watershed, and the equilibrium pastoral load and the actual load. The evolution of overgrazing between 2005 and 2016 was evaluated at the level of each rural area, and on the whole watershed, based on the evolution of all these usual indicators.

2. Data and methods

2.1. *Physical conditions and land use in the Tigrigra watershed*

The Tigrigra watershed drains an area of 1029.4 km² of the western part of the central Middle Atlas. It extends between longitudes 5°05'58" and 5°34'10" West and latitudes 33°10'42" and 33°36'20" North (Figure 1). It is a mountainous watershed straddling the plateaus of the Tabular Middle Atlas and those of the Western Meseta. The altitudes vary between 778 m and 2165 m.

The lithological distribution of the watershed shows that the Paleozoic shales cover 49% of its surface, the Quaternary basalts 26.13%, the limestones and dolomites of the Lias 21%, the doleritic basalts and red clays of the Triassic cover only 2.47%.

The climate of the watershed is Mediterranean, where the summer is very dry, and both winter and spring are wet and cold. Annual precipitation often exceeds 1000 mm in the upper part of the watershed, with a seasonal maximum recorded in winter. The average annual temperature is about 12.3°C with a maximum of 32.7°C in August and a minimum of 2°C in January.

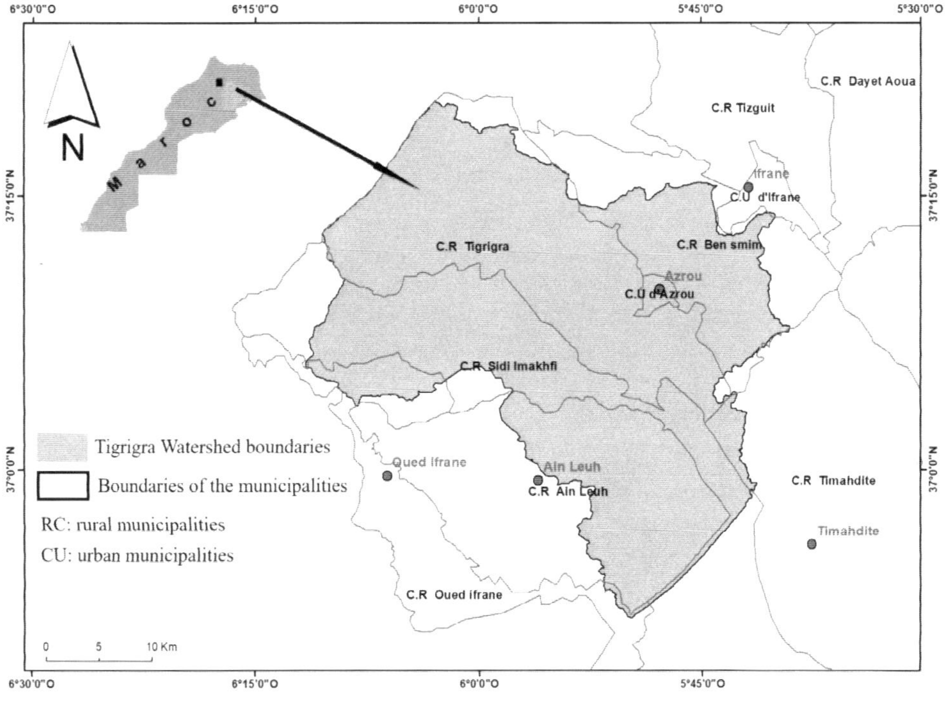

Figure 1. Location of the Tigrigra watershed and limits of the municipalities concerned by the study.

The land use map of the Tigrigra watershed (Figure 2) shows that agricultural land occupies 38.7%, forest 31.5% and rangeland 27.6%. The *Cedrus atlantica*, represents 47% of the forest cover; it is found in a pure state on basalts and in association with *Quercus Rotundifolia*, or with *Quercus faginea* and *Crataegus laevigata* on carbonate formations. The *Quercus Rotundifolia* covers 35% of the forest surface; it is found in coppice or in high forest, pure or mixed with other forest species. The *Quercus faginea* covers 11% of the forest; it occupies mainly the rich and deep soils of the cool areas. The *Genista Quadriflora* covers 6% of the forest area of the watershed; it grows mainly on the schist of the Meseta.

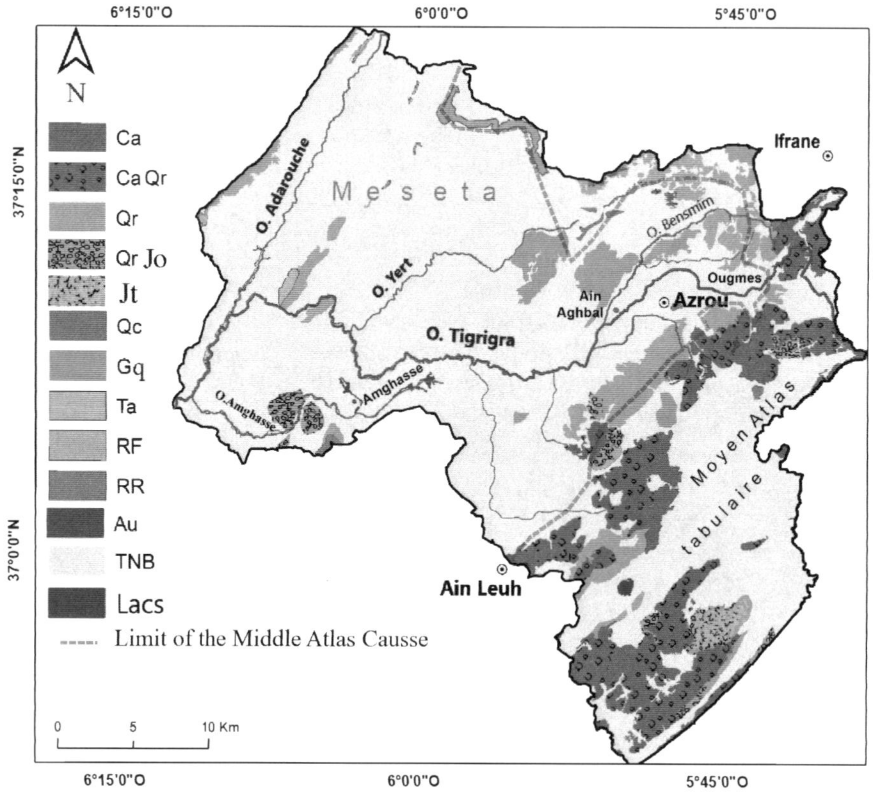

Figure 2. Biogeographic map of the Tigrigra watershed. Ca: *Cedrus atlantica*, Qr: *Quercus rotundifolia*; Jo: *Juniperus oxycedrus*; Jt: *Juniperus thurifera*; Qc: *Quercus faginea*; Gq: *Genista Quadriflora*; Ta: *Tetraclinis articulata*; RF: Reforestation of hardwoods; RR: Reforestation of softwoods; Au: *Crataegus laevigata*, TNB: Unforested land [4].

2.2. *Data*

The statistics used in this study come from the National Association of Sheep and Goat Breeders (ANOC) and the Agricultural Advisory Center of Azrou. The data used show that the livestock of most of the rural communities of the Tigrigra watershed, have experienced strong increases between 2005 and 2016 (Table 1). Indeed, the total evolution is 30%, with maximum value of evolution is 58.3%, recorded in Timahdite, while the only regression is recorded in the territory of Oued Ifrane, with a value of -5%.

Table 1. Evolution of livestock numbers between 2005 and 2016 in the rural municipalities, of the Tigrigra watershed (Ov : sheep, Bov : cattle, Cap : goats).

Rural Municipality	2005			2016			Evolution in %
	Ov	Bov	Cap	Ov	Bov	Cap	
Bensmim	31878	660	2199	46719	1044	3298	47
Tigrigra	77265	7632	5182	94226	7067	7538	20,8
Timahdite	4144	27	233	3975	48	2948	58,3
Sidi El Makhfi	45680	989	4548	71190	1028	7910	56,4
Ain Leuh	42034	317	6005	47038	701	5004	9,1
Oued Ifrane	5467	102	1311	5370	144	1023	-5

The total annual production of the different formations is obtained by multiplying the area of each rangeland by its average annual production (Table. 2).

Table 2. Annual fodder production by type of vegetation [5]

Type of vegetation	Forage Production (FU/ha/year)
Cedrus atlantica	300
Mixture of *Cedrus atlantica*, *Quercus rotundifolia* and *Quercus faginea*	450
Mixture of *Cedrus atlantica* and *Quercus rotundifolia*	450
Quercus rotundifolia	420
Quercus faginea	320
Quercus rotundifolia mixed with the *Juniperus thurifera*	410
Cedrus atlantica mixed with the *Juniperus thurifera*	410
Mixture of *Cedrus atlantica*, *Quercus rotundifolia* and the *Crataegus laevigata*	370
Mixture of *Juniperus oxycedrus* and *Quercus rotundifolia*	300
Genista quadriflora	310
Reforestation	394
Tetraclinis articulata	290
Unforested land	394

2.3. Methods

The study of the evolution of the fodder demand, at the level of each rural community, was carried out on the basis of the numbers of its livestock in 2005 and in 2016. The evaluation of the fodder needs for the whole Tigrigra watershed, was carried out, based on the synthesis of all the rural communities, while taking into account the rate of contribution of each community in the total surface of the watershed.

- *The evolution of forage demand:* The annual forage requirements are estimated in forage units (FU), based on 300 units of forage per year and per small livestock unit (FU/year/SLU), (With 1 SLU = 1 sheep; 1 cattle = 5 SLU; 1 goat = 0.8 SLU) [6].
- *The evaluation of pastoral pressure:* The forage deficit is expressed as a percentage of the total livestock requirement. It was calculated based on the following equation:

$$DF = 100*(RF- SF)/R \qquad (1)$$

where DF: forage deficit; RF: forage requirement, and SF: forage supply.
- *The equilibrium load (El):* This is the number of sheep that a hectare of rangeland can support without compromising its sustainability. It takes into account the current forage production of the rangelands. Its value is defined as the ratio between the average annual production of the rangelands per hectare (AP in forage unit) and the annual requirement of one small livestock unit (SLU):

$$El = AP \text{ per hectare (in FU)} / AR \text{ of 1 SLU (in FU)} \qquad (2)$$

- *The real load (Rl):* This is the load imposed on the range. In other words, it is the animal load that actually exists and uses the rangeland during the year in question. It is equal to the ratio between the number of small livestock units (SLU) and the area of the rangeland (S in ha), taking into account the livestock's stay:

$$Rl = \text{number of SLU}/S/yr \qquad (3)$$

Where: SLU = small livestock unit; 1 sheep = 1 SLU; 1 goat = 0.8 SLU and 1 cattle = 1 LBU = 5 SLU.
- *The overgrazing coefficient (Oc):* It is defined as the ratio of the equilibrium load (El) to the real load (Rl). It is expressed in % by the relationship:

$$Oc = 1- (El/Rl)*100 \qquad (4)$$

3. Results and discussion

3.1. Evolution of forage demand

The results obtained show that the total need is estimated at 91.88 million forage units in 2016 against only 70.7 million in 2005, an increase of 30%. The distribution of needs at the level of each municipality (figure 3) highlights a strong heterogeneity of fodder demand among the rural municipalities around the watershed. It also shows that the rate of change between 2005 and 2016 is not homogeneous from one territory to another. It varies from -4.99% for Oued Ifrane, to over 58.3% in Timahdite.

The increase in the demand for fodder for the majority of the communal territories of the watershed is mainly related to the growth of livestock. Indeed, the population growth and the stability of the breeders as well as the free access to the rangelands, favored the orientation of the population towards the breeding activity, which led to an excessive increase in the number of livestock.

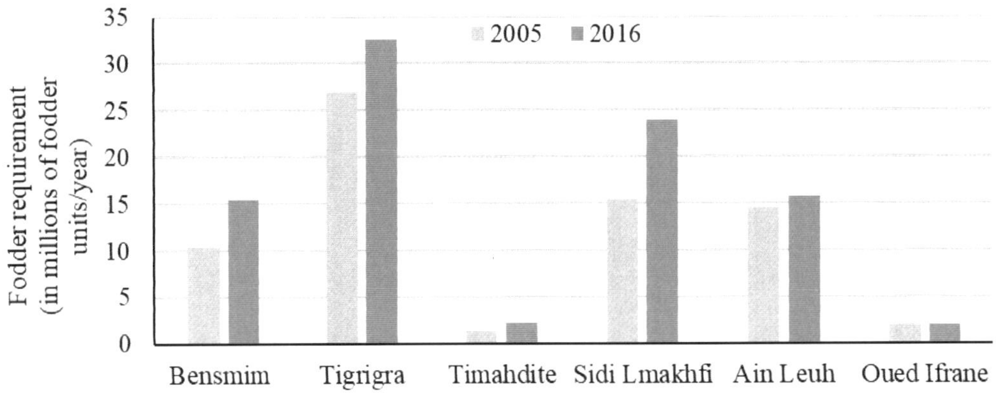

Figure 3. Supply and needs of fodder in the different rural territories of the Tigrigra watershed

3.2. Assessment of forage supply

It should be remembered that there is a close relationship between the type of vegetation and forage production in FU/year [7]. The general assessment of forage potentials was identified on the basis of the results of phyto-ecological surveys and pastoral inventory provided by the High Commission for Water and Forests [4] and [8].

The total annual production of different vegetation species is obtained by multiplying the area of each type by its average annual production (Table 2). The results obtained (Figure 4) show that the forage supply is very heterogeneous from one territory to another. It varies between 0.29 million units in the Timahdite territory and 7.57 million units in the Tigrigra territory. This heterogeneity is mainly related to the amount of forest cover and vegetation variation. The evaluation of the total forage supply in the whole Tigrigra watershed shows that the total annual forage production is about 24.19 million forage units. The spatialization of this supply in the whole watershed (Figure 5) clearly shows that the Causse area has a great capacity of fodder production in comparison with the Meseta. Indeed, in the Causse the climatic and edaphic conditions are favorable, while in the Meseta these conditions are unfavorable [9].

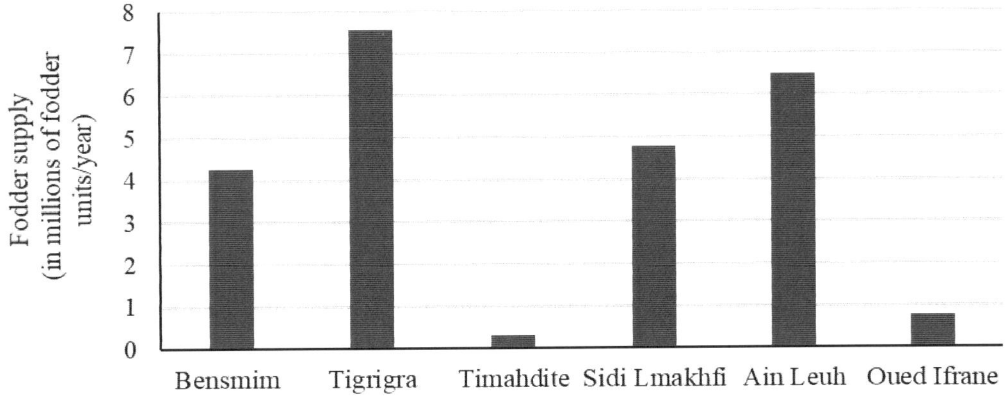

Figure 4. Fodder supply by community at the Tigrigra watershed

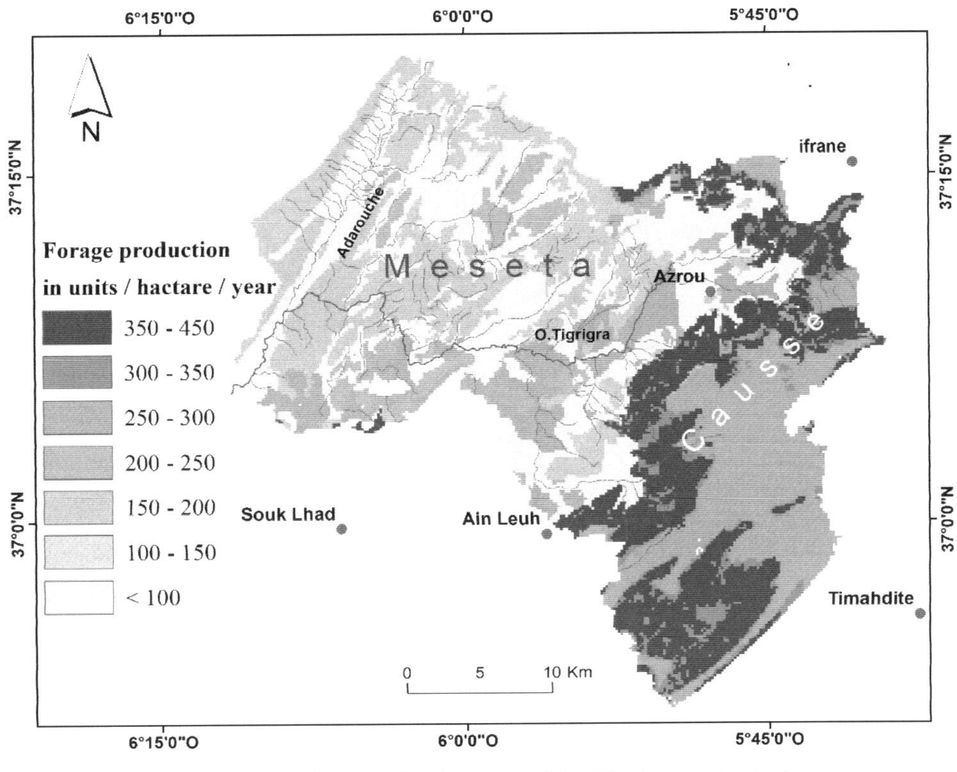

Figure 5. Fodder production map of the Tigrigra watershed

3.3. Assessment of forage deficit and pastoral pressure

The evaluation of the forage balance allowed us to assess the intensity of the forage deficit and overgrazing in the forest and peri-forest rangelands of the studied watershed. In order to evaluate the forage balances of the pastures, our investigations focused on the determination of the annual forage supply and on the quantification of the annual needs of the livestock. The determination of significant numerical indicators, which take into account the fodder balance and the theoretical and real loads of the pastures, made it possible to specify the state of the pastoral pressure at the level of different rural communities of the watershed.

IOP Conf. Series: Earth and Environmental Science 1090 (2022) 012005 doi:10.1088/1755-1315/1090/1/012005

3.3.1. The forage deficit of rangelands. The values of the forage deficit of the Tigrigra watershed rangelands for the years of 2005 and 2016 (figure 6), show that the forage balances are characterized by the increase in needs compared to the available supply and that all the rangelands of the watershed are threatened by degradation. Indeed, the global deficit at the scale of the watershed has increased from 66% in 2005 to 73% in 2016, an increase of 10.7%. At the scale of the communities, the deficit situation is alarming; it varies between 55.8% in Ain Leuh and 86% in Timahdite.

In terms of evolution of the deficit between 2005 and 2016, and with the exception of the territory of Oued Ifrane which recorded a regression of -3.1%, (63% in 2005 against 61% in 2016), all territories have recorded increases more or less accentuated. The courses of Bensmim and Sidi El Makhfi have recorded large increases in forage deficits (22.2% and 16.3%) while those of Tigrigra, Ain Leuh and Timahdite have recorded relatively moderate increases (6.7%, 6.8% and 10.6%). These results clearly show that the demand far exceeds the available potential in terms of fodder resources, which reflects a trend towards the degradation of the fodder capital of the watershed. This situation is mainly linked to the increase in the size of the livestock in the majority of rural communes.

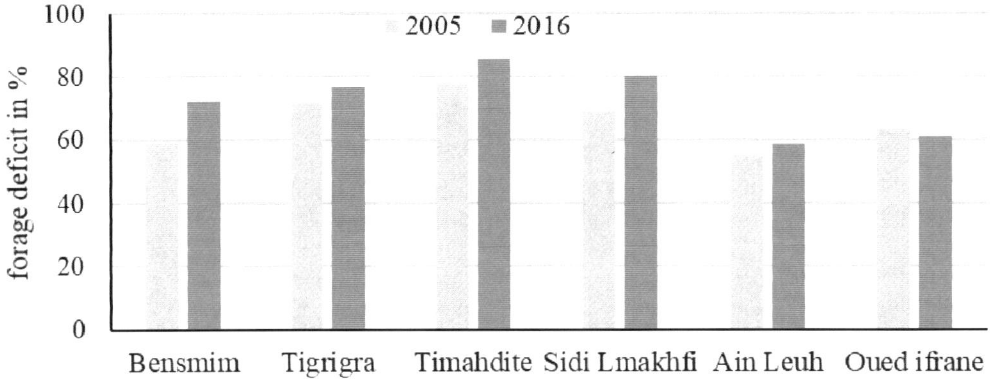

Figure 6. Assessment of the forage deficit in 2005 and 2016 in the different rural territories of the Tigrigra watershed.

3.3.2 The evaluation of pastoral pressure

To identify the pastoral pressure in different rural territories of the watershed, this study focused on determining the equilibrium load (El), the real load (Rl) and the overgrazing coefficient (OC).

3.3.2.1 The equilibrium load (El)

The results obtained for the whole watershed (Figure 7), show that the equilibrium load is not homogeneous between the different territories of the studied watershed. It varies between 1.1 SLU/ha in Oued Ifrane and 1.4 SLU/ha in Ain Leuh. The highest values are recorded in the Causse areas, where the forest cover is dense, while the lowest values are recorded in the Meseta where the pastoral potential is low.

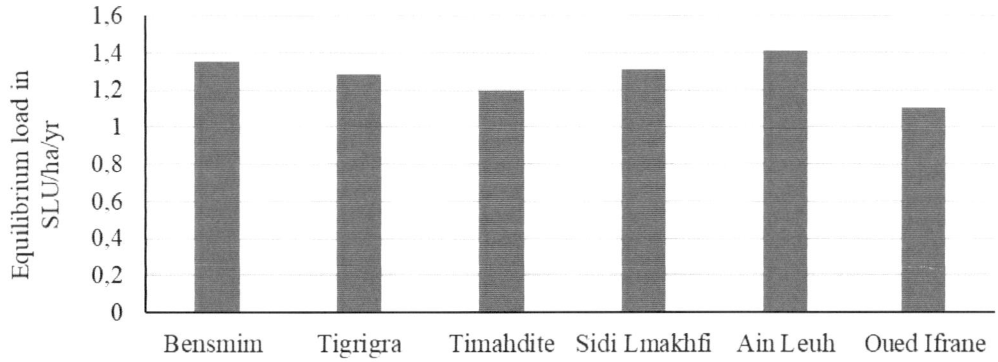

Figure 7. Balance load in the different rural territories of the Tigrigra watershed.

3.3.2.2. The real load (Rl).

The results obtained (Figure 8) indicate an increase in the real load between 2005 and 2016 in all rural territories of the watershed, except in Oued Ifrane. Indeed, the average value of this index increased from 2.08 SLU/ha in 2005 to 2.72 SLU/ha in 2016. This situation is alarming, especially in Timahdite and Sidi El Makhfi where the real loads have increased by 58.2% and 57.5% respectively. At the level of Oued Ifrane, the decrease in livestock has led to a drop of 5.7% in the real load.

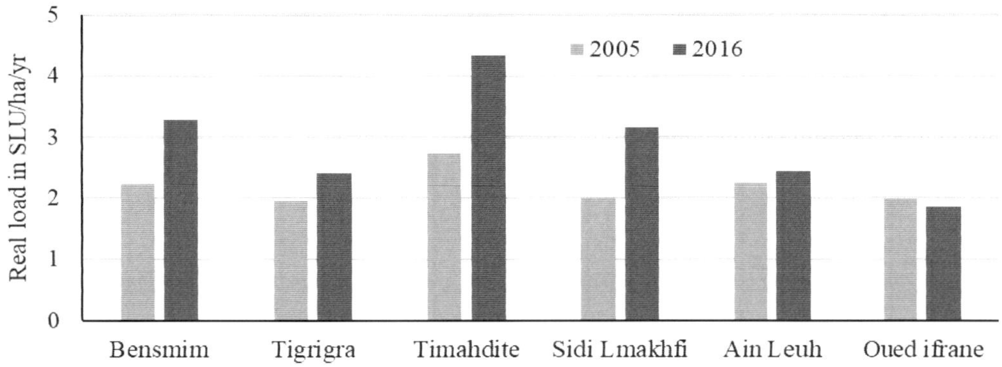

Figure 8. Evolution of the real load of the rangelands between 2005 and 2016, in the rural territories of the Tigrigra watershed

3.3.2.3. The overgrazing coefficient (Oc)

Overgrazing refers to a situation in which pastoral demand exceeds forage production. Theoretically, the intensity of overgrazing is proportional to the forage deficit: when this difference is zero, the range is in a state of load equilibrium; it is in a state of undergrazing when it is negative, and overgrazing when it is positive [6] and [10].

The overgrazing coefficient (Oc) is an adequate criterion for estimating pastoral pressure, as well as the state of degradation of rangelands [11]. It combines data on the number of livestock and the production of pastoral resources.

The results obtained (Figure 9), show that the coefficient of overgrazing is high, which confirms the strong pastoral pressure on forest and peri-forest ecosystems. Indeed, the average value of this coefficient over the whole watershed increased from 40.95% in 2005 to 53.18% in 2016. While, the maximum values are recorded in Timahdite, while the minimum values are recorded in Tigrigra in 2005 and Oued Ifrane in 2016. The increase in overgrazing shows variable levels, with a maximum of 23.9% recorded in Sidi El Makhfi, and a decrease of -3.35% recorded in Oued Ifrane.

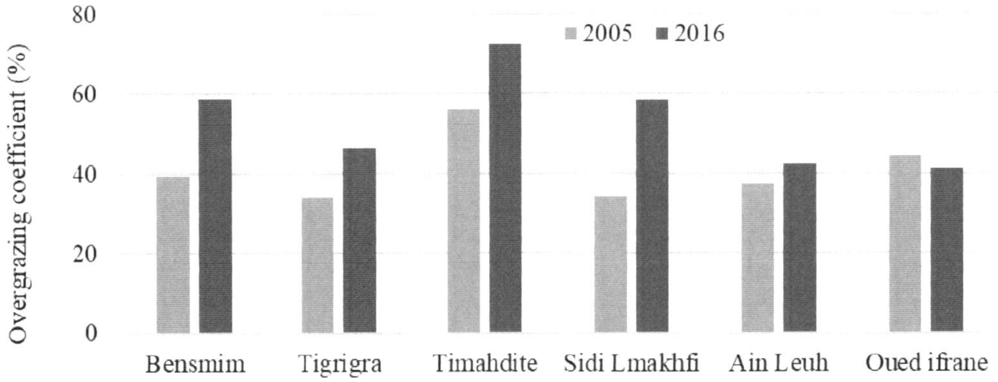

Figure 9. Evolution of the overgrazing coefficient between 2005 and 2016, in the rangelands of the rural territories of the Tigrigra watershed.

3.4. Current situation of rangeland degradation

The situation of overgrazing in the Tigrigra watershed is well confirmed by field observations. Indeed, the peri-forest areas are clearly marked by the effects of pastoral overload. In the same way, the forest rangelands with high pastoral use are degraded. In addition, to compensate for the fodder deficit, especially in winter, shepherds attack the holm oaks and Atlas cedars, by cutting the branches and tops. In many places, these recurrent actions have caused severe damage, reducing the ability of plants to regenerate naturally and affects production and fodder.

On the other hand, and in the absence of a rational management system for the pastoral space, based on rotation, the permanent stay of animals on the same grazing areas compromises seed production and limits the vegetative cycle of pastoral species. Indeed, the consumption of plants during the reproduction period affects seed stocks, and trampling of the soil prevents the burial of seeds, and limits their ability to germinate.

In the long term, this reduction in vegetation stock leads to a loss of biodiversity of the flora and affects mainly perennial pastoral species. As a result, rangelands are often invaded by weed species to the detriment of other species of good pastoral value. This state of imbalance compromises the possibility of establishing the balance and sustainability of the ecosystem.

4. Conclusion

The mode of exploitation of the pastoral space at the level of the Tigrigra watershed, is traditionally based on migratory movements between the winter pastures located on the Meseta (Azaghar) and those of summer located on the Causse du Moyen Atlas (Jbel). This mode of breeding has undergone a profound change over the past 5 decades. Indeed, and because of the stability of the breeders, many lands of course were transformed into lands of arboriculture and market gardening. In parallel with this evolution, the number of livestock has increased considerably, thus exerting increased pressure on the forest rangelands.

According to the data obtained by this study, it was shown that pastoral pressure was already significant in 2005. The comparison between the forage balances in 2005 and 2016 indicates that the rangelands are under increasing pressure, while the forage potential of the grazing areas cannot meet the needs of livestock. Indeed, the average value of the overgrazing coefficient of the watershed increased from 36.1% in 2005 to 50.3% in 2016. This pressure is alarming in Timahdite, where the overgrazing coefficient exceeds 72%.

In a Mediterranean climatic context, the situation of imbalance of forest and pre-forest rangelands of the Tigrigra watershed reflects the strong pressure on these ecosystems.

This situation implies the need to take urgent measures to preserve forest and pastoral resources, while taking into account the needs of the population. In this regard, we propose the following recommendations:

- To Limit the pastoral load by shifting from extensive to intensive and semi-extensive grazing.
- To set up compensations adapted to the pastoral potentialities of the protected forests.
- To augment the quantity and improve the quality of the local pastoral species.
- Establish a rotation system to ensure the regeneration and sustainability of pastoral species.
- Strictly apply the legislative laws relating to the conservation and exploitation of forests, and ensure compliance with the regulations of forest rangelands, in order to prevent illegal users.

5. References

[1] Jennane L 1986 Mutation récente des campagnes du Moyen Atlas et de ses bordures. *Méditerranée* **4** 49-62.

[2] Qarro M 1982 *Étude quantitative et qualitative de la végétation herbacée de la zone d'Ain Leuh, Moyen Atlas-plateau central (Maroc).* Doctoral thesis (FST St-Jérôme MarseilleFrance) p 200.

[3] Narjisse H, M'hirit O, Askarn O, Benzyane M and EL Omeran A 2001 Le couvert végétal marocain : Etat de dégradation, acquis et perspectives en matière de conservation et de mise en valeur. *Revue H.T.E.* **119** 25-29.

[4] HCEFLCD 2014 *Cartographie forestière et statistiques forestières : Mise à jour des cartes numériques et analogiques de l'inventaire forestier national dans les régions administratives de Meknès – Tafilalt et Tadla – Azilal* (Rabat : HCEFLCD) p 117.

[5] El Mderssa M, Achiban H, Benjelloun H, Zennouhi O, Nassiri L andIbijbijen J 2018. Evaluation of the contribution of forest rangelands in the forage balance in thewestern part of the Central Middle Atlas: Case of the forests of Azrou, Jbel AouaSouth and Sidi M'guild. *International Journal of Environmental & Agriculture Research* **4** 17-23.

[6] Laaribya S, Alaoui A, Gmira N and Nassim G2014 Contribution à l'évaluation de la pression pastorale dans la forêt de la Maamora. Parcours forestiers et surpâturage. *Revue Nature & Technologie, C- Sciences de l'Environnement* **10** 39-49.

[7] DagetPh, Poissonet J 1972 Un procédé d'estimation de la valeur pastorale des pâturages. *Revue Fourrages* **49** 31-39.

[8] HCEFLCD 2004 *Études d'aménagement concerté des forêts et des parcours collectifs de la province d'Ifrane. Rapport n° 5 – Étude pastorale de la zone Nord* (Rabat : HCEFLCD) p 101.

[9] AchibanH, Taous A, El-Khantoury I, El Mderssa M, Amechrouq A 2018 Quantification of soil loss in various lithological areas of the western Middle Atlas Central: application to the Ras-Elma, Tamelalet and Sebab watershed (Tigrigra watershed, Morocco). *E3S Web of Conferences* **37** 04003.

[10] Le Houérou H.N 1980 L'impact de l'homme et de ses animaux sur la forêt méditerranéenne. *Forêt Méditerranéenne* **II** 31-44.

[11] HCEFLCD 2013 *Le Programme d'Action National de Lutte Contre la Désertification : Actualisation et adaptation aux spécificités zonales* (Rabat : HCEFLCD) p 52.

A Framework Payment for ecosystem services to reconstitute forest areas closed to grazing in the Ifrane National Park (Morocco).

H Ismaili Alaoui[1]*, B El Asri[1], S Ghazi[1], N Brhadda[1], R Ziri[1]

[1]Plant, Animal and Agro-industry Production Laboratory, Faculty of Sciences, Ibn Tofail University, BP.133, University Campus, Kenitra 14 000, Morocco.

* E-mail: hassana.ismailalaoui@uit.ac.ma

Abstract. Overgrazing constitutes the major constraint threatening the sustainability of the forest ecosystems in the Ifrane National Park. The operations of regeneration and afforestation impose to close areas to grazing over one period exceeding ten years, generally lead to the opposition of the local population to the programs of reconstitution of the forest ecosystems. To solve these problems, a text of compensation for closed perimeters of afforestation or regeneration to the grazing was established in 2002. The main aim of this study is to evaluate the role of the compensation mechanism described as a payment tool for forest ecosystem restoration programs. For this purpose, an analysis of quantitative and qualitative indicators, based on individual and semi-structure interviews, before and after the introduction of the compensation mechanism is used. The GIS mapping approach is also used to visualize some results spatially. Based on results, this mechanism contributed to conserving the forest ecosystem and enhancing rural livelihoods at the level of Ifrane National Park, resulting in the regression of the forest infractions with a rate of 48%. On the socio-economic level, this mechanism contributed to initiate a climate of trust resulting in the collaboration of the population with the foresters and, to carry out many projects of community interest generated socio-economic benefits relevant for the users.

Keywords: Ifrane National Park, compensation mechanism, Payment for ecosystem services, conserving forest ecosystem, enhancing rural livelihoods.

1. Introduction

The environment is commonly considered as a form of economic capital whose integration ensures sustainable development and good politics [1-2]. It deserves special attention in the natural wealth management of all countries [3-4]. Indeed, it is largely recognized that natural ecosystems contribute to human well-being by providing society with a diverse range of ecosystem services. This multifunctionality is expressed in particular in Mediterranean-type forest ecosystems playing a crucial role in maintaining the components that ensure the well-being of local populations. Forest ecosystems provide the most fundamental ecosystem services (ES) that are important for human wellbeing [5]. However, today these ecosystems are certainly renewable resources, but with limited capacities, they can only be able to maintain themselves and illustrate their full socio-economic and environmental roles if they are managed sustainably [6-7].

To highlight society's dependence on natural ecosystems and raise public interest in their conservation, several authors have started to formulate ecological concerns in socio-economic terms. In response, Payments for Ecosystem Services (PES) have received increasing attention as a means of internalizing environmental externalities by creating appropriate markets [8]. The main objective is notto put a price on the environment but to estimate the value of qualitative and quantitative marginal changes in the availability of ecosystem services [9] by estimating the willingness of individuals to payto avoid change or the willingness to accept to make change.

Today, local socio-economic development strategies are increasingly encouraging to consider the capital value of ecosystem services in local policies and programs in a way that conservation strategies and socio-economic objectives could be harmonized, and environmental degradation reduced [10]. In this context, several works have been designed to find possibilities for win-win strategies that combine the conservation of these ecosystem services with the socio-economic development of user populations. Within that framework, at least 56 countries have adopted policies or programs on payments for ecosystem services [11], which have emerged as promoting tools for economic efficiency and environmental effectiveness in contrast to other integrated development projects [12-13].

Recognized as an efficacious and innovative policy approach to improve natural resource management [14 -15 -16] that is increasingly applied worldwide [17], PES is defined as a tool for the sustainable conservation of ES through contractual arrangements between producers and beneficiaries in which a producer of ecosystem service(s) is paid to implement practices that maintain the ecosystem service(s) [16].

PES is increasingly used for deforestation reduction, climate change mitigation, and watershed conservation. They also include policies to encourage carbon sequestration, and landscape embellishment [18]. PES schemes have also been linked to global poverty reduction efforts [19- 20] bycontributing to financial gains and income stability [21-20]. They are seen as a reward from donors to the provider of generally poor ES, implying a synergy between socio-economic and environmental impacts [19]. In China, PES is considered as one of the five main approaches to poverty reduction [22]. The PES mechanism is based on the principle of supply and demand. The demand is formulated by potential beneficiaries of ES which are disposed to pay for the maintenance of the ES. The supply side is formed by ES providers who are owners or managers whose interventions can ensure the continuance of the ES supply. This definition is based on the existence of a well-defined ecosystem service, a voluntary transaction between at least one buyer and one provider, and the condition that the provider guarantees the provision of the ecosystem service [12]. This last condition is one of the strengths of PES compared to conventional mechanisms that do not link payment to outcomes.

For the types of compensation for PES, the literature has revealed the existence of three methods, namely cash, in-kind, and capacity compensation [17]. Cash compensation allows compensated people to freely use the funds and purchase corresponding goods and services according to their preferences and needs. In-kind compensation is where the compensation provider creates output in the form of goods and services to meet the needs of the compensated individuals.

Capacity compensation means that organizations providing compensation offer free technical advice and training to compensated people to raise their level of knowledge and skills, to help them change and improve their livelihoods.

The first two sorts of compensation are less sustainable and less effective but are more applicable and easier to put.

In Moroccan, the forests ecosystems are a real illustration of the range of services provided and the problems faced. It provides a range of ecosystem services and suffers from various forms of degradation described as negative externalities [23]. Forest ecosystems are currently being degraded under the impact of many anthropogenic factors, the main one is overgrazing [24-25-26-27] which harms the natural regeneration of forests. It is considered the major factor of change and land degradation, affecting both soil and ecosystem components [25-28].

Therefore, to reduce the effect of overgrazing and to succeed the operations of regeneration and afforestation, foresters impose to close these areas for a period of sometimes more than twenty years often lead to opposition from users to forest ecosystems reconstitution programs. To solve these problems, the Department of Water and Forests seted up in 2002 a decree establishing compensation for afforestation and regeneration sites.

This mechanism of compensation for forest area closed to grazingis considered as a payment system for the grazing service avoided by the beneficiaries. It was set up to establish a voluntary conservation mechanism offering cash benefits to the beneficiaries of the local population to lead to a positive change in the use of forest areas. However, there are no studies assessingthe contribution of this mechanism to the achievement of this objective.

IOP Conf. Series: Earth and Environmental Science 1090 (2022) 012006 doi:10.1088/1755-1315/1090/1/012006

In this paper, we try to contribute to the enhancement of knowledge about the concept of PES as a compensation tool of the forest area closed to grazing. The objective is to study the role of this participative management mechanism to reconstitute forest areas and to generate socio-economic impacts for users. We also discuss some factors to ensure the success of this mechanism for other contexts of sustainable forest management.

Our study area is the Ifrane National Park (INP) when the first silvopastorals management associations are created in Morocco. Moreover, in this park, it was shown that pressure grazing is the most common cause of forest cover loss and land degradation [29].

To deal with this research question and to assess the effectiveness and the impacts of this mechanism through various spatial and temporal scales, we based on many factors which allowed us to analyze those impacts on user behaviors in an adequate way. We start with a description of the PES mechanism,then analyze how the compensation mechanism could support the sustainable management of forest resources at Ifrane National Park. We then discuss how to make PES successful in the national contextthrough a synthesis of all the key variables identified that can be applied to PES on both theoretical and practical dimensions. We conclude with some future directions for successful PES in both environmental and socio-economic models.

2. Materials and Methods

2.1. Study area

Our study area is classified under Moroccan law 22-07 as a national park. It was created in 2004 and initially covered 53,800 ha, in 2008 the national park was extended to cover 125,000 ha, this area covering almost all the forests of the province of Ifrane with an area of 65 290 ha (figure1).

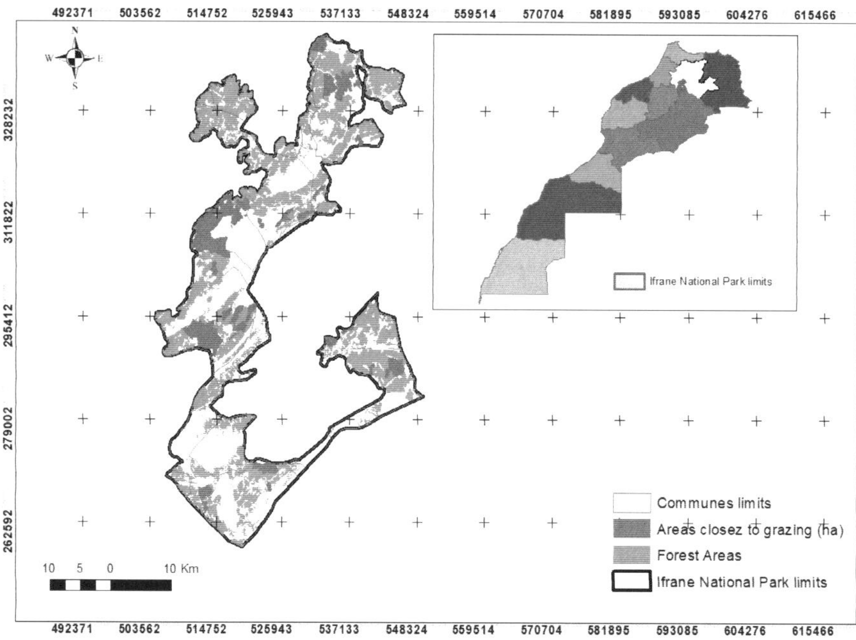

Figure 1. Mapping of forests stand in the Ifrane National Park

The Ifrane National Park is located at the central part of the Middle Atlas Mountains, almost entirely within the Sebou watershed [29]. It is characterized by very diverse flora among 1015 species of spontaneous plants recorded in the park, more than 250 taxa representing 47 families are endemic [29].In terms of altitude, it is situated between 1300 m and 2400 m, including all the cedar forests of the province. It contains, thus, 1/10 of the world surface of the Atlas cedar which classifies it in the heart of the Atlas Cedar Biosphere Reserve.

Since its creation, the National Park has aimed to conserve biodiversity and sustainably manage forest ecosystems to safeguard, restore the great natural balances, and enhance the educational, cultural, and socio-economic functions of the park [29].

The choice of Ifrane National Park as the study area was informed by the fact that for forest area closed to grazing programme has been implemented in the park since 2007.

Moreover, the area also experiences increased dependence on forests by poor local communities which influences their reconstitution. These contextual factors would have implications on the success or otherwise of PES.

2.2 Data collection

To analyze the effectiveness of this PES mechanism, we started with the mapping of the different areas closed to grazing for each association, and then we completed the data characterizing each association by semi-structured interviews, free listing and focus group, with members of the Executive Board of association, and forest managers.

Furthermore, to determine the users' opinion on the compensation mechanism, and its impact on their lives, individual interviews were carried out using stratified sampling based on the distance of the interviewee's localities from the closed grazing area. Thereafter, users were randomly selected Three classes were identified:

Class 1: distance < 10km; **Class 2:** 10 km < distance < 20km, and **Class 3**: distance > 20km. The intervals of the class are defined to highlight the degree of the impact of the closed areas on thepastoral activity of neighboring users, according to the distance separating them from these perimetersclosed to grazing.

The number of interviewees was set, depending on the availability of users around the perimeters. In total 140 users are interviewed, 54 users from **class 1**, 40 from **class 2,** and from **class 3**. The survey collected information on participation in the PES program as well as on socioeconomic related to the users interviewed.

Other quantitative data collected by consulting documents related to the water and forestry administration between 2003 to 2019 has made it possible to build a database on forest infractions and success rates of plantation operations in the forest areas closed to grazing, which is used to analyse the compensation mechanism.

The analysis of quantitative data was based mainly on the comparison of the averages of the different indicators before and after the year of the introduction of the compensation mechanism between 2003 and 2019. The determination of the statistical significance of the recorded differences was carried out by an independent samples test using the SPSS tool. The GIS mapping approach was also used to show some results spatially.

2.3 Context of compensation for forest area closed to grazing

The compensation mechanism for the forest area closed to grazing responds perfectly to the definition of a collective PES [30]. It is considered a payment system for the avoided grazing service [31]. This mechanism consists of a contract by which a grazing service avoided by forest rights holders is remunerated by the Department of Water and Forests, which is considered to be the purchaser of the environmental service, on condition of a voluntary commitment by the users [32]. The mechanism is designed to adapt the contract to communal land tenure and collectively managed resource systems.

The forest area closed to grazing allows the rights holders of the local population living near public forests and organized within silvopastoral associations to benefit from monetary compensation of 250 MAD/ha/year for a minimum area of 300 ha in exchange for sacrificing the right to use rangelands. The amount is fixed based on an average production in fodder units (FU) of one ha of forests which is 200

FU and counting 1.25MAD/FU. The monetary reward is conditional on compliance with the clauses of the collective PES contract. The granting of compensation for forest area closed to grazing is subject to the presentation of a file comprising a technical and an administrative part. The granting of the first compensation starts 12 months after the date of closing the sites to grazing, providing that during this period the following conditions are respected.

Given that forest area closed to grazing is considered as a loss of income in terms of grazing area for the rights holders, buying the right to graze until the age of defensibility of the plantations is a solution shared between the administration and the rights holders of the forest areas.

The areas concerned by the compensation include those to be reforested or recently exploited and whose density is considered very low, burned areas, or those in an advanced state of degradation. The administration retains the right to use the land that has been closed to grazing, where afforestation and regeneration operations can be planned.

To evaluate the impact of this mechanism on the reconstitution of forest areas through changes in the behavior sof users and the improvement of their living conditions, the analysis of the following indicators was used as discussed in the methodology.

3. Results

3.1. Situation of silvopastoral management associations in Ifrane National Park

The Ifrane National Park currently has twelve functional silvopastoral management associations, with two new associations created in 2020. Only the ten associations created before 2020, the year when this study was carried out, were considered in this work. This chapter aims to identify and present the ten silvopastoral management associations. It first places them in their spatial context in the Park. It then places them in the socio-economic context, as assessed by the poverty rate of the different communes in which the associations are located (Figure 2).

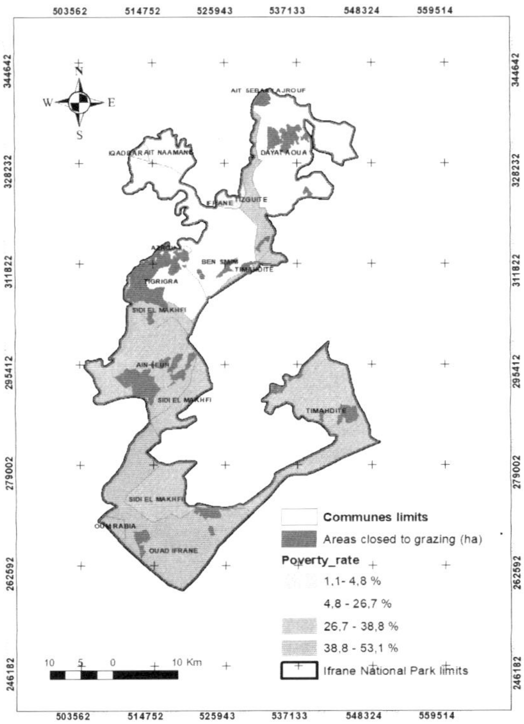

Figure 2. Localization of silvopastoral Management Association SMA)

The associations are located in territories with very high poverty rates compared to the national average of 4.8%, exceeding in some cases 53%. This puts into question the socio-economic conditions of the local communities, influencing their dependence on the natural resources of the Park, especially forest resources.

The associations are distributed over an area of 8215 ha, in 8 communes which represents 10.07% of the park's forest area and receives compensation of more than 2,053,750 MAD/year. These areas correspond to afforestation and regeneration perimeters realized by the forestry administration for which protection is essential to ensure their success.

The first associations started benefitting from compensation in 2007, four years after the initiative of the forestry administration under the project for the management and protection of the forest massifs of the Ifrane province was implemented in 2002. This project has fixed a method of working based on the realization of investments and accompanying actions in the profile of the users to encourage them to establish compensation contracts with the forest administration. But since 2010, the date of the end of the project, the process of contractualization process has continued at the initiative of the population. The most recent associations were created between 2013 and 2015 inspired by the accumulated experience of previous associations and the awareness of the state of degradation of their forests.

The time between the year of the association creation and the year of contractualization has varied between one year and nine years. This variation is caused by problems with the local population not being convinced of the interest of compensation, internalproblems between members, and the constraint of the minimum total area required which is 300ha.

According to the legislation, these associations are made up of the users of forest areas, breeders, and residents bordering on the forest areas to be rehabilitated or regenerated. Thus, the membership of the ten associations is 1770 distributed over 96 households, with men representing 96% of them

3.1.1. Evolution of forest infractions

One of the most factors contributing to the non-success of plantation operations and forest cover degradation is the infractions of cutting wood and overgrazing. Thus, to determine the trend of forest infractions in the forests where the compensation areas are located, the annual statements of infractions established by the forest managers were consulted for the period between 2003 and 2019. The data were filtered and analyzed by year, type, and localization of the infraction. The distribution was carried out by two types of infractions cutting of wood and grazing in closed areas, as these two types are the most frequent in the study area. The infractions considered are those committed on the same forest where the association benefits from compensation on the closed areas. Figure 3 shows the results obtained.

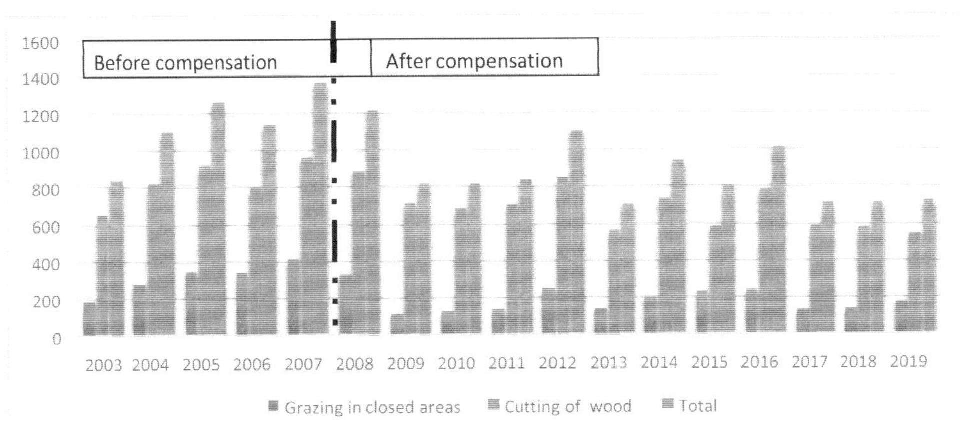

Figure 3. Distribution of forest infractions by year and type

During 2003-2019 period, a number of 15531 infractions were recorded in the INP. The year 2007 sawa maximum of 1369 infractions spread over the forests of study area. This could be explained by the sabotage actions carried out by people who are opposed to these associations and whose main concern is to graze liberally. However, a decrease was observed in 2013 (703 infractions) as reduction of 48%, which shows a positive change in the behaviour of users who are convinced of the socio-economic benefits of this PES tool.

The role of the guards distributed on both sides of the forest also played a very important role. These guards although they do not have the right to arrest offenders. They act as informers for the foresters at the appropriate time.

This regressive change in the number of infractions in the forests where the associations are located is very apparent in figure 4, which demonstrates the change in the number of infractions from one class toa lower class. The number of infractions is proportionally negative in relation to the forest areas closedto grazing.

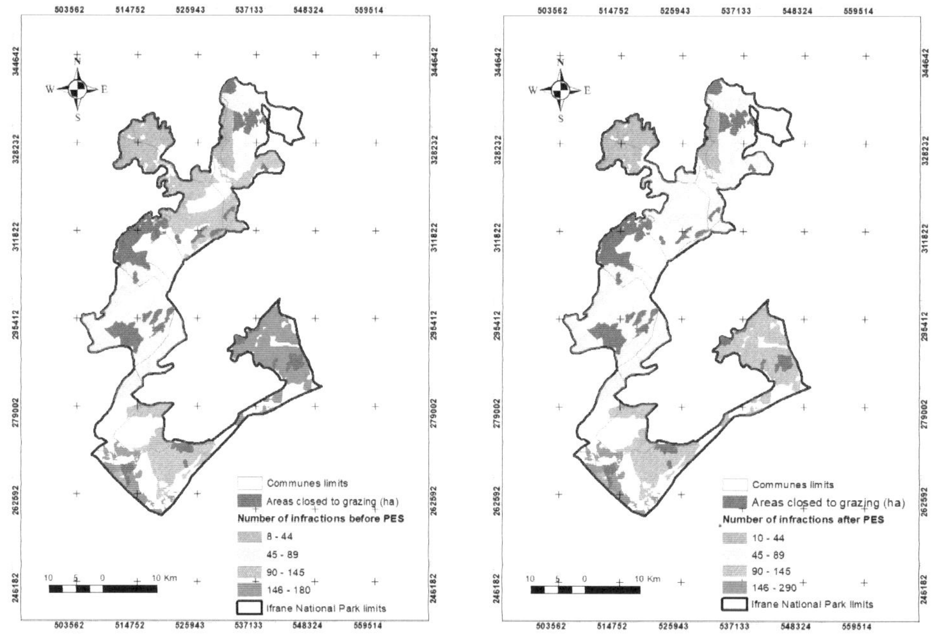

Figure 4. Mapping of total infractions before and after installation of PES.

An independent samples test using the SPSS tool was used to check if there was a statistically significant difference between the average number of infractions before and after the introduction of the compensation mechanism. The following table 1 shows the results obtained.

Table 1. independent samples test result before and after compensation

Nature of infractions	Compensation	Duration in years	Average	Standard Deviation	Sig.
Grazing in closed areas	Before	5	185,4	50,322	0,02<0,05
	After	12	184,67	67,396	
Cutting of wood	Before	5	684	138,402	0,058>0,05
	After	12	682,92	114,740	
Total infractions	Before	5	1140,20	201,396	0,045<0,05
	After	12	867,5833	166,597	

The average number of infractions decreased after the introduction of the compensation mechanism in 2007. Comparing the averages for grazing and wood cutting infractions before and after compensation, we find that the average has decreased significantly for grazing and not significantly for wood cutting. However, this regression remains significant for the two categories of infractions at INP. This proves the positive impact of the compensation mechanism on the reduction of violation of area closed to grazing.

3.1.2. Evolution of planting success rates

The success rate of the afforested area during a planting season is calculated after the summer season has passed, i.e., at the end of November of year n+1. When this rate is higher than 80%, it implies a good success of the plantation. Whereas, a rate of less than 60% is considered a plantation failure.

To monitor the evolution of afforestation success rates before and after the introduction of the compensation mechanism, the afforestation reports from the forestry administration have been consultedfor each campaign between 2003 and 2019 to obtain the information mentioned in figure 5.

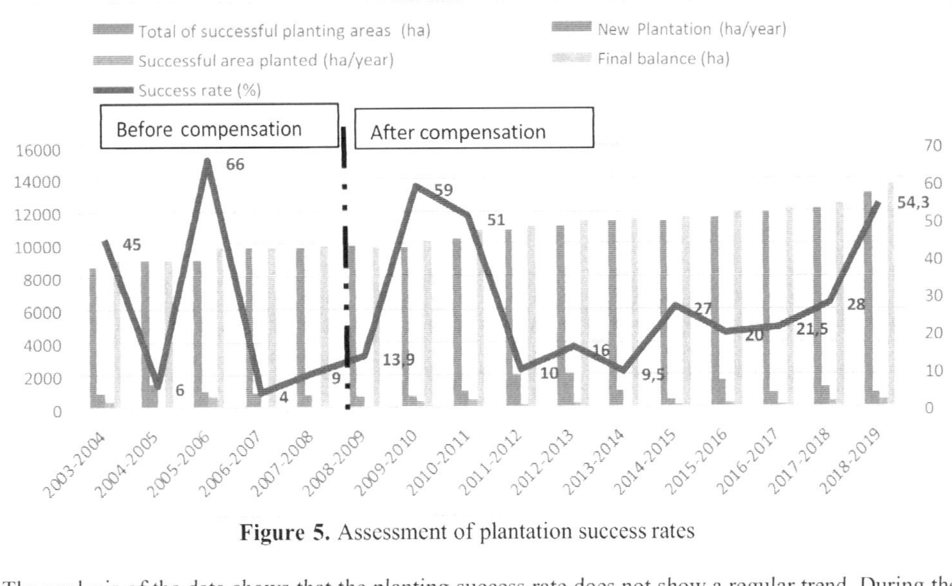

Figure 5. Assessment of plantation success rates

The analysis of the data shows that the planting success rate does not show a regular trend. During the period between 2003 and 2019 a total area of 18252 ha was planted, divided between afforestation, regeneration, and silvopastoral improvement. A maximum of 2115 ha was reforested during the 2012-2013 campaign, while a minimum of 551 ha was reforested in 2014-2015. Only 4346ha recorded successful with an average rate of 23.81%. The comparison of the average success rates of reforestation work in the study area before and after the introduction of the compensation mechanism is shown in the following table 2.

Table 2. Result of comparing the average success rates of plantations

Compensation	Duration in year	Mean	Standard Deviation	Sig.
Before	5	26	27,99	
After	10	25,59	16,82	0,977>0,05

Comparing the average area successfully reforested before and after the compensation mechanism can be seen that this average decreases after this date, but this decrease is not statistically significant.

Thus, the creation of associations at the INP has not improved the success of the annual planted areas in this area. Indeed, the compensation mechanism for forest area closed to grazing has a positive effect onthe reduction of infractions thanks to the associations' role in raising awareness and assigning guards. However, this role cannot contribute to improving the success rate of afforestation which is dependent on factors other than anthropogenic ones (climatic, edaphic, technical, etc.). Indeed, afforestation is a complex operation, particularly in the Mediterranean climate, of which its success depends on several factors.

3.1.3. Appreciation and collaboration with forest managers and associations

The households surveyed were all men (98%), 83% of whom were over 55 years old, 11.4% were between 35 and 54 years old, and only 5.6% were under 35 years old, which shows that the young peoplein the population were going to look for work in the city to help their families. 89.2% of the respondents are illiterate, the rest have primary education, 92.7% are married, 4% are
divorced, and the rest are single. This shows that most users are fathers of families that they have to support.

Most of the respondents stated that livestock (68%) and agriculture (29%) are their main activities, others were still waiting for the contribution of their sons who work in town.

According to Figure 6, all the users interviewed (100%) were informed of the existence of a silvopastoral associations and the areas closed to grazing, but only an average of 6% of respondents knew about the legislation (minimum acceptable area, value of the compensation per hectare, total amount of compensation paid to the association).

Following the interviews with users, the perception of the areas closed to grazing is not the same for allthe interviewees, nor all the associations or even within the same association.

In general, 61.8% of users accepted this mechanism, including 89% of association members, but this measure is rejected by more than 38,2%, especially by those with households less than ten kilometers away from these areas (47,8%) who were not members of the associations and who declared that this measure deprived them of their right to grazing without any compensation (figure 6). In other cases, this opposition is expressed even if the area is far from their households, and although the respondents are members of the association, which represents personal opinions that do not respect the collective commitment made in their name by the association. This disengagement reflects in some way the results of the conflicts between the rightful owners on the creation of the association

Indeed, the users accepting this process noted their collaboration which is reflected in the complaints about infractions within the forest area closed to grazing. These are meant to maintain the forest area closed to grazing to ensure the continuity of the association and the compensation allowance.

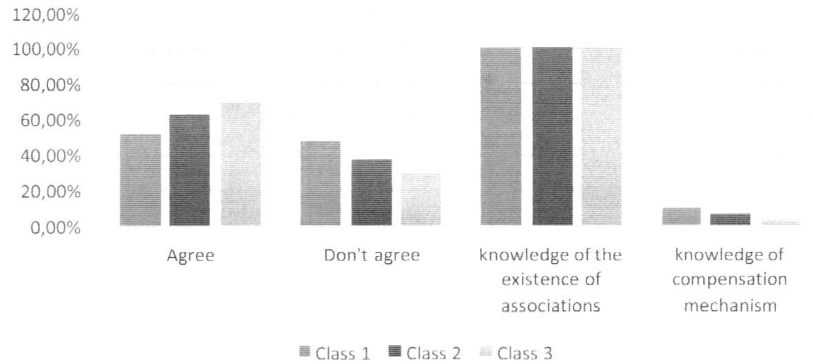

Figure 6. Users' appreciation of the compensation mechanism

For the aspects of collaboration with forest managers, gardening comes in first position. In addition to their role as informants on infractions committed inside and outside the areas area closed to grazing, some guards participate in surveillance rounds and tree marking operations organized by the forest managers. They are involved in the maintenance of the fences of the forest area closed to grazing.

The association also plays the role of interlocutor between the forestry administration and the local population. It intervenes particularly in afforestation project to convince the local population to avoid any opposition.

The establishment of the associations has also shown these roles for some traditional community action, which has allowed members of the rural community to meet them periodically to discuss collective matters relating to the problems and needs of the community members (tracks, drinking water, farm machinery, etc.) as well as how to manage the compensation amounts.

3.1.4. Contribution of the associations to improve the living conditions users through the compensation amounts

The figure 7 shows the use of the compensation amounts, these amounts represent the sum of the compensation paid to each association according to its annual closed grazing area, during the partnership period (2007-2019).

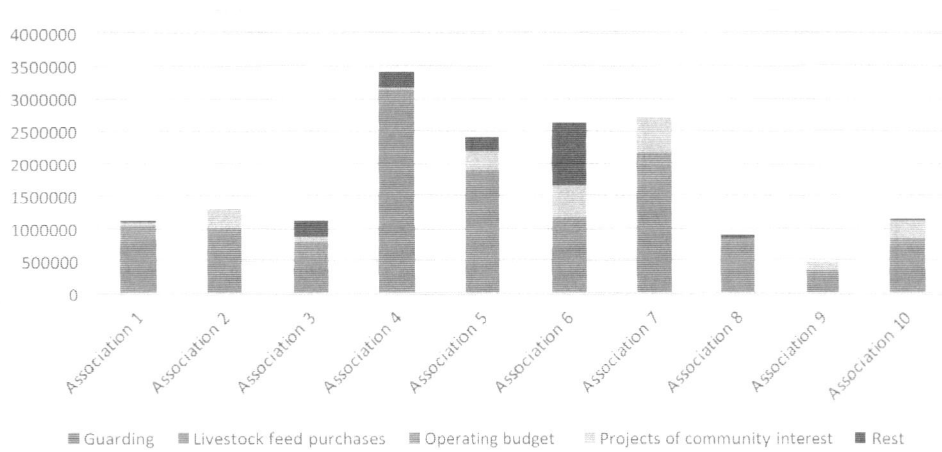

Figure 7. Use of compensation amounts (MAD)

The total compensation granted to the silvopastoral associations is 17,163,250 MAD, it is used mainly under four headings, namely the purchase of cattle feed (34,8%), guarding (31,33%), financing ofprojects of community interest (12,52%), and operating budget (11,19%). The results show that all theassociations, without exception, invest in guarding the guarded areas and buying supplements, with different percentages.

The guarding service is an income-generating job for about forty people, who are paid a monthly salary ranging from 1000Dhs to 2000Dhs, depending on the financial means and the number of guards available to the association. Some associations recruit permanent guards, while others change them to allow other members of the ethnic-spatial unit benefit from the job.

To compensate for fodder deficits during the lean season, most associations invest in the purchase of supplements for livestock to reduce the cost of supplements for breeders. Indeed, each member generally benefits from 1000Kg to 5000Kg per year (Bran, Barley, Maize, Compound feed). This quantity differs according to the amount dedicated by the associations to this type of project and according to the numberof beneficiaries of the compensation.

Eight out of ten associations invest in community projects (drilling and managing wells, distributing fruit plants, maintaining tracks, purchasing agricultural machinery, etc.), for a total percentage of 13%. Several actions conducted by the associations involved about a hundred local workers at a rate of 100Dhs/worker/day. These include the opening and maintenance of tracks, maintenance of fences in the forest area closed to grazing, drilling and equipping wells, water connection, developing water points, and cleaning campaigns.

The results also show that only two associations have not benefited from the full compensation during their operations because of non-compliance with the forest area closed to grazing and that no association shares the compensation amounts among its members.

The compensation for forest area closed to grazing is helped to reduce opposition to forest restoration and rehabilitation programs and consequently had led to a decrease in the number of fines for violationof forest area closed to grazing. The mechanism is also contributed to financing socio-economic projects of collective interest, promoting solidarity among groups of rights-holders by inviting them to organize themselves in pastoral associations considered as an interlocutor with the forestry administration. This mechanism is also helped to develop and strengthen local skills in organisational and financial management and in project planning.

In this context, the development of compensation PES could be seen as a renewal of a_common agreement between the public administration and the traditional community, as well as a tool for collective management of Moroccan forests. It can be seen as a collective PES that allows for attitudes change, transparency, and equitable distribution of costs and benefits among stakeholders [33]. This mechanism reflects those collective contracts that are the most feasible option when land is collectively owned [31 30].

Compensated land can also form the basis of a new PES mode, REDD, for the allocation of subsidies orcarbon credits which could provide additional income to local people who already offer a service of respect for the forest area closed to grazing.

However, this mechanism is hampered by the fact that the sum of 250MAD/ha granted as payment for the ecosystem service (250MAD/ha) is not discounted in line with the changing socio-economic realities. This value is still low compared to the high number of members of the associations who remain unrepresentative of all users of forest areas closed to grazing. In addition, there are other problems related to the lack of financial and administrative management of association members.

4. Discussions

Given that designing instruments capable of protecting forests and these ecosystem services on which societies depend is one of the greatest social policy challenges for Morocco and in general for developing countries. and considering that, the conception of PES should be implemented in circumstances where sensitivity to social equity is matched by concern for ecological efficiency [34].

It is noteworthy that the PES for closed areas to grazing, reviewed here have successfully generated both ecological and livelihood benefit, but with different degrees for beneficiaries, making it a source of conflict in some cases.

In fact, this initiative showed that integrating livelihoods into the PES approach promotes ecological outcomes, especially in the context where local people's livelihoods are largely dependent on natural resources. The success or otherwise of incentive forestry programs is influenced by the degree to which local communities participate in and benefit from the programs

Several research studies have shown that if PES programs does not take into consideration this socio-ecological complexity, negative effects may occur [35-36]. Indeed, the design of a PES must balance the opportunity costs and the benefits associated with ES [37-20] with equilibrium in terms of information between sellers and buyers [12-38-39]. Before the implementation of a PES, all parties involved in the scheme should have full knowledge of the contractual conditions to avoid the PES posing an additional social dilemma in the community. Designing a PES that tightly dictates the types of actions that are eligible for payment on one-size-fits-all basis risks demotivating potential participants. in our case study, some respondents showed that they were not aware of the approaches of this compensation mechanism and how the compensation amounts are fixed, which prevented them from participating effectively in the process. This proves that there is an effort to be made by the purchasers of the service, which are the forestry administration, to disseminate more information on this compensation system to all users.

Therefore, to improve this program, it should be adapted to the perception and culture of the local people [40]. ES providers need to maintain intrinsic motivations to justify participation [41].

The success of this program depends therefore on mastering the socio-cultural, institutional, economic, and ecological data that characterize each context [42]. Our case study shows that users who possess more livestock and whose collective land is poor in fodder units are the most interested in integrating this PES system.

A PES can only be assessed on three dimensions, namely its environmental effectiveness, its economic efficiency, and its impacts on social well-being [43]. In Morocco, where poverty is the main burden of the forest, and where the poor population considers the forests a gift from nature, it was difficult to adopt tools to make this population responsible without improving their living conditions. Indeed, local populations have a key role to play in the management process and the preservation of forest areas and their reconstitution. The utility of PES to support both conservation and poverty alleviation is attractivein places where the two are often considered incompatible as PES can provide new and/or additional income-generating opportunities for rights holders who have few other means of livelihood.

Thus, social capital will remain an essential element for the success of any payment for ecosystem services mechanism that aims to collectively manage common resources by the moral rules of local societies, while allowing local people to make decisions themselves that are adapted to their individual and socio-cultural needs and specificities. Payment schemes that are based on contracts adapted to communal land, and collectively managed resource systems, and that are structured to facilitate intrinsic incentives are, therefore, more likely to succeed [44], as endogenous rules are quite likely to be respected and understood by local people, compared to externally imposed rules [45].

In the absence of a belief in the benefits of a given process, externally imposed rules and incentives may even crowd out locally developed rules and social norms [46-45-47].

A collective PES model should also provide safeguards to avoid that some participants might receive compensation even if they pursue activities that degrade the environment [48]. The model should also ensure intra-community equity through the distribution of benefits in proportion to the contribution.

Analyzing the PES for closed area for grazing, we can show that this system has demonstrated its benefits in terms of infractions control, the establishment of some socio-economic projects using the compensation amounts, and creating a climate trust between the forestry administration and the users. However, the contracts of compensation are not clear enough. It did not mention the criteria forevaluating the respect of forest areas.

Moreover, some anomalies requiring improvement were identified, the mechanism does not specify the categories of infractions for which compensation may be suspended or canceled, either the surface areato be subtracted or the degree of damage requiring the termination of the contract.

The conditions of adhesion to the silvopastoral associations also pose problems. The eligibility conditions specified that the association must be composed of all the users residing in the areas bordering the forest to be rehabilitated or regenerated. However, the distance of proximity is not specified which influences user adherence as seen above.

Adding the non-specification of the management methods for the compensation amounts allocated. Thus, the associations are not supervised and each one manages the compensation payments in a way that is appropriate to it.

External problems with the local population and other associations can also hamper the smooth running of the association due to sabotage actions by users whose who claim that they were not implicated in the drafting of this mechanism of PES and that the administration deprives them of their rights in exchange for its inadequate sums of money.

To improve this experience and encourage other users to join this mechanism, the PES can also be used for other ecological functions related to water on the closed areas. PES can play important roles in building an upstream-downstream connection for ameliorating the quality and/or quantity of the water resource in the downstream of the catchment area by encouraging upstream users to adopt sustainable practices [42]. Different studies have shown the role of forests management and afforestation/reforestation practices to generate trade-offs within the provision of watershed-related ecosystem services of supply and regulation [49-50].

The areas closed to grazing can also be an ideal land to ensure sustainable tourism development based on landscape. it can be beneficial to wildlife re-introduction and consequently to develop ecological ecotourism for community tourism companies [51]. This can be achieved through economic incentives that can have very positive effects on conservation effortsand management of these areas. These incentives can come from entrance fees and taxes on tourism activities [52]. Thus, it is very important to capitalize on this compensation process to create other PES to serve forest users.

5. Conclusion

Payments for ecosystem services represent a mechanism to promote sustainable management of ecosystem services and can also be useful to support rural development and poverty mitigation of local communities.

Many models are observed in the literature, but none of these models has emerged as a standard approach for successful PES which proves that these systems are flexible and constantly evolving.

Indeed, a PES needs to balance environmental and socio-economic concerns through a community institution that guarantees the consultation process. Thus, particular attention should be paid to projects that focus on improving ecosystem services and providing local support to participating communities based on inclusive community governance models involving the government, the private sector, and local communities.

Undeniably, to date, the public sector has been the largest manager of ecosystem services. The government can strengthen its role in natural resources management through various regulatory measures. Additionally, public-private partnerships are also an important means for promoting PES.

This does not mean privatization or reduction of state efforts, but it does rather mean the gain of additional financial resources.

Integrated management has become an essential approach for the sustainable management of natural ecosystems. Although PES are innovative and promising, they are just one alternative among others to achieve optimal management of the environmental sphere, in particular, the flows of environmental services invisible on the markets in a more sustainable way. PES is not consistent enough if they are not supported by other legislative and economic instruments of environmental policy. Therefore, the success of such a mechanism is only possible if all interested parties are involved.

6. References

[1] Millennium Ecosystem Assessment Ecosystems and Human Well-being (MEA) 2005 *Synthesis* (Washington : Island Press) p 151.
[2] Barbier BD 2019 The concept of natural capital J. *Oxford Review of Economic Policy* **35** 14–36
[3] World Bank 2011 *The changing wealth of nations measuring sustainable development in the new millennium* (Washington, World Bank) P 224.
[4] World wildlife fund (WWF) 2018 Report living planet (Gland, Switzerland) p 75.
[5] Cudlín P Seják J Pokorný J Albrechtová J Bastian O and Marek M 2013 Forest Ecosystem Services Under Climate Change and Air Pollution J. *Dev. Environ. Sci.* **13** 521–546.
[6] Norgaard RB 2010. Ecosystem services: From eye-opening metaphor to complexity blinder J. *Ecol. Econ.* **69 (6)** 1219-1227.
[7] Barua SK Animon I and Matta R 2020 Valuing forest-based ecosystem services in Bangladesh: Implications for research and policies J. *Ecosyst. Serv.* **42** 101069.
[8] Pearce DW and Turner RK 1990 *Economics of Natural Resources and the Environment* Johns Hopkins University Press Baltimore (MD, USA) P 378.
[9] Turner BL Kasperson RE Matson PA McCarthy JJ Corell RW Christensen L Eckley N Kasperson JX Luers A Martello ML Polsky C Pulsipher A and Schiller A 2003 A framework for vulnerability analysis in sustainability science J. *PNAS.* **14** 8074–8079.
[10] Schmidt S Manceur AM and Seppelt R 2016 Uncertainty of Monetary Valued Ecosystem Services-Value Transfer Functions for Global Mapping J. *PLOS ONE.* **11** e0148524.
[11] Shang W Gong Y Wang Z and Stewardson MJ 2018 Eco-compensation in China: theory, practices and suggestions for the future. J. *Environ. Manag.* **210** 162–170.
[12] Wunder S 2006 The efficiency of payments for environmental services in tropical conservation J. *Conserv. Biol.* **21** 45–58.
[13] Pascual U Muradian R Brander Gómez-Baggethun E Martín-López B Verma M Armsworth P Christie M Cornelissen H Eppink F Farley J Loomis J Pearson L Perrings C Polasky S McNeely JA Norgaard R Siddiqui R David Simpson R ...and Simpson RD 2012 The economics of valuing ecosystem services and biodiversity Ecological and Economic Foundations (Taylor and Francis) chapter 6 pp 183-256.
[14] Ferraro PJ and Kiss A 2002 Direct payments to conserve biodiversity J. *Science* **298** 1718–1719.
[15] Pagiola S 2008 Payments for environmental services in Costa Rica J. *Ecol. Econ.* **65** 712–724.
[16] Wunder S 2015 Revisiting the concept of payments for environmental services J. *Ecol. Econ* .**117** *234–243.*
[17] Song X Liu Y Zhong F Deng X Qi Y Zhang J and Zhang Y 2020 Payment Criteria and Mode for Watershed Ecosystem Services : A Case Study of the Heihe River Basin, Northwest China J. *Sustainability* **12** 6177.
[18] Landell-Mills N and Poras I 2002 *Silver bullet or fool's gold ? A global review of markets for environmental services and their impact on the poor* IIED (London, UK) p 272.
[19] Bulte E Lipper L Stringer R and Zilberman D 2008 Payments for ecosystem services and poverty reduction: concepts, issues and empirical perspectives J. *Environ. Dev. Econ.* **13** 245–254.
[20] Wunder S 2008 Payments for environmental services and the poor : Concepts and preliminary evidence J. *Environ. Dev. Econ.* **13** 279–297.
[21] Pagiola S Agostini P Gobbi J Haan C Ibrahim M Murgueitio E Ramirez E Rosales M and Ruiz JP 2005 Paying for biodiversity conservation services experience in Colombia, Costa Rica, and Nicaragua J. *M. Res. Dev.* **25** 206–211.
[22] Le W and Leshan J 2020 How eco-compensation contribute to poverty reduction: A perspective from different income group of rural households in Guizhou, China J. *Cleaner Production* **275** 122962.
[23] Ellatifi M 2012 L'économie de la forêt et des produits forestiers au Maroc : bilan et perspectives Thèse de doctorat : Sciences économiques (Bordeaux) p 424.

[24] Poupon J 1980 L'aménagement et l'amélioration des parcours forestiers au Maroc 1ére Partie *forêt méditerranéenne* T II (1) pp141-150.
[25] Angassa A 2012 Effects of grazing intensity and bush encroachment on herbaceous species and rangeland condition in Southern Ethiopia J. *Land Degradation and Development* **25** 438- 451
[26] Harris SE 2012 Cyprus as a degraded landscape or resilient environment in the wake of colonial intrusion J. *Proc. Natl. Acad. Sci.* **109** 3670-3675.
[27] Moukrim S Lahssini S Naggar M Lahlaoi H Rifai N Arahou M et al., 2019 Local community involvement in forest rangeland management: case study of compensation on forest area closed to grazing in Morocco J. *RangelandJournal* **41(1)** 43-53.
[28] Lu X Yan Y Sun J Zhang X Chen Y Wang X and Cheng G 2015 Shortterm grazing exclusion has no impact on soil properties and nutrients of degraded alpine grassland in Tibet, China J. *SolidEarth* **6** 1195-1205.
[29] Haut-Commissariat aux Eaux et Forêts et à la Lutte Contre la Désertification (HCEFLCD) 2007 Plan d'aménagement et de gestion du Parc National d'Ifrane p 66 http://homepage.mac.com/jmdelacre/PNIfrane_rapport-_final strategie_V_07.pdf.
[30] Hayes T Grillos T Bremer LL Murtinho F and Shapiro E 2019 Collective PES : More than the sum of individual incentives J. *Environ. Sci. Policy.* **102** 1–8.
[31] Vanuxem S and Lahssini S 2016 Les Paiements pour Services Écologiques ou la reconquête de la Mâamora par ses habitants ? : Enquête sur une forêt emblématique du littoral méditerranéen marocain J. *Développement durable et territoires* **7** 11165.
[32] Vanuxem S 2016 Des paiements pour services environnementaux en faveur des populations locales ? Regards de l'administration forestière sur le mécanisme de compensations pour mise en défens marocain J. *VertigO* **16.**
[33] Milne S and Adams B 2012 Market masquerades: uncovering the politics of communitylevel payments for environmental services in Cambodia J. *Dev. Chang.* **43** 133–158.
[34] Liu J Li S Ouyang Z Tam C and Chen X 2008 Ecological and socioeconomic effects of China's policies for ecosystem services J. *Proc. Natl. Acad. Sci.* **105** 9477–9482.
[35] Miller BW Caplow SC and Leslie PW 2011 Feedbacks between conservation and social–ecological systems J. *Conserv. Biol.* **26** 218–227.
[36] Ban NC Mills M Tam J Hicks CC Klain S Stoeckl N and Chan KM 2013A social–ecological approach to conservation planning: embedding social considerations J. Front. Ecol. Environ. **11** 194–202.
[37] Grieg-Gran M Porras I and Wunder S 2005 How can market mechanisms for forest environmental services help the poor? Preliminary lessons from Latin America J. *World Dev.* **33** 1511–1527.
[38] Engel S Pagiola S and Wunder S 2008 Designing payments for environmental services in theory and practice: an overview of the issues *J. Ecol. Econ.* **65** 663–674.
[39] Kroeger T 2013 The quest for the optimal payment for environmental services program: Ambition meets reality, with useful lessons J. *For. Policy Econ.* **37** 65–74.
[40] Wynne-Jones S 2012 Negotiating neoliberalism: Conservationists' role in the development of payments for ecosystem services J. *Geoforum.* **43** 1035–1044.
[41] Kosoy N Martinez-Tuna M Muradian R and Martinez-Alier J 2007 Payments for environmental services in watersheds: Insights from a comparative study of three cases in Central America J. *Ecol. Econ.* **61** 446-455.
[42] Wunder S 2013 When Payments for Environmental Services Will Work for Conservation J. *Conserv. Lett.* **6** 230–237.
[43] Yu B and Xu L 2012 Study of eco-compensation in hydropower development in China J. *Procedia Environ. Sci.* **13 (10)** 891e898.
[44] Kaczan D Pfaff A Rodriguez L and Shapiro-Garza E 2017 Increasing the impact of collective incentives in payments for ecosystem services J. *Environ J. Econ. Manage.* **86** 48–67.
[45] Cardenas JC Stranlund J and Willis C 2000 Local environmental control and institutional crowding out J. *World Dev.* **28** 1719–1733.
[46] Bowles S 2008 Policies designed for self-interested citizens may undermine "the moral sentiments": evidence from economic experiments J. *Science* **320** 1605–1609.

[47] Vatn A 2010 An Institutional Analysis of Payments for Environmental Services J. *Ecol. Econ.* 69 1245–1252.
[48] Gatiso TT Vollan B Vimal R and Kühl HS 2018.If possible, incentivize individuals not roups: evidence from lab-in-the-Field experiments on forest conservation in Rural Uganda J. *Conserv. Lett.* 11 e12387.
[49] Hanson C Talberth J Yonavjak L 2011. Forests for Water: Exploring Payments for Watershed Services in the U.S. South WRI Issue Brief, Southern Forests for the Future Incentives Series Issue Brief 2 World Resources Institute (Washington DC, USA).
[50] UNECE Forests and Water 2018. Valuation and Payments for Forest Ecosystem Services United Nation as Economic Commission for Europe (Geneva, Switzerland) p 97.
[51] Ingram JC Wilkie D Clements T McNab RB Nelson F Baur EH Foley CA H 2014. Evidence of Payments for Ecosystem Services as a mechanism for supporting biodiversity conservation and rural livelihoods *J. Ecosystem Services* **7** 10–21.
[52] Iranah P Lal P Wolde BT and Burli P 2018 Valuing visitor access to forested areas and exploring willingness to pay for forest conservation and restoration finance: The case of small island developing state of Mauritius J. *Environ. Manage.* **223** 868-877.

Water quality cartography of the upstream part of the Oum Er Rabia hydraulic basin (Morocco)

O Boumalkha[1,2]*; T Hachi[2]; H Chorfi[3]; H Essabiri[1,2]; Y El Hammioui[1,2]; C Laiboud[1,2] and E H Abba[1,2]

[1] Biological engineering laboratory, Faculty of Sciences and Technology, Sultan Moulay Slimane University. B.P: 523 Béni – Mellal, Morocco.
[2] Heigher School of Technology. Sultan Moulay Slimane University B.P. 170, 54000 Khénifra, Morocco.
[3] Water treatment station. National Drinking Water Board. Khénifra. Morocco

*Corresponding author: b.oumayma021@gmail.com

Abstract. Water resources constitute the basis of human existence; however, they are becoming increasingly scarce under conditions of population increase, economic development, climate change and especially the impact of pollution due to human activities such as the discharge of wastewater without treatment in rivers, lakes ... Also, in Morocco, water resources are becoming scarce as in the global what makes this country among the areas under water stress. The main objective of this study is the evaluation of the water quality of the upper part of Oum Er Rabia. The results of bacteriological analyzes from the various stations showed an increase in total coliforms and feacal coliforms downstream of the Khenifra wastewater treatment plant and at the upstream station of the El Hansali dam. Regarding the physicochemical quality, the water of the majority of the controlled stations is of average to poor quality except one station which has a very poor quality according to the Moroccan standard relating to surface water.

Keywords: Water quality, upper Oum Er-Rbia, surface water, Morocco

1, Introduction

Water is the basis of human existence, with growing demography, socio-economic development and climate change, water resources are becoming scarcer globally [1]. Coupled with the impacts of climate change [2] and the various pollution generated [3] by human activities such as the discharge of untreated water in several regions, water resources in Morocco and around the world are becoming scarce and classify the country among water-stressed areas [7]. Ecosystem functions in general and aquatic ecosystems in particular provide a number of important services for humans [8]. The degradation of the water quality of aquatic ecosystems impacts the functioning of receiving environments (Bassem, 2020), all trophic levels and consequently the abundance of different animal and plant species and can lead to the disappearance of certain polysensitive species and ecosystem loss [8]. The Oum Er Rabia watershed, with an area of 50,000 km2 and a well-developed hydrological network, is also subject to various human activities that lead to the degradation of the quality of its water resources. The main objective of this study is the evaluation of the quality of the water of the upper part of Oum Er Rabia by physico-chemical and bacteriological parameters to put the point on the degree of impact of the anthropic activities following the reasons: presence of anthropogenic activities, impact of central demineralization and domestic water discharged directly without treatment into the waters of Oum Er Rabia, the lack of descriptive and evaluative studies of the impact of these activities as well as the impact of river flowing into the Oum Er Rabia as in the case of the Oued Srou which constitutes one of its effluents.

2, Materials and Methods

2.1. Presentation of the study area

2.1.1. Geological and structural context. The Upper Oum Er-Rbia watershed is part of the Middle Atlas. The powerful stratigraphic series of the studied basin is composed at its base by a Paleozoic basement on which rests, in discordance, a sedimentary cover ranging from the Upper Triassic to the Quaternary. The carbonate formations (limestone and dolomite) of the lower and middle Lias form the backbone of the Middle Atlas Causse. The Triassic, which rests in unconformity with the Paleozoic, is formed of red ocher clays, which are generally in direct contact with the plinth and contain several lenses of gypsum and salt. However, in the valley of Srou Wadi, these clays rest on a conglomerate sandstone formation marking the base of the Moroccan Triassic terrigenous outflow. The clay and saline formation show significant thicknesses (more than 800 m) in the center of the basin east of El Kbab [10].

2.1.2. Hydrological context. Its hydrological origin is made up of ten sources, which are of two types, salty on the right side of the river and soft on the left side. On its upper course, the mountainous massifs receive relatively abundant precipitations which can reach 1000 mm / year. The Oum Er Rabia river is characterized in its upper part by an irregular longitudinal profile and steep slopes [11]. The flow rate favored by a flow rate of up to 105 m3 / S [12], and its steep slope gives it significant self-purification power in places. Downstream from Khénifra, Oum Er Rabia receives on its left bank one of its first tributaries, the Srou Wadi.

2.2. Choice of sampling stations

The choice of 8 sampling stations (figure 1) was made, taking into account the lithological diversity of the basin, the distribution of anthropogenic activities along the upper Oum Er- Rbia, as well as the areas of its confluence with other Wadis (Srou Wadi).

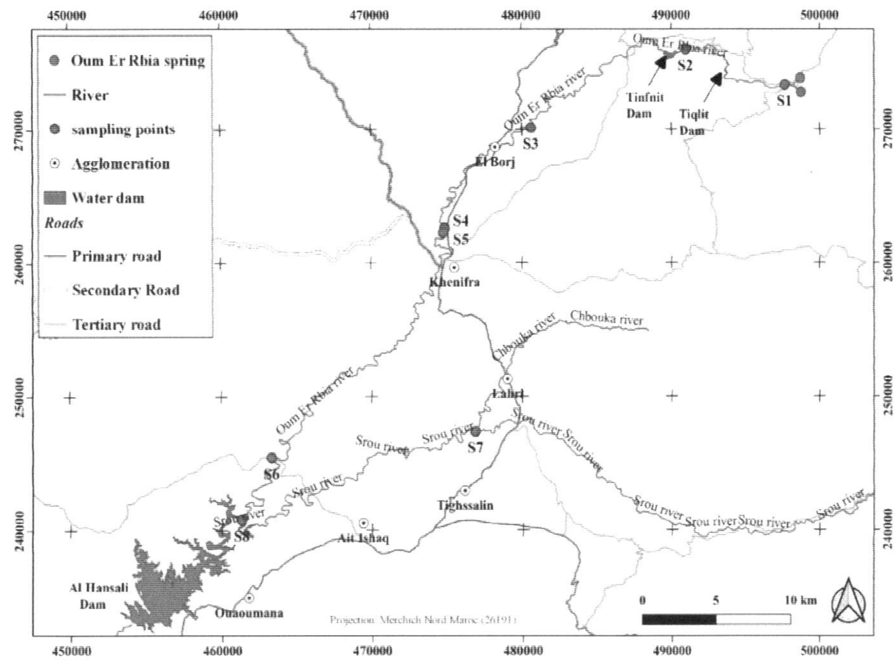

Figure 1. Study aera and different sampling station

IOP Conf. Series: Earth and Environmental Science 1090 (2022) 012007 doi:10.1088/1755-1315/1090/1/012007

Samples were taken at 8 sampling stations which reflect the actual characteristics of surface water throughout the study area. These are 3 stations upstream from the town of khenifra (St1, St2 and St3); two stations at the level of khenifra (St4 and St5), one station downstream from the town of khenifra (St6) and one after the confluence of Srou Wadi with Oum Er-Rbia river in addition to another at the level of Srou for get an idea of the influence of human activities installed on this tributary. The table 1 shows the different stations and their geographical coordinates.

Table 1. Location of different Sampling station.

Stations	Description	Latitude	Longitude
ST1	At the level of the Oum Er-Rbia sources (reference station)	33,056652	-5,424178
ST2	between hydroelectrical station and Tanfnite dam	33,080536	-5,4951
ST3	Upstream of hydroelectrical station El Borj	33,027596	-5,606111
ST4	Next to the old water intake of the treatment plant of khenifra	32,959996	-5,667207
ST5	300m downstream of treatment plant of Khenifra	32956737	-5,668391
ST6	Near the bridge of Oum Er-Rbia river, Road Aourach	32,804562	-5,790065
ST7	On Srou river just after the confluence of Echbouka river	32,822195	-5,645728
ST8	After confluence Srou and Oued Oum Er Rabia	32,762628	-5,811493

2.3. Parameters investigated

Rivers are ecosystems whose functioning and composition reflect the interaction with geological, climatic and anthropogenic conditions in the catchment area. Physico-chemical characteristics are among the key factors structuring the functioning of aquatic ecosystems.

Physico-chemical parameters [13] such as Hydrogen Potential (pH), Dissolved Oxygen (O2), Temperature (T °), Chloride (Cl-), turbidity (SM) and electrical conductivity (EC) are the most important parameters that have influenced aquatic biodiversity as well as the behavior and fate of contaminants in the water column. These parameters also largely control the behavior and mobility of contaminants between the different compartments of the aquatic ecosystem. The frequency of temporal monitoring of the water column at the different sampling sites was chosen according to the Moroccan standard on water quality and the inventory of the degree of water pollution [15]. In the present study, in addition to the six parameters mentioned (pH, O2, EC, Cl-, SM and T °) we followed the chemical parameters indicative of organic pollution (COD, BOD5, NH4 +, NO2-, NO3- and Phosphorus total (PT)) and bacteriological quality of the study area, over a period of 6 months (February to June 2020), (Table 2) presents the various parameters and analysis methods adopted in this study.

Table 2. Parameters studied and different methods of analysis

	Parameters analysed	Unit	Methods of analysis monitored
1	Temperature	°C	In situ
2	pH	-	Electro-metric measurement
3	Conductivity	µS/cm	NM ISO 7888 :2001: Water quality - Determination of electrical conductivity
4	Dissolved oxygen O2	mg/ L	NF EN 25814 / 1993: Water quality - Dissolved oxygen determination - Electrochemical probe method
5	Suspended matter	mg/L	
6	COD	mg/l d'O2	NF T 90 101 /2001: water quality- Determination of COD
7	BOD_5	mg/l d'O2	NF EN 1899 part 2 / 1998: Determination of BOD after n Days (BODn) - Part 2: method for non-diluted samples
8	Chloride: Cl^-	mg/L	Titrimetry (mercuric nitrates)

9	Nitrate NO_3^-	mg/L	ISO 789063/1988(8): Water quality: dosage de determination of nitrates-spectrometric method with sulfo-salicylic acid
10	Nitrite: NO_2^-	mg/L	NM ISO 6777/1999: Water quality: determination of nitrites-methods by molecular absorption spectrometry.
11	Ammonium NH_4^+	mg/L de NH4+	Molecular absorption spectrometry
12	Total Phosphorus	mg/L Pt	ISO 6878 :2004/ Water quality: phosphorus measurement - ammonium molybdate spectrometric method
13	Total and fecal Coliforms (TC and FC)	UFC/100mL	NF ISO 9308-2 /2012: Water Quality-Escherichia coli and Coliform Enumeration-Part 2: Most Probable Number Method.

2.4. *Sampling campaign*

The frequency of temporal monitoring of the water column on the different sampling sites was chosen while drawing inspiration from the Moroccan standard relating to water quality and the inventory of the degree of water pollution (Memcat, 2002). This law stipulates in its article 4 that "The minimum number of samples on the basis of which the quality of the water is assessed is:

- 12 per year for places that are subject to the influence of pollution sources, at the rate of 1 per month;
- 4 per year for places that are not affected by pollution, at the rate of 1 per season.

According to this Moroccan normative reference and given that our study area may be subject to pollutant discharges, therefore, 3 surface water sampling campaigns from upstream Oum Er Rbia, were carried out during the period February -June of the year 2020 at the level of 8 sampling stations whose positioning is presented in figure 1.

Particular attention has been paid to the nature and cleanliness of the equipment used for the storage of the samples in order to minimize any risk of contamination. The water samples were collected in polyethylene bottles previously cleaned by shaking for 24 h with 10% HNO3.

In the field, 4 bottles (1L; 500ml; 250ml and 250ml of river water sample were taken manually, at a depth of 20 cm to 50 cm in bottles previously rinsed 3 times with river water, stored by the rest in a cooler at 5°C until they are processed in the laboratory. Each bottle bears a label mentioning the date and the sampling site. The measurement of the physico-chemical parameters (pH, temperature, dissolved O2 and conductivity) was also carried out in the field, directly in the river using a multi-probe device (HACHE).

3, Results and discussion

3.1. *Temperature*

In the study area, the mean temperature values vary between 9 ° C and 26.80 ° C.

The figure 2 shows an increase in water temperature according to the time of sampling and vice versa at the altitude of the station from upstream to downstream of the watercourse, with the exception of the drop in temperature at the level of the St4 and a remarkable increase at the St3 (a great exchange with the ambient temperature) and this can be explained by the conduction of the majority of flow via the channel to join its natural course just after the St3.

The waters of station St7 (Srou Wadi) are lukewarm and relatively influence the temperature of station St8.

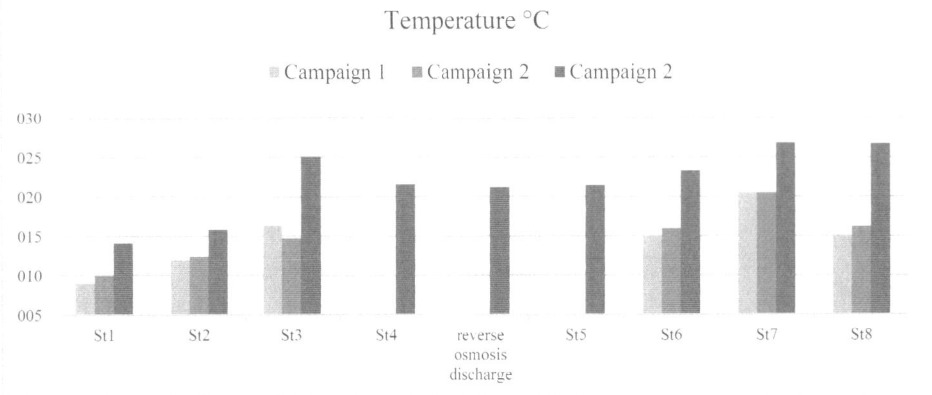

Figure 2. Spatiotemporal temperature variation in the study area.

3.2. *Hydrogen potential*

This parameter revealed in all the stations studied values between 7.25 (St 1) and 8.50 (St 8) (figure 3).

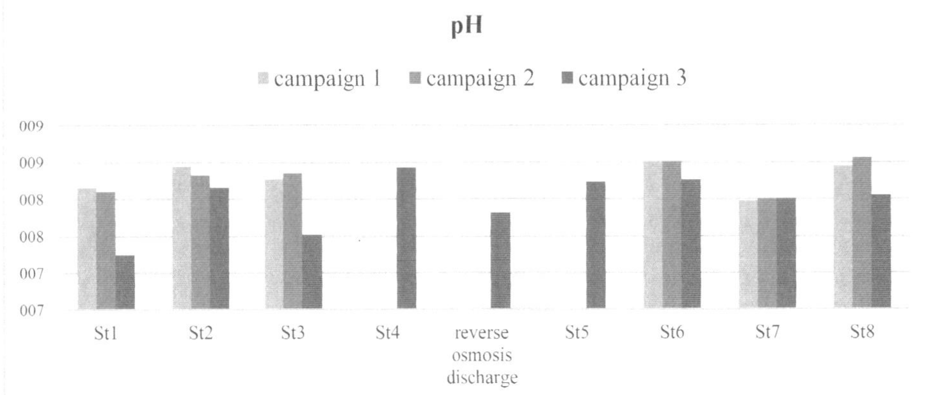

Figure 3. Spatiotemporal variation in pH in the study area

This relative pH stability is due to the buffer system developed by the carbonates and bicarbonates which are important in this medium. Likewise, it should be noted that the geological formations at the top of the Oum Er-Rbia basin, especially at the level of the sources, are limestone in nature.

The rejection of the demineralization station has a pH (7.8), it is a water that has undergone a conventional treatment with alumina sulphate and the one that influences this drop and it will have a relative impact on the St5 (pHSt4 = 8.42 and pHSt5 = 8.23).

3.3. *Conductivity*

The values of the conductivity of the stations studied are very variable and between 2070 (St 1) and 6700 µS / cm (St 7).

From the figure 4, the conductivity of the water of Oum Er-Rbia are, in general, increasing from upstream (St1) to downstream (St8) following brackish runoff from the Triassic lands in addition to its content of other mineral ions.

The relatively high conductivity at St3 is explained by its low flow to which is added salts and brackish runoff.

The St6 located downstream of the wastewater treatment plant includes a slight increase in conductivity and this may be due to the discharge of the latter. While the St7 includes very high values and influencing the conductivity of the St8.

IOP Conf. Series: Earth and Environmental Science 1090 (2022) 012007 doi:10.1088/1755-1315/1090/1/012007

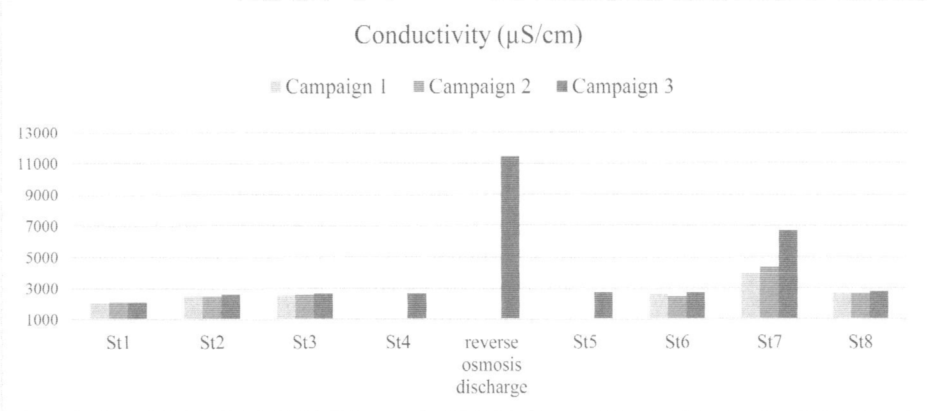

Figure 4. Spatiotemporal variation in conductivity in the study area.

3.4. *Chloride*

The chloride ion values recorded in the water at the top of Oum Er-Rbia vary between 568 (St 1) and 1100.5 mg / l (St 8) (Figure 5).

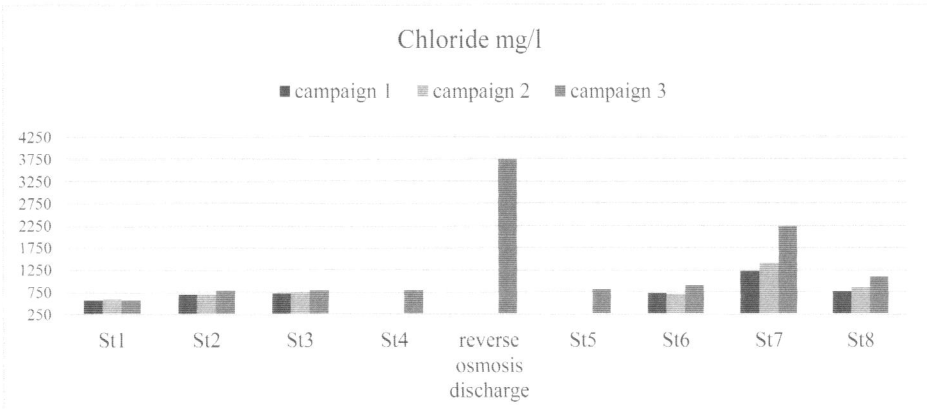

Figure 5. Spatiotemporal variation in the chloride content in the study area.

The chloride ion values increase going from upstream to downstream, under the influence of runoff from Triassic lands to which is added brackish water from discharges from the Khenifra demineralization station (St5) and discharges from the Khenifra wastewater treatment plant for St6 and that of Srou Wadi (St8).

3.5. *Suspended matter (turbidity)*

The increase in turbidity seems normal (dry weather) going from upstream to downstream and it can be explained by the presence of suspended matter of mineral origin which comes from lithology or organic (chlorophyll and micro- algae) (figure 6).

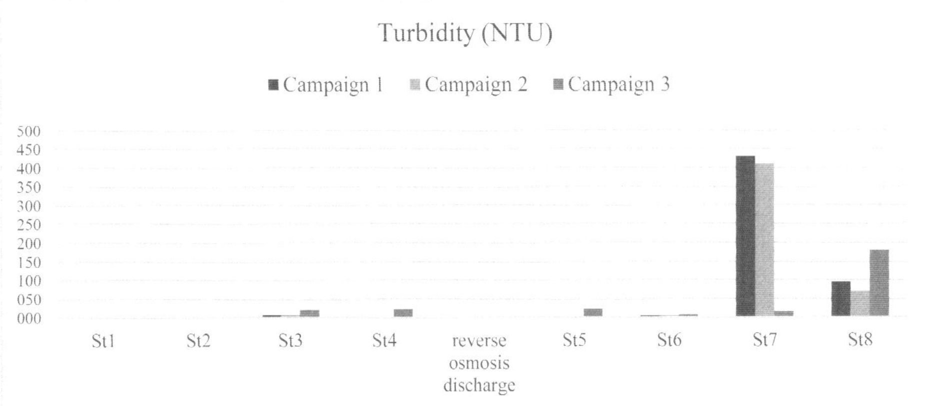

Figure 6. Spatiotemporal variation of suspended matter (turbidity) in the study area

The water of the bottom of Oum Er-Rbia include a high load of suspended matter at the level of the St7 which originates from Srou Wadi which is the subject of a strong pollution of suspended matter by the fact of the existence of sand extraction quarries (+410 NTU), while at the level of St6 only reaches a maximum of 8.70 NTU in the three companies.

During the third campaign, the turbidity of Srou Wadi (St7) recorded a very minimal value (16.80 NTU) knowing that the mission was carried out only one week after the flood, this flood can be the explanation of the strong turbidity experienced by the St8 due to sludge stagnation at the mouth of the Elhansali dam.

3.6. Dissolved oxygen

Dissolved Oxygen values are between 7.05 (St8) and 9 mg / l (St2) (figure 7).

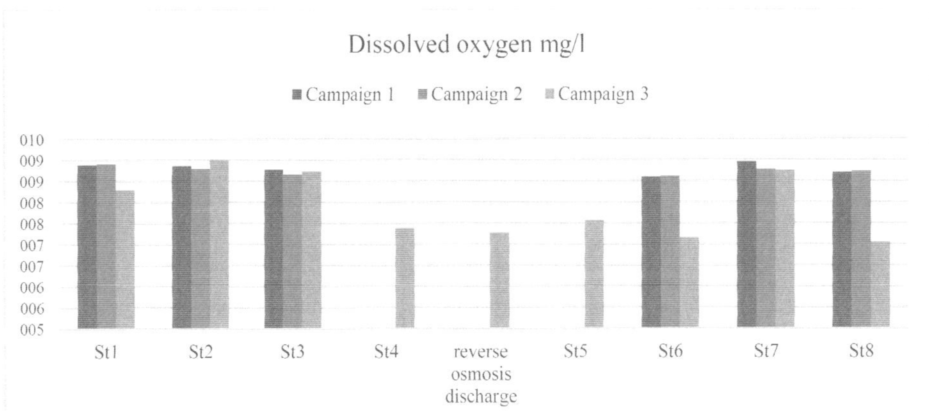

Figure 7. Spatiotemporal variation of dissolved oxygen in the study area

Indeed, the stations which are located downstream of the wastewater treatment plant have a relatively low dissolved oxygen content, this can be explained by the load of biodegradable organic matter of industrial and domestic origin which constitute major consumers of O_2 from the aquatic environment. The decrease in the concentration of dissolved oxygen also depends on the temperature, which increases from upstream to downstream of the river.

The O_2 content in St2 remains high compared to St1 and this is due to the mixing created by the turbining of the hydroelectric station even if its temperature is relatively higher than that of St1.

The dissolved oxygen content of Srou Wadi St7 still remains high in the samples of the three companions, this can be explained by the large contact surface of the river with the atmosphere and by the mixing of the water during the flow.

3.7. Chemical oxygen demand (COD)

From the following graph (figure 8); the COD is important at the level of the St1 and it is decreasing from upstream to downstream for all the stations, except the St7 (Srou) and St8, thus in progressive decrease from the campaign 1 to the campaign 3, this may be due to a decrease in biodegradable releases caused by the limitation of human activities imposed by the Covid-19 circumstances.

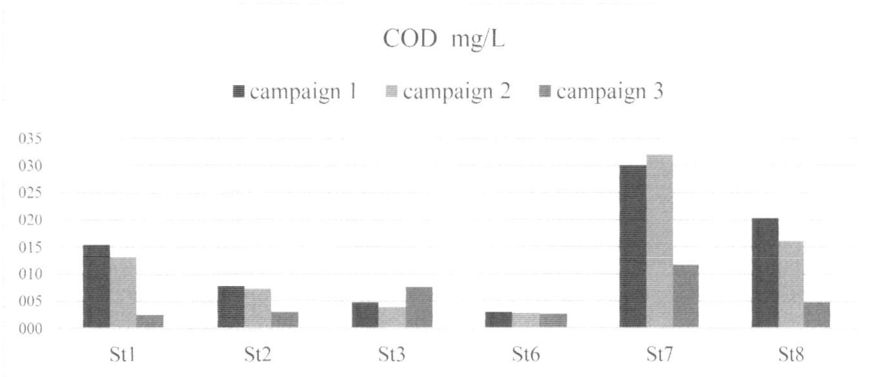

Figure 8. Spatiotemporal variation of COD in the study area.

3.8. Biological oxygen demand BOD_5

Depending on the evolution of this parameter, the stations can be classified into 4 categories (figure 9):

- Stable BOD_5: St1 which can be explained by a biodegradation of organic matter of faunistic and floristic origin in the ecosystem itself plus the discharges linked to the presence of an upstream agglomeration.
- BOD_5 increasing according to the campaigns, case of stations: St2 and St3 this can be explained by the development of bacteria following the increase in temperature as observed in (figure 2).
- Decreasing BOD_5: St6 and St8, the values in these stations are very high but their decrease during the campaigns can be explained by the positive effect of the temperature on the treatment of wastewater in the Khenifra station.
- BOD_5 increases then decreases in St7 by recording maximum values 7-8-5mg / l which can be linked to domestic and industrial discharges at the level of Srou Wadi.

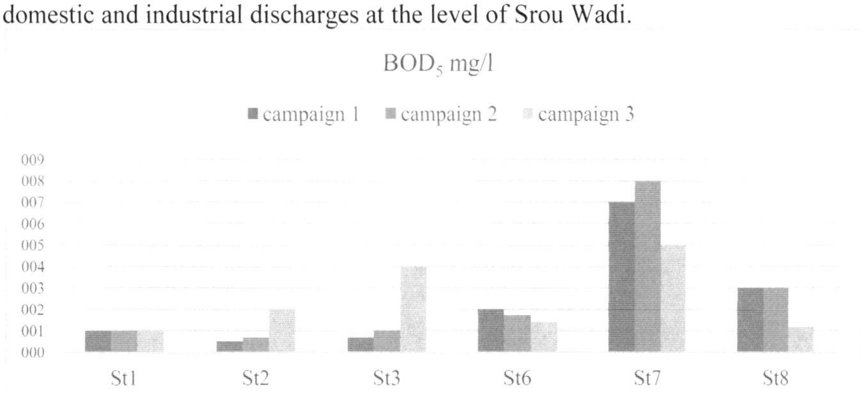

Figure 9. Spatiotemporal variation of BOD5 in the study area.

The increase in BOD_5 levels can be explained by the establishment of conditions for degradation of organic matter by microorganisms whose activity intensifies with warming water.

3.9. Ammonium NH4 $^+$

According to the figure 10, during campaign 1 and campaign 2 we observe that the Ammonium content is in general increasing from station St1 to station St6 to decrease relatively in station St8. Its content at St7 is the same as that of St8. We can say that this is because of the presence of decaying organic matter which may be of faunistic or floristic origin in the ecosystem or it is about discharges from neighboring inhabitants.

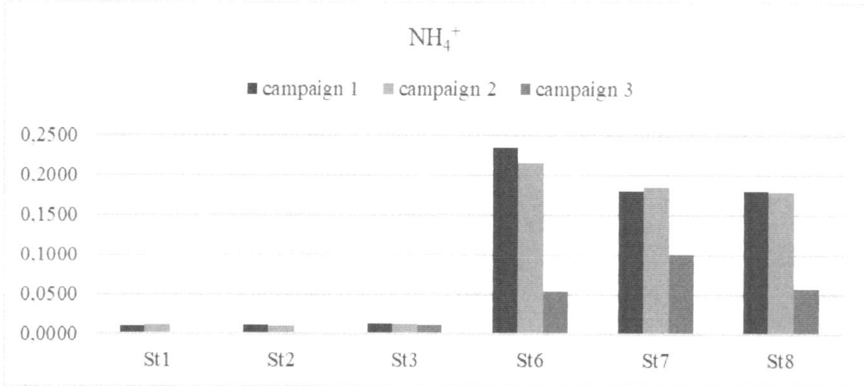

Figure 10. Spatiotemporal variation of ammonium in the study area.

However, during campaign 3 we notice the absence of NH_4^+ in St1 and St2 with a decrease, compared to previous campaigns at St3, St6 and St8 as well as at St7 located on Srou Wadi this can be explained by a strong bacterial oxidation of the NH_4^+ element which may be linked to the change of the season, therefore an abundance of nitrifying and nitric bacteria.

3.10. Nitrates (NO_3^-) and Nitrites (NO_2^-)

There is an increase in nitrates in Station St2 which can be explained by the location of this station around the Tanfnite settlements and agricultural activities.

The increase observed at the St6 station is because it is located downstream from the Khenifra wastewater treatment plant, while the values reached at the St8 can be explained by the runoff from the irrigation of neighboring land (figure 11).

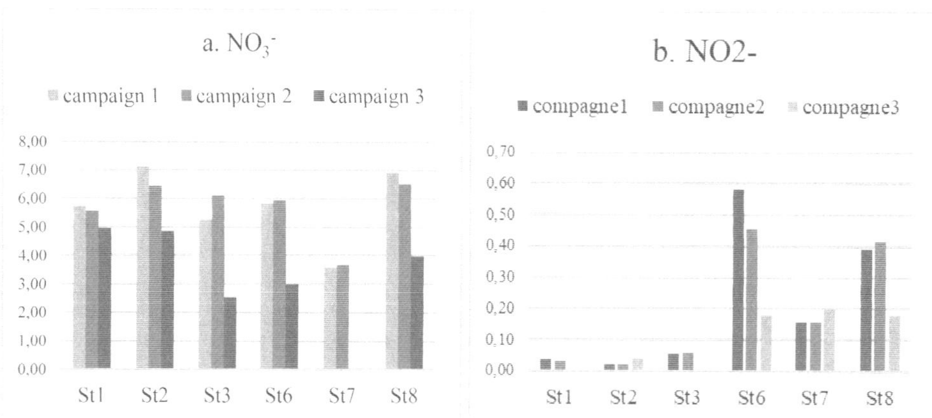

Figure 11. Spatiotemporal variation of nitrates (a) and nitrites (b) in the study area.

The content of nitrite NO_2^- can be explained by the rate of oxidation of NH_4^+ is more important than the rate of oxidation of NO_2^- to NO_3^-, indeed we notice that the concentrations of the elements: $[NH_4^+] < [NO_2^-]$ whatever $[NO_2^-] << [NO_3^-]$.

3.11. Total phosphorus (tP)

The analysis of the results of the figure 12 shows that the concentration of total phosphorus in the water of the stations decreases going from the station St1 towards the St3 during the three campaigns and this can be explained by the assimilation of phosphorus by the plants and aquatic algae while we notice an increase in phosphorus content during the 3rd campaign for the first three stations and this can only be due to the runoff of fertilizers with the irrigation water of the neighboring land.

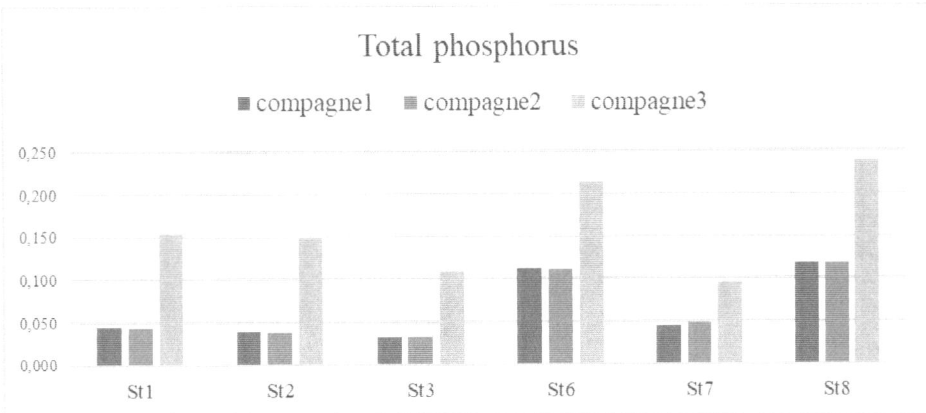

Figure 12. Spatiotemporal variation of total phosphorus in the study area.

At station St6, this content becomes higher given the discharges from the upstream purification station (detergents) in addition to phosphorus of agricultural origin.

A remarkable increase in the Total phosphorus content going from St6 to St8, and this is due to agriculture since Srou Wadi is subject to a very significant impact from agricultural activities.

3.12. Total and faecal coliforms

This bacteriological parameter showed very variable values which are between 360 / 100ml and 110,000 / 100ml (figure 13).

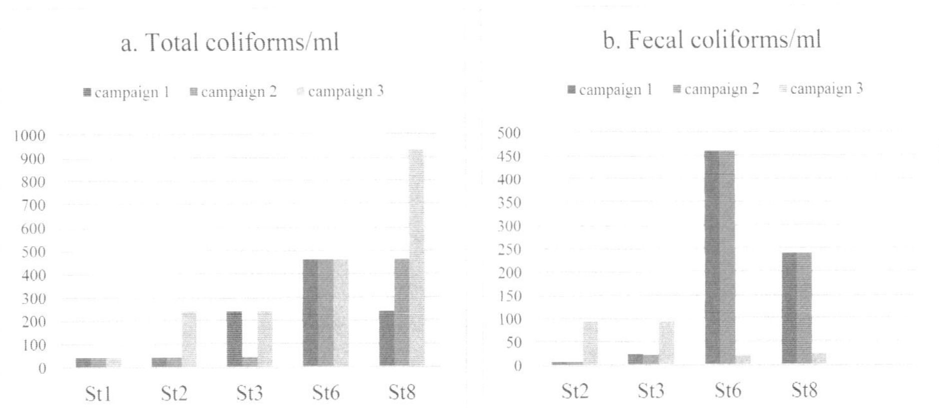

Figure 13. Spatiotemporal variation of total (a) and fecal (b) coliforms in the study area.

Total and fecal coliforms are minimum in the first two stations St1, St2 and slightly significant at St3 which can be infected by direct discharge or by leaching of animal fecal matter. They increase remarkably at St6, the latter is contaminated by discharges from the Khenifra wastewater treatment plant.

The concentration of these bacteria at St8 has decreased. It seems that the conditions of the richness of Oum Er-Rbia river in mineral salts play a limiting factor of bacterial development which may originate from domestic waste from neighboring villages.

3.13. Spatiotemporal typology of the environment by statistical study: Principal Component Analysis

The PCA; Principal Component Analysis, is a method of processing data that explores the connections between variables and the similarities between individuals [16].

So, we used this method in order to establish a relationship between the different physicochemical parameters and to make the distinctions / similarities of the sampling stations in the form of groups (water quality) in order to visualize the effect of anthropogenic activities on the quality of surface water of the Oum Er-Rbia river (case of the upper Oum Er-Rbia).

Indeed, a statistical treatment NPCA (Normalized Principal Component Analysis) was applied to all the parameters of the 3 campaigns which were carried out during the practical period. The PCA was carried out on a data matrix consisting of 5 samples (5 stations x 1 February campaign), 5 samples (5 stations x 1 March campaign) and 5 samples (5 stations x 1 June campaign). During these samples, the 14 variables were measured, but for reasons related to the application of this method, only 8 parameters were retained (water temperature, conductivity, dissolved oxygen, biochemical oxygen demand, chlorides, total phosphorus, Ammonium, nitrates).

The PCA was only feasible by these variables (physicochemical parameters) and these individuals (the sampling stations) mentioned above which satisfy the three conditions of such a PCA:

- The correlation of the majority of variables must be greater than or equal to 0.5.
- KMO index> 0.5 and Bartlett's test tends towards zero (table 3).

Table 3 : KMO index and Bartlett test.

Kaiser-Meyer-Olkin index for measuring sampling quality.		0,531
Bartlett's sphericity test	Approximate Khi-square	96,366
	ddl	28
	Signification	0,000

So, this PCA will be done with a data loss of 37.37% (100-62.622), while the data image will only represent 62.622% of the data extraction quality (see table 4 representation quality).

Table 4 : Total Variance Explained.

Component	Sums extracted from the load square	Sums of rotation of the load square		
	Cumulative %	Total	% of variance	Cumulative %
1	39,146	4,554	37,949	37,949
2	62,622	2,961	24,673	62,622

In the Factorial plane F1x F2 the correlation circle formed by the axes F1 and F2 (figure 14) and (Table 4) giving 62.622% of the total information shows:

According to the F1 axis (37.949%), on the one hand highly mineralized water rich in chlorides and on the other hand, it is associated with variables such as, BOD_5, NH_4^+ and COD which are indicators of domestic pollution, agricultural and industrial. This axis then defines a gradient of mineralization and organic pollution going from the left to the right of the F1 axis

The F2 axis (24.673%) is associated with variables such as: Pt, NO_3^- defines an axis of organic pollution of agricultural origin and domestic discharges. We note that the DO is negatively correlated with Pt and positively with NO_3^- this can be explained by the fact of the aquatic plants which assimilate the Pt and release oxygen by the phenomenon of photosynthesis in addition to the mixing of water knowing that the top of Oum Er Rbia is characterized by a significant slope (loss of 1000m / 50km or an average slope of 2%); we also notice that the DO is negatively correlated with the temperature which is very logical since the dissolved oxygen content increases with the decrease in temperature.

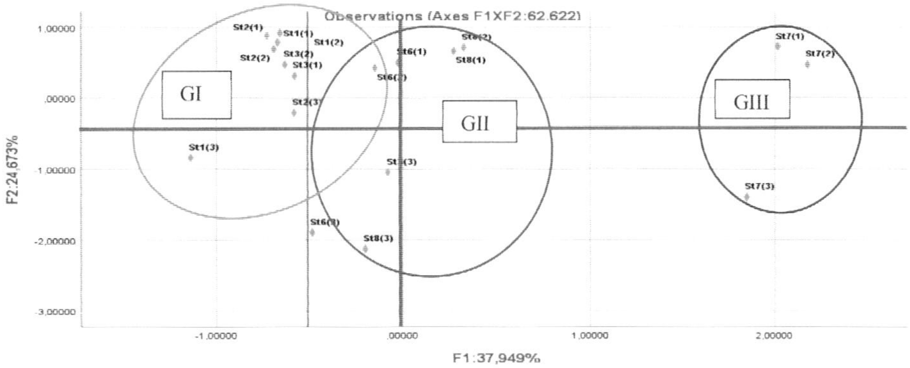

Figure 14. Graphical approach of the PCA of the physicochemical parameters in water according to the F1x F2 plan. C, factorial map of the stations.

The global analysis allows us to define a typology dominated by the individualization of three groups of stations GI, GII and GIII (figure 14).

• The GI group includes the stations located upstream from the city of Khenifra (St 1, St2 and St 3) characterized by less polluted water.

• The GII group is made up of the St6 and St8 stations, they are located respectively after the Khenifra wastewater treatment plant and after the confluence of Srou Wadi, their water is of intermediate quality of the GI and GII.

• The GIII group contains the St7 station characterized by strong mineralization and pollution of agricultural and industrial origin.

• St4 and St5 stations are not involved in this statistic given their incomplete data due to the sanitary conditions of the COVID 19 pandemic.

3.14. *Quality assessment of upstream d'oum Er rbia according to moroccan standards*

The table above summarizes the water quality at each station and during each of the three campaigns with reference to the Moroccan standard relating to the quality of river water (table 5 and table 6).

Table 5. the quality of the stations by campaign for each parameter according to the Moroccan standard for the quality of river water.

Parameter	Campaign	St1	St2	St3	St4	Reverse osmosis discharge	St5	St6	St7	St8
T°C	1	9	12	16,4	....	....	....	15	20,5	15,1
	2	10	12,5	14,7	....	....	....	16	20,5	16,2
	3	14,2	15,8	25,1	21,6	21,2	21,5	23,3	26,8	26,7
Conductivity	1	2070	2480	2552	....	....	....	2600	3960	2630
	2	2120	2490	2600	....	....	....	2500	4380	2680
	3	2080	2570	2620	2690	11480	2720	2730	6700	2800
pH	1	8,16	8,44	8,27	....	....	....	8,5	7,97	8,44
	2	8,1	8,32	8,35	....	....	....	8,5	8	8,5
	3	7,25	8,16	7,52	8,42	7,81	8,23	8,26	8	8,05

Chloride	1	568	692,25	720,7	....	....	....	718	1219	760
	2	585,75	702,9	738,4	....	....	....	710	1380	860
	3	568	788,7	790,4	795	3745	805,9	901	2229	1101
OD (mg/l)	1	8,89	8,86	8,78	....	....	....	8,6	8,95	8,7
	2	8,9	8,8	8,66	....	....	....	8,62	8,78	8,72
	3	8,3	9	8,72	7,38	7,27	7,58	7,17	8,75	7,05
BOD_5 (mg/l)	1	1	0,5	0,7	....	....	....	2	7	3
	2	1	0,7	1	....	....	....	1,7	8	3
	3	1	2	4	....	....	....	1,4	5	1,2
COD (mg/l)	1	15,4	7,73	4,71	....	....	....	3	30	20,3
	2	13	7,21	3,9	....	....	....	2,8	32	16
	3	2,46	3	7,58	....	....	....	2,59	11,65	4,68
Nitrate (mg/l)	1	5,75	7,11	5,26	....	....	....	5,82	3,56	6,93
	2	5,58	6,43	6,12	....	....	....	5,93	3,69	6,51
	3	4,96	4,84	2,52	....	....	....	3	0,05	3,97
NH_4+ (mg/l)	1	0,0092	0,0108	0,013	....	....	....	0,2349	0,181	0,181
	2	0,011	0,0098	0,013	....	....	....	0,2154	0,185	0,179
	3	0	0	0,011	....	....	....	0,0535	0,101	0,057
Pt(mg/l)	1	0,045	0,039	0,033	....	....	....	0,113	0,045	0,119
	2	0,044	0,037	0,032	....	....	....	0,111	0,049	0,119
	3	0,153	0,148	0,109	....	....	....	0,214	0,095	0,238
CF /1ml	1	3,6	7,4	23	....	....	....	460	460	240
	2	3,6	7,4	21	....	....	....	460	460	240
	3	23	93	93	....	....	....	19	23	23
CT/1 ml	1	43	43	240	....	....	....	460	460	240
	2	43	43	43	....	....	....	460	460	460
	3	43	240	240	....	....	....	460	1100	930

So, in order to attribute an overall quality to each station that is part of the study area, the classification is made according to the average of each parameter analysed from each station during the three campaigns.

According to the same standard, the overall water quality takes the highest class presented by the parameters analysed (table 6).

Table 6. the surface water quality grid (Moroccan standard)

		Class 1	Class 2	Class 3	Class 4	Class 5
Parameters	Units	Excellent	Good	Average	Poor	Very Poor
Temperature	°C	<20	20-25	25-30	30-35	>35
PH		6,5-8,5	6,5-8,5	6,5-9,2	<6 ou >9,2	<6 ou >9,2
Conductivity in 20°C	µS/cm	<750	750-1300	1300-2700	2700-3000	>3000
Chloride (Cl-)	mg/l	<200	200-300	300 -750	750 -1000	>1000
Dissolved O_2	mg/l	>7	7-5	5-3	3-1	<1
BOD5	mg/l	<3	3-5	5-10	10-25	>25
COD	mg/l	<30	30-35	35-40	40-80	>80
Nitrates (NO_3-)	mg/l	≤10	10-25	25-50	>50	
Ammonium	mg/l	≤0,1	0,1-0,5	0,5-2	2-8	>8
P total (PT)	mg/l	≤0,1	0,1-0,3	0,3-0,5	0,5-3	>3
Fecal coliforms	/100mL	≤20	20-2000	2000- 20.000		
Total coliforms	/100mL	≤50	50-5000	5000- 50.000		

The table 7 and figure 15 summarizes the quality of the stations in the study area. Stations St4 and St5 are stations that were only introduced into the study area in June to find out the degree of impact of the discharges from the demineralization station on the quality of Oum Er-Rbia in terms of water salinity (chlorides).

Table 7. Assessment of the average water quality of the stations studied according to each parameter according to the Moroccan standard for river water

Parameters	St1	St2	St3	St4	Reverse osmosis discharge	St5	St6	St7	St8
T°C	11,07	13,433	18,733	21,6	21,2	21,5	18,1	22,6	19,333
Conductivity	2090	2513,3	2590,7	2690	11480	2720	2610	5013,3	2703,3
pH	7,837	8,3067	8,0467	8,42	7,81	8,23	8,42	7,99	8,33
Chloride	573,9	727,95	749,8	795	3745	805,85	776,33	1609,3	906,83
OD(mg/l)	8,697	8,8867	8,72	7,38	7,27	7,58	8,13	8,8267	8,1567
BOD_5 (mg/l)	1	1,0667	1,9	....	....	....	1,7	6,6667	2,4
COD (mg/l)	10,29	5,98	5,3967	....	....	....	2,7967	24,55	13,66
Nitrate (mg/l)	5,428	6,1245	4,6349	....	....	....	4,9139	2,4325	5,8039
NH4+ (mg/l)	0,007	0,0069	0,0125	....	....	....	0,1679	0,1557	0,1387
tP(mg/l)	0,081	0,0749	0,058	....	....	....	0,1462	0,063	0,1585
FC /1ml	10,07	35,933	45,667	....	....	....	313	337.66	167,67
TC/1ml	43	108,67	174,33	....	....	....	460	673.33	543.33
Global quality	Average	Average	Average	Poor		Poor	Poor	Very Poor	Poor

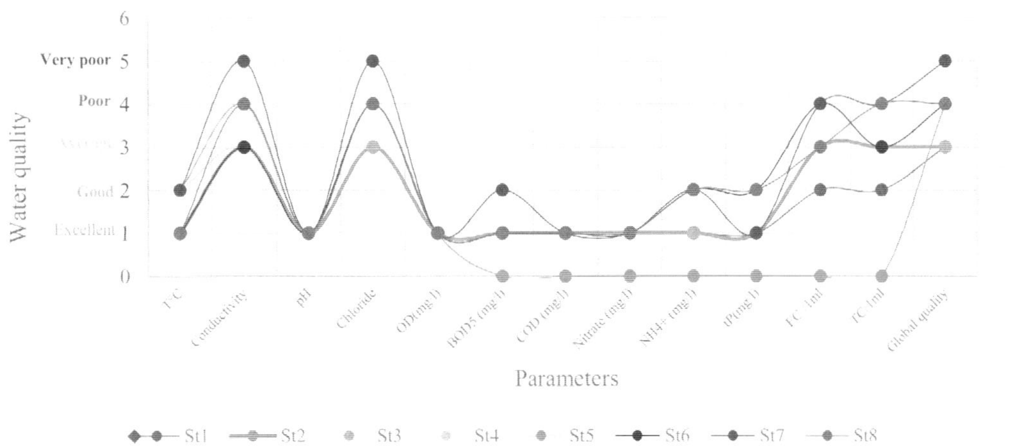

Figure 15 : Assessment of the average water quality of the stations studied according to each parameter

Three stations have an average overall quality. These are the stations considered as reference stations (St1, St2 and St3) these stations have taken this class following three parameters: chloride, conductivity and bacterial load. Four stations (St4, St5, St6 and St8) whose quality is within the limit of bad classes mainly due to the chloride content and their conductivity. As for the 7th station (St7) represents a very poor quality (class V) due to its chloride content and high conductivity (figure 16).

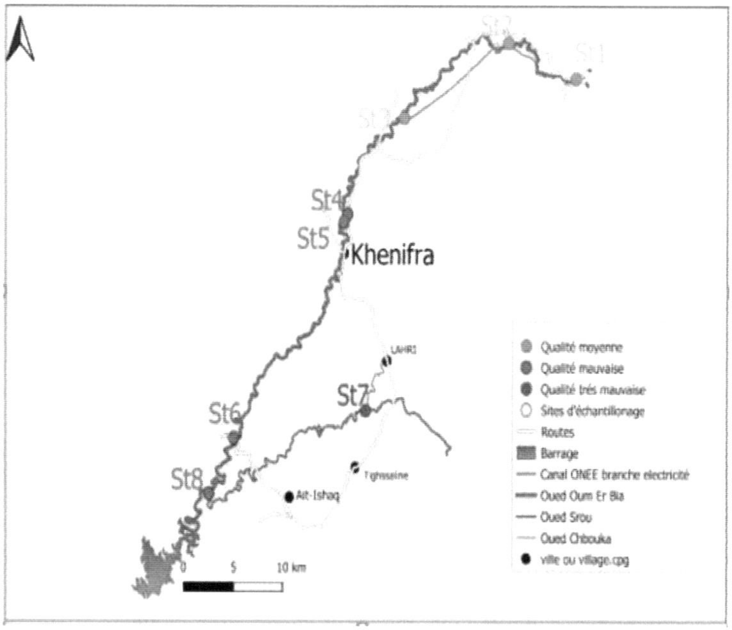

Figure 16. Global distribution of water quality in the study area

4, Conclusion and recommendations

Spatiotemporal monitoring of different physicochemical and bacteriological parameters, such as temperature, pH, conductivity, SM, dissolved O_2, COD, BOD_5, total phosphorus, NO_3^-, NO_2^-, NH_4^+, Cl, fecal and total coliforms, showed a deterioration in water quality, particularly in the area subject to wastewater discharges from the town of Khenifra respectively at the St 6 stations, and downstream of the confluence from Srou Wadi (St 8). The Normalized Principal Component Analysis of the physicochemical data allowed us to highlight the existing correlation between the different parameters and to differentiate a zonality of water quality in the studied region.

The results of the water analysis allow us to conclude that the quality of the stations located directly on Oued Oum Er Rbia (7 stations) are distributed as follows: 42.86% of the upstream stations have an average quality or 57.14 % of other stations have poor quality.

Reasoning on all the stations we will have: 37. 50% of the stations are of average quality, 50% of poor quality (St4, St5, St6 and St8) and 12.50% of very poor quality represented by the St7 upstream of the confluence of Srou Wadi with Oued Oum Er Rbia which explains the intense anthropization experienced by Srou Wadi in addition to its high salt content (Chlorides). In addition, we were able to identify 4 industrial units installed at the level of Srou Wadi, they are totally concentrated in a very small area, and a wastewater treatment station that does not treat all of the wastewater from the area Khenifra city since we have not yet managed to collect a significant part of these.

Among the means of combating pollution of the Wadis we suggest:

- A census of all polluting activities;
- Ensure the collection and treatment of all wastewater from rural centers as well as the collection and treatment of waste from all subscribers of the city of khenifra;
- Raising the awareness of owners of traditional as well as modern oil mills;
- Characterization of the liquid discharge from industrial units and oil mills located in the Oum Er Rbia basin;
- The evaluation of the purification yields of the purification stations in other towns and rural centers for the parts of Oued Oum Er-Rbia downstream of the study area;

- Biological control against the development of algae in dam reservoirs (Tiklit and Tanfnite dams, etc.) Introduction of carpillons (common Chinese carp and herbivores).

Confirming via these results the significant impact due to human activities on the water quality of Oued Oum Er-Rbia, it is therefore necessary to install a program of control and continuous monitoring of the various sources of pollution and their effect on the environment at regional level, by obliging the various existing industrialists to equip themselves with an adequate treatment system for their products, residual water before its evacuation into the natural environment, as well as the recycling and reuse of their purified wastewater in certain type of irrigation. It is in this context that our future study fits into the objective of quality assessment water from the part of Oued Oum Er-Rbia existing in the vicinity of the controlled landfill of the city of Khénifra.

5, References

[1] Li X, Liu C, Wang G, Bao Z, Diao Y, Liu J. Evaluating the Collaborative Security of Water–Energy–Food in China on the Basis of Symbiotic System Theory. *Water*. 2021 Apr 17;13(8):1112.

[2] Ouhamdouch, S., Bahir, M. P. and Carreira, M. (2017). Climate change impact on water resource in semi-arid environment: example of Essaouira basin (Morocco). *J. Water Sci.*, 31(1), 13-27.

[3] Zhang, M. (2006). Assessing the Impact of Leather Industry to Water Quality in the Aojing Watershed in Zhejiang Province, China. Environ. Monit. Assess., 115, 321–333.

[4] Echihabi, L., Foutlane, L., Bourchich, L. and Zeitschrift, L. (2002). Problèmes majeurs de pollution et leurs impacts sur les ressources en eau au Maroc Revue économique et sociale : bulletin de la Société d'Etudes Economiques et Sociales Band (Jahr): 60.

[5] Hahid, A. and Saba, I. (2018). Water Pollution and its Sources, Effects & Management: A Case Study of Delhi. Int. J. Curr. Adv. Res., 07(2), pp. 10436-10442.

[6] Kubicz, J., Pawelczy, A. and Lochynski., P. (2018). Environmental Health Risk Posed by Contamination of the Individual Water Wells. Chemosphere, 208, 247-256

[7] Ceoworld magazine (2019).> The Latest > Stats Gate > Most Water-Stressed Countries in The World For 2019.

[8] Doi H, Katano I, Negishi JN, Sanada S, Kayaba Y. Effects of biodiversity, habitat structure, and water quality on recreational use of rivers. Ecosphere. 2013 Aug;4(8): art102.

[9] Bassem SM. Water pollution and aquatic biodiversity. Biodiversity Int J. 2020;4(1):10–16. DOI: 10.15406/bij.2020.04.00159.

[10] Khalil N, Er Rouane S, Mania J, Mudry J. Essai de bilan hydrochimique sur les eaux du Haut Bassin de l'Oum Er Rbia (Moyen Atlas, Maroc). Rev Fr Geotech. 2004;(109) :75–85.

[11] Loup J. - L'Oum Er Rbia. Contribution à l'étude hydrogéologique d'un fleuve marocain. Institut scientifique chérifien, Service de géographie physique, n° 9, Rabat, Maroc, 1960.

[12] Benamar, A., Mahjoub, F.Z., Kzaiber, F. and Oussama, A. (2019). Evaluation of water quality of Oum Er Rbia River (Morocco) using water quality index method. J. Appl. Surf. Interfaces, 5(13) 1-12.

[13] Ampon Payakka And Taeng-On Promm. 2014. Aquatic Insects Biodiversity and Water Quality Parameters of Receiving Water body.Current World Environment Vol. 9(1), 53-58 (2014)

[14] Chinche Sylvie Belengfe, Zebaze Togouet Serge Hubert, Moanono Patrick,Georges Thiery, Pountougnigni Oumarou F, Tuekam Kayo Raoul, Polycarpe and Fomena Abraham. 2019. Impact of physicochemical parameters on biodiversity and groundwater quality in Tiko, Cameroon. International Journal of Fisheries and Aquatic Studies 2019; 7(6): 39-46

[15] Abba EH, Idrissi I, Bennani Y, El Yaacoubi A, Cherroud S, Ainane, Mhamed Khaffou. 2021. Assessment of water quality of Oum Er Rabia River by Microbiological Quality Index and Water Quality Index. Pollution [Internet]. 2021 Jul [cited 2021 Dec 5];7(3). Available from:

[16] Sabatier R, Rymer JC, Pau B. Standardisation en immuno-analyse : 6rude multicentrique de la dispersion des résultats par analyse en composantes principales. :6 pp 270, 271.

[17] Norme Marocaine de Qualité de l'Eau (NMQE) (2002). Rabat, Bulletin Officiel Maroc, N°5062.

Structural and Mining Geology of a Copper site with Epigenetic-type auriferous levels in the region of Imi Nisk in M'goun, Central High Atlas, Rural town of Abachkou, Azilal. Morocco

M Alaoui[1], A Wafik[2], M Khalid[3], A Boutsougame[4] and T Saki[5]

[1]Multidisciplinary of Technology and Social Sciences Laboratory. EST-Khenifra. Morocco
[2]Dynamic Lithosphere Laboratory and Genesis of Resources, DLGR, Cadi Ayyad University FSSM. Marrakech. Morocco
[3]Marine Geosciences and Soil Sciences Laboratory. CNRST Associate Research Unit (LGMSS). FS-El Jadida. Morocco
[4]Environment and Natural Resources Management Team. EST-Khenifra. Morocco
[5]Topography and Civil Engineering Unit. OFPPT. Meknes. Morocco

E-mail: alaouimustapha0@gmail.com

Abstract. Our initial intention was to use point geochemical data in relation to regional and/or local geology to identify and quantify the mineral potential of these Paleozoic formations, and the nature of the soil in relation to weathering by external climatic agents. The mineralogical study of the substrates proved to be essential to confirm the results of the spot sampling by channel sampling of the mineralized structures. From there, it was decided to focus on structural geology, regional tectonics, geochemistry, geomorphology, and the extension of veins from other deposits around this site. Because it is located at the Ordovician-Devonian contact, the copper mineralization at the future Imi Nisk mine site in the Central High Atlas has been interpreted in terms of sulfides-rich cluster deposits (Chalcopyrite, Azurite and Covellite) in Ankerite or siderite, and quartzite. The new structural and mineralogical data presented in this work favor the hypothesis of epigenetic mineralization, meaning that it was formed after the host rock. The present work consists in establishing a preliminary research in order to give a general idea and a very precise overview of the gitology, geology and geochemistry of the mineralized structures and veins located in this research permit in mining exploration of Imi Nisk. Also, no study has been published in this thematic on this site.

Keywords: Ordovician, Paleozoic, Central High Atlas, Epigenetic, Mineralization, Copper, Imi Nisk.

IOP Conf. Series: Earth and Environmental Science 1090 (2022) 012008 doi:10.1088/1755-1315/1090/1/012008

1. Introduction

Deposits and mining indices in Morocco are grouped within provinces metallogenic established in connection with major geodynamic events which have characterized the geology of Morocco [1-2]. The mineral paragenesis is fairly constant: chalcopyrite, bornite, Covellite, chalcocite, malachite and rarely pyrite in a gangue of chlorite, barite and orthosis [3]. Many research works were carried out during the 1940 is by the Issougrid Mines Company, which was partially taken over in 1967 by the former BRPM. In accordance with current regulations, the future Imi Nisk mine was established as a research permit (PR 2437860) in November 2008. The permit was renewed on November 27, 2011 and then transformed into an operating permit in 2013. After having exhausted the various renewals, the operating permit benefited from the exceptional extension in 2015. This exceptional extension is valid for 4 years but the holder must comply with the provisions of the new law relating to mines 33/13. On the site, old mining works were carried out there including a main gallery having reached a depth of 5 meters. On the surface, trenches have been dug but these are mainly scrapings to recognize the quality of the quartz structures visible on the surface especially when these include oxides of sulphides (Chalcopyrite) and iron (Limonite) which constitutes the main guide. for the operator. Tertiary copper-lead epigenetic mineralization exists in the Mesozoic carbonate cover of the Eastern High Atlas [4]. Morocco, while stibnite, barite and copper mineralization are located in the Precambrian to Paleozoic age basement [5]. It is estimated that sulfur from bacteriologically reduced syngenetic sulfides and Mesozoic marine sulfates remobilized during tertiary orogenesis are the sources of H_2S and SO_4 [6]. It is estimated that Cambro-Ordovician sulphates constitute the source for the Jbel Zelmou deposits in the basement, of a segment of the Hercynian chain crossed by the Menhouhou fault [7-8]. The isotopic composition of galena lead in the tertiary deposits of the Mesozoic cover is rigorously homogeneous. The ages vary from 230 to 180 million years. It is considered that significant remobilization processes are at the origin of tertiary metallogenic in Morocco [9-10-11-12].

Although, in the study area, only the land allocated to the Ordovician-Silurian-Devonian is exposed, we will give a brief overview of the geology of the area, as a whole, taking particular inspiration from the work of the Azilal map at 1/100,000. Historically, undated land had been allocated to the Ordovician-Devonian-Silurian. For some authors, this term includes all the shale-quartzite formations located under the graptolite formations of the Silurian [13-14-15-16-17]. Mining remains an essential and growing part of the modern industry. By some estimates, it makes up nearly 45% of the total global economy, and mineral production continues to increase as demand for raw materials grows around the world. However, many mining techniques still in use can have serious impacts on both the mining site itself and the surrounding environment [18].

2. Materiel and methods

First of all, it is necessary to choose the site with important potential in targeted ore, to make visits and to take samples to determine the clark of the minimum of exploitation. The methodology applied in this research work is purely geological, based on the collection of field data by the mineralogy, geochemistry, and geological to choose the appropriate mining method and define the exploitable content based on the samples analysed (Figure 1). Geological mapping is an essential tool for mining exploration work; it is performed at different scales depending on the state of progress of the work and led to the creation of geological maps indicating the lithology of the outcrops and the structures with the location of mineralization and mineral showings. A great mastery in the conduct of geochemistry campaigns for the search for precious metals and base metals in various geological contexts. Rock lithogeochemistry or geochemistry is the final phase of a geochemistry program to determine the primary anomaly on the parent rock. Also used for the geochemical characterization of surface alterations, in particular the exploration of sulphides and oxides in ancient soils. For the characterization of the mineralogical composition of the mineral species, we proceeded on the one hand to the study by sampling and then to the geochemical analysis of thirteen (13) samples and

IOP Conf. Series: Earth and Environmental Science 1090 (2022) 012008 doi:10.1088/1755-1315/1090/1/012008

finally to the determination of the average contents. A structural and tectonic cartography proves to be essential to know the process of deposition of these indices.

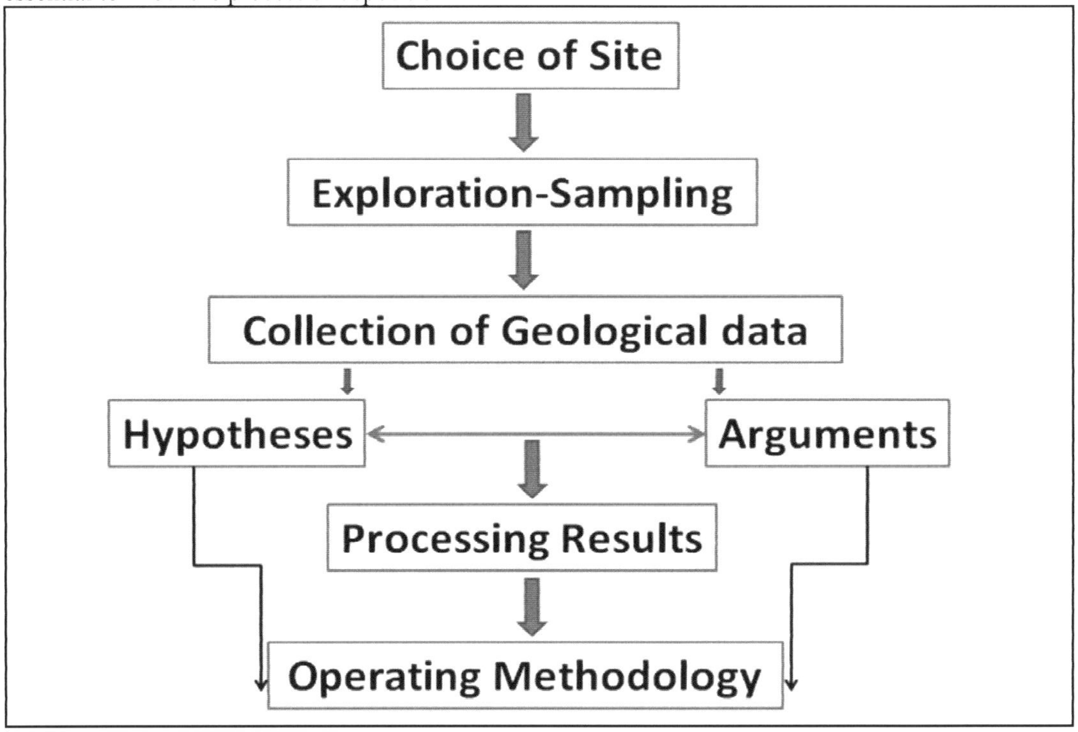

Figure 1. Methodology of Mining prospection field

3. Geology setting

The Moroccan High Atlas, to the south of the Marrakech city, is the highest part of the chain, and it contains its highest points including the Jbel Toubkal 4167 m and the Jbel Ouankrim 4089 m, in the same way the High Atlas Central, the summit of M'goun peaks at 4070m above sea level. Most of the chain is occupied by ancient, Paleozoic but also Precambrian lands, which outcrop in several buttonholes, the largest of which is that of the old Block, overlapping on secondary and tertiary lands. Reverse Horst faults reflecting the overlap of the Paleozoic on the secondary, consequence of the opening of the Atlantic Ocean after filling of the Jurassic basin. Indeed, the faults that played in normal breaks in Graben during the Hercynian, after they replayed in reverse faults during the tertiary (Figure 2a) and (Figure 2b). These reverse structures have been filled by mineralization such as Zn, Lead, Barite by circulation of hydrothermal waters. (Figure 3a). The High Central Atlas in Morocco is a segment of a vast intracontinental chain, which extends from the Atlantic Ocean in the region of Agadir in Morocco to the Gulf of Gabes in Tunisia.

Nowhere in these massifs is the shortening due to Atlas tectonics such that it can mask ancient structures; in Paleozoic terrains, we can therefore recognize hercynian deformations and compare them to that of the Mesetian domain, such as the N70 faults which have taken an N120 direction.

The ancient Bloc, for example, with its schistose, metamorphic and granitized domains, extends the central areas of the Meseta, from the buttonhole of Ait Tamlil, with its carboniferous gravity layers, corresponds to the eastern areas of the Jbilet. The Western High Atlas is separated from the Atlas of Marrakech by the Triassic corridor of Argana.

From a structural point of view, Imi Nisk's substratum reveals the presence of two phases of brittle and plicatives deformations relating to the Hercynian and Atlas Metallogenically, a single mineralized vein was well distinguished on the three hills. (Figure 3b), (Figure 4) and (Figure 5).

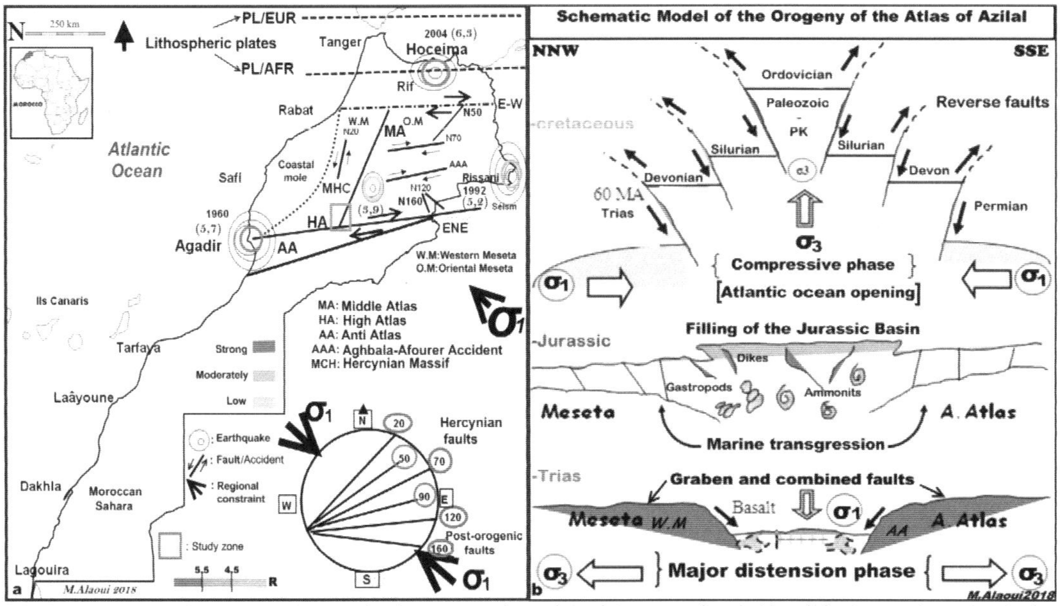

Figure 2. Schematic Structural and Sismic Map and Model of Orogenesis; a) Simplified tectonic and principal Sismic Mapping of Morocco; b) Synthetic Model of the Central Atlas orogeny.

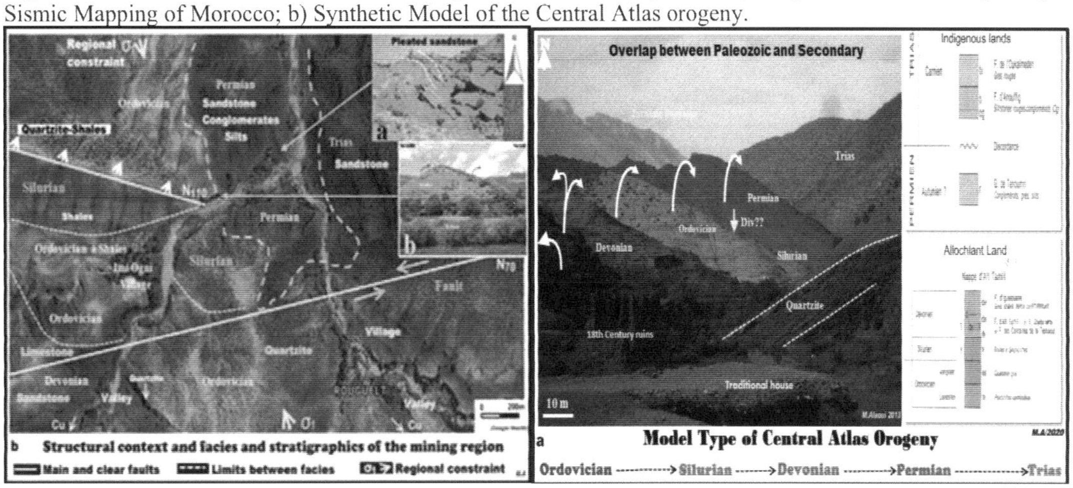

Figure 3. Panoramic view and satellite photo of Paleozoic overlap on the secondary field,
a) Folding of the Permian gres; **b)** Side view of the overlap between Ordovician and Permian

The Imi Nisk deposit is located in the province of Azilal, 90 km SW of the municipality of Abachkou on the road to Ait Abbas and Ait Bouguemaz. GPS Navigation
Geographic Coordinates of vein mineralized: [31.533104 N] & [-6.639867 E] or [N31°31'59.14''] & [W6°38'15.44''] and regional coordinates "Lambert": X=381000, Y= 105000 (More or less 500)
Altitude: 2298 meters, elevation site of river: 278 meters

Explored Substance: is the Copper (Cu).

Research keys: "Epigenetic" type Copper mine, the mineralization is hosted in greyish quartzite and brownish ankerite (Iron carbonates) between the Devonian and the Ordovician with ductile tectonic in faulty overlap in places. It is a vein which extends over more than one km, oriented NE-SW/N55 and dipping 65°-75° towards the NNW.

The central High Atlas, in Morocco, is a segment of a vast intracontinental Mesozoic mountain ranges intracontinental chain, individualized in the Triassic the Triassic in the northern part of the African Plate African plate (Laville et al., 1994).

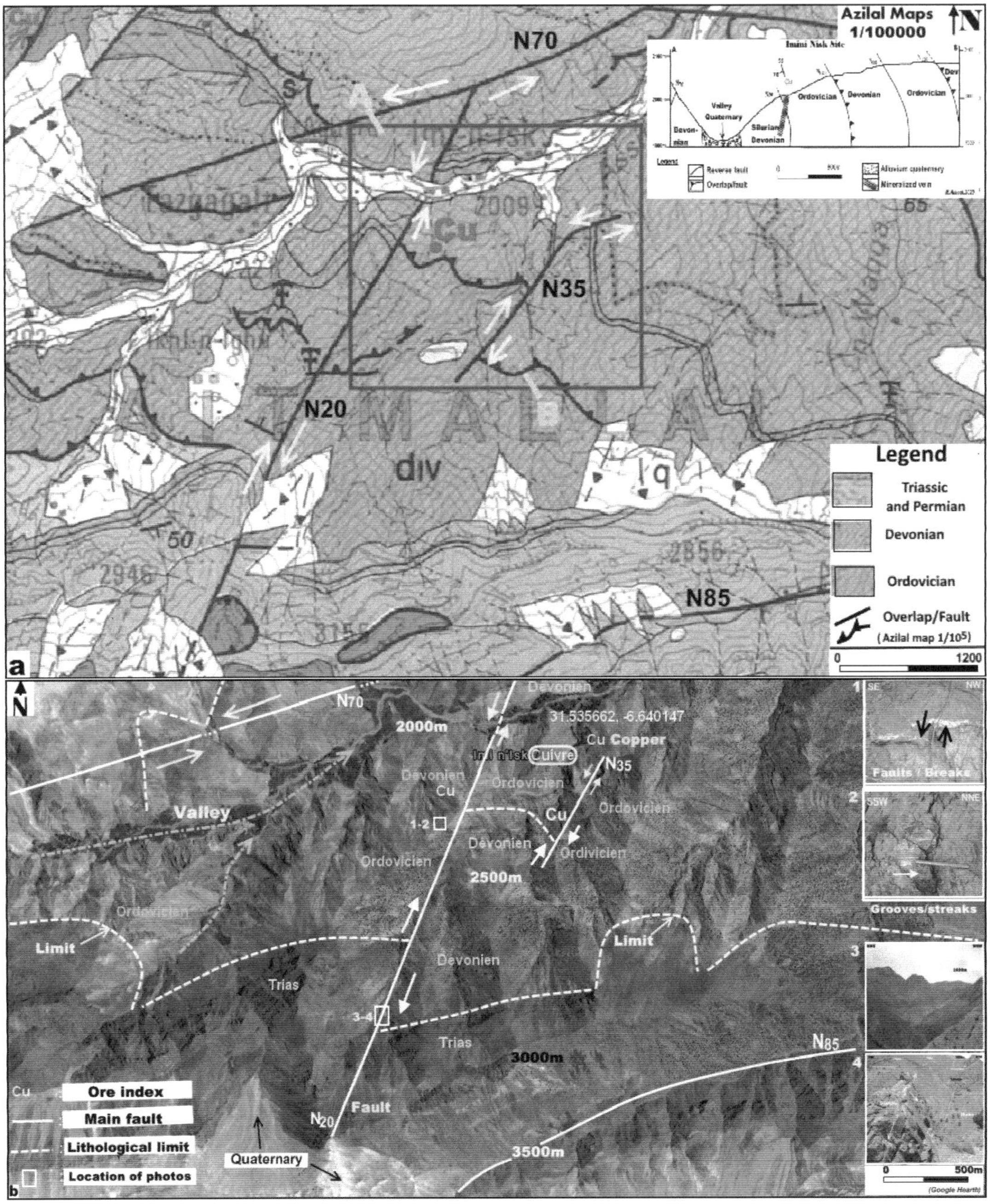

Figure 4. a) Extract from the geological map of the area and structural section [AB]. [NNW-SSE] synthetic of the copper hill; b) Aerial lithochronological and structural view; b1) Micro-fault in a greyish quartzite, b2) Streaks on the plane of a shear, b3) Triassic field extending towards the South, b4) Plan of the fault N20 and its extension towards the SSW.

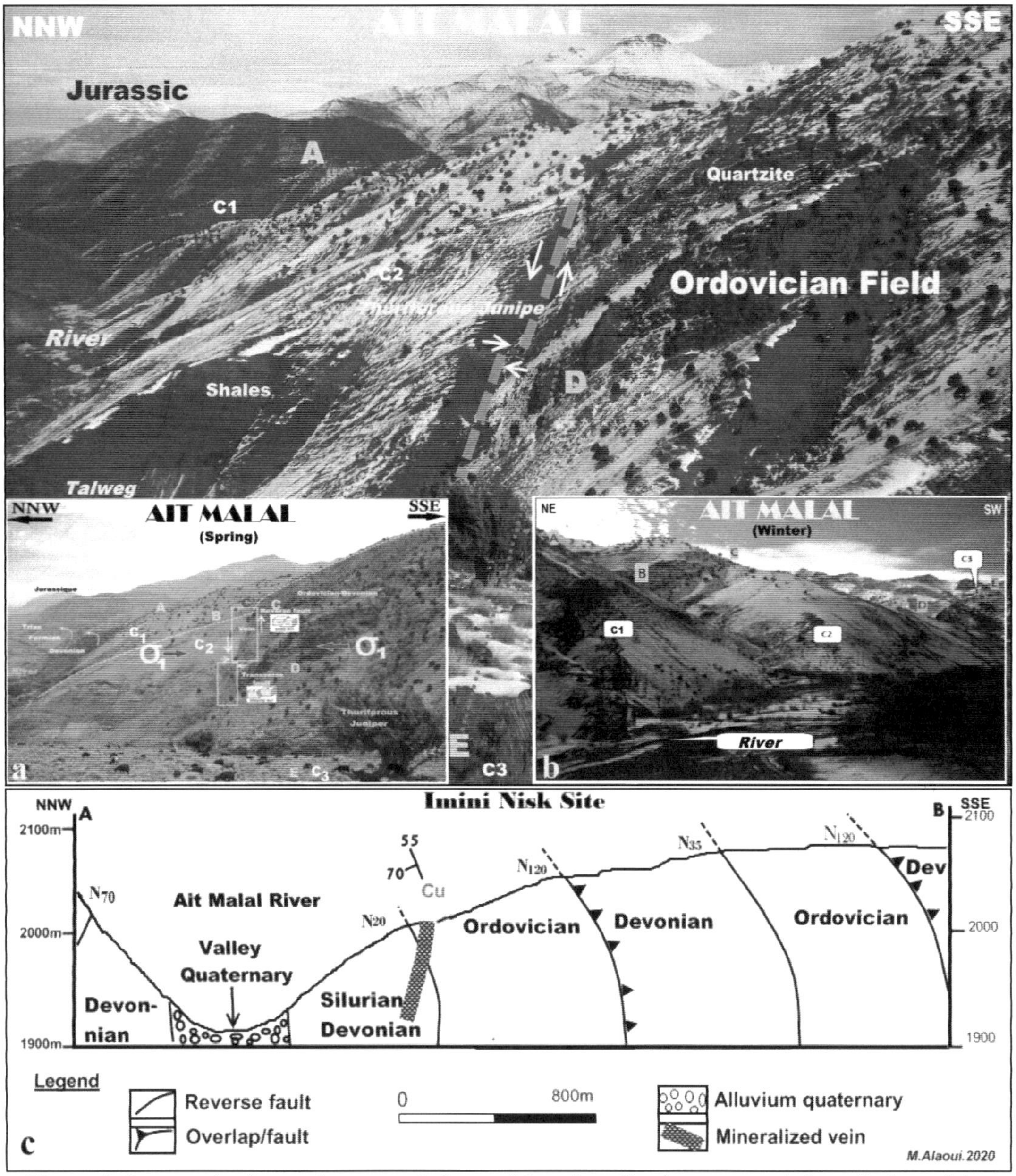

Figure 5. Panoramic view and geomorphology of the mineralized structure; **a)** Overall view of the mineralization valley during the spring towards the WSW; **b)** General view of the snow-capped hills during the winter carrying the mineralized vein towards the NNE; **c)** Simplified schematic section NNW-SSE of the copper mineralization site.

It is a vein mineralization linked to the pre-Atlas phase; this brittle phase is responsible for many vein mineralizations, low temperatures hosted either in the Paleozoic basement or in the covers of the Triassic-Jurassic. We mainly report the lead veins of Tafilalet and Adana, the [Pb-¬Cu-Zn] veins of the Jbilet and the central High Atlas and the barite veins of the western Jbilet and the High Atlas. We can speak of the stratiform mineralization at the base of the transgressions, for this type of mineralization formed in a sedimentary context; the base-cover contact played a very important paleogeographic role in the concentration of metals. In the Triassic, the main mineralizations are those of the lead-red Zayda type, occurrences of copper from Argana and concentrations of rock salt, gypsum and potash from Mohamadia and Safi. In the Jurassic there are stratiform deposits of manganese from Bouarfa and Tiharatine, the red-bed at [Pb-Cu] from Sidi-Rahmoune and the gypsum and anhydrite deposits from the Safi basin. In the Cretaceous, the most important mineralizations are the manganese concentrations of Imini, the lead, zinc and copper mineralizations of Merija and Tansrift, [3] (Beni Mellal Atlas), the phosphate mineralizations in the Khouribga, Gantour, Oulad Abdoun and Bou Kraa basins and the marls and oil shales in the Tarfaya, Boujdour, Timahdit and Tangier sectors [4].

The thickness of the vein varies from a few centimeters to several meters (power of 8 meters have been observed in two places), the average mineralized net thickness of the vein is around one meter, it is hosted in shales and attributed quartzite to the Ordovician. Note the alignment of the trees growing along the outcrop (Below the line in red dashes) ; these trees (Thuriferous Juniper) grow well on the laterites of the Gossan. The points A, B, C, D, E mark the visible extension of the vein between hill C1 and hill C3 via hill C2 (Figure. 6). A total of thirteen 13 samples were taken along the seam over a length of about 1Km, per trench.

The Imi Nisk site is characterized by a fault system between N20 and N80, which gave this sinuousness of the vein along the hills. On hill C1, its direction is between N110 and N55, at hill C2, its direction is almost East-West, and on hill C3 its direction varies between N20 and N50. These games are directly responsible for the second degree ore crushing zones. (Table.1).

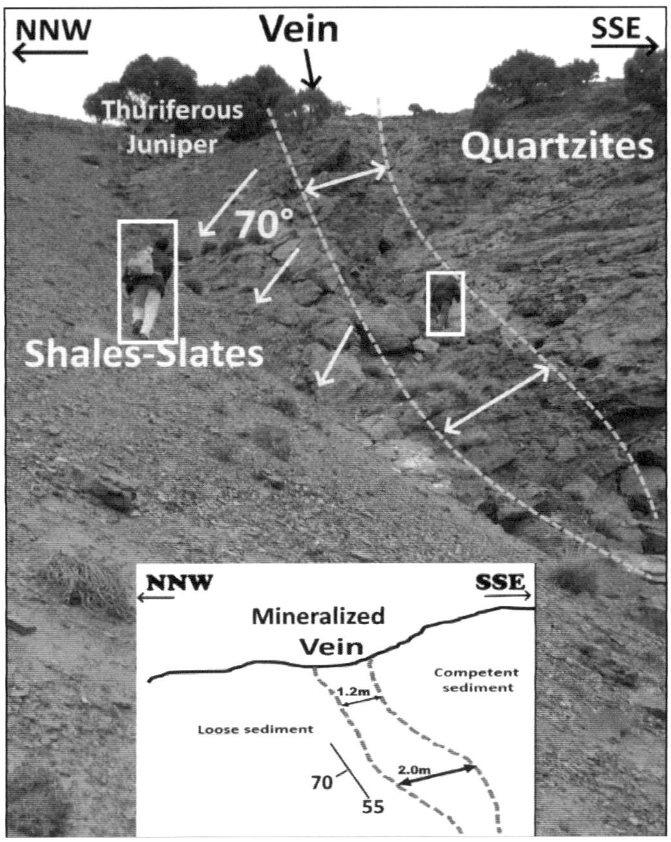

Figure 6. Close-up view of the mineralized vein with variation in actual thickness

Table 1. Location of real and apparent stations and thicknesses at the level of hill "C1-C2-C3"

N° Station	Coord - Lamb			Oxidized area	
	X	Y	Z	Apparent thickness	Actual thickness
Station1	381465	104827	2330	0,8	0,6
Station2	381440	104766	2334	2	1
Station3	381411	104747	2353	1	1,2
Station4	381445	104799	2345	1	2
Station5	381504	104895	2299	1	2
Station6	381532	104967	2285	2	3
Station7	**100m N10 from the last station**			0,6	1

4. Results and discussion

The Central High Atlas is bounded to the south by the southern Atlas accident and to the north by the overlapping fault. This area is characterized by a structure mainly formed of faults oriented North-East /South-West to East-West inherited from the Orogeny of Hercynian age or even Precambrian age (Laville E *et al.* (1978). The Central High Atlas shows a symmetrical morphology and double spill. In this vast structural domain there are various Paleozoic and Jurassic litho stratigraphic formations.

At the level of hill C3 extending towards the SSW the vein has changed direction, a few degrees to the South, this is explained by the play of the fault which gave the Talwegs (Figure.7) (Figure.8b) and (Figure. 9f). The difference between the apparent and real thicknesses varies from 0.2m to 1m, we can argue by the following hypothesis, it is that the vein has become subvertical, so the core samples must be between 70° and 80° for more information on the vertical extension of the mineralization.

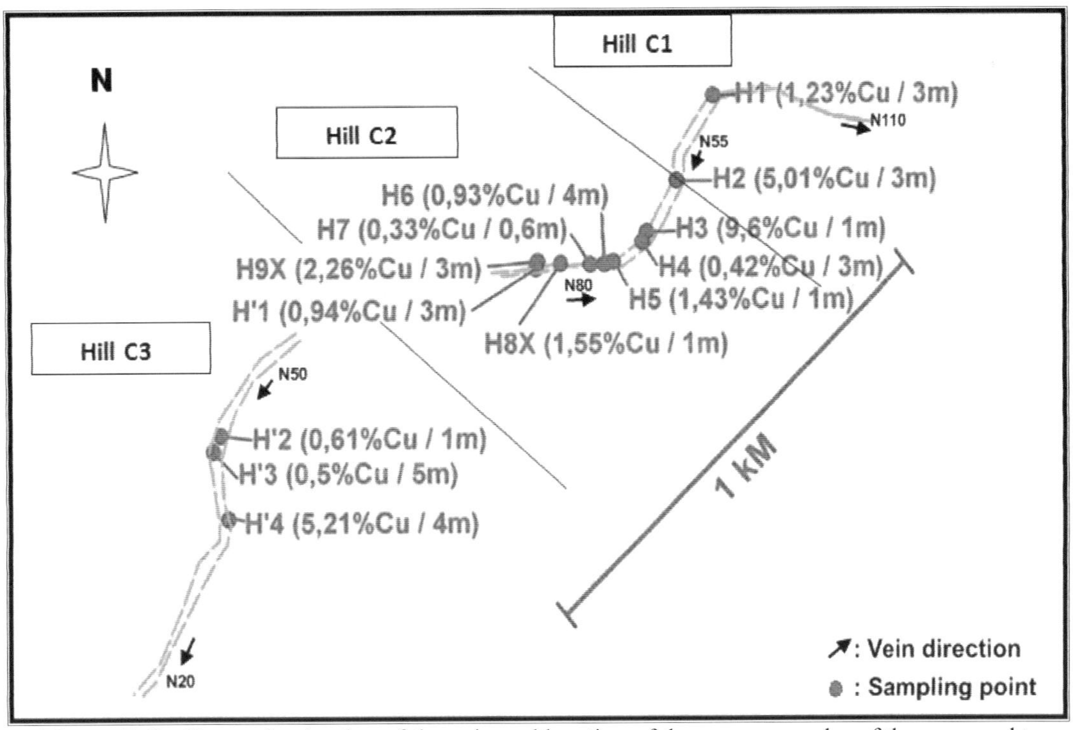

Figure 7. Profile-section in plan of the vein and location of the average grades of the ore sought

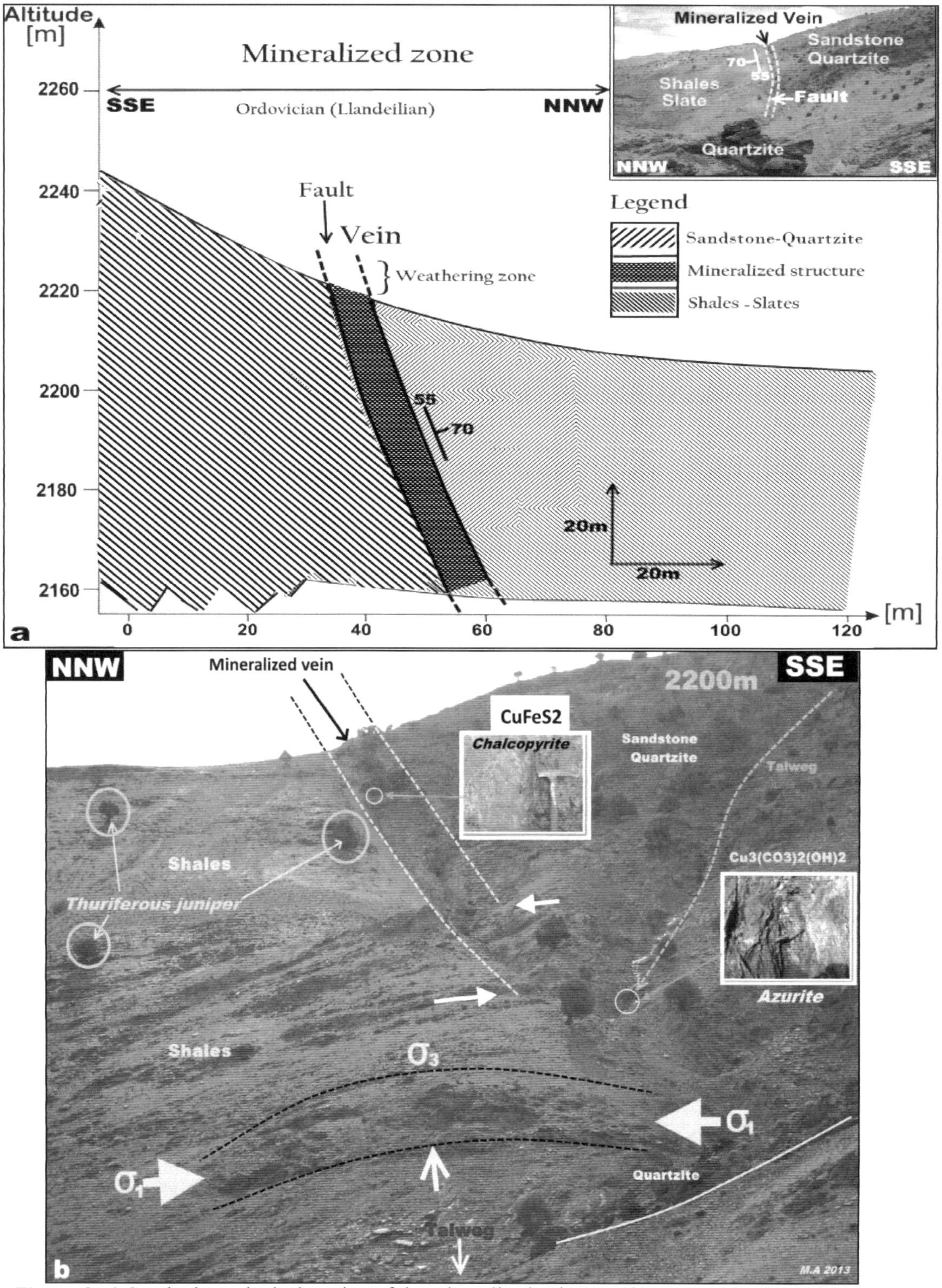

Figure 8. a) Synthetic geological section of the mineralized vein pass; b) Panoramic view of the left flank of the hill with a folding zone reflecting the regional constraint of the orogenesis (σ1–σ3)

IOP Conf. Series: Earth and Environmental Science 1090 (2022) 012008 doi:10.1088/1755-1315/1090/1/012008

Some rare geological phenomena in nature were encountered on the study site, a rosacea of schistosity linked to local constraints (Figure. 9a), and a phenomenon of geological competence, or rheology of the facies, between quartzite's and shales compacted, we speak of the enclaves, which gave gray areas, we often see them between the garnet and its gangue under the microscope (Figure. 9c).

Traces of Gold (Au) were marked on all the hills explored in four spaced samples. The presence of CaO lime is due to iron carbonates or siderites (Crude form $FeCO_3$ with traces of Mn ; Mg ; Ca ; Zn ; Co), also called ankerite (Ca (Fe, Mg, Mn) $(CO_3)_2$), at the level of hill C1 and hill C2. Silica SiO_2 is remarkable on most of the vein samples, this is explained by the shale cover. Iron oxides Fe_2O_3 characterize the C2 hill where the iron carbonates contain the mineralization on apparent thicknesses exceeding 10 meters in some places. (Figure. 9d). Exploitation work on the Imi Nisk site was carried out in a traditional and archaic manner, galleries and trenches were found in the heart of the vein at 2100m above sea level (Figure. 9e), but given the conditions which were difficult in the past, to transport the ore, the craftsmen abandoned the deposit. At 2013, the exploration and exploitation license was sold for a Moroccan mining company. Exploitation works have been started in order to

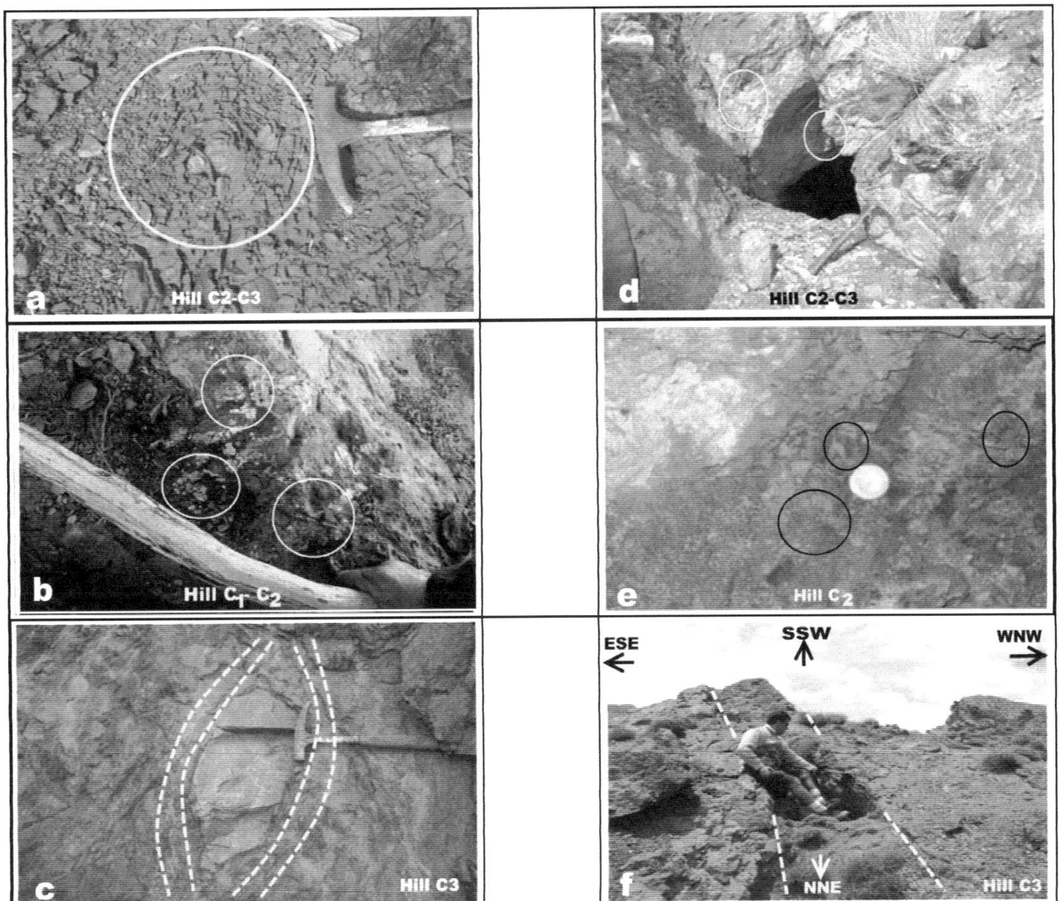

Figure 9. Some geological, tectonic, and mineralogical aspects of the site and their position on the three hill; a) Rose of schistosity in a circle with a hard radius; b) Chalcopyrite, Covellite, azurite and limonite in siderite, c) Phenomenon of competence creating a shadow zone of a quartz enclave; d) Galleries of the old artisanal exploitations; e) Indices of mineralization as azurite, malachite, and limonite between a siderite (Ankerite) and shales; f) South extension of the hill trench vein (C3).

produce the materials while coming in copper ore (Product is in bulk on the vein hill) for flotation stations installed in the region. Currently, this mine is temporarily shut down, but it will certainly restart after the circumstances of the Covid19. Although significant, the analyzes obtained correspond to the outcropping areas exposed directly to surface alterations : oxidation, leaching, etc.

The vein extends over more than one km, it is oriented NE-SW (N55) and inclined between 65° and 75° to NNW. The structure is hosted in shales and quartzites attributed to the Ordovician.

Obtaining such copper contents and traces of gold in these levels is a rarity given the situation of this vein in a region where climatic agents are more aggressive. In such climatic conditions and when the host is "acidic": shales, granites ... the sulphides dissociate and transform into secondary minerals: oxides (iron caps) and /or carbonates (ankerite or siderite) which arch and fill the top parts, while the leached cations such as Cu, Zn, Au, Ag, migrate to a relatively deeper zone marking the transition from an oxidizing medium to a non-oxidizing medium called the cementation zone where they concentrate to form economic mineral masses (we speak of a mineral deposit). In this hypothetical model, value minerals such as ; copper, gold (from altered sulphides), dissolved metal from the oxidized zones reconcentrate in the carburizing zone. According to the chemistry of the medium, the oxidation zone is characterized by a set of free cations such as Cu^{2+}, Zn^{2+}, and anions such as $SO4^{2-}$, HCO^{-3}, with a pH which varies between 4 and 8. (V. Saegart, et al. La Caridad deposit, Mexico). The copper which comes from the dissolution of sulphides has a great analogy with gold in the deposit area, except that it deposits a little lower than this, in the basal part of the cementation zone. The contents of copper vary over the entire extension of the vein depending on the host, at hill level (C2) the traces of copper appear in chalcopyrite and Covellite located in iron carbonates or siderite, also called ankerite, impregnated in shales.At the level of the hill (C3) the host is a greyish to brownish quartzite where the chalcopyrite is disseminated in centimeter pallets and the azurite at the heart of the Talwegs in the presence of iron oxides (Limonite) (figure. 9b). At the level of the old galleries of archaic exploitation, the indications of copper are manifested in a light greenish or malachite tint. Traces of gold (Au) exist along the mineralized structure, in four samples, its content can vary depending on the depth of the structure of a pyrite heap in the sulphide zones. The content of silica SiO_2 exceeds, in places, 87% this is explained by the abundance of shales in contact with the mineralized veins. The CaO lime content exceeds 31%, it is, certainly, of origin the carbonates of iron from the northeast part ofthe, vein. At the level of the hill (C3) the host is a greyish to brownish quartzite where the chalcopyrite is disseminated in centimeter pallets and the azurite at the heart of the Talwegs in the presence of iron oxides (Limonite). At the level of the old galleries of archaic exploitation, the indications of copper are manifested in a light greenish or malachite tint (Figure 9d). According to the local geological map and the projection on the satellite outlet of the site, we can speak of a mineralization linked to the ductile shear between the two dextral faults which affected the Ordovician in overlap on the Devonian, by creating zones of grinding.

The mineralized hills, in particular, going southwest, show a system or dextral and senestial decays that enter into the development of crushing and folding zones generated by the regional stress, as shown in the 3D aerial photo. (Figure. 10).

Table.2 the average content of the 13 samples along the vein is 2.23%, according to the copper Clark, which varies between 1,0 and 6,0%, we can qualify the deposit of Imi Nisk of a future copper mine ultimate. Thus, and in general, the deposits [Cu-Au] of the atlas chain in Morocco are mainly controlled by faults or shear zones of regional scale and / or associated secondary faults / shear zones and are set up later in the grinding zones of corridors with regional metamorphism.

Table 2. Content of chemical elements along the vein (Σ Cu = 2.23%)

Slampes	Trench length (m)	SiO2 (%)	Fe2O3 (%)	CaO (%)	MnO (%)	Cu (%)	Au (G/T)
H 1	3	3,16	32,42	31,41	2,79	1,24	<0.05
H 2	3	23,74	32,6	15,86	3	5,01	<0.05

H 3	1		Ankerite			9,60	
H 4	3					0,42	
H 5	0,7	7,85	63,82	1,04	4,76	1,43	
H 6	4		Iron Hat			0,93	
H 7	0,6					0,33	
H 8 X	1					1,55	
H 9 X	3	7,67	26,37	27,32	2,93	2,26	<0.05
H' 1	5	81,97	9,65	0,95	0,79	0,94	
H' 2	0,7	Quartzite				0,61	
H' 3	1,5	87,15	6,52	0,82	0,3	0,49	<0.05
H' 4	0,7	Shales				5,21	

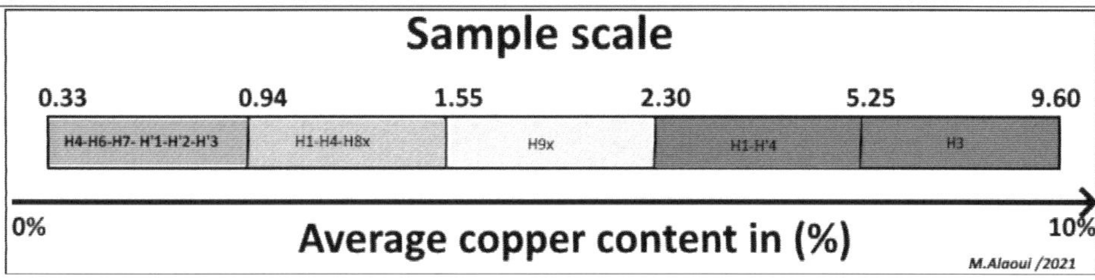

5. Conclusion

The reason which pushed us to study this field in mining research on the one hand, it's fascinating geological structures which attracted us and which can contain many interesting associated mineralizations, on the other hand to give a socio-economic development and also to explore and exploit its mineral potential by opening a copper-mine.

Sampling difficulties :

- Stripping of the snow cover to find the vein and sample;
- Dispersion and burial of sampling products in the snow;
- Difficulty in collecting the sampling products to put them in bags.

Analysis shows that the only economic metal that is expressed in significant quantities in oxidized areas is copper. It could be assumed that the primary mineralization (before alteration) would consist of iron sulphides (pyrite and / or pyrrhotite) and copper (Chalcopyrite, bornite, Covellite ...) hosted in a gangue quartz.

The structure in question would therefore correspond to a copper vein par excellence. The objective of the research to be undertaken would be, in particular, to try to locate and determine the mineralogical and mining importance of the cementation zone. The extent of copper mineralized outcrops over more than one kilometer demonstrates the importance of the mining potential of this sector.

In terms of infrastructure ; we note the existence and proximity of :

- Electricity:(medium voltage :22KV) ;
- The abundance of water : (river with permanent flow) ;
- The telephone network ;

The critical effects of mining on the environment, the impact of mining methods and the ways in which the industry tries to make itself more environmentally friendly.

- Air impact

The dust from the mined ore and the gases released by the mining equipment are bad for the health of the miners as well as for the environment. Over time, exposure to dust generated by mining operations can lead to disease and the build-up of scar tissue in the lungs.

- Water impact

Material from after mining can easily drain into local water supply systems, leading to increased acidity and heavy metal contamination which can destroy flora and fauna and render the water unsuitable.

Some forms of mining also require the drainage of water reservoirs from aquifers, which can have serious consequences, such as the drying up of springs, the cutting of rivers and the degradation of local ecosystems.

-Access by track during redevelopment (bridges and rafts under construction);

-The favourable and cooperative social.

- Soil erosion

Surface mining, one of the most common techniques, damages land to exploit raw materials. It dredges the earth and strips the vegetation, leaving the area vulnerable to soil erosion and topsoil wear and tear. Topsoil is necessary for plant growth, and without it, mining sites cannot fully recover.

Worse, soil erosion can often spread, meaning that mining can have effects on soil beyond the site.

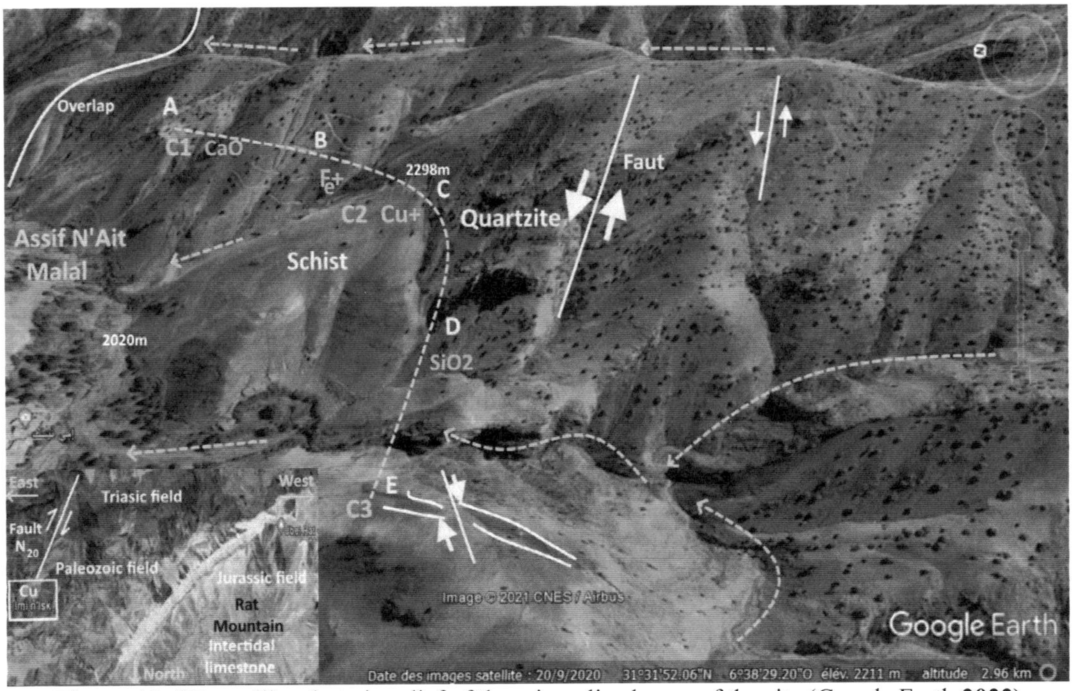

Figure 10: 3D satellite photo in relief of the mineralized zone of the site (Google Earth **2022)**

6. References

[1]. E. Laville 1978. Indice des jeux successifs d'un accident synsédimentaires sur les structures plicatives du versant nord du Haut Atlas central, (Maroc). Bull. Soc. Géol. Fr., (7), 20, 329-337.

[2]. A Gaouzi, A Chauvet, L Barbanson, L Badra, J C, Touray, S Oukaroud and M El Wartitib. Septembre 2001. Mise en place syntectonique des minéralisations cuprifères du gîte d'Ifri (district du Haut Seksaoua, Haut Atlas occidental, Maroc). Jean-Claude Touray ISTO, UMR 6113, université d'Orléans, bât. Géosciences, BP 6759, 45067 Orléans, France. Volume 333, Issue 5, 15 September 2001, Pages 277-284.

[3]. H, Ibouh, A. Michard, M. Hibti, Kh. Elamari, 2011. Le cuivre des Couches Rouges de Tansrift (Atlas d'Azilal) / The Copper Deposit of the Tansrift Red Beds (Azilal Atlas)

[4]. A A., Barodi E.B., Maacha L. & Zinbi Y. (1998): Les minéralisations cuprifères du Maroc. Chron. Rech. Min531-532, 29-44.

[5]. J. Caïa 1969. Les minéralisations plombo-cupro-zincifères stratiformes de la région des plis marginaux du Haut Atlas oriental: un exemple de relations entre des minéralisations et une sédimentation détritique continentale. Notes Serv. Géol. Maroc
[6]. A.S. Ayangbenro, O.S. Olanrewaju, & O.O. Babalola 2018. Sulfate-reducing bacteria as an effective tool for sustainable acid mine bioremediation. Frontiers in microbiology, 1986.https://doi.org/10.3389/fmicb.2018.01986
[7]. E Pelleter A Cheilletza D Gasquet A Mouttaqi M Annichd A El Hakourd G Féraude 2008. The Variscan Tamlalt–Menhouhou gold deposit, Eastern High-Atlas. Bolata Meriem (1995): Etude géologique de la région de Zelmou (Haut atlas oriental, Maroc)
[8]. P. Rajlich, J. Legierski & V. Šmejkal 1983. Stable isotope study of base metal deposits from the Eastern High Atlas, Morocco. Mineralium Deposita, International Journal for Geology, Mineralogy and Geochemistry of Mineral Deposits volume 18, pages161–171 (1983)
[9]. R. Brede 1992. Structural aspects of the Middle and High Atlas (Morocco), phenomena and causalities. Géol. Rdsch, 81, 171-184.
[10]. A. Michard, H. Ibouh & A. Charrière 2011. Syncline-topped anticlinalridges from the High Atlas: a Moroccan conundrum, and inspiring structures from the Syrian Arc. Terra Nova, Volume23, Issue 5, 314–323. doi: 10.1111/j.1365-3121.2011.01016.
[11]. R. Moussu 1952. Géologie des Gîtes Minéraux Marocains (zones française du Maroc); Chap. 7, Le Cuivre. Notes et Mém. Serv. Géol. Maroc, n°87, 157-177
[12] A. Subra 1980. Emersions anté-crétacées et minéralisations dans l'Atlas de Béni-Mellal. Le gisement de Tabaroucht, un remplissage paléo karstique cuprifère. Mines Géol. Energie, n°47, 43-70.
[13]. A. Subra 1972. Inventaire gîtologique de la région de Ouaouizerht (Atlas de Béni-Mellal, Maroc). Rapp. SEGM 940: 22.
[14]. A. Subra 1970. Contribution à l'étude métallogénique du cuivre: étude d'un gisement filonien (Padern) et d'un gisement stratiforme (Tansrift). Thèse 3ème cycle, Univ. Paul Sabatier, Toulouse, inédit, 170 p.
[15]. A. Azza. (2019): Géologie et métallogénie des minéralisations à or et Antimoine du Jbel Haouanite (Haut Atlas oriental, Maroc). Thèse de doctorat - Université Sidi Mohamed Ben Abdellah. USMBA
[16]. H. EL Hnot, M Zahraoui. 2005. Origine et mode de mise en place de la barytine du gisement de Bou Ouzzal (Meseta marocaine). Bulletin de l'Institut Scientifique, Rabat, section Sciences de la Terre, 2005, n°27, 47-54.
[17]. J. Guezal, M. El Baghdadi et A. Barakat. 2014. The Jurassic–Cretaceous volcanism of the Atlas of Beni-Mellal (Central High Atlas, Morocco): evidence from clinopyroxene composition. Arabian Journal of Geosciences volume 8, pages977–986 (2015)
[18]. J. Tsui 2020. The Environmental Impact of Mining (Different Mining Methods Compared). Environment March 4, 2020/https://get-green-now.com/

The potential role of sediments in nutrients (N, P) cycle in Marchica lagoon (Mediterranean Sea, Morocco)

Abderrahmane RAHHOU [1,*] , Mostafa LAYACHI [2] , Mustapha AKODAD [1] , Najib EL OUAMARI [2] , Asmae AKNAF [1] , Hicham GUEDDARI [1] ,Brahim OUDRA [3] and Mourad BAGHOUR [1]

[1] Laboratory of Biology, Geosciences, Physics and Environment (OLMAN-LBGPE) Multidisciplinary Faculty, Nador, Morocco
[2] Regional Center of INRH-Nador, Morocco
[3] Water, Biodiversity and Climate Change Laboratory, Faculty of Sciences Semlalia, Cadi Ayyad University, Marrakech, Morocco

[*] corresponding author: Abderahmane.rahhou@ump.ac.ma

Abstract. The aim of this study is to present the distribution of nutrients in water and sediment of Marchica lagoon, a post-restorated lagoon located in Mediterranean Sea (NE Morocco), to estimate the impact of bottom sediments on the water quality of this lagoon, and to identify the areas more affected by eutrophication risk. Nutrient concentrations in the water column and sediments were determined in thirteen samples taken from different station of the lagoon in October 2019. Nutrient concentrations in sediment were found higher than those in surface water; very high levels of nitrogenous elements showed in all sediments samples and mainly nitrogenous components (averages of 43.58 mg of N-NH4; 0.827 mg of N-NO3; 0.054 mg of N-NO2 and 773 mg of TN per kg of dry sediment) and high levels of phosphorus with average of 620 mg P/kg of dry sediment. A strong positive correlation showed between sediment and water parameters in 0.01 and 0.05 levels. High concentrations were recorded near the nutrient-rich wastewater treatment plant effluent discharges, watershed, and groundwater input. The lowest concentrations are near the new pass, this is due to the low continental inputs and the renewal of the lagoon waters by the Mediterranean Sea. Generally, the distribution of nutrient levels is influenced by continental inputs, hydrodynamics, and organic matter content in the sediment. These sediments are considered as a reservoir of nutrient pollutants, which will influence water quality by releasing sedimentary materials in the water.

Keywords: Marchica Lagoon, Pollution, Sediment, Nitrogen, Phosphorus, Eutrophication.

1. Introduction

The Coastal lagoons, the interface between the continent and the sea, are the most productive ecosystems in the biosphere [1]. In addition, they show great spatial and temporal variability in the physicochemical characteristics of their waters due to the mixing of freshwater with marine waters [2].

Most of the nitrogen load is provided by rivers and wastewater treatment plants, but precipitation also has an important contribution [3]. The large amounts of nitrogen can lead to a progressive eutrophication of lagoons [4]. The main source of nutrients comes from human activities such as agriculture, urbanization, industry, and aquaculture [5-7].

In lagoons, sediments transform deposited materials via degradation and dissolution processes and return transformation products, e.g., nutrients, to the water column, which can control the functioning of the upper water column [8]. Sediments can act as sinks with potential implications for nutrient limitation in the water column [9].

For several years, the daily discharge of domestic, agricultural, and industrial waste into the Nador lagoon has stimulated algal blooms. This increases the consumption of oxygen, causing an imbalance in this ecosystem, which leads to eutrophication [10]. Numerous studies on eutrophication, bathymetry, phytoplankton, fauna, and fisheries in the Nador lagoon have been carried out before the restoration of the lagoon [11, 12], and other studies on water quality during the post-restoration period, the variation of physicochemical parameters and nutrients in the water was studied [13-15]. All these works have been conducted on water, and no studies have been conducted on the contribution of sediments in the supply of nutrients.

Since 2010, two wastewater treatment plants have been set up in order to preserve the biological and ecological qualities and the development of tourist activities around this ecosystem (Nador wastewater treatment plant in 2010 and Arekmane wastewater treatment plant in 2013) [15], the opening of a new pass between the lagoon and the Mediterranean Sea in 2011 to exchange water, which has improved water quality in the lagoon [16]. These measures represent the restoration plan for the Nador lagoon.

The objective of this study is to determine the spatial variation of physicochemical parameters of water and sediments and to study the relationships between these variables and their possible distribution, as well as to understand the sedimentary role in the recycling of nitrogen and phosphate nutrients and their potential role in the eutrophication process.

2. Materials and methods

2.1. Study area

Marchica lagoon is one of the largest coastal lagoons in the Mediterranean Sea, with a length of 25 km, a width of 7.5 km and an estimated area of 115 km^2. It is located on the north-eastern coast of Morocco between the "Cap des trios fourches" and the "Cap de l'eau", more precisely between latitudes 35°05'N -35°14'N and longitudes 2°44'W - 2°56'W (Figure 1). It is separated from the sea by a coastal dune (25 km). In order to allow a better exchange of the lagoon's waters with the Mediterranean Sea waters, a new inlet was opened in 2010, about 300 m wide and 6.5 m deep.

2.2. Sampling strategy and analysis methods

A total of 13 sampling sites were selected for the study of water and sediment quality in the Nador lagoon (Figure 1). The samples were taken in October 2019 at surface layer of water and sediments (20 cm). pH, dissolved oxygen, temperature, turbidity, salinity, and conductivity were measured *in situ* using portable devices with special probes for each parameter. Water samples collected were subject to other analyses: ammonium NH4 (NF T90-015), total phosphorus (TP), nitrite NO2 and nitrate NO3 [17].

The dried sediment samples were analyzed for NO3 and NO2 after extraction with a 2M KCl solution, total nitrogen (TN) and total phosphorus (TP) by oxidation with Potassium persulfate [18,

19] and then analyzed using methods described for water. Organic matter in sediments (OM) determination is carried out in an oven at 550 °C [17].

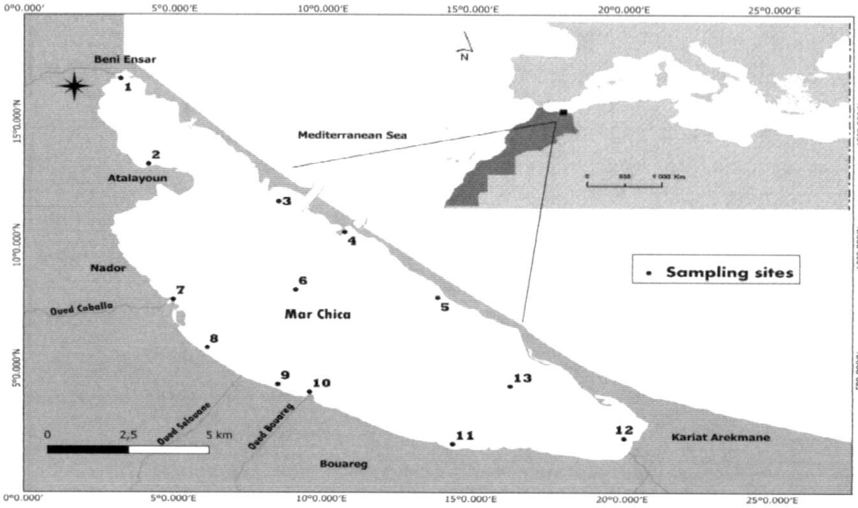

Figure 1: Location of the study area and sampling stations

2.3. Statistical analysis
Statistical analysis was used to detect significant correlations between the different parameters (Spearman's coefficient). To study the relationships between the variables and their possible distribution pattern, Principal Component Analysis (PCA) was employed, as described in other studies [20-24]. Several studies have combined Geographic Information System (GIS) techniques and PCA to understand the anthropogenic impact in the different areas [22, 25, 26]. The data were processed using SPSS IBM 26 software.

3. Results and discussions

3.1. Temperature, salinity, electrical conductivity, pH, O2 and turbidity in wate

Regarding dissolved oxygen, the results did not show the presence of anoxia (Figure 2). The minimum value (5.58 mg O_2/l) was recorded in the area near the Nador wastewater treatment plant and the maximum value (12.5) was recorded in the north-eastern area. The minimum and maximum values of pH (7.82-8.21), salinity (37.53-39.03 ‰) and temperature (20.10-21.6 °C) are homogeneous throughout the lagoon and this is due to the strong hydrodynamics and circulation that homogenizes the lagoon waters with sea waters after the opening of the new pass [27].

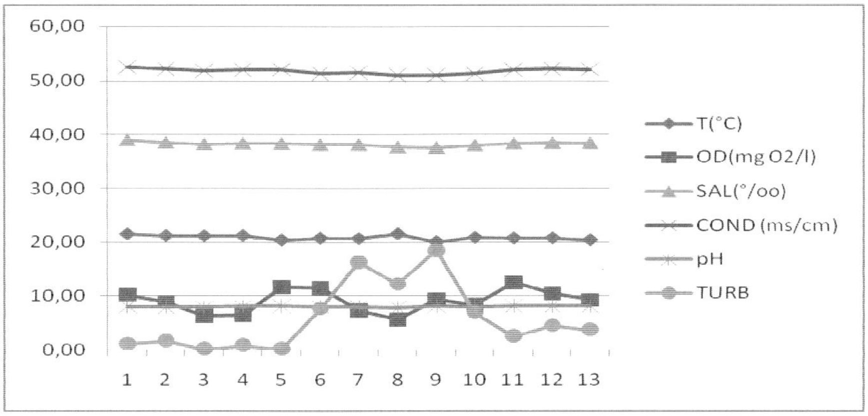

Figure 2. Physicochemical parameters of the surface waters at the sampling stations

3.2. Ammonium (NNH4-W), Nitrite (NNO2-W), Nitrate (NNO3-W), and Total Phosphorus (PT-W) in water.

The results show concentrations of nutrients responsible for pollution and eutrophication and high values of dissolved oxygen, in comparison with previous study [13]; water quality along the lagoon is generally improved, except for some areas that receive discharges of treated wastewater and the mouths of Wadis [13, 15]. The values for nitrates and nitrites in the lagoon are between (0 and 0.03 mg/l); ammonium in the lagoon presents maximum values of 0.25, 0.104 and 0.155 mg/l respectively in Nador wastewater treatment plant area, Cabaillo and Kariate Arekmane Wadis). A clear improvement in phosphorus concentration was recorded in the lagoon, with the exception of the area receiving wastewater from Nador wastewater treatment plant area. Indeed, after the opening of the new inlet in 2011, water exchanges are indeed better between the lagoon and the Mediterranean Sea and the developments around the lagoon have improved the oxygenation of surface waters and reduced the eutrophication of the lagoon according to [13-15].

3.3. Ammonium (NNH4-SED), Nitrite (NNO2-SED), Nitrate (NNO3-SED), Total Nitrogen (NT-SED), Total Phosphorus (PT-SED) and Organic Matter (MO-SED) in sediments

In addition to water analyses, sediments of the Nador lagoon are very rich in nitrogen and phosphorus, the maximum values of these two nutrients were recorded in the areas known by a strong eutrophication and an important anthropogenic pressure, in particular the discharges of wastewater from the cities of Beninsar, Arkmane and Nador wastewater treatment plant. The low concentrations recorded in the area near the new pass (table 1).

Table 1. Results of nitrogenous components and phosphorus in water (mg/l) and sediment (mg/kg dry weight)

	NNO2_W	NNO3_W	NNH4_W	PT_W	NNO2_SED	NNO3_SED	NNH4_SED	PT-SED	NT_SED	MO_SED
S1	0	0	0,01	0,04	0,02	0,04	11,5	670	300	2,65%
S2	0,001	0,002	0,002	0	0,01	0,12	103	980	1900	12,40%
S3	0	0	0,004	0	0,01	0,08	9,5	370	400	2,00%
S4	0	0	0,005	0	0,008	0,48	15,5	240	300	4,20%
S5	0,001	0,004	0,054	0	0,008	0,6	13,5	300	200	3,00%
S6	0,002	0,003	0,079	0	0,07	1	42,6	930	1100	16,20%
S7	0,001	0,001	0,104	0	0,09	1,36	23,7	840	250	2,60%
S8	0,022	0,028	0,25	0,112	0,09	1,4	110	970	2450	9,50%
S9	0	0	0,016	0	0,07	0,84	19,2	330	200	2,50%

S10	0,003	0,003	0,028	0,05	0,08	1,48	17,6	300	200	2,00%
S11	0,008	0,012	0,014	0	0,08	1,08	33	650	900	12,30%
S12	0,009	0,014	0,155	0,09	0,09	1	98,5	720	1300	10,90%
S13	0,001	0,002	0,006	0,03	0,08	1,28	68	880	550	14,70%

Spearman correlation analysis was carried out to study the relationship between physico-chemical parameters of surface water with sediment parameters (Table 2); the correlation is highly significant at the 0.05 and 0.01 level between water parameters. A strong positive correlation showed between organic matter, nitrogen and phosphorus in sediments. For phosphorus in sediment and water, no significant correlation found, while for nitrogenous there is a high significant correlation at the 0.05 and 0.01 levels.

It has been shown that the sediment surface in the Venice lagoon plays an important role in regeneration processes [28], and that the calculated overall amounts of nutrients released from the sediments are of the same order of magnitude as the nutrients entering the lagoon from external sources [29]. In this context, a study of the mechanisms of nutrient release from sediments is important for understanding eutrophication processes.

Sediments can act as sinks with potential implications for nutrient limitation in the water column [9]. Establishing the magnitude of benthic fluxes is important for understanding geochemical budgets and for determining the role of sediment in nutrient recycling. Nutrients at the sediment-water interface are known to contribute to biogeochemical cycles in most shallow water systems [30, 31].

Table 2. Spearman correlation between sediment and water parameters

	NNO2_W	**NNO3_W**	**NNH4_W**	**PT_W**	**NNO2_SED**	**NNO3_SED**	**NNH4_SED**	**PT_SED**	**NT_SED**	**MO_SED**
NNO2_W	1									
NNO3_W	,951**	1								
NNH4_W	,661*	,643*	1							
TP_W	0,52	0,45	0,44	1						
NNO2_SED	,693**	0,52	,675*	,579*	1					
NNO3_SED	,716**	,583*	,627*	0,41	,786**	1				
NNH4_SED	,698**	,623*	0,35	0,34	,589*	0,5	1			
TP_SED	0,41	0,32	0,17	0,22	0,44	0,17	,770**	1		
NT_SED	0,52	0,48	0,07	0,28	0,3	0,04	,747**	,776**	1	
MO_SED	0,41	0,44	0,02	0,01	0,11	0,08	,693**	,624*	,706**	1

**. The correlation is significant at the 0.01 level. *. The correlation is significant at the 0.05 level

The PCA was applied to the NH4, NO2, NO3 and TP contents of the waters of the Nador lagoon and to the variables TN, TP, MO, NH4, NO2 and NO3 in the sediments. The inertia carried by the two factors is 76.4%, i.e. 40.4% for factor PC1 and 30% for factor PC2. Nutrients in water and oxidized forms of sediment nitrogen contribute positively to the PC1 axis. The PC2 axis, on the other hand, is defined mainly by the sediment-related elements that contribute positively to it (Figure 3).

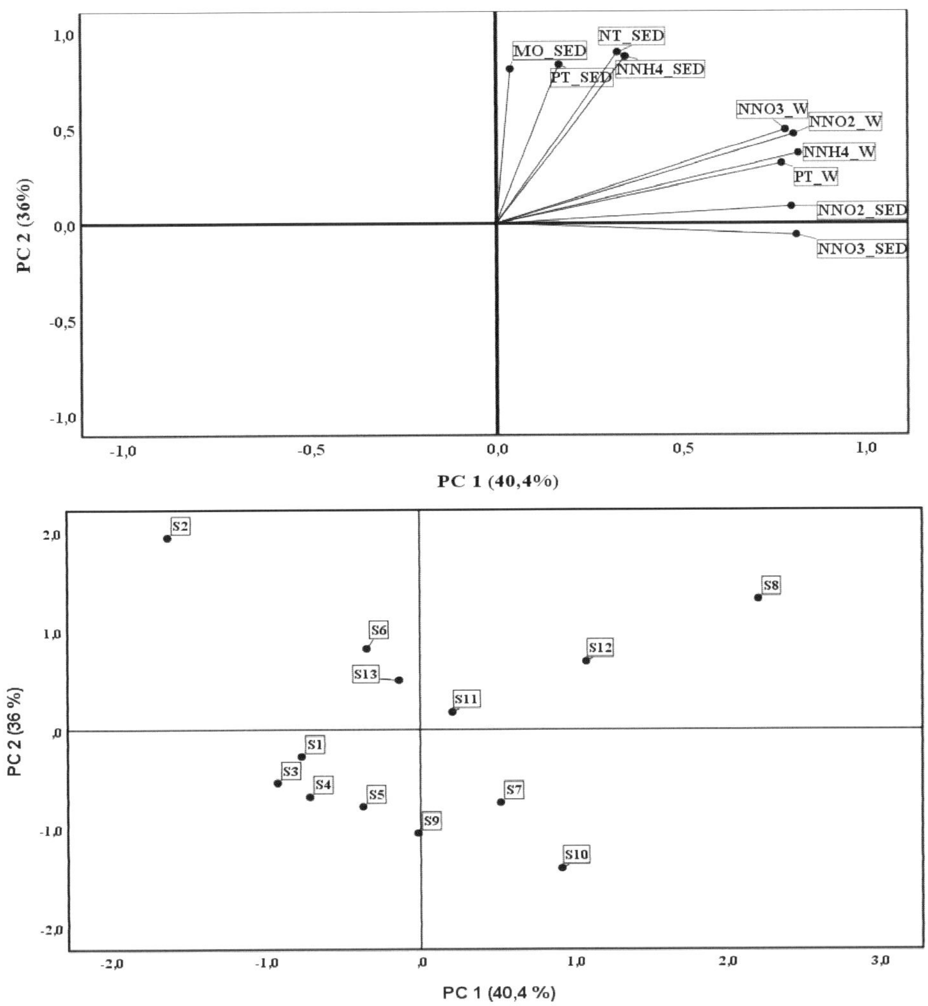

Figure 3: PCA of variables and observations

The spatial distribution of the PC1 and PC2 factors using GIS allows the determination of the area of impact by nutrients in water and sediments (Figure 4).

The results show four areas influenced by sediments (North-East zone, North-West zone, the lagoon center and the zone close to the Nador wastewater treatment plant), these zones have been impacted by important inputs of wastewater highly loaded in organic matter and nutrients, with low hydrodynamism and exchange with the Mediterranean Sea which decreases the renewal rate of the lagoon waters [27]. The area of PC1 values represents, on the other hand, the lagoon surface most influenced by nutrients in the surface water column and is located mainly on the extended fringe of the discharge point of the Nador wastewater treatment plant and the Bouareg coast characterized by agriculture activities and watershed inputs.

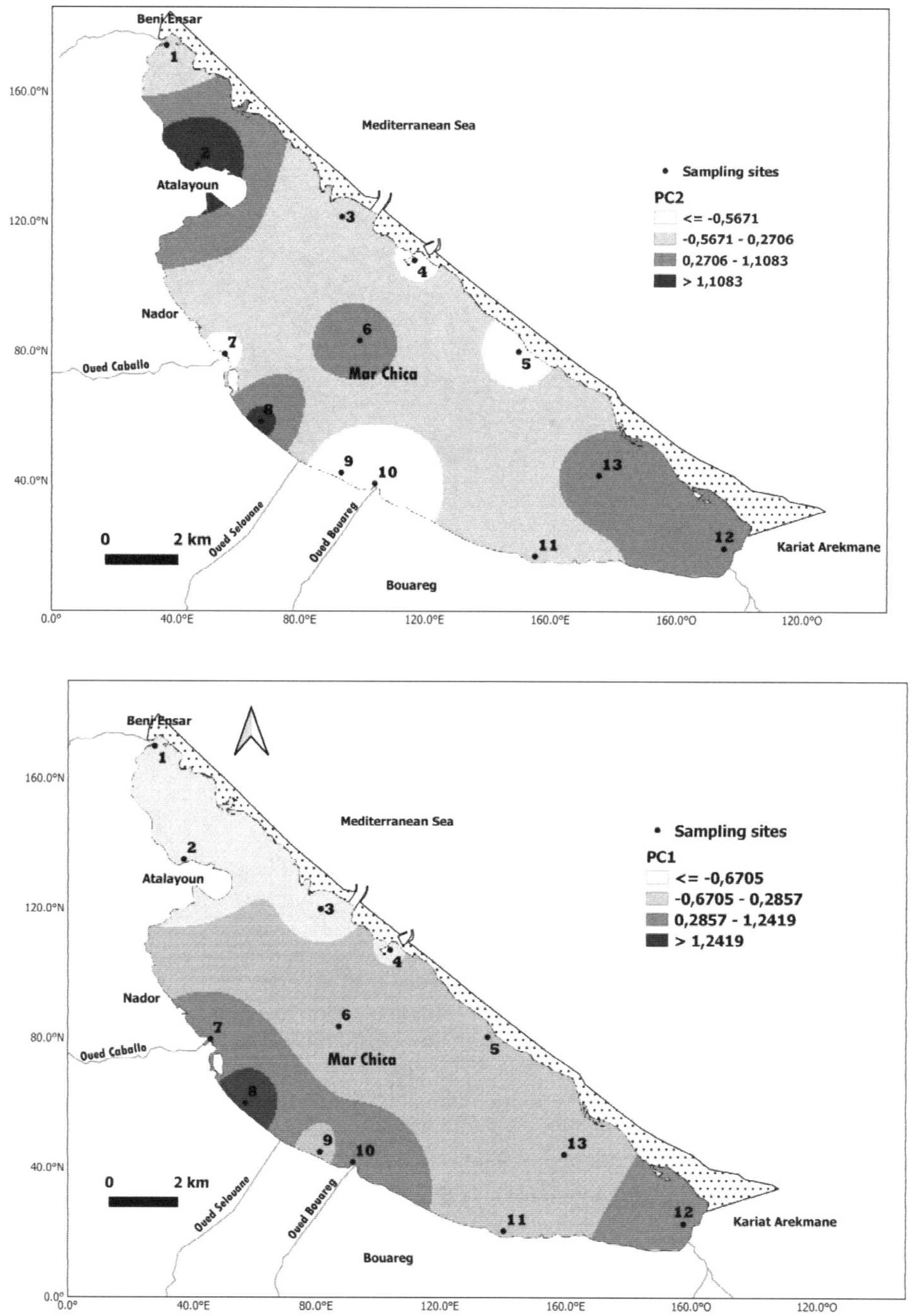

Figure 4: spatial distribution of PC1 and PC2 factors.

4. Conclusion

The Results show low concentrations of nutrients responsible for pollution and eutrophication, high dissolved oxygen values, which implies the improvement of the overall water quality, except for a few points that receive discharges of treated wastewater and the mouths of Wadis. Compared to the previous studies, the quality of the lagoon water is influenced by the terrestrial input of pollutants, sediments, and hydrodynamics. Overall, the water quality of the lagoon is improving due to the renewal of the water through the new Mediterranean channel and the liquid and solid wastewater treatment of the city of Nador. The highest concentrations of nitrogen, phosphorus and organic matter are found near the discharge of nutrient-rich wastewater effluent. The lowest concentrations are found near the new pass. The distribution of nutrient levels is influenced by continental inputs, hydrodynamics, and sediments; these sediments are considered as a reservoir of nutrient pollutants, which will influence water quality through the release and resuspension of sedimentary material in the water. To understand the dominant processes leading to eutrophication, a research program focused on identifying the sources of eutrophic nutrients (both direct and through regeneration processes), the rates of input and the fluxes between the water and sediment components is required. Only then can the resulting eutrophication be understood in terms of increased rates of primary production and of phytoplankton and benthic algae.

5. References

[1] Abreu PC, Odebrecht C, González A. 1994 *J Plankton Res*. 16 (7):737–53.
[2] Piccini C, Conde D, Alonso C, Sommaruga R, Pernthaler J. 2006 *Appl Environ Microbiol*. 72 (10):6560–8.
[3] Tournoud M-G, Payraudeau S, Cernesson F, Salles C. 2006 *Ecol Model*. 193 (1):19–33.
[4] Kjerfve B. Chapter 1 Coastal Lagoons.1994 In: Kjerfve B, editor. Elsevier Oceanography Serie. *Coastal Lagoon Processes*; vol. 60. 1–8.
[5] Affian K, Robin M, Maanan M, Digbehi B, Djagoua EV, Kouamé F. 2008 *Environ Monit Assess*. 2008 159 (1):531.
[6] Allouche N, Maanan M, Gontara M, Rollo N, Jmal I, Bouri S. 2017 *Environ Model Softw*. 88:168–82.
[7] Roselli L, Fabbrocini A, Manzo C, D'Adamo R. 2009 *Estuar Coast Shelf Sci*. 84(4):539–52.
[8] Soetaer K, Herman PMJ, Middelburg JJ. 1996 *Limnol Oceanogr*. 41(8):1651–68.
[9] Gunnars A, Blomqvist S. 1997 *Biogeochemistry*. 37 (3):203–26.
[10] Ruiz F, Abad M, Olías M, Galán E, González I, Aguilá E, N.Hamoumic N, Pulidod I, Cantanoa M. 2006 *Environ Res*. 102 (2):215–29.
[11] Karim Bloundi M, Faure P, Duplay J. 2008 *Comptes Rendus Geosci*. 340 (12):840–9.
[12] Kada O, Abdellaoui S, Ramdani M, Nachit D. 2009 *Bulletin de l'Institut Scientifique, Rabat, section Sciences de la Vie*, 31 (2), 91-98.
[13] Aknaf A, Akodad M, Layachi M, Madani FE, Jaddar A, Mesfioui A, Baghour, M. 2017 J. Mater. Environ. Sci. 8 (9), 3216-3225.
[14] Mostarih MMM, Madani FE, Ali HS, Hachemi OE, Abdellaoui S, Chafi A. 2016 *J. Mater. Environ. Sci.* 7 (12) 4795-5809.
[15] Oujidi B, Tahri M, Layachi M, Abid A, Bouchnan R, Selfati M, Bounakhla M, Mohamed Maanan M, Bazairia H, Snoussi M. 2020 *Reg Stud Mar Sci*. 35:101127.
[16] Karim, H., Makaoui Ahmed, D., Mohammed, I., & Abdellaoui Benyounes, D. 2015 *Eur. Sci. J*, 11, 418-428.
[17] Rodier J, Legube B, Merlet N. 2009 Dunod Paris.
[18] Smart MM, Rada RG, Donnermeyer GN. 1983. *Water Res*.17 (9):1207–11.
[19] Ostrofsky ML. 2012. *Hydrobiologia*. 696 (1):199–203.
[20] Mendiguchía C, Moreno C, García-Vargas M. 2007. *Chemosphere*. 69 (10):1509–17.
[21] Daniel A. Réseau Hydrologique Littoral Normand (RHLN) : cycles annuels 2001-2003 et proposition d'indicateurs d'eutrophisation. 2004 Dec 1 [cited 2021 Dec 5]; Available from: https://archimer.ifremer.fr/doc/00020/13148/
[22] Zhou F, Guo H, Hao Z. 2007. *Mar Pollut Bull*. 54 (9):1372–84.
[23] Simeonov V, Stanimirova I, Tsakovski S. 2001. *Fresenius J Anal Chem*. 370 (6):719–22.
[24] Rigollet V, Sfriso A, Marcomini A, De Casabianca ML. 2004 *Bioresour Technol*. 95(2):159–67.

[25] Zerrouqi Z, Sbaa M, Oujidi M, Elkharmouz M, Bengamra S, Zerrouqi A. 2008. *Int J Environ Sci Technol.* 5(1):125–34.
[26] Zerrouqi Z, Sbaa M, Chafi A, Aqil H. 2013. *Bulletin de l'Institut Scientifique, Section Sciences de la Vie, 35*, 51-59.
[27] Maicu F, Abdellaoui B, Bajo M, Chair A, Hilmi K, Umgiesser G. 2021 *Cont Shelf Res.* 228:104535.
[28] Degobbis D, Homme-Maslowska E, Orio AA, Donazzolo R, Pavoni B. 1986 *Estuar Coast Shelf Sci.* 22 (4):425–37.
[29] Facco S, Degobbis D, Sfriso A, Orio AA. 1986 *Estuarine Variability Academic Press* 307–18.
[30] Smetacek VS. 1985. *Mar Biol.* 84 (3):239–51.
[31] Graf G. 1989 *Nature.*341 (6241):437–9.

Permian volcanism of the Khenifra Massif (Central Morocco) Petrography and geochemical affinity

A Boutsougame[1,*], H Ouazzani[2], E H Abba[1] and M Alaoui[3]

[1]: Higher School of Technology- Khénifra-, P.B: 170-54000, Khénifra, Morocco.
[2]: Faculty of Sciences Meknes, Department of Geology, B.P11201 Zitoune, Meknes, Morocco.
[3]: Sciences and Technologies Faculty of Sultan Moulay Slimane University, Beni Mellal, Morocco.

[*] E-mail: a.boutsougame@usms.ma

Abstract: The Permian volcanic complex of Khenifra (Moroccan Central Hercynian Massif) manifests itself mainly by volcanic rocks of andesite, rhyolite, dacite, rhyodacite, ignimbrite and trachyte nature. These are rocks with a microlitic to porphyritic texture. They are formed by a primary paragenesis made of plagioclase, pyroxene, olivine, feldspar, biotite, amphibole and quartz and a secondary mineral association made of augite, calcite, chlorite and opaque. Intermediate and acidic rocks are widely represented in Khenifra basin. These facies are of calco-alkaline affinity.

Keywords: Petrography, Geochemical affinity, Permian Volcanism, Khenifra, Moroccan Central Hercynian Massif.

1. Introduction

In Morocco, the Permian series are conserved in isolated intramontane basins and distant from each other. Most of the Permian eruptions occur in the Moroccan Central Hercynian Massif. The placement of magmatic rocks, of very varied nature, is controlled by Hercynian tectonics [1]. The main Permian volcanisms of the Moroccan Central Hercynian Massif are; Khénifra volcanism, Chougrane El Had Bouhsoussene volcanism, Bou Achouche volcanism and Tiddas-Sebt Ait Ikkou volcanism.

The Khenifra Basin is located at the SSE of the MCHM. It is a continental basin, individualized at the limit between the SW end of the tabular Middle Atlas and the SE fringe of the MCHM. At the Outcrop, it is the largest of the Permian basins with an area of around 100 km^2.

The Permian volcanic complex of Khenifra studied here has rarely been the subject of detailed geological surveys. The facies of this volcanic complex are andesite, rhyolite, dacite, rhyodacite, ignimbrite and trachyte.

To carry out this petrographic and geochemical study, we sampled all the facies encountered in the field.

2. Materials and Methods

2.1. Lithostratigraphy

The main Permian volcanisms of the Moroccan Central Hercynian Massif are; Khénifra volcanism, Chougrane-El Had Bouhsoussene volcanism, Bou Achouche volcanism and Tiddas-Sebt Ait Ikkou volcanism (figure 1).

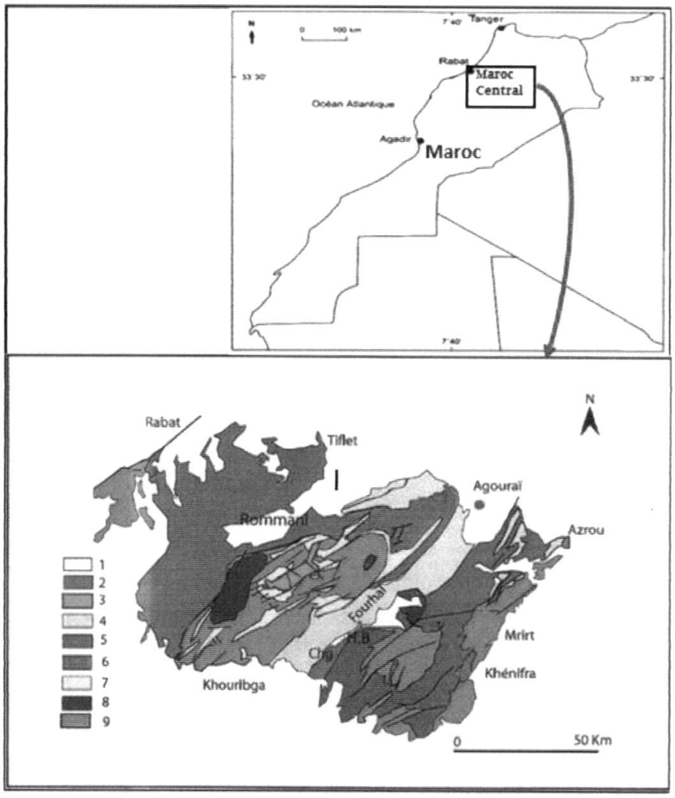

Figure 1. Geological map of the Massif Central [2] with location of the Permian volcanic complex of the Chougrane basin (Chg) - El Had des Bouhsoussene (H.B). 1) Post-Paleozoic Formation, 2) Permian, 3) Namurian-Westphalian, 4) Lower Namurian, 5) Visean, 6) Devono-Dinantian, 7) Silurian, 8) Late-Hercynian granite and 9) Lower Paleozoic (Cambrian to Silurian).

The geological deposits of Khenifra are represented by the deposits sandstone with shales of Cambro-Ordovician age, surmounted in discordance by transgressive limestones of the higher Viséan [3]. The allochthonous deposits consist of sandstone and schist facies of higher Ordovician age. These two sets, folded and schistosed during the Hercynian orogeny, are surmounted by Autunian formations represented by red continental deposits (figure 2), mainly made up of conglomerates (figure 3) with some bands of sandstone and siltstone [4], intercalated with calc-alkaline synsedimentary volcanic rocks [5], [6], [7]. These lands are surmounted by Triassic deposits of the Middle Atlas region, made up of clays and evaporites intercalated with basaltic flows ([8], [9].

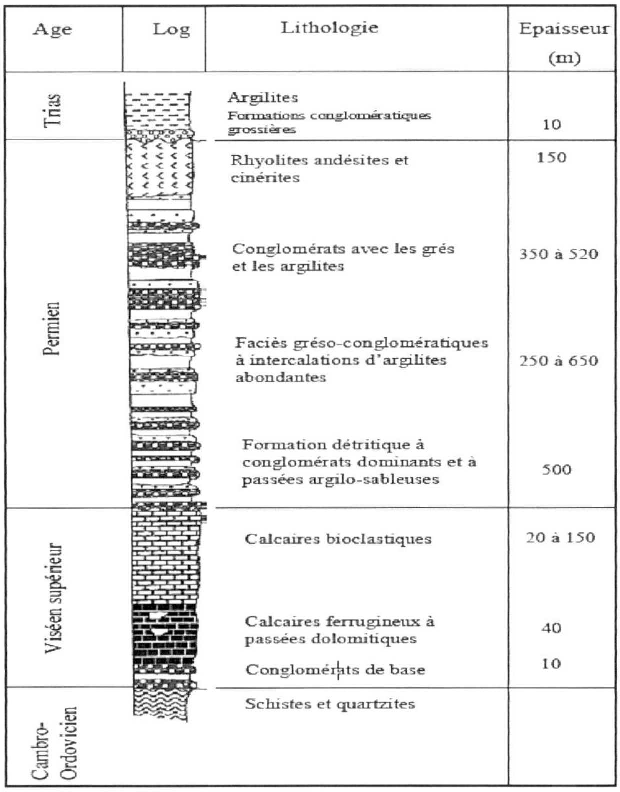

Figure 2. Synthetic stratigraphic log of the Khenifra basin [5] variable scale column.

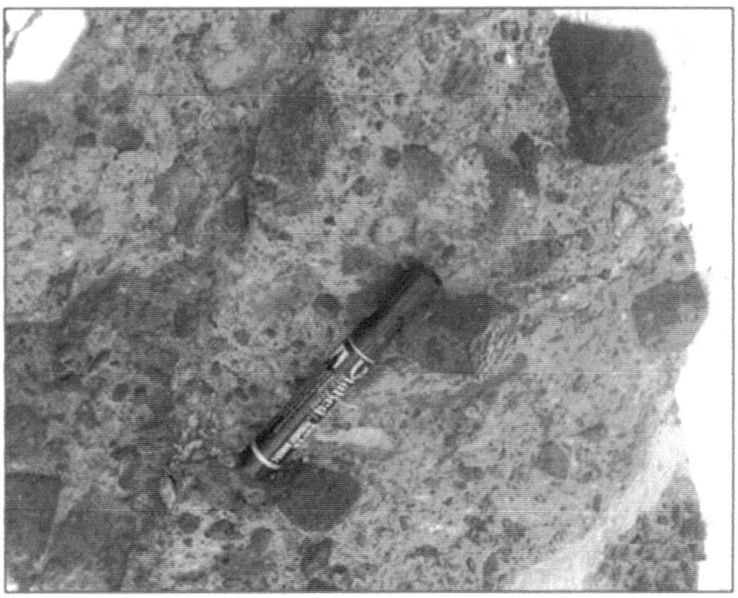

Figure 3. Polygenic conglomerates of Khenifra composed by coarse heterometric terrigenous material assembled by a sandy clay matrix.

The volcanic outcrops are located in the NNW periphery of the Permian Basin of Khenifra where significant volcanic manifestations are noted:
The first activity dated on the basis of structural arguments is of Permian age [10].

The majority of the volcanic rocks of these manifestations, which are rhyolites, andesites, dacites, cinerites, rhyolites-ignimbritic and pyroclastites, are found on the NW fringe of the basin at the level of Jbel Bou Hyyati, Bou Tmlaline, Bou Wadou, Aghnja, Moucharki, Bou Tajat, Sidi Tir, Fas and Akellalou.

The majority of the volcanic rocks of these manifestations which are rhyolites, andesites, dacites, cinerites, rhyolites-ignimbritic and pyroclastites, is found on the NW fringe of the basin at the level of Jbel Bou Hyyati, Bou Tmlaline, Bou Wadou, Aghnja, Moucharki, Bou Tajat, Sidi Tir, Fas and Akellalou (figure 4). In addition to these places, other activities of the same generation exist inside the basin, at Jbel Taghat [6].

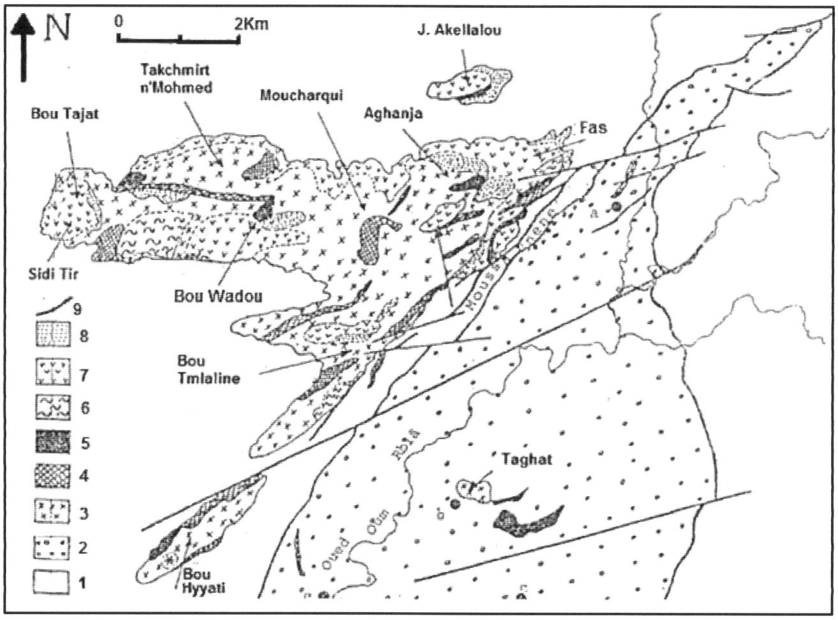

Figure 4. Simplified geological map of the Permian volcanic stacking of Khenifra [11].1: pre-Permian formations; 2: Sedimentary Permian with Autunian fossiliferous levels; 3: rhyolites; 4: igimbrites; 5: cinerites; 6: dacites; 7: Andesites; 8: Pyroclastites; 9: abnormal contacts.

The second volcanic manifestation is that of plio-quaternary basaltic flows [12] which take place along a large fault (SSW-NNE) followed by Oum Rabiâ River. It is a fissured volcanism which progressed along a large tear having been reactivated during the movements of Quaternary uprisings of central Morocco [4]. The same manifestations are found in the northern fallout from the central Moroccan plateau [12]; they have been described as closely related to the tectonic instability of the MCHM during the Quaternary [13].

The volcanism dated back to 264 ± 10Ma, K / Ar Jebrak [14], is marked mainly by two eruption episodes [15], 1995): (i) calc-alkaline eruptions dominant with andesites and rhyolites; (ii) alkaline or "mixed" eruptions with basalts, gabbros and dolerites, microdiorites and lamprophyres. These volcanic rocks, which accompany the deposition of the Autunian late-orogenic continental molasse, have long been considered as probable Pliocene ejectas [12].

Khenifra volcanic rocks are dated back to the Permian on the basis of structural arguments [10], [16], [11], [6], [17], [5], because they are intercalated in Permian detrital formations [4], [18]. The estimated thickness for all the volcanic rocks is around 250m ([19].

2.2. *Petrographic and mineralogical characteristics*

The tops of the reliefs formed by the volcanic rocks are made up of massive rocks, especially rhyolites and andesites, while the depressions are made up of dacites, pyroclastics and ignimbritic rhyolites.
The petrographic study has identified rhyolites, andesites and dacites which represent the majority of volcanics in the Khenifra volcanic complex.

2.2.1. Rhyolites. The rocks take the form of balls of a few meters to a few tens of meters. The thickness can reach up to 160m. These rocks are yellow, brick red, and rarely gray (figure 5).

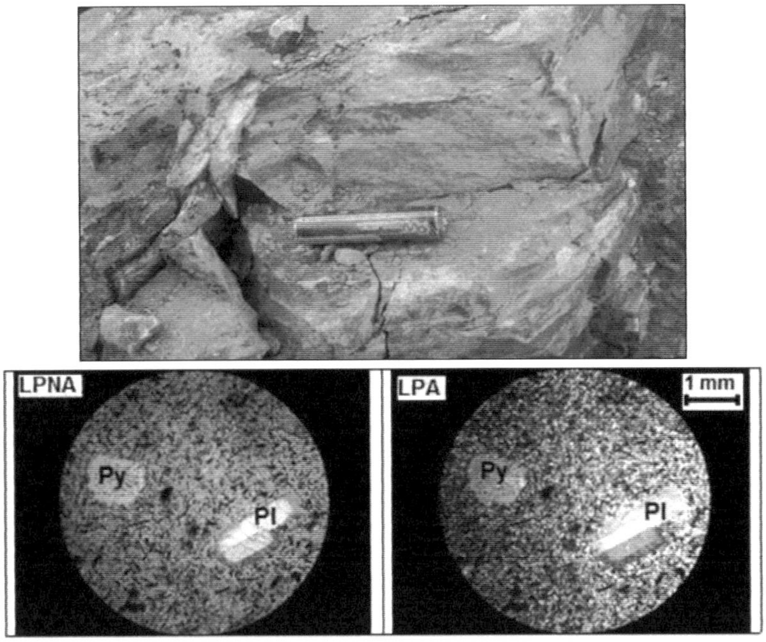

Figure 5. Photographs of Khenifra rhyolite (Bou Hyyati rhyolite).

Rhyolites are formed from plagioclase, quartz, amphibole and biotite. Their textures are porphyritic microlitic.

- Plagioclase (Pl) exists in the form of automorphic to subautomorphic crystals. It is most often fractured and weathered into an assemblage of calcite and kaolinite. It sometimes shows oscillatory zonations.
- Quartz (millimeter dimensions) is in the form of isolated crystals with irregular contours.
- Pyroxene (Py): medium in size and in small quantities.
- Biotite (very scarce) is in the form of small tablets. It is slightly chloritized.

Secondary minerals (calcite, chlorite and oxides) are widely represented.

2.2.2. *Andesites.* They are located in the northern fringe of volcanic massifs, in the summit of Jbel Sidi Tir, Fas and Akellalou (figure 4).

The geographic distribution of these andesitic facies is governed by the extent of fracturing which shows an increasing gradient from West to East of the entire volcanic sector [4].
Andesites are made up of plagioclase, quartz, amphibole and pyroxene arranged in a porphyry microlitic texture (figure 6).

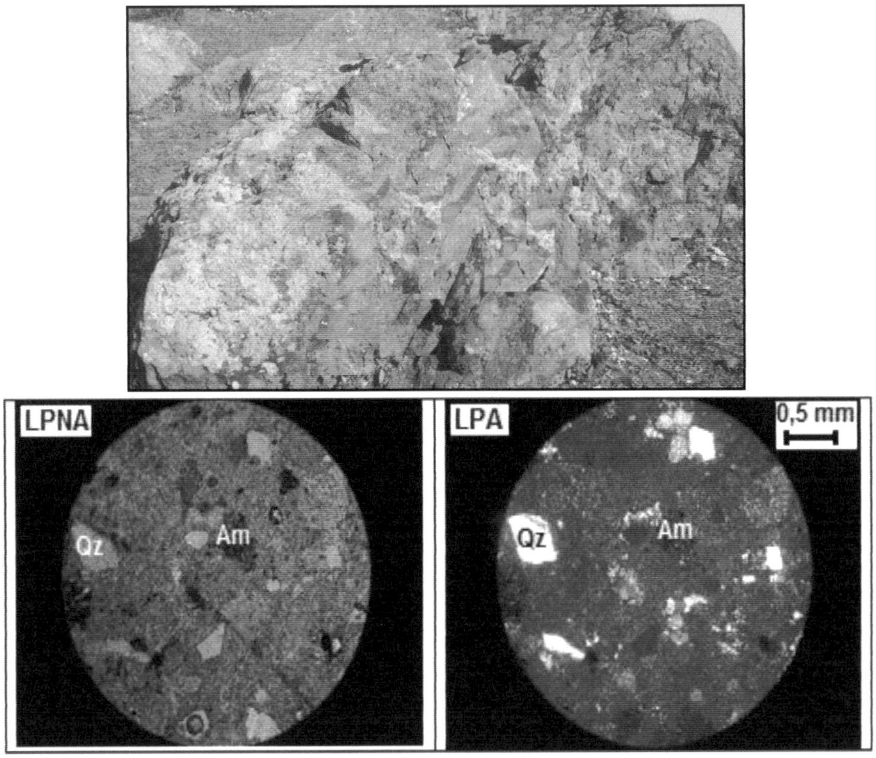

Figure 6. Photographs of the andesite of Khénifra (Andesite from Jbel Sidi Tir).

- Plagioclase (Pl) occurs in the form of automorphic to subautomorphic crystals. It is most often altered and fractured; the fractures are filled with oxides. This plagioclase is zoned.
- Quartz (dimensions 0.5 to 1mm) is in the form of isolated globular crystals.
- Amphibole (Am), recognizable by its characteristic habit. It is epigenized as a mixture of chlorite and oxides.
- Biotite (very scarce) is in the form of small tablets. It is slightly chloritized.

The weathering minerals, calcite and oxides are widely present in rocks.

2.2.3. Dacites. Black or greenish in color and massive in structure, they emerge as veins of limited extent. These rocks are found east of Sidi Tir, at Jbel Fas and south-east of Bou Tmlaline.

Dacites are made up of plagioclase, quartz, amphibole and biotite. Their texture is a porphyritic (figure 7).

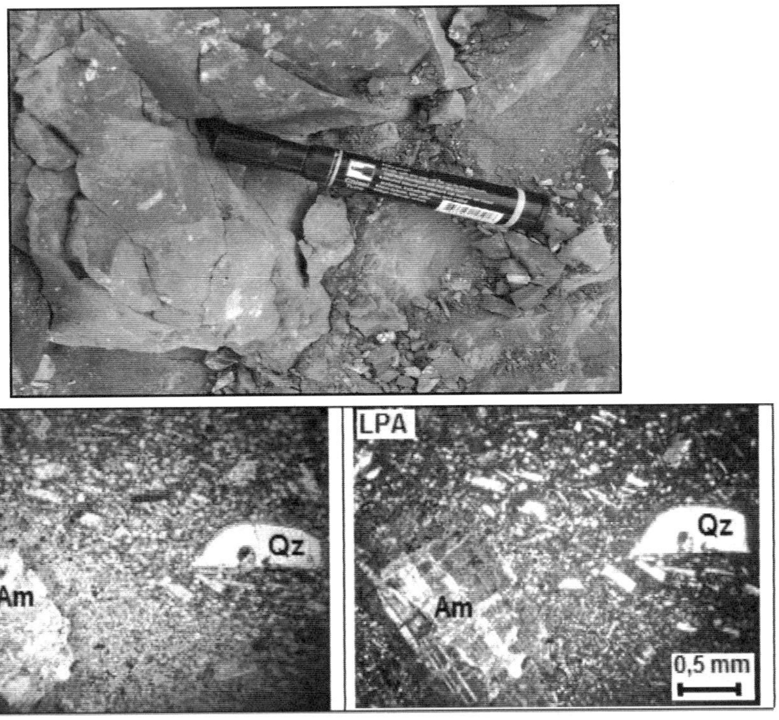

Figure 7. Photographs of dacites from Khenifra (dacite from BouTmlaline)

- Plagioclase (Pl) is in the form of automorphic to subautomorphic crystals, sometimes grouped in glomeruli. It sometimes shows oscillatory zonations or double-twinned.
- Quartz (millimeter dimensions) is in the form of isolated crystals with irregular contours.
- Biotite (very scarce) is in the form of tablets. It is partially chloritized; the latter is associated with grains of oxides in the cleavage planes.
- Amphibole is in the form of pleochroic phenocrysts, recognizable by its characteristic habitus. It is largely replaced by an assemblage of chlorite and oxides.

Secondary minerals are represented by calcite, chlorite and oxides.

2.3. Tectonics

In their structural interpretation of the Permian basins [10], link the opening of the Khenifra Basin to a sinister unhooking clearance along the NE-SW accidents, during submeridian compression, accompanied by distension following direction EW. Allary [20] pointed out the existence of these accidents, but for this author, the fault is constantly destral strike-slip. These accidents would have a Permian [20] or post-Permian-anté-Triassic game [4].

A late-Hercynian NE-SW tightening gives the opening of the Autunian Basin of Khenifra in a NNE-SSW direction ([10].

2.4. Geochemistry of rocks of the Khenifra volcanic complex

The chemical analysis of the samples studied is given in the table 1.

Table 1: chemical composition of the volcanic rocks of Khenifra. Oxides in%, trace elements in ppm. Andesites and rhyolites are analyzed by Youbi [6].

N° éch	YT1	YT154a	Kh11	Kh12	Kh8	YT51	YT26
Facies	Andesite		Trachydacite			Rhyolite	
SiO_2	58,02	61,37	64,2	65,2	65,5	67,89	72,26
Al_2O_3	15,95	15,76	15,1	16	15,3	14,83	14,97
Na_2O	3,78	4,23	5,78	5,8	6,04	2,32	1,24
Fe_2O_3	6,75	5	3,96	3,22	3,1	3,84	1,79
MgO	3,69	3,62	3,19	2,01	2,49	0,68	0,4
K_2O	2,02	1,1	2,95	2,8	2,54	7,96	8,01
MnO	0,07	0,08	0,0758	0,0508	0,0367	0,03	0,03
CaO	5,03	4,91	1,79	2,35	2,6	0,66	0,3
TiO_2	0,95	0,68	0,473	0,415	0,381	0,43	0,32
P_2O_5	0,31	0,26	0,19	0,195	0,208	0,19	0,07
P.F	4,01	2,59	1,88	1,69	1,52	1,73	1,22
Total	100,58	99,6	100	100	100	100,2	100,61
Rb	46	50	144	143	192	145	138
Ba	566	653	653	638	599	1043	799
Sr	389	471	254	251	383	109	60
Zr	202	173	196	197	322	199	149
Nb	-	8	-	-	-	6	-
Y	23,19	16,94	-	46,2	147	14,61	-

3. Results and discussion

3.1. Petrographic determinations

In the classification of volcanic rocks [21] (figure 8), the rocks analyzed are positioned in the field of trachydacites. The rocks analyzed by Youbi [11] correspond to andesites or rhyolites.

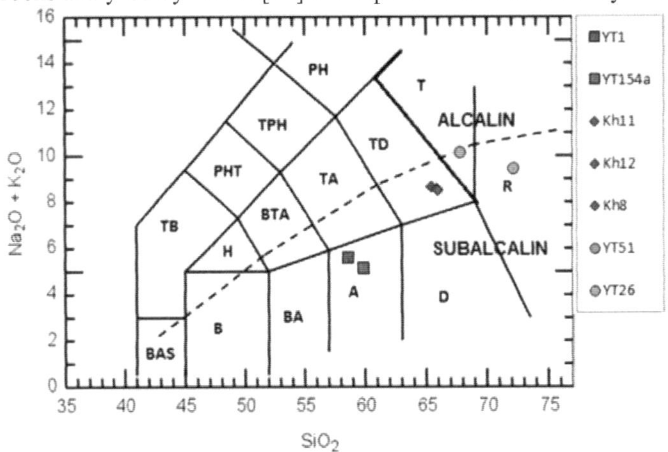

Figure 8. projection of the volcanic rocks of Khenifra in the $Na_2O + K_2O$ vs SiO_2 diagram (in%) of Le Bas et al., [14]. YT1 and YT154a: andesites; Kh8, Kh11 and Kh12: trachydacites; YT51 and YT26: rhyolites. Legend: BAS: basanites; B: basalts; BA: basaltic andesites; D: dacites; H: trachybasalts; BTA: andesites trachybasaltes; TA: trachyandesites; TD: trachydacites; T: trachytes; R: rhyolites; TB: basanites and tephritis; PHT: phonolitic tephritis; TPH: tephritic phonolites; PH: phonolites. The broken line separates the domain of alkaline lavas from the domain of subalkaline lavas.

3.2. Variation diagrams of major elements

In the diagrams of Harker [22], (figure 9), we see a decrease in the oxides Fe_2O_3, MnO, MgO, CaO, Al_2O_3 and P_2O_5 and an increase in K_2O with the increase in silica taken as an index of differentiation from andesites to rhyolites through the trachydacites. The evolution of Na_2O shows an increase since then, from andesites to trachydacites, then a drop from trachydacites to rhyolites; the Na_2O could have been leached into the rhyolites. These trends of the major elements are compatible with an evolution of the three facies from each other by a phenomenon of fractional crystallization.

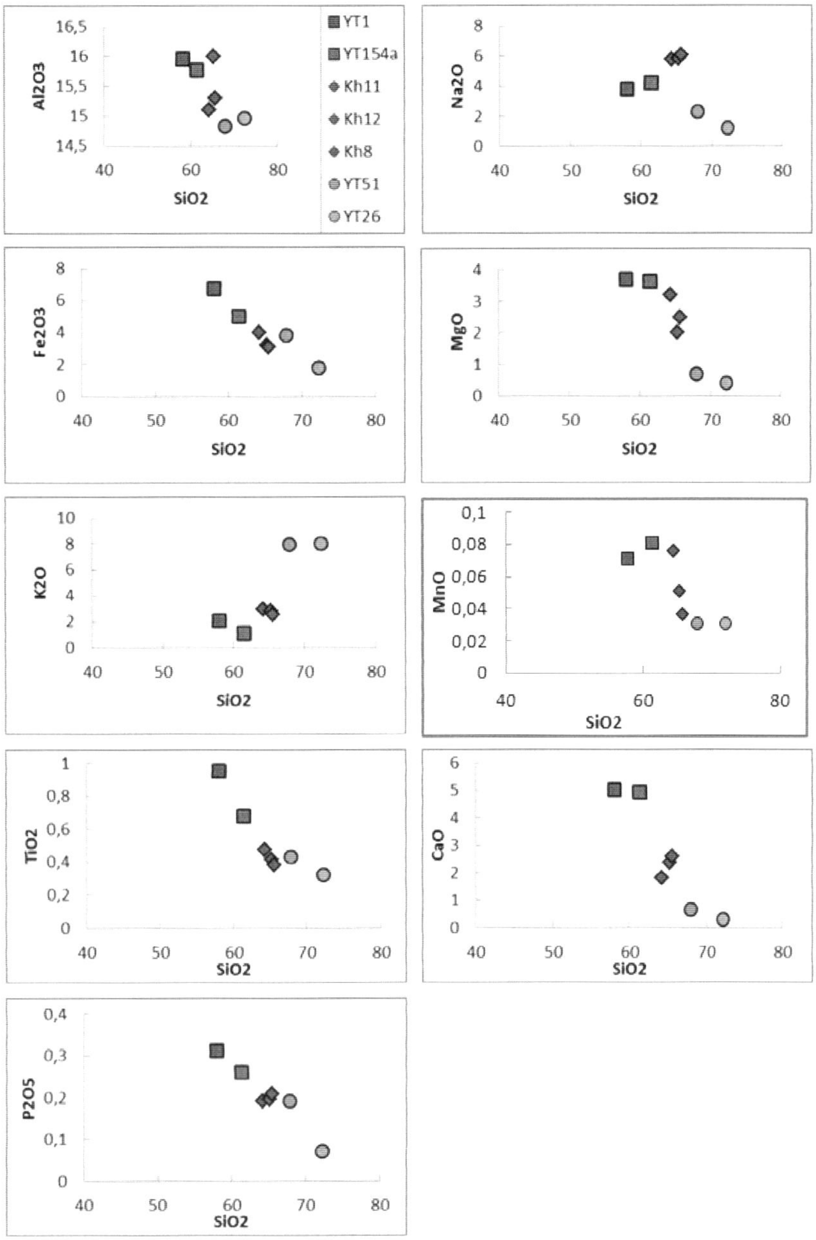

Figure 9. diagrams of variations of oxides (in%) as a function of SiO_2 (in%) for the volcanic rocks of Khenifra. YT1 and YT154a: Andesites; Kh8, Kh11 and Kh12: Trachydacites; YT51 and YT26: Rhyolites.

3.3. Variation diagrams of trace elements

In the diagrams of variation of trace elements in SiO_2 (figure 10), Sr levels continuously decrease during differentiation from andesite, trachydacites to rhyolites. This decay is the result of the extensive fractionation of plagioclase feldspar. The Sr content is high in andesites and trachydacites (251 ppm to 471 ppm).

Ba and Rb contents increase during differentiation of andesite, trachydacites to rhyolites. The increase in Ba is due to the fractionation of K feldspars and biotite. Whereas the increase in Rb, suggests a fractionation of plagioclase and clinopyroxene.

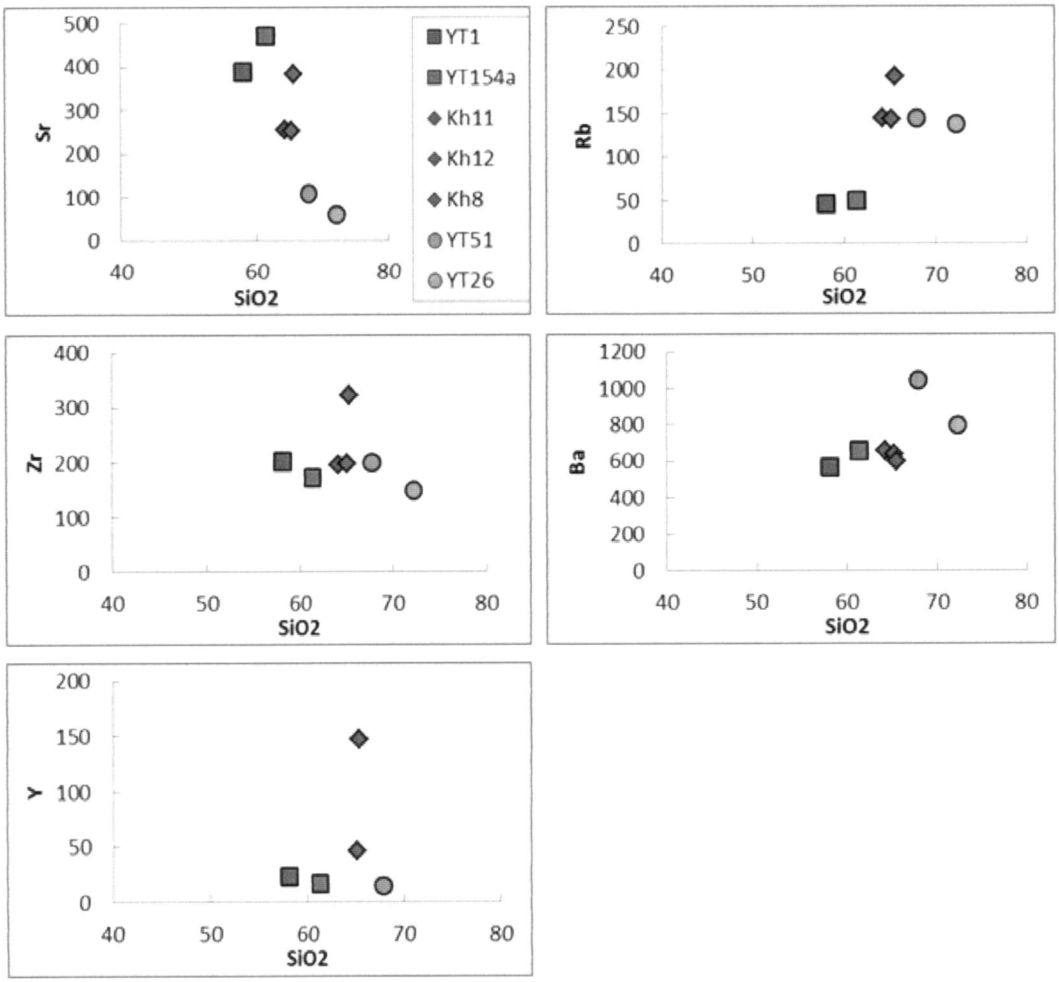

Figure 10. diagrams of variations of trace elements (in ppm) as a function of SiO_2 (in%) for the volcanic rocks studied. YT1 and YT154a: Andesites; Kh8, Kh11 and Kh12: Trachydacites; YT51 and YT26: Rhyolites.

3.4. Magmatic affinity

The representation of the samples in the $Na_2O + K_2O$ vs SiO_2 diagram of Le Bas et al., [21] shows the projection of these samples in the sub-alkaline field (figure 8).

3.4.1. Diagram of SiO_2 vs FeOt / MgO

The rocks analyzed are distributed in the SiO_2 vs FeOt / MgO diagram of Miyashiro [23] in the typical calc-alkaline field and also the intermediate field (figure 11), which confirms the work of [14] and [13].

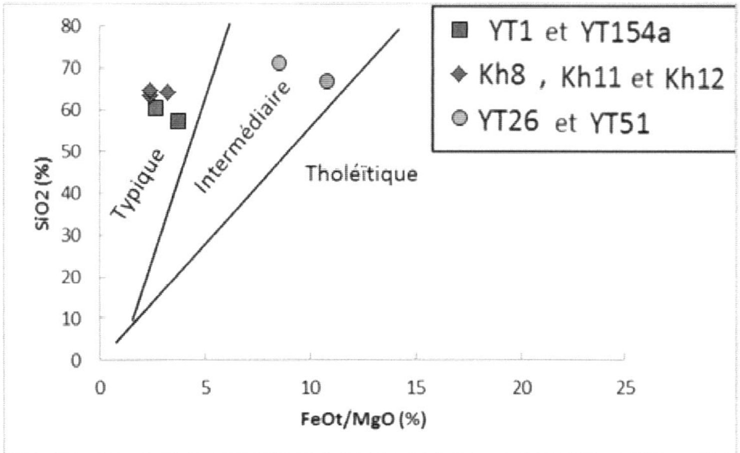

Figure 11. Distribution of andesites (YT1 and YT154a), trachydacites (Kh8, Kh11 and Kh12) and rhyolites (YT51 and YT26) from Khenifra in the SiO_2 vs FeOt / MgO diagram from Miyashiro [23].

4. Conclusion

The petrographic and geochemical studies of the Permian volcanic rocks of Khenifra made it possible to complete the existing data and determine the petrographic and geochemical characteristics of the main volcanic rocks of the Permian basin of the massif studied.

From a geochemical point of view, the intermediate and acid terms are widely represented in the Khenifra basin.

Most major and trace elements show correlations with silica as a differentiation index. This reflects the important role played by petrogenetic processes such as fractional crystallization. The characterization of these elements makes it possible to highlight that these rocks belong to a calc-alkaline series.

In general, the study of Permian volcanisms in the Khenifra basins has provided new data, petrographic and geochemical, which will allow us to contribute to a discussion on the geotectonic significance of this volcanism.

5. References

[1] Cailleux Y, Delouche C, Gonord H et Zouine E.M 1986 Synthèse sur le volcanisme Permien du Maroc. Son insertion dans le contexte géodynamique ouest-méditerranéen. *111e congrès national des sociétés savantes* 1, 221-235.

[2] Bouya N 2014 Utilisation des données géophysiques aéroportées magnétiques et spectrométriques dans la cartographie géologique de la partie NE du massif central marocain et des causses moyen atlasiques. *Thèse Doctorat, Univ. Moulay Ismail, Fac. Sciences*, 190p

[3] Allary A, Andrieux J, Lavenu A et Ribeyrolles M 1972 Les nappes hercyniennes de la Méséta sud orientale (Maroc Central). *C. R. Acad. Sci.*, Paris, t. **274**, D, 16, pp. 2284-2287.

[4] EL Wartiti M 1990 Le Permien du Maroc mésétien : étude géologique et implications paléogéographiques. *Thèse Sciences, Univ. Mohammed V*, Rabat, 501p.

[5] Youbi N, Cabanis B, Chalot-Prat F et Cailleux Y 1995 Histoire volcano-tectonique du massif permien de Khénifra (Sud-Est du Maroc central). *Geodinamica Acta*, 8, 3, 158– 172.

[6] Youbi N 1998 Le volcanisme "post-collisionnel" : un magmatisme intraplaque relié à des panaches

mantelliques. Etude volcanologique et géochimique. Exemples d'application dans le Néoprotérozoïque terminal (PIII) de l'Anti-Atlas et le Permien du Maroc. *Thèse Sciences, Univ. Cadi Ayyad*, Marrakech. 660p.

[7] Boutsougame A, 2017 Volcanisme permien de massifs de Chougrane-El Had des Bouhsoussene, Bou Achouche et Khénifra (Massif Hercynien Central Marocain), pétrographie, minéralogie et géochimie : implications géodynamiques. *Thèse Sciences, Univ. Moulay Ismail*-Méknès- *120p.*

[8] Charroud A, Charroud M, Fedan B, Laville E, Rioult M, Piqué A et Du Dresnay R 1996 Dynamique sédimentaire des formations triasiques du Moyen Atlas méridional. In : Medina F. (Ed) - Le Permien et le Trias du Maroc : état des connaissances, *PUMAG,* Marrakech, pp. 269-289.

[9] Ouarhache D 2002 Sédimentation et volcanismes (effusif et explosif) associés au rifting triasique et infraliasique dans le Moyen Atlas sud-occidental et la Haute Moulouya. *Thèse d'Etat, Univ. Mohammed V*, Rabat Agdal, 264 p.

[10] Cailleux Y, Gonord H, LE Guern M et Sauvage M 1982 Sur l'âge permien du volcanisme acide tertiaire dans le Maroc central. *9éme R.A.S.T.* Paris.

[11] Youbi N 1990 Le complexe volcanique permien de Khénifra (SE du Maroc Central). Cartographie, volcanologie, géochimie, minéralogie, considérations structurales et implications géodynamiques. *Thèse de doctorat de 3ème Cycle, Univ. S. M. B. A* Fès, 341p.

[12] Termier H 1936 Etude géologique sur le Maroc central et le Moyen-Atlas septentrional. *Notes et mém. Serv. Mines et carte géol.* Maroc, n°33, t.**1**, 743p.

[13] El Wartiti M, Broutin J et Freytet P 1986 Première découverte paléontologiques dans les séries rouges carbonatées permiennes du bassin de Tiddas (Maroc central). *C. R. Acad. Sci.* Paris, 303, II, p. 263-268.

[14] Jebrak M. (1985) - Contribution à l'histoire naturelle des filons (F, Ba) du domaine varisque français et marocain. Essai de caractérisation structurale et géochimique des filons en extension et en décrochement. Document du B.R.G.M., no 99, 510p.

[15] Youbi N et Cabanis B 1995 Le volcanisme permien du Maroc : un magmatisme relié à une subduction de type B associé à des failles transformantes. *In* : A. El Boukhari et al. (Eds.) - Résumés de la Réunion extraordinaire de la Société géologique de France sur le thème : Accrétion continentale au Protérozoïque supérieur et au Paléozoïque du Nord Ouest africain. 84p.

[16] Youbi N 1987 La branche nord du massif volcanique permien de la région de Khénifra (S-E. du Maroc Central). *Rapp. Inédit, faculté des Sciences.* Fès. 58p.

[17] Youbi N, Chalot-Prat F et Cailleux Y 1989 Données préliminaires sur le volcanisme permien de Khénifra. *Collogue Géologique franco-marocain, Strasbourg, résumé* : 203.

[18] Assoumi H, Broutin J, Youbi N, El Wartiti M et Galtier J 1995 Découverte de Scleromedulloxylon Cf. Aveyronense, bois fossile de Coniférophyte, dans le complexe volcanique permien du bassin de Khénifra (Maroc Central). Intérêt stratigraphique et floristique. Annales de Paléontologie, volume 81, *Fasicule**1***, pp. 1-16. Paris.

[19] Boushaba A, Cailleux Y et El Wartiti M 1987 Les granitoïdes hercyniens et le volcanisme permien du domaine mésétien marocain. *Bull. Sci. Terre,* Rabat, n°3, pp. 104-138.

[20] Allary A, Lavenu A et Ribeyrolles M 1976 Etude tectonique et microtectonique d'un segment de chaîne hercynienne dans la partie sud orientale du Maroc central. *Notes &Mém. Serv.géol.* Maroc, 261, 169p.

[21] Le Bas MJ, Le Maitre RW, Streckeisen A and Zantti B 1986 A chemical classification of volcanique rocks basid on the total alkali-silica diagram. *J. Petrol*, Oxford. 27, **3**, pp.745- 750.

[22] Harker A 1909 The natural history of igneous rocks. *Mac millon, New York*, 384p.

[23]Miyashiro A. (1974) - Volcanic rock series in island arcs and active continental margins. American Journal of Science, Vol. 274, p. 321-355.

Sustainable Sanitation: an appropriate solution in rural areas

A Taouraout[1,2*], A Chahlaoui[2], M Sadki[2], A Maliki[2], M Khaffou[3], D Belghyti[1]

[1] Laboratory of Agrophysiology, Biotechnology, Environment and Quality, Faculty of Sciences. Ibn Tofail University, kenitra. Morocco.
[2] Laboratory of Natural Resources Management and Development Team, Health and Environment, Faculty of Science, Moulay Ismail University Meknes, Morocco.
[3] High School of Technology of Khenifra.

[*]**Corresponding author:** aziz.taouraout@uit.ac.ma

Abstract. In Morocco, reuse of treated wastewater can reduce the negative effects of drought in the agricultural sector. However, it is necessary to adopt a sustainable sanitation in order to reduce the risk of this type of water on public health and environment. This new sanitation concept is based on the principle of separation, recovery and reuse of wastewater and taking in consideration the social, economic, environmental, and institutional of the local conditions. The aim of this article is: to give the main results of a survey conducted in 2017 at Dayet Ifrah village which housed the first pilot ecological sanitation project in Morocco concerning appreciation and attitude of the population towards this new system of sanitation; and on the other hand to study the efficiency of a pilot scale system composed on three filters used to treat domestic wastewater coming from a single house in Meknes. In general, ecological sanitation is appreciated by the population of the village and showed its effectiveness to improve sanitation in Moroccan context. The pilot scale system proved his effectiveness to trait organic pollution and nutrients, faecal pollution and parasites.

Keywords: Reuse, Sustainable sanitation, Wastewater, Water scarcity, Morocco.

1. Introduction

Currently, rural areas of Morocco face different challenges such as water stress [1], degradation of quality of renewable water resources and environment [2-3]. Indeed, rural sanitation of Morocco is lagging far behind, given the absence of adequate wastewater collection, evacuation and treatment infrastructure. The constraints that hamper the development of sanitation in these areas are of several aspects (institutional, legal, financial, strategic, technical, etc.) [4]. Aware of this situation, Morocco has invested in the field of the rationalization of the management of water resources by the regularization of the use of conventional and unconventional water (law 36-15). Furthermore, the Rural Sanitation National Program sets as an objective to be achieve by 2040 [5]: eradication of defecation in the wild; 100% households will be equipped by adequate sanitary structures and 50% of Wastewater will be depolluted. To do this, separation and treatment of excreta and gray water at the source is the strong point of this new type of sanitation, since it is in the 1% of wastewater (excreta) that is concentrated almost all of the most harmful pollutants (pathogens, parasites and nitrates...) for the environment and public health [6]. So, by separating and treating excreta out of wastewater, using dry toilets, all of these problems are virtually eliminated from economical and efficient way [7]. So it becomes obvious that the dry toilet solution is significantly less problematic in its impact on the environment and increases protection of public health [8,9].

There are many ecological sanitation techniques ranging from simple dry toilets, Urine Diversion Dehydration Toilets (UDDT) to much more complex systems (vacuum flushes coupled with a methanizer producing biogas). The choice of the technique to be used will then depend on the geographical conditions, economic situation, their use on a large or small scale, the involvement of users, technical skills, financial resources, location of the project and especially its acceptability by the population [10]. According to the WHO [11], it is recommended to apply a system with a concept of multiple barriers to preserve public health in the reuse of wastewater, urine and faeces.

It is within this framework that our study comes to strengthen the Rural Sanitation National Program by ecological, economical and sustainable techniques to fill the deficit of which hits the Moroccan rural sanitation sector. To do this, a survey was conducted at Dayet Ifrah village in the Middle Atlas which housed the first pilot ecological sanitation project in Morocco to study the appreciation and attitude of the population towards this new system of sanitation, on the other hand, an experimental study on three filters (Vertical Constructed Wetland (VCW), Vertical Multi-Soil-Layering (V-MSL) and Horizontal Multi-Soil-Layering (H-MSL)) was carried out on a pilot scale to treat domestic wastewater.

2 Materiel and methods

2.1. Survey

In order to judge perception and attitude of the population towards Ecological Sanitation (EcoSan) structures installed in Dayet Ifrah village, a survey was carried out in 2017. Several techniques and data collection tools were used in this study. The qualitative and quantitative data were collected through field observation, a survey, a group discussion and interviews [12].

2.2. Treatment of wastewater

The characteristics of the system used in this study are summarized in Table 1 and Figure 1.

Table 1. Main characteristics of the pilot unit.

Caracteristics	VCW	V-MSL	H-MSL
Area (m^2)	0.24	0.12	0.15
Dimensions (LxWxH)	0.40x0.60x0.78	0.20x0.60x0.78	0.30x0.50x0.55
Feeding type	Alternate and sequenced	Discontinued	Continued
Flow direction	Vertical	Vertical	Horizontal
Aeration	Yes	Yes	No
Medium	Unsaturated and aerobic	Aerobic / anaerobic	anaerobic
Plantation	Yes	No	No

The choice of substrates used in our system, is mainly related to their availability and their cheap cost in Morocco. The V-MSL is composed by six permeable layers (PLs) with gravel (diameter between 3 to 5 mm), five Soil-Mixture-Blocs (SMBs) which are deposited vertically; unlike H-MSL their layers are horizontal. The SMBs measuring 20 cm (L) x 16.5cm (W) x 4 cm (H), are composed by 60% of soil, 10% of charcoal, 10% of iron sawdust and 20% of wood sawdust. On the other hand, VCW is composed by gravel, iron sawdust and crushed concrete. Crushed concrete and iron sawdust were used for the first time in constructed wetland filter in order to enhance the adsorption of phosphate [13].

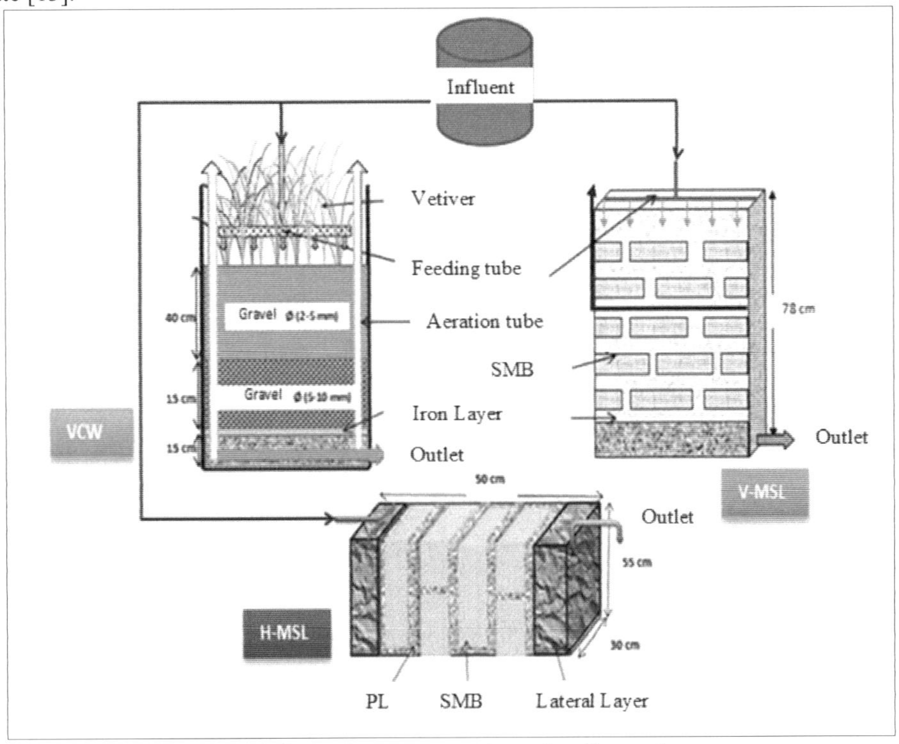

Figure 1. Schematic representation of a pilot scale system

2.3 Analyzes carried out

Samples of raw and treated water were collected at the same time each month (from June until November 2017) and analyzed for different water quality parameters using the French standard methods. Biochemical oxygen demand (BOD_5) was determined by the BOD Sensor (VELP SCIENTIFICA), chemical oxygen demand (COD) was analyzed according to the dichromate open reflux method and Total Suspended Solids (TSS) concentration was determined by the filtration method. Ammonium (NH_4^+) concentration was determined by the indophenol method, and orthophosphate (PO_4^{3-}) concentration by the molybdate and ascorbic acid method. In addition, parasitological analysis was performed using Arther Fizerald Fox flotation concentration technique [14]. Concerning fecal pollution indicators analysis, Table 2 summarizes the various bacteriological parameters analyzed as well as the method undertaken.

Table 2. Bacteriological parameters and methods of analysis

Parameter	Technique	Reference
Total coliforms (TC)	Count by membrane filtration (0.45 μm) on Tergitol agar with TTC 7 24 hour Agar at 37 °C.	ISO 9308-1:2014 [15]
Fecal coliforms (FC)	Count by membrane filtration (0.45 μm) on Tergitol agar with TTC 7 24 hour Agar at 44°C	ISO 9308-1:2014

2.4 Data analysis

The collected data were analyzed using software: Microsoft Excel and SPSS (Analysis of variance test (ANOVA)).

3. Results and discussions

3.1 Survey results

The EcoSan structures installed at Dayet Ifrah were:

- 7 bathrooms with Urine Diversion Dehydration toilets (UDDT);
- 1 biogas digester for wastewater and manure treatment;
- Horizontal constructed wetlands to treat domestic wastewater,
- Demonstration gardens for reusing organic fertilizers (excreta after treatment) in Agriculture.

The survey showed that 95% of users are positively appreciate these ecological structures installed in their village, with the exception of the biogas digester which was badly perceived by the sole beneficiary, for maintenance and profitability reasons. In addition, more than 90% of households without toilets showed a real interest to acquire this new type of toilet.

It is clear that Eco-toilets and horizontally constructed wetlands meet the expectations of beneficiaries since they contribute to create a healthy and clean life and contribute to protect groundwater and Dayet Ifrah aquatic ecosystem (water surface).

However, the population did not accept the reuse of excreta after treatment in agriculture (psychic effect). In order to overcome this challenge, raising awareness through scientific days, channels conventional (media, relays, etc.), associative actors, elected representatives of the region and representatives of administrations, etc., remain the most appropriate means.

3.2 Domestic wastewater treatment

3.2.1 Elimination of organic pollution and nutrients

The BOD_5, COD and TSS concentrations before and after treatment were respectively, (186.7±42.9) mg / L, (386.8±84.1) mg / L and (371.3±102.7) mg / L for influent; (40.3±25.3) mg / L, (80±47.9) mg / L and (61.3±42.6) mg / L for VCW; (28.1±12.3) mg / L, (59.1±24.3) mg / L and (48±32.3) mg / L for V-MSL and (65.5±31.3) mg / L, (123.7±52.4) mg / L and (87.3±43.3) mg / L for H-MSL (Table 3). These results showed clearly that the three filters recorded an good ability to reduce these parameters with superiority given by V-MSL filter with an average of 84.3% followed by VCW filter with 81% while H-MSL filter did not exceed 75%.

Concerning the elimination of ammonium, the V-MSL filter was able to remove (83.1 ± 6.1)% followed by VCW with (49.8 ± 18.3)% and only (29.2 ± 12.6)% for the H MSL.

Regarding orthophosphate reduction, V-MSL and VCW gave good results with (72.5 ± 8.8)% and (68.6 ± 9.4)%, respectively; however, H-MSL filter has recorded a moderate reduction with only (45.2 ± 29.4)%.

Table 3. Average concentrations of organic pollution in influent and effluent

	Influent	VCW	V-MSL	H-MSL
BOD	186.7±42.9	40.3±25.3	28.1±12.3	65.5±31.3
COD	386.8±84	80±47.9	59.1±24.3	123.7±52.4
TSS	371.3±102	61.3±42.6	48±32.3	87.3±43.3

Significant difference of the performance between H-MSL filter and the other two filters (VCW and V-MSL) is obtained by ANOVA test concerning BOD_5 and COD removal at $p < 0.05$. The V-MSL filter showed superior performance regarding the removal rate of organic pollution followed by VCW filter.

3.2.2 Reduction of fecal pollution

Influent average concentration of total coliforms was 6.11 log units CFU / 100 mL with a maximum of 6.19 and a minimum of 6.01. Concerning fecal coliforms, it was 5.99 log units CFU / 100mL with a maximum of 6.12 and a minimum of 5.83.

The average reductions of CT and CF were respectively, 1.48and 1.50 for the VCW; 1.61 and 1.64 for V-MSL and 0.62 and 0.55 for H-MSL. These results showed clearly that V-MSL and VCW (aerobic filters) gave the good reduction rates with an average of 95%; against only 76% for H-MSL (anaerobic filter). Significant difference between influent and effluent was observed concerning bacterial loads with $p \leq 0.001$. We can therefore conclude that unsaturated aerobic media favor the elimination of pathogens compared to saturated anaerobic ones.

3.2.3 Reduction of parasites

An average parasite load of 12.7 eggs / 2L is detected in influent during the study period (June to March) with a maximum of 47 eggs / 2L and a minimum of 0 eggs / 2L. Parasitological monitoring revealed the presence of eggs in 70% of raw wastewater samples and 10%, 30% and 20% at the outlet of VCW, V-MSL and H-MSL, respectively.

In raw waters, helminths are most abundant with a total of 113 eggs / 2L followed by protozoa with 14 eggs / 2L. Four species of parasites have been identified: *Ancylostoma sp.* , *Ascaris sp.*), *Hymenolepis diminuta*) and *Blastocystis hominis* (table 4).

Significant difference between influent and effluent was observed concerning the number of eggs / 2litres with $p \leq 0.01$ (ANOVA test).

Table 4. Average concentrations (Number of eggs / 2L) of parasites before and after treatment

		Helminthes			Protozoa
		Nematodes		Cestodes	
		Ancylostoma spp	*Ascaris spp*	*Hymenolepis diminuta*	*Blastocystis hominis*
Number of eggs/2litres	Influent	38	70	5	14
	VCW	0	4	0	0
	V-MSL	1	7	0	1
	H-MSL	1	8	0	0

Figure 2 reports the average performance to reduce parasites by the three filters. VCW filter showed a good performance, because it was able to eliminate more than 97% of nematodes (100% of *Ancylostoma spp* and 94.3% of *Ascaris spp*), 100% of cestodes (*Hymenolepis diminuta*) and 100% protozoa (*Blastocystis hominis*); followed by V-MSL and H-MSL with respectively, 94% and 93% for nematodes, 100% for cestodes and 100% and 92.9% for protozoa.

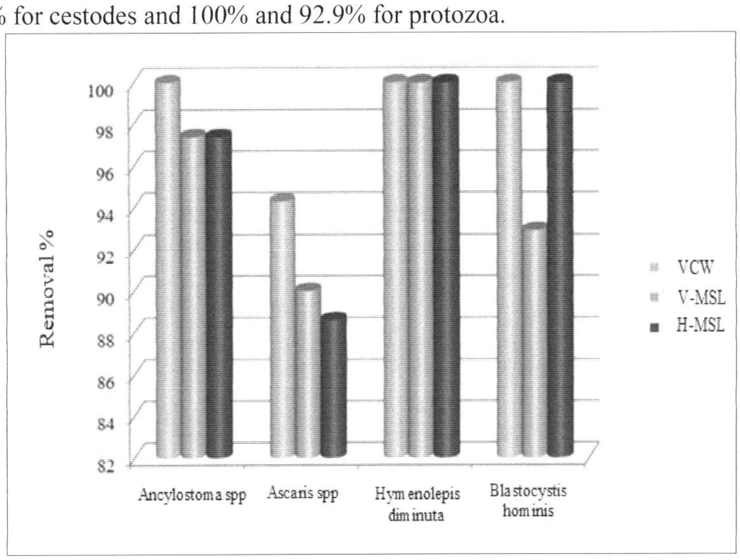

Figure 2. Average reduction of parasites by the three filters

The domestic water treated by the three filters meets the discharge standards adopted by the Moroccan legislator concerning organic pollution, bacterial load; however, for a possible reuse in irrigation, the WHO standard requires the absence of helminths and a fecal bacteria less than 1000 CFU/100 ml. This is very difficult to achieve even with sophisticated filters. Separation and treatment at the source is the best approach to reduce bacterial pollution and parasites, since it is in the 1% of wastewater (excreta) that is concentrated almost all of the most harmful pollutants (pathogens, parasites and nitrates...). So, by separating and treating excreta out of wastewater, by using dry toilets, all of these problems are virtually eliminated from economical and efficient way. It becomes obvious that the dry toilet solution is significantly less problematic in its impact on the environment and increased protection of public health. On the other hand, gray water must be considered as a precious resource and not as waste. Despite the described risks of inadequate gray water management, gray water nonetheless has great potential to reduce the water stress currently facing regions of the world. The reuse of gray water to irrigate the garden or for flashing toilet is an effective measure to save water at the domestic level [16, 17].

Gray water treatment techniques are less documented. However, some techniques have been used in pilot projects in some of the countries. The criteria for evaluating and making decisions about the sanitation system to be adopted should not be limited to finding an effective system, but other criteria, such as adaptation to local conditions, social assessment, risks on the health and the environment, economic and financial considerations, the possibility of reusing effluents after treatment, the energy requirement for operation, etc.

4. Conclusion

As result of this study, the pilot scale system installed showed an efficiency to remove organic pollution and nutrients but less performed to eliminate fecal pollution bacteria and parasites. This is why, Ecological Sanitation comes to overcome this problem by separating and treating these pollutions at source using eco-toilets and treating gray water alone (low concentration of organic pollutants and less contaminated by fecal pollution). So, if we were interested to treat gray water alone by the filters (VCW, V-MSL and H-MSL) the results will be better.

Regarding the survey conducted at Dayet Ifrah village, EcoSan structures installed were appreciated by almost all people related on their benefits such as: protects against cold, rainy and dark conditions, promotes health and environment.

5. References

[1] Sinan M, Belhouji A, 2016. Impact of the Climate Change on the climate and the water resources of Morocco on horizons 2020, 2050 and 2080 and measures of adaptation. Houille Blanche - August(4), 32–39 (2016). https://doi.org/10.1051/lhb/2016037
[2] High Commissioner of Planning: The Morocco between millennium goals for the development and objectives of sustainable development achievements and challenges. National Report, Aug 2015 13. 4.
[3] WHO/UNICEF, 2017. Drinking Water, Sanitation and Hygiene Update and SDG Baselines
[4] PNAR, 2013. Programme National d'Assainissement en milieu Rural au Maroc ... www.agire-maroc.org/activites/assainisse...rier-2013-rabat.html
[5] DGCL/DEA, 2015. Présentation du Ministère de l'intérieur direction générale des collectivités locales direction de l'eau et de l'assainissement. Atelier sur le PNAR.
[6] TDM, 2009. Toilettes Du Monde – Guide toilettes sèches 2009. Source: TDM 2009 Des toilettes sèches à la maison.
[7] Taouraout A, Chahlaoui A, Belghyti D, Najy M, Taha I, and Kharroubi A, 2019. Perceptions and Attitudes of the Rural Population of Morocco Towards EcoSan Latrines UDDTs. © Springer Nature Switzerland AG 2019 (Eds.): SCA 2018, LNITI, pp. 104–117, 2019. https://doi.org/10.1007/978-3-030-11196-0_11

[8] Ferguson G, Dakers A, and Gunn I, 2003. Sustainable wastewater management. A handbook for smaller communities. Published by the ministry for the environment Manatu Mo Te Taiao, New Zealand. ISBN: 0-478-24094-5

[9] Taouraout A, Chahlaoui A, Belghyti D, Sammoudi R, Najy M, 2020a. How to Improve Wastewater Treatment in Smart City. © Springer Nature Switzerland AG 2020 (Eds.): SCA 2019, LNITI, pp. 1141–1153. https://doi.org/10.1007/978-3-030-37629-1_82

[10] Naramabuye FX, Haynes RJ, Modi AT, 2008. Cattle manure and grass residues as liming materials in a semi-subsistence farming system. Agric Ecosyst Environ 2008, 124:136-141

[11] OMS, 2006. Directives de l'OMS sur la réutilisation des eaux usées, des excréta et des eaux grises.

[12] Taouraout A, Chahlaoui A, Belghity D, Najy M, Kharroubi A, 2018. *The socio-cultural acceptance of EcoSan latrines (UDDTs) in rural areas of Morocco*. SCA '18, October 10–11, 2018, Tetouan, Morocco © 2018 Association for Computing Machinery. ACM ISBN 978-1 -4503-6562-8/18/10…$15.00. https://doi.org/10.1145/3286606.3286815

[13] GuanY, Zhang Y, Huang X, & Fu J, 2015. *Effect of operating factors on the contaminants removal of a soil filter: multisoil-layering system.* Environmental Earth Sciences, 74(3), 2679 2686 (2015). https://doi. org/10.1007/s12665-015-4288-8.

[14] Arther RG, Fizgerald PR, & Fox JC, 1981 .Parasits ova in anaerobically digested sludge. J. Wat. Pollut. Cont. Fed., 53, 1333-1338.

[15] ISO 9308-1:2014, Qualité de l'eau - Dénombrement des Escherichia coli et des bactéries coliformes (Référence pour CT, CF et E. coli) — Partie 1: Méthode par filtration sur membrane pour les eaux à faible teneur en bactéries.

[16] Eriksson E, Auffarth K, Henze M, Ledin A, 2002. Characteristics of grey wastewater. Urban Water 4(1), 85–104 (2002)

[17] Saumya S, Akansha S, Rinaldo J, Jayasri MA, Suthindhiran K, 2015. Construction and evaluation of prototype subsurface flow wetland planted with Heliconia angusta for the treatment of synthetic greywater. J. Clean. Prod. 91, 235–240 (2015) Improving Safe and Sustainable Gray Water Reuse 1177

IOP Conf. Series: Earth and Environmental Science 1090 (2022) 012012 doi:10.1088/1755-1315/1090/1/012012

Alfalfa in arid and semi-arid regions Taroudant as an example, good and bad cultural practices on the environment: A statistical study

H E Moussaoui[1]*, L F Z Ainlhout[2], A Bourezi[3] and L Bouqbis[2]*

[1] Faculty of Sciences, Laboratory of Biotechnology, Materials and Environment, Ibn Zohr University, Agadir, Morocco
[2] Faculty of Applied Sciences, Laboratory of Biotechnology, Materials and Environment, Ibn Zohr University, Agadir, Morocco
[3] Faculty of Legal, Economic and Social Sciences, Ibn Zohr University, Ait Melloul, Morocco

*E-mail: hassan.elmoussaoui@edu.uiz.ac.ma / l.bouqbis@uiz.ac.ma

Abstract: A field study was made on the cultivation of alfalfa and their different components of this crop in the Taroudant region in the south of Morocco, the questions of this study based on the cultivation and fertilization methods, irrigation, and finally prospects. This study showed that the majority will continue to cultivate alfalfa her because the use of alfalfa is mainly as fodder food for their own herd, the whole population does the plowing before the crops, the varieties used are mainly from USA. The population uses manure with different quantities, it depends on the cultivated area and on the use of chemical fertilization. This study showed high consumption of water and chemical fertilizers with a means between 44066.8 m^3/year/ha is 62399.9 m^3/year/ha, and between 165.6 kg/ha/year and 205.7 Kg/ha/year respectively with a 95% confidence interval. What knocks the alarm danger in environmental matters.

Keywords: Alfalfa, Environment, Fertilization, Irrigation, Taroudant, Morocco.

1. Introduction

Agriculture in Morocco is the fastest growing sector in the country, this sector including livestock currently represents about 15% of the gross domestic product of Morocco, and almost a quarter of its exports [1], which requires providing suitable feed to guarantee the success of the overall ration in order to maintain stable animal production [2], for that a diet which is constituted a proportion of alfalfa increases the value of this diet taking into account its low production inputs, its environmental and nutritional benefits and its continuous production for four to five years [3].

The problem of drought and the increased water needs by agriculture in Morocco [4], requires to characterize this primordial crop in order to solve it, indeed the crop issue, the needs hydric, nutritional and pedoclimatic, are the main causes for a bad cultural practice which destroyed the soil. Furthermore, the environment and the natural resources used in the region decreased the quality of soil as a result the decrease of plant growth. Our study focused to draw the next perspectives related to the continuity of the crop of alfalfa in the conditions current by a questionnaire.

2. Material and methods

2.1. Sampling method and sample size

The survey carried out in the south of Morocco in region of Taroudant which are recognized by agriculture. The survey was carried out on a total sample of one hundred farms. In fact; we worked on a five regions with a sample size of thirteen to fifteen percent of the total population growing alfalfa.

2.2. Questionnaire and survey compositions

The questionnaire was in the form of a survey focused on the components, objectives, and prospects for growing alfalfa. It was divided into three parts. Firstly, the questions describe the population, the land and the objectives of cultivation. Secondly, the method of growing alfalfa by the description of the tillage practice, the amendments, the sowing, the irrigation as well as the harvest and the diseases of this crop. Finally, the perspectives in the form of questions on the continuity of cultivation of alfalfa.

2.3. Data analysis

The statistical analyses were performed using XLSTAT.v2018. The direction of different relationships existing between the different nominal qualitative variables is done using the MCA method and for the quantitative variables we used the PCA method. The modes are also calculated for all the components, the quantiles and the medians are schematized in the form of a Boxplot. The estimate of the average differences is made by a 95% confidence interval by IBM SPSS Statistics 23.

3. Results and discussion

3.1. Response rates

The results of the survey by which we have based in this study were only eighty-five agricultures of one hundred surveys; indeed, four farmers refused the response of the survey. While eleven questionnaires were eliminated since it was no answer on certain questions of our survey.

3.2. Description of the population

The results of the descriptive statistics Table 1 show that alfalfa is mainly used as a fodder food for herds this represents 96%, and also shows that the most dominant complementary food ratio is Corn silage + Straw + Cereals which represents more than 56% of all dietary ratios, farmers do not depend on pasture in diets since this represents a small proportion.

The diversification of supplementary fodder regime for alfalfa in the population due to the availability of these foods for agriculture and the food needs of the herd. The nutrient content of many alternative foods varies widely, making it necessary to analyze or assess the value of the food. Producers need to know the energy, protein and key levels of minerals in these foods to develop a balanced, low-cost diet for livestock to meet the needs of livestock and the objectives of producers, the ration must be balanced [5]. From a commercial point of view, alfalfa is among the largest forage crops in many countries. For all that, forage legumes can also be high quality animal feed. But its growth is threatened by changing climatic conditions, which indicates the need to cultivate varieties capable of resisting and also adapting to the negative effects of climate change [6].

Table 1. The descriptive statistics of the farming practice of alfalfa.

Variable\Statistics	Nb. of observations	Mode	Mode (effective)	Modalities	Proportion by modality
Use_of_alfalfa	85	fodder for personal herd	82	Sale	0.035
				fodder for personal herd	0.965
Complementary_ food	85	Corn silage+Straw+Cereals	48	Corn silage+Green beans+Straw+Cereals	0.153
				Corn silage+Straw+ Cereals	0.565
				Corn silage+Straw+Cereals +Pasture	0.188
				Cereals	0.024
				Pasture	0.024
				Pasture+Green beans+Cereals	0.012
				Sale	0.035

3.3. Alfalfa cultivation method

The variables analyzed by the MCA are Type of manure; Kind of fertilizers; Crop rotation. The symmetric graphic outputs of the variables (Figure 1) visualize two principals groups of the methods of cultivation of alfalfa the first is based on the use of multiple types of manure, mineral fertilizers and multiple crop rotations at the same time, the second is based on the use of 33% ammonitrate, the use of a mixture of manure from two third cattle and one third sheep and a crop rotation based on cereals and vegetables, The other two modes are very rare either do not use any fertilizer and no crop roation or a fertilizer with widely different NPK proportions to the first two modes, with the incorporation of herbs in the cereal crop rotations.

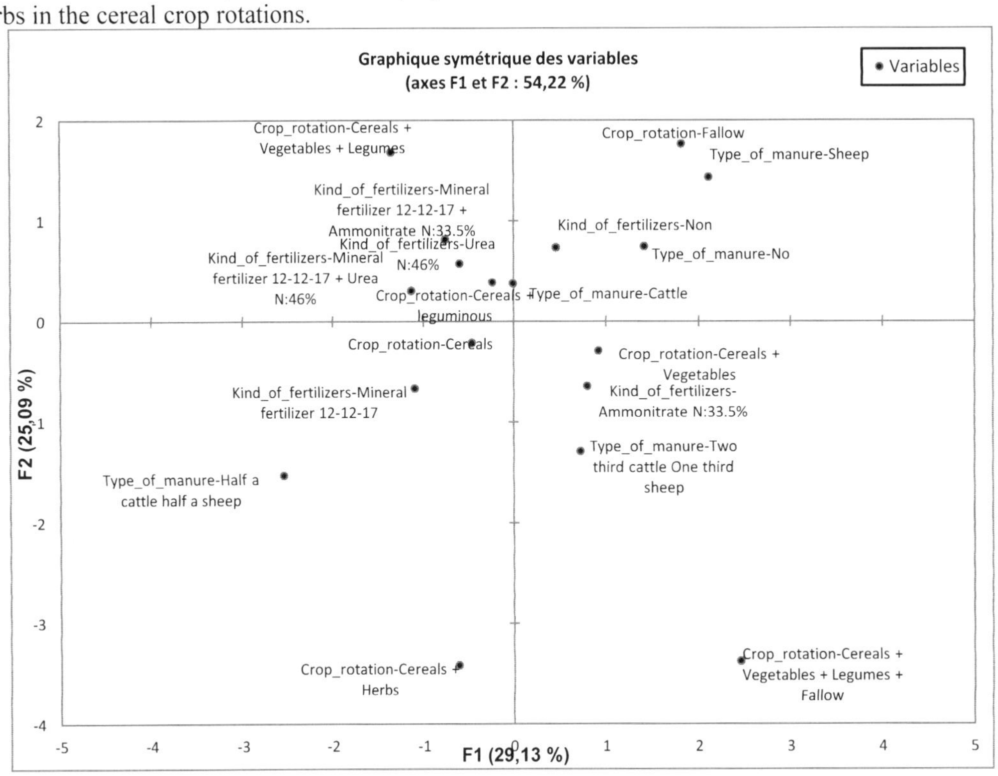

Figure 1. The symmetric graphic outputs of the Multiple Correspondence Analysis (MCA) for the variation of method of cultivation of alfalfa by agricultures of region. The variables analyzed: Type of manure; Kind of fertilizers; Crop rotation. The black dots indicate the variables studied.

The results of the descriptive statistics Table 2 show that the majority of the population uses Monosoc Brabant plow and Cover crops V, and anomalies that Two-soc Brabant plow and Cover crops PXL use in plowing, the entire population studied applied plows deeply followed by plowing superficial. Cattle manure occupies a large part of the organic amendments of this crop it is generally applied by the spreader followed by the proportions of cattle and sheep manure a very small proportion which uses sheep manure alone or does not use the manure, the unavailability of sheep manure is the reason for this low proportion of use, almost a third of the population studied uses Ammonitrate N: 33.5% as chemical fertilizers each quarter, the use of the Mineral fertilizer 12-12-17 when applying manure remains high compared to the use of Urea N: 46%, the use of Mineral fertilizer 12-12-17 + Ammonitrate N: 33.5% or Mineral fertilizer 12-12 -17 + Urea N: 46% do not even represent 5% of the population, almost 30% of the population does not use fertilizers.

Cereals and also cereals are found in all combinations of rotation crops except a proportion of 3% which leaves the land fallow, more than half of the fields studied suffer from diseases (Butterfly of Cydia medicaginis - Tychius aureolus - Acyrtosiphon pisum; Aphid - Uromyces striatus; rust) and 84% of the agricultural crops treat with chemicals these pests the remaining part either it does not treat or it uses conventional method to reduce the impact of pests on the crop (irrigation by the Water Diffuser or reduce the time between harvests).

Table 2. Descriptive statistics for the description of the population

Variable\ Statistics	Nb. of observations	Mode	Mode (effective)	Modality	Proportion by modality
Deep_plowing	85	Monosoc Brabant plow	82	Monosoc Brabant plow	0.965
				Two-soc Brabant plow	0.035
Superficial_ plowing	85	Cover crops V	82	Cover crops PXL	0.035
				Cover crops V	0.965
Type_of_ manure	85	Cattle	59	Cattle	0.694
				Half a cattle half a sheep	0.035
				No	0.035
				Sheep	0.024
				Two third cattle One third sheep	0.212
Kind_of_ fertilizers	85	Ammonitrate N:33.5%	25	Ammonitrate N:33.5%	0.294
				Mineral fertilizer 12-12-17	0.212
				Mineral fertilizer 12-12-17 + Ammonitrate N:33.5%	0.024
				Mineral fertilizer 12-12-17 + Urea N:46%	0.024
				Non	0.294
				Urea N:46%	0.153
Crop_ rotation	85	Cereals	37	Cereals	0.435
				Cereals + Herbs	0.012
				Cereals + Vegetables	0.165
				Cereals + Vegetables + Legumes	0.024
				Cereals + Vegetables + Legumes + Fallow	0.012
				Cereals + leguminous	0.318
				Fallow	0.035
Chemical_treatment_against_alfalfa_ diseases	85	Yes	38	No	0.156
				Yes	0.844

Traditional and modern agriculture in the Moroccan agronomic system are undergoing degradation processes due to improper use of tillage practices tools. In contrast the transient beneficial effect of tillage, the sustainability of the adverse effects of conventional tillage systems destroys the environment [7]. The application of cattle manure to acid soils increases the pH, as it is attributed to buffering of organic acids and bicarbonates in cattle manure. The available mineral P, N, K, Mg and Ca increased immediately after manure application, and the available P and K remained significantly higher in modified soil by manure compared to unmodified soil [8]. The cattle manure increases in pH, contents of organic C, and total N and P. Also an increased microbial biomass and dehydrogenase and phosphatase activities of soil which gives an indication of the increase in soil fauna activity [9]. Alfalfa which is cultivated in soils polluted by heavy metals could tolerate high concentrations of Cd, Zn and Pb in the soil and had very good growth performance. Also, sheep manure also increases phytostabilization. Application of sheep manure can be helpful in increasing electrical conductivity and reducing the concentrations of DTPA-extractable metals in the soil [10], and can also reduce the effects of saline irrigation and increase the productivity of alfalfa [11]. The alfalfa and clover fix nitrogen and can reduce dependence on organic fertilizers and improve soil structure, for example the rotation between wheat and alfalfa [12]. Mineral nitrogen fertilization can be abandoned and arriving at a good yield of wheat by the cultivation of alfalfa for a period more than 3 years before the cultivation of wheat [13].

In order to meet the nutritional needs of humans to grow with quality and efficiency, it is known that the chemicals used in plant nutrition as well to increase soil efficiency and pest treatment are the most common practices, important for controlling the production and the quality of plant production, this non-reasonable use causes damage to the soil as the salinity, the accumulation of heavy metals, the eutrophication of water and the accumulation of nitrate and also pollution air by sulfur and nitrogen which they have dangerous effects on the environment like the greenhouse effect and consequently of the effects on the health and the existence of the human being [14]. The groundwater is exposed to contamination from pesticides and fertilizers used in agriculture [15]. Among strategies useful in reducing the use of fertilizers and pesticides are the productions of legumes, which fix atmospheric nitrogen by symbiosis with rhizobium nitrogen-fixing bacteria, in rotation with non-legume crops [16]. Also the combination of manure and straw increases both the yield and the microbiological activity of the soil also minimize of production input costs [17].

3.4. Alfalfa cultivation Perspective

According to eighty-five questionnaires, only one person has another source of income, and four people will give up growing alfalfa, no one has answered yes regarding land abandonment or land conversion to real estate.

3.5. Quantitative survey parameters

In order to conclude the different correlations between the different quantitative variables we analyzed all the quantitative variables by PCA analysis (Figure 2). This analysis showed a strong positive correlation between the flock size and the area of cultivated alfalfa also between the quantity of fertilizers and with the numbers of the years of the successive harvest this graphical representation shows a negative correlation between the flock size and the incorporation of alfalfa in the feed ratio, that means when the size of the herd increases the quantity of alfalfa incorporated into the feeding system, other foods decreased in favor, also the graph shows a positive correlation between water consumption ,cultivated area and the price. Although the price of m^3 changes according to the irrigation method. The PCA analysis shows that there is no relationship between water consumption and the use of organic or mineral fertilizers. Regarding the analysis of the range results of the individuals, it is observed that a group based mainly on a large ration of alfalfa compared to the other feed diet foods, this group is distinguished by a small herd size and an average consumption at low water since the cultivated area is low, the second group is characterized by a high consumption of water and a low use of fertilizers, the third group is not based on alfalfa despite the cultivated area with alfalfa which is important but they do not support the large size of the herds.

The area cultivated alfalfa in Morocco is appreciated by a quarter of the areas cultivated by fodder food in Morocco [18]. The incorporation of wheat straw in a diet which constitutes the corn silage acts better than a diet which is based on alfalfa straw since it has improved food intake and ruminal pH and also improves undigested neutral detergent fiber after 30 hours (uNDF30) and the ruminal function of high-yielding dairy cows, but may decrease feed efficiency [19]. The diet and its impact is not influenced by the change in composition but it is even influenced by the size of the same food for example dry matter intake (DMI), average daily gain (ADG) and the carcass weight[20].

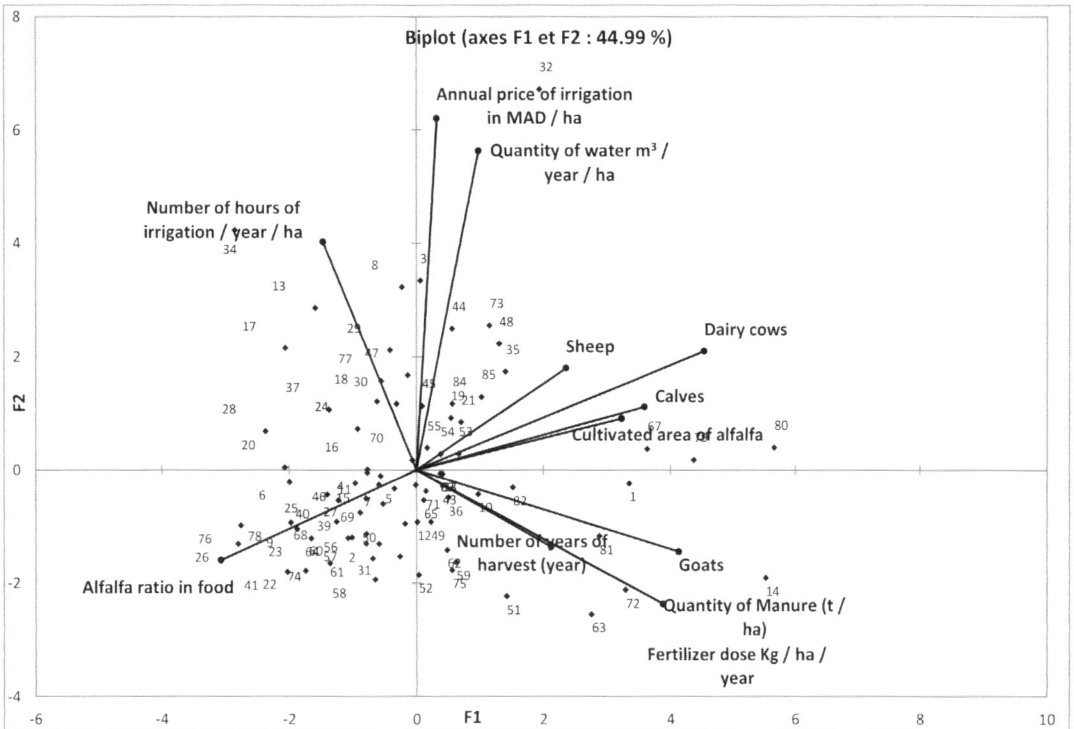

Figure 2. Biplot graph of the principal component analysis (PCA) for the variation of following quantitative variables: Alfalfa ratio in food (%); Quantity of Manure (t/ha); Fertilizer dose (Kg/ha/year) ; Number of hours of irrigation (year/ha) ; Quantity of water (m3/year/ha) ; Annual price of irrigation in (MAD/ha) ; Number of years of harvest (year). ● Variable; ♦ Individual.

Box plots Figure 3. showed values of the application of extreme manure of the sample between 83t/ha and 96t/ha, the average use is about 37.13t/ha. Reconciling the median and the average shows statistical homogeneity and non-concentration of the data. The results of Fertilizer dose (Kg/ha/year) have shown strong annual uses of chemical fertilizers, 75% of the population uses an annual dose of chemical fertilizers greater than 120 Kg/ha/year. We also observe that 75% of the population with a number of hours less than 1128 hours/year/ha, a significant dispersion of the data for 25% of the individuals who stay. In accordance with the results of Box plots Quantity of water (m3/year/ha) and Annual price (MAD/ha), the statistical data are concentrated between 40000 m3/ha/year and 3600 m3/ha/year and 29680 MAD/ha/year and 1800 MAD/ha/year which represents 50% of the population, the other 50% of the population presents a significant dispersion of the data. The entire population decreased irrigation in winter with proportions that have been taken into account in this present study. These observations can be explained by the heterogeneity in prices and debits according to the regions. Estimating means with a 95% confidence interval Table 3 showed the following results: The means of using chemical fertilizers is between 165.6 Kg/ha/year and 205.7 Kg/ha/year and Truncated mean at 5% 182.1 Kg/ha/year, and the means of the amount of Irrigation water is between 44066.8 m3/year/ha is 62399.9 m3/year/ha and Truncated mean at 5% is 49356.9 m3/year/ha.

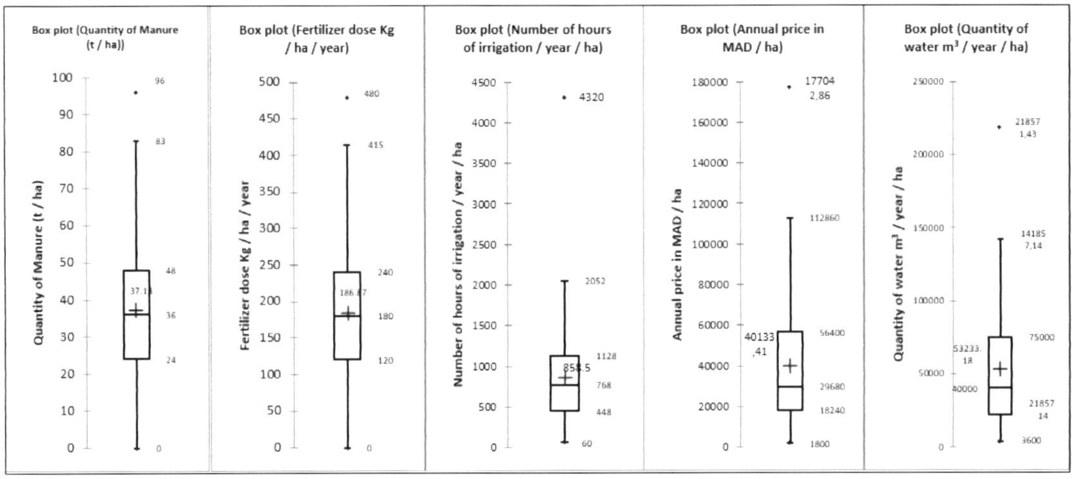

Figure 3. BoxPlots which describe the parameters of irrigation and quantity of the organics and chemicals fertilizer

The water consumption by various plants is influenced by climate, water supply, soil and topography [21], [22]. This consumption is very important and very expensive hydraulically and economically for alfalfa given its continuous production and the long evapotranspiration times [23]. Insufficient irrigation negatively influences the quality and quantity of alfalfa since increases NDF and decreases the total productivity of alfalfa [24]. The increase in the number of years in the production of alfalfa is accompanied by a high consumption of water consumed from deep soil. Four years is the good duration of the cultivation of alfalfa which keeps both good production and low impact on the consumption of water from the deep soil [25]. The three-year alfalfa rotation followed by two Maize years increased the production of Maize and corn leaf area index, biomass, and yield by comparison with the Maize rotation alone and also decreased total evapotranspiration [26]. For continuous cultivation of alfalfa, it should be an additional of 4 to 5 tonnes / acre of farmyard manure to each cut. For the improvement of yield, it requires about 2 to 3 tonnes / acre of superphosphate per alternate year [27]. One-third-ton livestock manure and ammonitrate 33.5% fertilizer at 75 Kg/ha per application are most used on vegetable and leguminous crops in the Gharb region in Morocco [15]. The spread of pre-planned manure between 9 and 40 t / ha does not systematically result in more nitrogen in the soil or in the plant after 2 years of alfalfa [28].

Table 3. Estimate of the quantity of mineral fertilizers (A) and the quantity of water consumed per year (B) with a confidence interval of 95% with Average truncated at 5%.

Feature A				
			Statistics	**Standard deviation**
Fertilizer dose (Kg/ha/year)	**Mean**		185,67	10,082
	95 % confidence interval for the Mean	**Lower bound**	165,62	
		Upper bound	205,72	
	5 % truncated average		182,05	

Feature B				
			Statistics	**Standard deviation**
Quantity of water (m^3/year/ha)	**Mean**		53233,17	4609,41
	95 % confidence Interval for the mean	**Lower bound**	44066,85	
		Upper bound	62399,48	
	5 % truncated average		49356,89	

The application of manure up to 62.1t / ha for 4 years having positive effects on the biomass and on the dry weight also on Density of root mass and specific root length and on Plant available nitrogen compared to mineral fertilization with all applied doses arriving at N 210 kg / ha, P 300 kg / ha and K 150 hg / ha [29]. The application of high concentration of manure (336 kg N ha- ~) had a negative effects on the productivity of alfalfa, Soil Nitrate and on nitrate in soil water over a long period [30]. The toxic effects of high doses of fertilizers reacted differently depending on the type of soil [31].

4. Conclusion and perspectives

This survey showed the importance of the cultivation of alfalfa for the agricultures of the region and the problems linked to this crop and the bad cultural practices applied by agriculture such as the heavy dependence on plowing and the unreasonable uses of water resources. and mineral fertilizers, which firstly requires a study of the soil and the remaining water resources in order to describe the situation and the current state of these resources and to make decisions in order to save the situation, secondly, to seek other crops which do not consume water and are not sensitive to pests with the same positive impact or find control methods and water management methods in order to continue the cultivation of this fodder food.

5. Acknowledgments

We thank the Laboratory of Biotechnology, Materials and Environment of the Faculty of Science of Agadir, the Polydisciplinary Faculty of Taroudant and the Faculty of Applied Sciences of Ait Melloul, University Ibn Zohr for the funding of this study.

6. References

[1] R. Harbouze, J.-P. Pellissier, J.-P. Rolland, and W. Khechimi, "Rapport de synthèse sur l ' agriculture au Maroc.," p. 105, 2019, [Online]. Available: https://hal.archives-ouvertes.fr/hal-02137637/document.

[2] E. Short, "The importance of forage," *Equine Heal.*, vol. 2018, no. 40, pp. 14–15, 2018, doi: 10.12968/eqhe.2018.40.14.

[3] B. Boller, U. K. Posselt, and F. Veronesi, *Fodder Crops and.* 2010.

[4] D. Verner *et al.*, "Climate Variability, Drought, and Drought Management in Morocco's Agricultural Sector," 2018.

[5] G. Lardy and V. Anderson, "Alternative Feeds for Ruminants," *Ndsu*, vol. 1182, no. May, pp. 1–24, 2003.

[6] K. P. Kulkarni, R. Tayade, S. Asekova, J. T. Song, J. G. Shannon, and J.-D. Lee, "Harnessing the potential of forage legumes, alfalfa, soybean, and cowpea for sustainable agriculture and global food security," *Front. Plant Sci.*, vol. 9, p. 1314, 2018.

[7] R. Mrabet, R. Moussadek, A. Fadlaoui, and E. van Ranst, "Conservation agriculture in dry areas of Morocco," *F. Crop. Res.*, vol. 132, pp. 84–94, 2012, doi: 10.1016/j.fcr.2011.11.017.

[8] J. K. Whalen, C. Chang, G. W. Clayton, and J. P. Carefoot, "Cattle Manure Amendments Can Increase the pH of Acid Soils," *Soil Sci. Soc. Am. J.*, vol. 64, no. 3, pp. 962–966, 2000, doi: 10.2136/sssaj2000.643962x.

[9] J. A. Parham, S. P. Deng, W. R. Raun, and G. V. Johnson, "Long-term cattle manure application in soil. I. Effect on soil phosphorus levels, microbial biomass C, and dehydrogenase

and phosphatase activities," *Biol. Fertil. Soils*, vol. 35, no. 5, pp. 328–337, 2002, doi: 10.1007/s00374-002-0476-2.

[10] Z. Elouear, F. Bouhamed, N. Boujelben, and J. Bouzid, "Application of sheep manure and potassium fertilizer to contaminated soil and its effect on zinc, cadmium and lead accumulation by alfalfa plants," *Sustain. Environ. Res.*, vol. 26, no. 3, pp. 131–135, 2016, doi: 10.1016/j.serj.2016.04.004.

[11] S. Mbarki *et al.*, "Performance of Medicago sativa Grown in Clay Soil Favored by Compost or Farmyard Manure to Mitigate Salt Stress," *Agronomy*, vol. 10, no. 1, pp. 1–14, 2020, doi: 10.3390/agronomy10010094.

[12] W. J. Nauta, G. J. Burgt, and T. Baars, "Partner farms: a participatory approach to collaboration between specialised organic farms.," in *Designing and testing crop rotations for organic farming. Proceedings from an international workshop*, 1999, pp. 149–158.

[13] A. R. Safahani Langeroodi, E. Campiglia, R. Mancinelli, and E. Radicetti, "Can biochar improve pumpkin productivity and its physiological characteristics under reduced irrigation regimes?," *Sci. Hortic. (Amsterdam).*, vol. 247, no. September 2018, pp. 195–204, 2019, doi: 10.1016/j.scienta.2018.11.059.

[14] S. Savci, "An agricultural pollutant: chemical fertilizer," *Int. J. Environ. Sci. Dev.*, vol. 3, no. 1, p. 73, 2012.

[15] B. Marouane *et al.*, "Impact of agricultural practices on groundwater quality: Case of Gharb region-Morocco," *J. Mater. Environ. Sci.*, vol. 5, no. Suppl. 1, pp. 2151–2155, 2014.

[16] J. E. Fox, J. Gulledge, E. Engelhaupt, M. E. Burow, and J. A. McLachlan, "Pesticides reduce symbiotic efficiency of nitrogen-fixing rhizobia and host plants," *Proc. Natl. Acad. Sci.*, vol. 104, no. 24, pp. 10282–10287, 2007.

[17] E. Liu *et al.*, "Long-term effect of chemical fertilizer, straw, and manure on soil chemical and biological properties in northwest China," *Geoderma*, vol. 158, no. 3–4, pp. 173–180, 2010, doi: 10.1016/j.geoderma.2010.04.029.

[18] A. Bouizgaren *et al.*, "Assessment of summer drought tolerance variability in Mediterranean alfalfa (Medicago sativa L.) cultivars under Moroccan fields conditions," *Arch. Agron. Soil Sci.*, vol. 59, no. 1, pp. 147–160, 2013, doi: 10.1080/03650340.2011.606216.

[19] A. Kahyani *et al.*, "Performance of dairy cows fed diets with similar proportions of undigested neutral detergent fiber with wheat straw substituted for alfalfa hay, corn silage, or both," *J. Dairy Sci.*, vol. 102, no. 12, pp. 10903–10915, 2019, doi: 10.3168/jds.2019-16869.

[20] A. E. Francisco *et al.*, "Effects of alfalfa particle size and starch content in diets on feeding behaviour, intake, rumen parameters, animal performance and meat quality of growing lambs," *Meat Sci.*, vol. 161, p. 107964, 2020, doi: 10.1016/j.meatsci.2019.107964.

[21] H. F. Blaney and W. D. Criddle, *Determining consumptive use and irrigation water requirements*, no. 1275. US Department of Agriculture, 1962.

[22] H. El Moussaoui and L. Bouqbis, "Interactive Effect of Biochar and Bio-Compost on Starting Growth and Physiologic Parameters of Argan," *Sustainability*, vol. 14, no. 12, p. 7270, 2022.

[23] C. C. Sheaffer, C. B. Tanner, and M. B. Kirkham, "Alfalfa Water Relations and Irrigation," no. 29, pp. 373–409, 2015, doi: 10.2134/agronmonogr29.c11.

[24] S. Jafarian, M. R. Chaichi, and M. Moradi-Ghahderijani, "Effects of surfactant and limited irrigation on forage yield and quality of alfalfa ('Medicago sativa'L.)," *Aust. J. Crop Sci.*, vol. 10, no. 1, pp. 76–80, 2016.

[25] H. Yang, S. Li, H. Sun, W. Wang, and F. Zhao, "Effects of substrate material on plant growth and nutrient loss," *Polish J. Environ. Stud.*, vol. 27, no. 6, pp. 2821–2832, 2018, doi: 10.15244/pjoes/81069.

[26] T. Sun and Z. Li, "Alfalfa-corn rotation and row placement affects yield, water use, and economic returns in Northeast China," *F. Crop. Res.*, vol. 241, no. June, p. 107558, 2019, doi: 10.1016/j.fcr.2019.107558.

[27] R. Rashmi and V. Manisha Sarkar, "Cultivation of alfalfa (Medicago sativa L)," *Anc. Sci. Life*,

vol. 17, no. 2, p. 117, 1997.
[28] M. A. Schmitt, C. C. Shaeffer, and G. W. Randall, "Preplant manure and commercial P and K fertilizer effects on alfalfa production," *J. Prod. Agric.*, vol. 6, no. 3, pp. 385–390, 1993.
[29] V. Vasileva and O. Kostov, "Effect of mineral and organic fertilization on alfalfa forage and soil fertility," *Emirates J. Food Agric.*, vol. 27, no. 9, pp. 678–686, 2015, doi: 10.9755/ejfa.2015.05.288.
[30] J. Daliparthy, S. J. Herbert, and P. L. Veneman, "Dairy manure applications to alfalfa: Crop response, soil nitrate, and nitrate in soil water," *Agron. J.*, vol. 86, no. 6, pp. 927–933, 1994, doi: 10.2134/agronj1994.00021962008600060001x.
[31] J. E. Ells and A. E. McSay, "Allelopathic Effects of Alfalfa Plant Residues on Emergence and Growth of Cucumber Seedlings," *HortScience*, vol. 26, no. 4, pp. 368–370, 2019, doi: 10.21273/hortsci.26.4.368.

Geoenvironmental characterization of two REE deposits: the Montviel carbonatites and Kipawa silicates, Quebec Canada

M EDAHBI[1,2*]; B PLANTE[2]; M BENZAAZOUA[2]; A CAYER[3]

[1] Ecole supérieure de Technologie (ESTFBS), Université Sultan Moulay Slimane, Maroc
[2] Université du Québec en Abitibi-Témiscamingue (UQAT, Rouyn-Noranda J9X 5E4, QC, Canada
[3] Ressources Géoméga, St-Lambert, Canada

*Corresponding author: Mohamed.edahbi@uqat.ca

Abstract. Drainage water quality is the significant environmental concern for the rare earth elements (REE) mining industry. REE deposits are associated with other metals and radioactive bearing minerals. REE mining and refining activities can generate significant quantities of liquid and solid wastes. Therefore, a long-term integrated approach covering the full mine-life cycle is required to mitigate possible environmental concerns. In the present study, two REE concentrates were prepared and all deposit lithologies of carbonatites and silicates sampled and investigated for their mineralogy, geochemistry, and their environmental behavior using kinetic testing. For the Montviel carbonatite (enriched in light rare earth elements, or LREE), the majority of REE-bearing minerals are associated with carbonates (i.e., monazite, kukharenkoite, burbankite, etc.), whereas the REE-bearing minerals associated with the Kipawa silicates (enriched in heavy rare earth elements, or HREE) are fluorbritholite, eudyalite, mosandrite, etc. The kinetic tests showed a neutral to alkaline pH of leachates and a low leachability of REE (carbonatites <140 μg/L; silicates <15 μg/L) with a higher mobility of HREE than LREE. The reactivity of REE carbonates are one to two orders of magnitude higher than REE silicates. For sustainable mineral development, geological and environmental data was integrated into the geometallurgical model to identify and control the environmental risks associated with mining those two deposits.
Keywords: rare earth elements (REE), carbonatites, silicates, geoenvironmental characterization, REE leachability, and REE mobility.

1. Introduction

In recent years, the government of Canada has taken a very strong interest in rare earth elements (REE) deposits [1, 2, 3, 4] leading to the production of promising results during the detection of various REE deposits [5, 6, 7]. For example, in Québec, several REE mining projects are in development (e.g., Montviel carbonatite, Kipawa silicates, etc.). The Montviel carbonatite and Kipawa silicates deposits are two REE-bearing ores located in Abitibi-Témiscamingue, Quebec, Canada [7, 8]. These REE deposits are geologically different, notably in terms of mineralization; Montviel being a classical carbonatite, which consists of a series of mafic/ultramafic rocks, enriched in light rare earth elements (LREE, La to Sm), whereas Kipawa is much more singular, with silicate minerals forming a sill with an enrichment in heavy rare earth elements (HREE, Eu to Lu). The mineral reserves were estimated at 250 million tonnes (Mt) of REE-Nb carbonatites with a REE content 1.47% of rare earth oxides (REO) and

19 Mt of material with an average grade of 0.41% REO for Montviel carbonatites [9] and Kipawa silicates [10] , respectively. The pre-feasibility study of these two REE ores projects indicates that their extraction involved an open-pit method due to ore body morphology, overburden height, low-grade, and large-tonnage REE ores [9, 10]. However, the mining and refining of these REE deposits could generate significant quantities of liquid and solid waste [11]. The solid wastes can be magnetic minerals, non-magnetic minerals, and/or minerals from fine-grained flotation tailings. These wastes are a principal environmental concern for the REE mining industry, as a result of its effect on drainage water quality. Moreover, these REE deposits are associated with some radioactive elements such as thorium (Th), and other elements recognized by their toxicity like arsenic (As), strontium (Sr), barium (Ba), nickel (Ni), and niobium (Nb), etc. Consequently, waste disposal and management strategies must take into account these environmental challenges to ensure the physical and chemical stability of the mine wastes. As a result, the mining and refining of these REE ores require environmental precautions by adapting a long-term integrated approach that could cover the full mine-life cycle.

Environmental contamination can occur when different contaminant (REE, Sr, Ba, etc.)-bearing minerals stored in the surface are in contact with mine drainage. Since the contaminants are associated with carbonates, sulfides, and silicates, the contaminants could be released in the leachates during water-rock interactions through dissolution and oxidation phenomena. For the Montviel carbonatites, a previous study revealed that Ba, Sr, and F show increased leaching trends with time [12]. Laboratory kinetics tests indicate that pH remains neutral to alkaline despite mineralogical variabilities among the humidity and weathering cells. Regarding the metals and metalloids, their concentrations are below the regulatory limits of Directive 019. In order to predict the environmental behavior of these two REE deposits, site characterization is crucial to localize the zone with high contamination potential and understand the type of contamination that can occur.

The Montviel and Kipawa REE deposits are composed of different lithologies characterized with a wide variety of REE content (0.12-3% Montviel carbonatites; 0.05-1.17% Kipawa silicates). For this reason, the geoenvironmental characterization is considered one of the most reliable approaches allowing to assess the early environmental impacts of REE mine wastes. This approach requires a combination of different methods of characterization (mineralogy and geochemistry) to investigate with more accuracy the mineralogy and the environmental geochemistry of the selected samples and identify their related environmental hazards.

In the present study, the geoenvironmental characterization approach of these two REE deposits was based on five main steps; 1) selection and preparation of representative samples from different lithologies, 2) preparation of REE concentrates using gravity separation and magnetic separation, 3) submission of REE concentrates to weathering cells to test their geochemical responses, 4) (geo)chemical and mineralogical characterization of the selected samples, and 5) integration of the obtained data into the geoenvironmental model to identify potential sources of REE for Montviel carbonatites.

2. Material and Methods

2.1. *Site study and sampling*

The Montviel site is located approximately 40km of Lebel Sur-Quevillon, Quebec, Canada. The REE carbonatites are associated with mafic/ultramafic rocks little metamorphozed and deformed [8,9], while the REE ores deposits of Kipawa are located at 150 km south of Rouyn-Noranda, Quebec, Canada. The mineralization is associated with a peralkaline intrusion and occurs in silicates minerals varying from meso to mafic syenites and impure marbles [4,13].

Three lithologies from the Kipawa silicates, called syenite (SY), calcite-poor (CP), calcite-rich (CR), and three lithologies from the Montviel carbonatites, called calciocarbonatites (CaC), ferrocarbonatites (FeC-LG), and ferrocarbonatites (FeC-HG) (Figure 1), were sampled using two sampling methods: drilling for Montviel carbonatites and bulk samples for Kipawa silicates. Approximately 10 kg from each lithology were collected and transferred into high density polyethylene (HDPE) pails. Due to the low amount of REE-bearing minerals within the samples, two REE concentrates, called magnetic concentrate form Montviel (MCM) and magnetic concentrate form Kipawa (MCK), were prepared using a combination of techniques (gravity and magnetic separation) to preconcentrate REE-bearing minerals and simulate the extreme scenario for REE leaching during rock-water interactions. The selected samples were dried at room temperature, crushed, homogenized, divided into different small aliquots, and stored in sealed plastic bags and analyzed for their mineralogy and geochemistry.

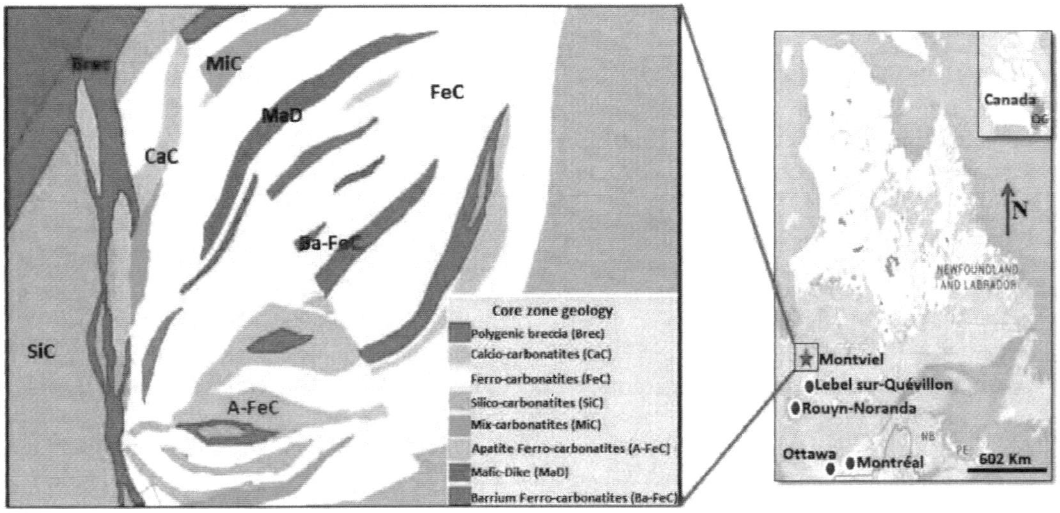

Figure 1. Location and geological setting of studied samples

2.2. Geochemical and mineralogical analyses

The bulk geochemical composition of the selected samples was determined using a combination of chemical techniques including Inductively Coupled Plasma Atomic Emission Spectroscopy and Mass Spectroscopy (ICP-AES/MS) (Perkin Elmer Optima 3100-RL ICP-AES and Agilent 7700 ICP-MS), an X-ray Fluorescence (XRF) (Bruker, Tiger Model) analysis for the whole rock, and an induction furnace analyzer (ELTRA CS-2000) for sulfur/carbon analysis. After multi-acid digestion using HCl, HNO_3, HF, and Br_2, the digests obtained were analyzed to determine metalloids and REE concentrations using ICP-AES/MS.

The mineralogical composition and textural properties of the studied samples were assessed using a couple of mineralogical methods such as scanning electron microscope equipped with a microanalysis system (SEM-EDS), electron probe micro-analyzer (EPMA), and quantitative mineralogy (Quantitative Evaluation of Materials by Scanning Electron Microscopy QEMSCAN). Overall, the use of these methods aims to determine mineral identification, mineral association, mineral texture, mineral liberation, and mineral quantification. All the mineralogical analyses were performed on polished sections prepared carefully to ensure homogeneity and avoid the preferential sedimentation of heavy minerals (e.g., REE-bearing minerals and sulfide minerals). All these mineralogical and geochemical data were used to build a geoenvironmental characterization model (Figure 2).

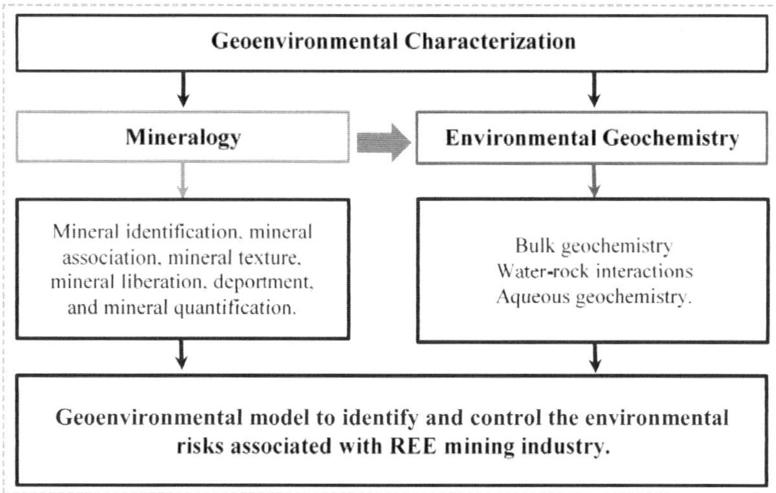

Figure 2. Geoenvironmental characterization methodology

3. Results and Discussion

3.1. *Bulk geochemistry of solid samples*

An overview of the geochemical composition of the studied samples is shown in figure 3. Total sulfur (ST) in Montviel lithologies is an average of 0.54% (Figure 3), whereas, in Kipawa lithologies, it is an average of 0.03%. These values suggest that sulfide minerals are present only as trace minerals in those REE deposits. Total carbon (CT) values for Montviel lithologies ranged from 5.7 to 9.77wt% with a maximum content in ferrocarbonatites lithologies. Unlike the Montviel lithologies, CT in the Kipawa silicates varied between 0.03wt% and 6.36wt% with maximum in CR. The static tests (net neutralization potential = neutralization potential – acid potential) calculated based on CT and ST values and previously published indicate that the lithologies form those REE deposits are not acid generating [7]. The CT comes mainly from calcite and dolomite in those two REE deposits. The CR lithology from Kipawa silicates show a similar CT content when compared to Montviel lithologies. This high CT content in CR lithology and Montviel lithologies possibly ensure a high neutralization potential to neutralize acidic mining effluents and favor the natural mitigation of the contaminants through various sorption phenomena.

The whole rock assays (Figure 4) show that all Kipawa lithologies have the same order of magnitude of SiO_2 (56-70%), except in the CR lithology, where SiO_2 exhibits the lowest content value (19%). Fe_2O_3 and MgO contents ranged from 0.5 to 9% and 0.8 to 24.8% respectively. The contents of Al_2O_3, Na_2O, and K_2O ranged from 0.6-13.7%, 0.18-7.84%, and 0.36-5.03%. Unlike the Kipawa silicates, the Montviel lithologies show high CaO content (19-30%) compared to other oxides. MgO varied from 3.8 to 8.5%. Fe2O3 content ranged from 11.5 to 23.5 % in all lithologies with a maximum in ferrocarbonatites LG and HG. The Na_2O (0.4-1.8%) and K2O (0.5-2.5%) abundance is similar in all Montviel lithologies. Contents of Cr_2O_3, V_2O_5, P_2O_5, and TiO_2 were less than 1% in all lithologies for those two deposits. These results indicate that the mineralogical compositions of the Montviel lithologies are associated mostly with carbonates, whereas the mineralogical compositions of the Kipawa lithologies are associated with aluminosilicates.

Regarding REE concentration, the whole rock analysis by ICP-MS indicates that LREE (0.8330-2.9158 % and HREE (0.0081.3-0.0113 %) concentrations varied widely within individual Montviel lithologies, with maximum in FeC-HG. LREE and HREE concentrations within MCM are 3.6882 mg/L and 0.0041.2 %, respectively. Regarding the Kipawa lithologies, LREE (0.0412-0.0865 %) and HREE (0.0260-0.0613 %) concentrations were relatively similar within individual SY and CP with yttrium (Y 0.0219-0.0450 %) is the most abundant element within all Kipawa lithologies, whereas cerium (Ce

0.4200-1.4 %) is the most abundant element within all Montviel lithologies. The enrichment factor (ΣLREE/ΣHREE) shows a huge difference between Montviel (36.5-257) and Kipawa (1.41-1.58) lithologies. This difference is probably due to the formation mode of Montviel and Kipawa REE deposits and the geochemical properties of LREE and HREE. During fractional crystallization, HREE have a tendency to concentrate in tardive magmatic fluid contrary to LREE, which fractioned in the early stages of fractional crystallization. This difference between LHREE and HREE is probably due to difference in ionic radius (ionic radius of LREE >> ionic radius of HREE) and geochemical affinity to other chemical elements [5,14,15].

The hydrometallurgical treatment and valorization processes of REE concentrates (MCM and MCK) would generate different types of tailings magnetic, non-magnetic, and flotation tailings. These tailings must be managed separately to ensure effective and long-term environmental management. Magnetic and flotation tailings are the most problematic in terms of their high content of contaminants-bearing minerals (Ba-minerals, Sr-minerals, Zr-minerals, and Th/U-minerals) and sulfide minerals, notably pyrrhotite that is considered the most reactive mineral in sulfides family. In contrast, non-magnetic tailings were composed mostly of paramagnetic minerals (i.e., calcite, dolomite, pyrite, and aluminosilicates minerals). These interpretations must be taken into account during the design of the environmental management approach of these tailings. The contaminants occur as micro-inclusions within crystal lattices of the minerals and consequently, the mineralogical investigations are critical for environmental geochemistry to increase our understanding of long-term environmental behavior of these tailings.

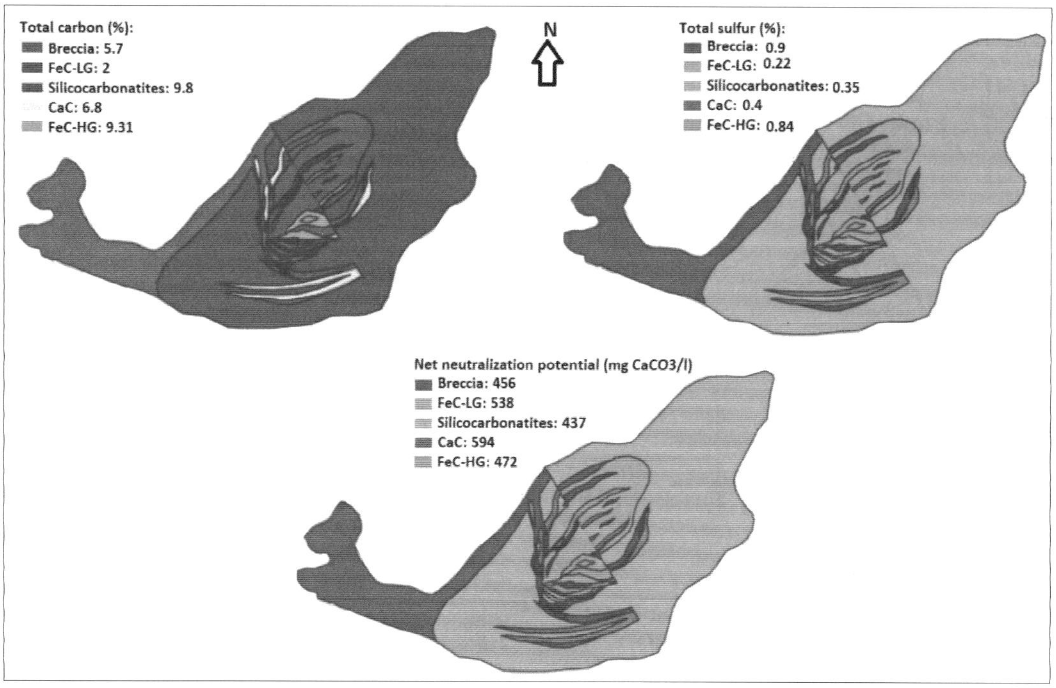

Figure 3. Distribution of total carbon, total sulfur, and net neutralization potential of Montviel lithologies.

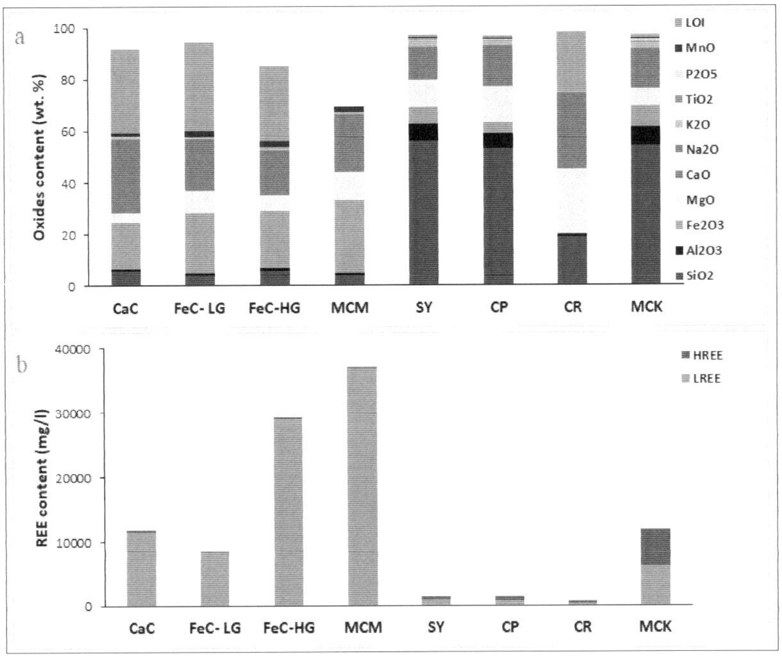

Figure 4. Bulk geochemistry of studied samples; loss on ignition (LOI) is not shown for MCM; a) represent distribution of minor and major elements and b) the distribution of REE within studied samples.

3.2. *Mineralogy and Environmental implications*

The mineralogical composition determined using QEMSCAN modal analysis shows that the Kipawa lithologies are mainly composed of aluminosilicates (Na-Mg Silicates, biotite, albite, and quartz), iron oxides, Fe-carbonates, and barite as trace minerals. REE minerals consist of mosandrite, fluorbritholite-Ce, monazite, and aqualite (Eudialyte Group Table 1). Moreover, low levels of REE are also detected in apatite, titanite and thorosteenstupine (Table 2). The mineral liberation data, calculated in the base of area percentage, shows that 26 to 48 % of combined REE Minerals are free (100%), except for the magnetic concentrate where the degree of liberation is around 58.5 %. The remaining minerals are associated with other minerals (36.5-69.6%) and/or encapsulated within other gangue minerals (4.3-10.3%).

In Montviel carbonatites, QEMSCAN analyses show that gangue minerals are dominated by carbonates including calcite (most common in CaC lithology), ankerite, siderite, Mn-siderite, and a (Ca, Mg, Fe)-carbonate. The main REE-bearing minerals detected in the studied samples are Kukharenkoite, burbankite, ancylite, carbocernaite, qaqrssukite, and apatite. The mineral liberation data of REE-bearing minerals shows that 15 to 26% is free (100%) and the remaining minerals are associated with other minerals (64-80%) and/or encapsulated within other gangue minerals (10-20%). Sulfide minerals are present as trace minerals in both those REE deposits (Table 1).

It is well known that environmental geochemistry is linked to the mineralogy of geological materials, particle size distribution, and the surface area [16] and mineralogical examinations can play key roles in understanding and managing lithologies under study. The mineralogical data of Montviel carbonatites lithologies suggest that qaqarssukite, norsethite, carbocernaite, olekminskite, pyrochlore, and barytocalcite are the main sources of REE but also of contaminants elements (i.e., Ba, Sr, Nb, and F). These elements can be released easily during natural processes (natural weathering) and they can have negative impacts for global and local ecosystems in case of the absence of careless waste disposal. Contrary to Montviel lithologies, the main detected contaminants within the Kipawa lithologies are Nb, Zr, Th, and U with their main sources being thorosteenstrupine, aqualite, fluorbritholite- (Ce), and mosandrite. These contaminants can be released in response to different natural processes (i.e., physical, chemical, and biological). REE waste management is very challenging due to the complexity of the associated mineralogy. A fundamental understanding of this mineralogical complexity requires further information about minerals (chemical crystal chemistry, structure, surface state, solubility, etc.).

Table 1. Mineralogy of Montviel carbonatites and Kipawa silicates using QEMSCAN modal analysis

	Montviel Carbonatites			
	CaC (%)	FeC-HG (%)	FeC-LG (%)	MCM (%)
Sulfides	0.7	1.5	1	7
Oxides (Fe, Mn, and Al oxides)	0.2	0.5	0.1	6
Carbonates	70.7	54	77.5	37.3
Silicates	13.9	18.5	12.3	7
Apatite	1.6	0.6	1.3	3.
Pyrochlore (Nb Ca Ox)	1.1	2.1	1.1	9.3
Norsethite (MgBa Carb)	0	0	0	1.1
Barytocalcite (Ba Ca Carb)	5	2.2	1.5	1.8
Monazite	0.04	0	0.03	0.9
Olekminskite (Ba Strontianite)	1.7	2.4	2	3.3
Qaqarssukite (Ba REE)	0.9	1.9	1.2	2.8
Kukharenkoite (Ce,La)	0.8	3.6	1.6	7.5
Burbankite	1.9	11	0.4	10.7
Ancylite	0.3	0	0.01	0
Carbocernaite (High Sr Ca Ba REE)	1	1.4	0.08	1.8
Other	0.02	0.01	0	0.06
Total	100	100	100	100
	Kipawa Silicates			
	SY (%)	CP (%)	CR (%)	MCK (%)
Sulfides	0.05	0.02	0.03	0.12
Barite	0.01	0.01	0.01	0.05
Fe Oxide/Metallic Fe	0.03	0.02	0.01	0.1
Carbonates	2.5	2.4	56.6	1.5
Quartz	0.3	0.1	0.08	0.7
Orthoclase	9.6	0.04	0	7
Albite	18	13.8	0	12.3
Other Plagioclase	0.1	0.09	0	1
Amphiboles/Pyroxenes	54.5	62	6.1	57.9
Biotite	10.6	15	3.8	8.9
Chondrodite	0	0	33.2	0
Titanite	0.1	0.5	0.02	0.6
Other Silicates	0.8	0.2	0.02	1.1
Pyrochlore	0.08	0.02	0.01	0.3
Apatite	0.3	1.5	0	0.4
Mosandrite	0.2	0.03	0	4.1
Fluorbritholite - (Ce)	0.09	0.07	0.06	0.5
Monazite	0	0.06	0	0
Aqualite	0.05	0.01	0	1.3
Thorosteenstrupine	0.08	0.01	0.01	0.1
Reidite (Zr Silicate)	1.2	3.8	0.08	1.2
Catapleiite	1.3	0.05	0	0.5
Total	100	100	100	100

To calculate accurate deportments of REE, EPMA compositional data was collected and recombined with the QEMSCAN mineralogy. More specifically, the EPMA allows for determination of the REE contents within small spots ($\simeq$ 20 μm) using wavelength-dispersive spectroscopy (WDS). In Montviel carbonatites, EPMA data show that all REE-bearing minerals contain only LREE (Ce, La, Sm, Nd, and Pr) with Ce is the most abundant element. Fe, Sr, Ba, and F are present as traces elements in all analyzed REE-bearing minerals, except for barytocalcite where Sr and F are under detection limits. Overall, the REE concentration range from 0.44 to 53.62 wt%, with Ce and La reaching up to about 28.1wt% and 13.47wt% in monazite. Nd concentration range from under detection limit in Olekminskite,

barytocalcite, and Pyrochlore) and 8.28wt% in monazite. Ce, La, Sm, Nd, and Pr are under detection limits in barytocalcite and Pyrochlore respectively.
Regarding the Kipawa silicates, EPMA data show that REE concentration range from 0.73wt% to 46.1wt% with maximum in monazite and minimum in titanite. Y, Zr, and Nb are detected in mosandrite, fluorbritholite, and titanite. Other contaminants (Th, U) occur as micro-inclusions within thorosteenstrupine, mosandrite, and fluorbritholite minerals. These contaminants are probably integrated into the crystal lattices of these minerals due to a cationic exchange process, notably between REE, Ca, and radioactive elements (Th, U).

Table 2. EPMA compositional data of Montviel carbonatites and silicates; values in grey are below the detection limits of the instrument; all values are in wt%.

Montviel Carbonatites												
	Fe	Ce	La	Ba	Sm	Nd	Pr	Ti	Nb	P	Sr	F
Detection Limits	*0.09*	*0.37*	*0.24*	*0.34*	*0.15*	*0.14*	*0.14*	*0.03*	*0.13*	*0.04*	*0.05*	*0.06*
Kukharenkoite	0.8	10.9	6	42.1	0.2	3.3	1.1	0.05	0.01	0.06	0.8	2.2
Burbankite	0.2	8	4.9	19	0.1	2.3	0.8	0.03	0.01	0.01	14.1	0.4
Qaqarssukite	1	16.6	8.5	20	0.5	5.6	1.7	0.02	0.03	0.01	3.5	2.4
Carbocernaite	1	9.3	4	5.5	0.2	3	1	0.01	0	0.01	28.1	0.5
Monazite	1.2	28.1	13.5	0.5	0.6	8.7	2.7	0.02	0	12.7	1.1	1.3
Ancylite	2.5	19.9	9	0.9	0.15	5.7	2	0.01	0.03	0.01	17.1	1.4
Olekminskite	0.5	0.3	0.11	3.2	0.05	0.08	0.09	0	0	0.01	50.2	0.2
Barytocalcite	0.2	0.15	0.05	47.2	0.07	0.06	0.11	0	0	0.02	0.04	0
Pyrochlore	0.4	0.25	0.1	0.05	0.05	0.11	0.05	1.8	47.4	0	0.8	6.3
Apatite	0.7	0.5	0.2	0.15	0.2	0.5	0.11	0	0	17.4	1	4.6
Kipawa Silicates												
	Ce	La	Sm	Nd	Pr	Ti	Nb	P	Y	Zr	Th	U
Detection Limits	*0.37*	*0.24*	*0.15*	*0.14*	*0.14*	*0.03*	*0.13*	*0.04*	*0.14*	*0.11*	*0.12*	*0.1*
Mosandrite	3.9	1.2	0.6	1.9	0.5	6	0.7	0	4.8	0.3	0.3	0.05
Monazite	28	14.2	1	9.4	2.8	0.02	0	12.5	0	0	0	0
Fluroapatite	1.8	0.9	0.13	0.8	0.2	0	0	17.1	0	0	0	0
Fluorbritholite	21.7	12.3	1.4	4.3	1.9	0.01	0.01	0.36	1.8	0	1.1	0.1
Aqualite	0.6	0.3	0.10	-0.02	0	0.2	0.5	0	1.3	9.5	0.01	0
Titanite	0.4	0	0.09	0.15	0.02	21.6	0.8	0	0.2	0.11	0.02	0
Thorosteenstrupine	0.5	0.21	0.07	0.11	0	0.01	0.02	0	0.2	0	28.2	0.7

The environmental behavior of Montviel carbonatites and Kipawa silicates are clearly different in the light of these mineralogical findings. The mineralogical composition is different and consequently, the mineral reactivity is different (Figure 5); however, the REE-bearing mineral liberation within all studied lithologies are comparable across lithologies (Montviel and Kipawa). To compare the reactivity of REE-carbonatites and that of REE silicates, a REE carbonate concentrate and a REE silicate concentrate were submitted to weathering cells. The results of leaching behavior revealed strong differences between REE carbonates and REE silicates in terms of normalized cumulative charges. The obtained values vary from one to two orders of magnitude higher than REE silicates. Since these two concentrates have the same characteristics (i.e., grain size distribution, the liquid / solid ratio, the degree of liberation, etc.), the differences in terms of normalized cumulative charges could be due mainly to the difference in reactivity between the two concentrates. The REE carbonate concentrate appears to be more reactive than REE silicate concentrate.

The geochemical behavior of studied Montviel lithologies were tested using humidity cells [5]. REE released concentrations were stabilized between 0.15 and 9 µg/L after 147 days of leaching [5,7]. Due to the low leachability of REE, only the results of the kinetic tests of REE concentrate are presented. The leachates of REE concentrate represent the worst possible scenario of REE leaching into natural landscapes. Previous studies indicate that sorption phenomena and precipitation of secondary minerals may control the fractionation of REE as suggested by thermodynamic equilibrium calculations [5]. The kinetic tests results of REE concentrates using weathering cells showed a neutral to alkaline pH of leachates (pH > 7.5) and a low leachability of REE (carbonatites <180 mg/Kg; silicates <95 mg/Kg) with a higher mobility of HREE than LREE. All elemental concentrations were under regulatory limits (directive 019). However, barium showed an increasing trend, which requires long-term monitoring of its environmental behavior. These results are coherent with previous results from humidity cells [5]. Regarding the radioactive elements, these elements are analyzed only for Montviel humidity cells leachates. The detected radioactive isotopes are radium (Ra-226 and Ra-228) and uranium (U-234, U-235, and U-238). The results show that the concentrations of these radioactive elements are below the Directive 019 and Canadian Guidelines for the Management of Naturally Occurring Radioactive Materials (NORM) limits (Ra 5 Bq/L; U 1 Bq/L). Several points such as low reactivity of radionuclides-hosting minerals and REE-bearing minerals and their fractionation with kinetics tests can explain their low concentration in the leachates. The mobility and bioavailability of REE and radionuclides can be controlled by pH-Eh conditions, temperature, mineral solubility, sorption phenomena [17, 15].

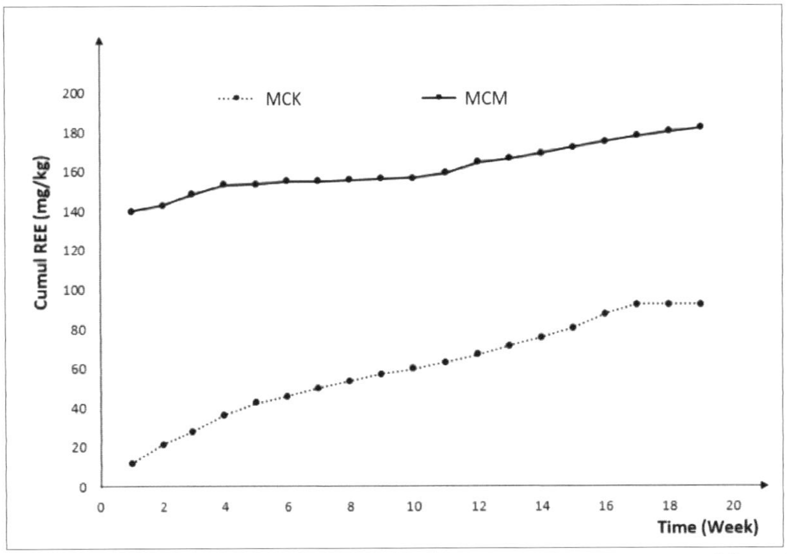

Figure 5. Comparison of leaching behavior of two REE concentrates prepared from Montviel carbonatites and Kipawa silicates.

3.3. *Waste management and disposal*

As demonstrated in section (bulk geochemistry), the REE mine wastes are not acid generating, but they could generate contaminated drainage (CND). Consequently, the restoration techniques of acid mining drainage (AMD) are not suitable for CND. Several options can be used for REE mine wastes disposal. The experimental design of the future REE waste rock dumps must be taken into account mineralogical, geochemical, ore morphology, and physical proprieties of the REE waste rocks. In this subsection, only the design of the Montviel lithologies dump will be shown (Figure 6). To increase the chemical and physical stability of the dump, the authors suggest to compact the REE waste rocks to favor runoff and limit water percolation. As previously demonstrated [5], the silicocarbonatites lithology is relatively characterized by low content of metal(loid)s-bearing minerals and high contents of carbonates compared to FeC-LG and FeC-HG lithologies. Moreover, during the REE ore exploitation, the mining operation will start by excavating this lithology (silicocarbonatites) surrounding the REE ore deposit, which will serve as the first layer of the dump (Figure 6). The excavated materials will be put at the base of the dump as a natural filter. Moreover, the authors suggest depositing the first layer of silicocarbonatites with an inclination slope 20-30% to accelerate the runoff of water.

The CaC lithology, which is closed to REE ore deposit, is also characterized by low content of sulfides, low content of REE minerals, low content of contaminants-bearing minerals. Moreover, this lithology (CaC) also contains a high content of calcite, which is known for its high potential neutralization and sorption capacity of metal(loid)s [18]. For these reasons, the authors suggest putting their materials as limestone drain to treat any residual contamination at the outlet of the waste rock pile before the effluents reach the natural environment. This passive treatment approach will cost less compared to the active treatment of drainage water.

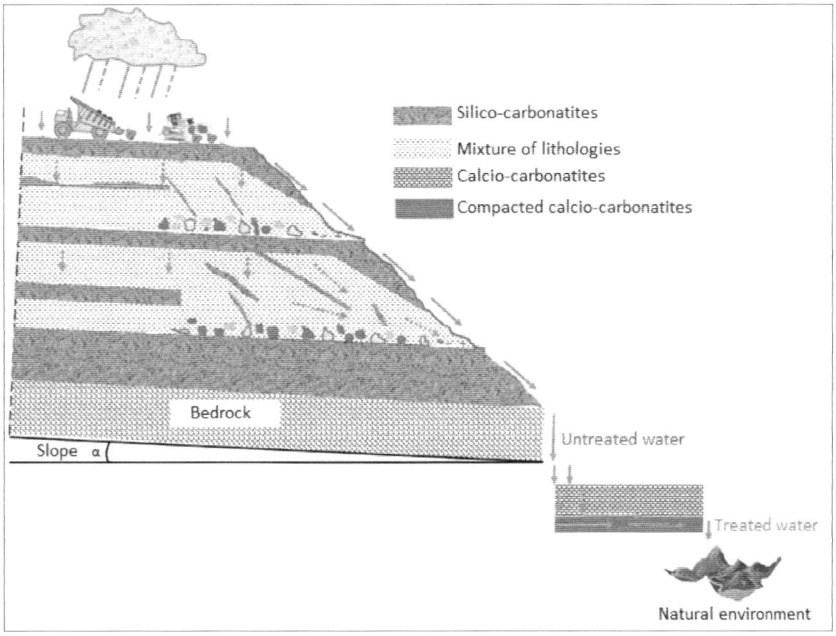

Figure 6. Conceptual model proposed for waste management and disposal of Montviel REE carbonatites.

4. Conclusion

Geoenvironmental characterization based on mineralogical investigations, bulk geochemistry, and prediction tests could be a robust approach allowing to understand and predict the long-term environmental behavior of REE mine wastes. Overall, the mineralogical assessment using QEMSCAN, EPMA analysis, and bulk geochemistry allow to determine several parameters such as modal mineralogy of REE-hosting minerals, mineral distribution, REE liberation, and deportment, which can control the environmental behavior of REE mine wastes. Studying all REE ores lithologies separately enhanced our understanding of the importance of REE mine wastes mineralogy and geochemistry. Consequently, this geoenvironmental characterization approach allowed for differentiation between problematic and non-problematic lithologies. This integrated approach can cover the full mine-life cycle starting from the preparatory stage of exploration to closure of mine. It can also help to ensure long-term environmental management of these tailings and identify their associated risks.

5. References

[1] Barakos, G. (2017). An assessment tool for the mineability of rare earth element deposits. Technische Universität Bergakademie.
[2] Barakos, G., Mischo, H., & Gutzmer, J. (2016). Legislation; challenges and policy strategies: in search for a regulatory framework for sustainable development in the rare earths mining industry. In Conference: SME Annual Meeting and Mining Engineering.–Conference Paper.–Project:

Mining Valuation of REE deposits. At: Phoenix, Arizona, USA.–February.
[3] Parliament, C. (2014). The rare earth elements industry in Canada– Summary of evidence. Standing Committee on Natural Resources. Retrieved October 14, 2015.
[4] Constantin M, Fleury F (2011) The Kipawa alkaline syenite complex and its rare earth element deposit. Abstract presented at the GAC-MAC Annual Meeting, Ottawa, Ontario, Canada.
[5] Edahbi, M., Plante, B., Benzaazoua, M., & Pelletier, M. (2018a). Geochemistry of rare earth elements within waste rocks from the Montviel carbonatite deposit, Québec, Canada. Environmental Science and Pollution Research, 25(11), 10997-11010.
[6] Edahbi, M., Plante, B., & Benzaazoua, M. (2019). Environmental challenges and identification of the knowledge gaps associated with REE mine wastes management. Journal of cleaner production, 212, 1232-1241.
[7] Edahbi, M., Benzaazoua, M., Plante, B., Doire, S., & Kormos, L. (2018b). Mineralogical characterization using QEMSCAN® and leaching potential study of REE within silicate ores: A case study of the Matamec project, Québec, Canada. Journal of Geochemical Exploration, 185, 64-73.
[8] Nadeau, O., Cayer, A., Pelletier, M., Stevenson, R., & Jébrak, M. (2015). The Paleoproterozoic Montviel carbonatite-hosted REE–Nb deposit, Abitibi, Canada: Geology, mineralogy, geochemistry and genesis. Ore Geology Reviews 67, 314-335.
[9] Desharnais, G., and Duplessis, C. (2011). Montviel core zone REE mineral resource estimate technical report, Quebec. Geomega Resources Inc. Member of SGS group (SGS SA).
[10] Roche in collaboration with SGS, Golder and Genivar. (2012). Feasibility study for Kipawa project, Temescamingue Area, Québec, Canada. http://www.matamec.com/vns-site/uploads/ documents/ matamecfs 2013sedar.pdf (accessed on August 15th, 2017).
[11] Hurst, C., (2010). China's rare earth elements industry: What can the west learn?, Institute for the Analysis of Global Security Washington DC.
[12] Edahbi, M. (2018). Prédiction du comportement géochimique des rejets miniers porteurs de terres rares (Doctoral dissertation, Université du Québec en Abitibi-Témiscamingue).
[13] Saucier, G., Noreau, C., Casgrain, P., Côté, P., Larochelle, E., Bilodeau, M., ... & Kissiova, M. (2013). NI-43-101 report-feasibility study for the Kipawa project is not represented area, Québec, Canada. Matamec Explorations Inc., Montréal.
[14] Fernández-Caliani, J. C., Barba-Brioso, C., & De la Rosa, J. D. (2009). Mobility and speciation of rare earth elements in acid minesoils and geochemical implications for river waters in the southwestern Iberian margin. Geoderma, 149(3-4), 393-401.
[15] Yusoff, Z. M., Ngwenya, B. T., & Parsons, I. (2013). Mobility and fractionation of REEs during deep weathering of geochemically contrasting granites in a tropical setting, Malaysia. Chemical Geology, 349, 71-86.
[16] Anawar, H. M. (2015). Sustainable rehabilitation of mining waste and acid mine drainage using geochemistry, mine type, mineralogy, texture, ore extraction and climate knowledge. Journal of environmental management, 158, 111-121Erguler, Z. A., & Erguler, G. K. (2015). The effect of particle size on acid mine drainage generation: Kinetic column tests. Minerals Engineering, 76, 154-167.
[17] Hierro, A., Bolivar, J. P., Vaca, F., & Borrego, J. (2012). Behavior of natural radionuclides in surficial sediments from an estuary impacted by acid mine discharge and industrial effluents in Southwest Spain. Journal of environmental radioactivity, 110, 13-23.
[18] Alkattan, M., Oelkers, E. H., Dandurand, J. L., & Schott, J. (1998). An experimental study of calcite and limestone dissolution rates as a function of pH from− 1 to 3 and temperature from 25 to 80 C. Chemical geology, 151(1-4), 199-214.

Energetic valorization of municipal solid waste in Morocco: financial feasibility of refuse derived fuel production

A Ouigmane[1,2*], O Boudouch[2], A Hasib[2], O Ouhsine[1], E Abba[3], R J Isaifan[4], M Aadraoui[5] and M Berkani[1]

[1] Team of Applied Spectro-Chemometry and Environment. University of Sultan MoulaySlimane, BeniMellal, Morocco
[2] Team of Agro-Industrial and Environmental Processes. University of Sultan of MoulaySlimane, BeniMellal, Morocco
[3] Team of Biotechnology, Analytical Sciences and Natural Resources Management,Higher School of Technology Khenifra, University of Sultan MoulaySlimane, BeniMellal, Morocco
[4] Division of Sustainable Development, Hamad Bin Khalifa University, Qatar Foundation, Education City, P.O. Box 5825, Doha, Qatar
[5] Georesources and Environment team, Faculty of Science and Technology,University of Sultan of Moulay Slimane, Beni Mellal, Morocco

*ouigmaneabdellah@gmail.com

Abstract. The management of municipal solid waste is considered as a problem of municipalities in developing countries due to its impact on the environment and on human health. Energy recovery is considered as a sustainable waste management solution. The objective of this paper is to carry out a financial feasibility study of refused derived fuel (RDF) recovery in a region in Morocco. The results show that the cement plant requires 24,000 tons of RDF to substitute 15% of the pet coke used to produce 1 million tons of clinker per year. Thus, 14,000 tons of pet coke will be saved annually in this case. The results of comparing the net present values (NPV) of three investment scenarios show that the project is profitable if the investor is the cement plants or the group of municipalities with a selling price of 36 USD/ton of RDF. On the other hand, if the landfill company is the investor, the project becomes profitable from a sale price of 39USD/ton of RDF.

Keywords: Municipal solid waste, RDF, Cement plant, NPV, Morocco

1. Introduction

The cement sector is considered as one of the most fossil fuel-intensive industries [1], and in order to achieve the 2030 sustainable development targets (SDS), fossil fuels must be replaced by alternative fuels and other renewable energy sources. Recent studies have reported on the economic and environmental benefits of the co-incineration of alternative fuels (RDF) in cement kilns [3-4]. In Morocco, the cement sector is undergoing an evolution in terms of cement production, with twelve production plants distributed throughout the country with an annual production capacity of 21 million tons [5], making it the industrial sector with the highest energy consumption at 30% [6]. Some cement plants use different types of waste to substitute pet coke, such as the cement plant in the city of Oujda that uses coal mine wastes, scrap rubber tires or the sludge from waste water treatment plants [7]. The majority of cement plants use pet coke to provide the thermal energy necessary for clinker production despite the pollution generated, thus, pet-coke imports are subject to an Internal Consumption Tax (ICT) of 8.73USD/ton, a Para fiscal tax on imports of 0.25% and value added tax (VAT) on imports of 20% [8].

On the other hand, the production of municipal solid waste (MSW) continues to increase as a result of improved living standards and population trends. According to the Word Bank, household waste generation will increase from 2.01 billion tons in 2016 to 3.40 billion tons in 2050 [9]. Thus, the most commonly used disposal method in developing countries is landfilling despite its negative impacts on the public health and ecosystems [10]. Waste to energy conversion shows several economic and environmental benefits [11-12]. In developed countries; studies on RDF have been carried out and have shown results encouraging benefits for the co-incineration of RDF. These include the reduction of waste volume, saving of fossil fuels, incorporation of residues in the composition of clinker and the reduction of CO_2 emitted by fossil fuels [13-14]. Hence, the use of RDF as an alternative fuel in cement kilns show many economic advantages for the cement factories [4, 15, 16]. However, before investing in an RDF production project, a financial study is necessary. The literature review on the financial aspects of RDF production for use in cement kilns in Morocco and Africa revealed a lack of data and research on this subject. The financial benefits of RDF depend on a number of factors such as the RDF production chain, waste composition, calorific value, etc. In addition, there are many significant financial unknowns in the production and use of RDF [17]. The most adopted method in order to conduct financial studies is the NPV calculation [18]. In research, the author conducted a financial study comparing several RDF production scenarios [19]. They found that only the plants directly using the produced RDF for energy production were found to be economically viable. Another financial study was conducted by [20] on the production of RDF in a region in Italy showed that the production cost of RDF per unit of mass was between 0.1 and 0.2 €/kg. These values can be reduced if the plant is able to treat also other wastes such as biomass.

Waste recovery can be carried out by the public sector or by the private sector specialized in waste management. Thus, cement plants can install RDF production units for energy recovery. Generally, privatization of waste management is more efficient in terms of service management [21].

The objective of this study is to carry out a comparative financial feasibility of RDF production in Morocco according to three different investment scenarios. The RDF produced will be co-incinerated in cement kilns by substituting petroleum coke. The method used is the calculation of NPVs and profitability indices for each scenario.

2. Materials and methods

2.1. Waste management in the study area

The province of Khenifra is located in the center of Morocco and is made up of 22 communes (Figure 1) with an estimated population of 315,552 inhabitants in 2020 [22].

The management of household waste is a major obstacle for the territorial municipalities and especially in the phase of elimination. Waste collection in the province is carried out by municipal services, with the exception of the cities of Khénifra and Mrirt, where the service has been delegated to a private operator. Regarding waste disposal, since 2017, an evolution in the mode of waste management in the province of Khénifra is noticed, moving from landfilling in dumps to controlled landfilling. The landfill and recovery center has a storage capacity of 51. 100 tons of household and similar waste per year, i.e. 140 tons per day, the center will have a positive social, environmental and economic impact. The center will allow to extract recycled materials as plastic, glass, aluminum, paper, scrap metal, reduce greenhouse gas emissions as well as the amount of landfilled waste, preserve natural resources, improve the working conditions of waste collection agents, while creating new employment opportunities for the local workforce, which currently numbers about 22 employees and 6 supervisors. Despite the efforts made by the municipalities and the Ministry of Environment, the current waste management has problems and does not meet expectations, as the energy and material contained in the garbage is lost underground. In addition, the center is poorly located, in fact its geographical location can cause several impacts on health and on environment. Therefore, it is necessary to look for sustainable and profitable solutions to take advantage of the material and energy without causing any impact on the environment.

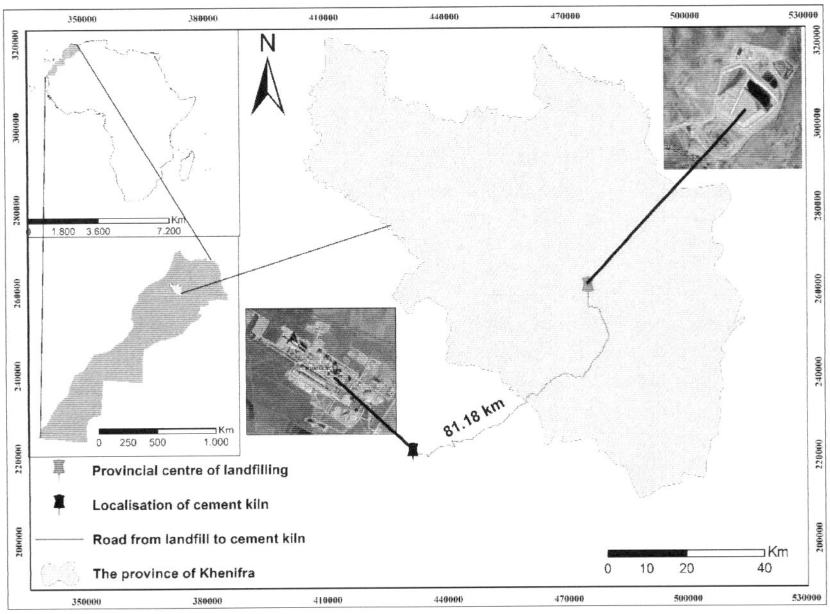

Figure 1. Situation of the study area and the cement plant

2.2. Characterization of RDF in the study area

The characteristics of RDF produced from municipal solid waste in the study area are presented in table 1 [23].

Table 1. Characteristics of RDF in the study area

Parameters	value	unit
LCV	18	MJ/kg
Moisture content	14	%
Ash content	6.5	%
Chlorine content	0.21	%
C	44	%
O	43	%
H	4	%
N	1	%
S	0.07	%
Pb	229	ppm
Cd	0.63	ppm
As	2.72	ppm
Cr	53.49	ppm

2.3. Calculation method

The purpose of the paper is to make a comparative study of three investment scenarios in RDF production for energy recovery in a cement kiln near the study area. For this purpose, NPV and profitability indices (PI) are calculated for each scenario using Eqs. (1) and (2) [19, 20]:

$$NPV= -I0+\sum_{i=1}^{n} CF(1+t)^{(-i)} \qquad (1)$$

$$PI=\sum_{i=1}^{n} CF(1+t)^{(-i)} / I \qquad (2)$$

Where:

I0: Initial investment; CF: Cash-flow; n: Lifetime, t: Discount rate; PI: profitability index.

NPV measures the absolute advantage that can be derived from an investment project. It constitutes:

- A rejection criterion for each project for which it is negative;
- A selection criterion between two projects (Retention of the one with the highest NPV).

The PI is a rejection criterion for any project whose index is less than 1. For two or more projects, the one with the highest profitability index will be chosen.

2.4. Scenario1

2.4.1. *Investment context.* The investor in the RDF production project in the first scenario is the landfill delegate. It will reserve a space within the landfill to build an RDF production unit. The product will be sold to the nearest cement plant at a competitive price compared to fossil fuels.

2.4.2. *Turnover*. The gain made by the delegate in this investment is the return from the sale of RDF to the cement plant. Since there is no market for the sale of RDF in Morocco, a price of 36 USD/ton of RDF has been estimated. Thus, the potential of combustible waste that could be used in the production of RDF in the study area is calculated using Eq.3.

$$QRDF=RP*Pm*(1+(PG/100))\ n\text{-}m\ *365*\ RRDF \qquad (3)$$

Where: QRDF is the quantity of RDF produced per year in all communes; RP: Ratio of daily waste production in Morocco 0.76 kg per capita per day; Pm: Population of year m, (as per population census in Morocco in 2014) [22]; PG: Population growth rate; RRDF: The percentage of RDF in the waste (0.28%) [24].

The turnover is deducted using Eq.4.

$$Turnover\ S1 = QRDF * the\ estimated\ price\ of\ RDF\ (USD/ton) \qquad (4)$$

2.5. Scenario 2

2.5.1. *Investment context.* The second scenario states that the cement plant produces RDF for use in its kiln as a substitute for pet coke. The investor buys a piece of land next to the landfill to build an RDF production plant.

2.5.2. *Turnover.* The percentage of substitution of petroleum coke by RDF used in the calculation of this scenario is 15%. This percentage is the most optimal from a technical and economic point of view according to several studies [4, 15, 25]. The cement plant's gain lies in the saving in pet coke following its substitution by 15% of RDF. The saved petroleum coke will be multiplied by 83USD/ton. The calculation data of the petroleum coke saving over ten years is shown in Table 2.

Table 2. Petroleum coke saving by the cement plant

Year	2020	2021	2022	2023	2024	2025	2026	2027	2028	2029
Annual clinker tonnage (T/year)	1000000	1010000	1020000	1030000	1040000	1050000	1060000	1070000	1080000	1090000
Quantity of RDF required at 15% (T/year) *	24000	24240	24480	24720	24960	25200	25440	25680	25920	26160
Quantity of petroleum coke required 85% (T/year)**	83000	83830	84660	85490	86320	87150	87980	88810	89640	90470
Quantity of petroleum coke required 100% (T/year)	97000	97970	98940	99910	100880	101850	102820	103790	104760	105730
Petroleum coke saved (T/year) ***	14000	14140	14280	14420	14560	14700	14840	14980	15120	15260

* The amount of RDF is calculated based on the energy required to produce one kilogram of cement (730 kcal/kg) [26] and on the RDF ICP of the study area determined in the laboratory according to EN 15400, 2011 [27] (4453 kcal/kg).

** The quantity of petroleum coke is calculated on the basis of the energy required to produce one kilogram of cement (730 kcal/kg)[26] and the ICP of petroleum coke (7500 kcal/kg).

*** The saving of petroleum coke is deduced by subtracting the amount of petroleum coke used without and with substitutions substitution

The turnover is deducted using Eq.5.

$$Turnover\ S2 = Petroleum\ coke\ savings * the\ price\ of\ petroleum\ coke \qquad (5)$$

The price of petroleum coke taken into consideration is 83 USD/ton which is variable and depends on the price of oil.

2.6. Scenario 3

2.6.1. *Investment context.* The investor in the RDF production project is the association of municipalities in the province. Land will be purchased for the construction of the RDF production plant. The combustible fractions will be used to produce RDF, while the non-combustible fractions will be sent to the landfill. The RDF produced will be sold to the nearest cement plant at a competitive fossil fuels price.

2.6.2. *Turnover*. The gain that the group of communities will make in this investment is the return from the sale of RDF to the cement plant and the gain in landfill costs. The sale price of RDF is 36 USD/ton and the landfilling price is 25 USD/ton. The same formula from Scenario 1 was used to estimate the annual RDF potential.
The turnover is deducted using Eq.6.

$$Turnover\ S3 = (QRDF * Price\ of\ RDF\ (USD/ton)) + (QRDFL * 25USD) \qquad (6)$$

Where: QRDF is the amount of RDF produced per year, QRDFI is the amount of waste minimized to landfill.

2.7. Estimated price of RDF

Due to the lack of a market for RDF sales in Morocco, the following formula was used to estimate the price per ton of RDF (Pedro, 2016).

$$RDF\ price\ (USD/ton) = FFP\ (USD/ton)*0.59 - OIC\ (USD/ton) \qquad (7)$$

Where:
PCF: Fossil fuel price, 83 USD/ton.
OIC: reflects the operating and investment costs resulting from technical adaptation to RDF use, equipment maintenance costs, waste co-incineration rate and investment amortization in infrastructure that consumer organizations are willing to pay.
0.59: is the ratio NCV RDF/ NCV pet coke.

2.8. Investment costs

Total capital investment is the sum of direct and indirect installation costs. It was calculated using the factor method [19, 28], as shown in Table 3. The investment I0 is the investment in the RDF production plant (Trommel sieve, conveyor, crusher, sorting equipment, small equipment...) and transport equipment (tipper and loader). The project life is 10 years.

Table 3. Factors for estimating the capital investment of the RDF unit

Cost element	Value
Cost of equipment purchased	A
Installation of equipment	0.3A
Instrumentation	0.15A
Piping	0.1A
Electrical Installations	0.15A
Buildings	0.25A
Services and site preparation	0.3A
Land *	0.05A
Total direct cost	**B=2.3A**
Engineering and supervisory manpower	0.1B
Construction Spending	0.1B
Overheads	0.1B
Total Capital Expenditures	**C=1.3B=2.99A**

* The price of land has not been taken into account for the first scenario

2.9. Operating costs

Operating and maintenance costs are generally divided into fixed operating costs (maintenance, operating labour, supervision, plant operation, laboratory costs, insurance and local taxes, etc.), variable operating costs (raw materials, miscellaneous operating equipment and transport, etc.), and general costs (administration, distribution and marketing costs, research and development costs, etc.). The annual operating costs were calculated using the factor method as shown in Table 4.

Table 4. Factors for estimating the operating costs of the RDF unit

Element of cost	Value
Energy consumption	D
Manpower	E
Maintenance manpower	0.1C
Equipment maintenance	0.015C
Local Tax	0.02C
Insurance	0.01C
Cost of treatment	F= (0.145C)+ D+(1E)
Cost of administration	(0.15 E)+ (0.15C)
Distribution and sales	0.05F
finance charges	0.05C
Transport of RDF	G
Disposal of process residues	H
Annual operating cost	**I= (0.352C)+(1.05D)+(1.2 E)+G+H**

3. Results and discussion

3.1. Price of RDF

The estimation of the price of RDF over 10 years of operation is shown in figure 2. The details of the calculations are presented in the appendix (Table A1).

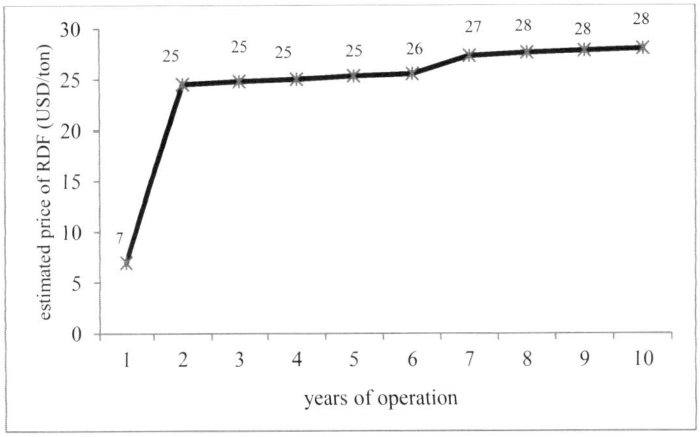

Figure 2. Results of price RDF estimation

It is found that the price of RDF increases with time, and could reach 28USD/ton in the tenth year of operation. A change in the quality of RDF and the price of petroleum coke can influence the selling price negatively or positively.

3.2. Turnover

The results of the turnover estimates for the three scenarios for 10 years are shown in figure 3.

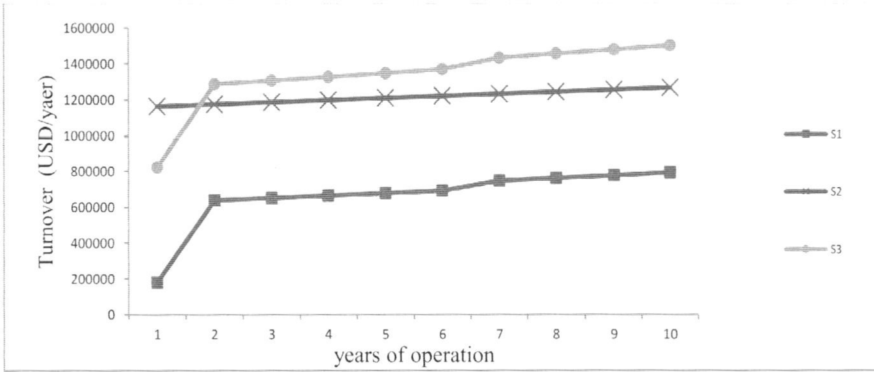

Figure 3. Turnover calculation results for the three scenarios

From the results of the calculation of turnover, it is noted that scenario 3 is the most advantageous; since the turnover could increase from 821,240 USD in the first year to 1,501,945 USD in the tenth year depending on the evolution of the selling price of RDF. On the other hand, the turnover of scenario 1 is small compared to the other scenarios with a figure of 179,102 USD in the first year that could reach 791,674 USD in the tenth year of operation.

3.3. *Capital and operating costs*

The results of the investment estimate and the results of the operating costs are shown in table 5.

Table 5. Estimation results of capital investment and investment in USD

	Scenario 1	Scenario 2	Scenario 3
Total Capital Expenditures	1,035,303	1,053,661	1,050,365
Annual operating cost	554,803	561,265	559,509

The capital and operating costs for the three scenarios are the same with the exception of Scenario 1, as the landfill delegate will use the landfill site to build the production unit. However, for scenarios 2 and 3, the land will be acquired by the investor. The detailed calculation of the investment and operating costs is shown in Tables A2 and A3 in the appendix.

3.4. *NPV and PI*

The comparative results of the three scenarios are shown in figure 4. The details of NPV calculations for the three scenarios are presented in tables A4, A5 and A6 in the appendix.

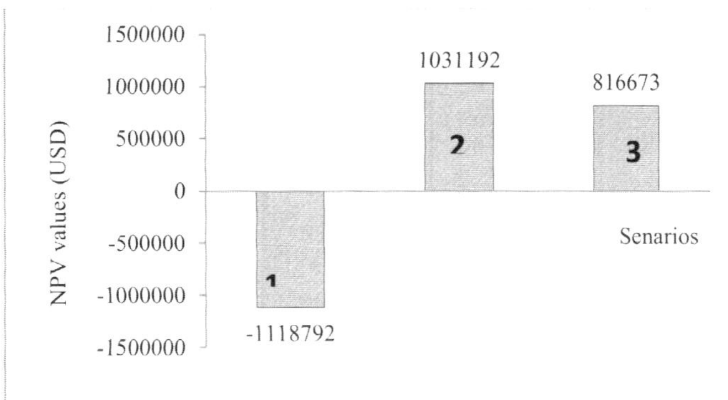

Figure 4. NPV comparison of the three scenarios

- The NPV found for Scenario 1 is -1,118,792 USD for a RDF selling price of 36USD/ton. The PI is not calculated since the NPV is negative ;
- The NPV calculated for scenario 2 is 1,031,191USD greater than zero, so the project is profitable. The calculated profitability index is 1.98 ;
- The NPV value for scenario 3 is 816,673 USD greater than zero, so the project is profitable at an RDF selling price of 36USD/ton. The calculated profitability index is 1.78.

From the comparative results, it can be seen that investment scenario 2 is the most favorable and advantageous with a higher NPV and a large PI due to the direct consumption of RDF produced by the cement plant without purchase. The third scenario is advantageous because the NPV value is relatively high with a profitability index that reaches 1.78. Hence, the new law of territorial communes in Morocco 113-14 [29] has given the possibility to communes to create companies. As a result, such an investment could help communes to alleviate the financial problems they face and at the same time manage waste in a sustainable way. On the other hand, scenario 1 is not profitable since the NPV found is less than zero which is due to the low selling price of RDF. In order to visualize the impact of the RDF price on the NPV and on the cement plant gain, a simulation was conducted with the pricing of RDF over 10 years of operation. The selling prices were varied from 3 USD/ton to 52 USD/ton of RDF. The results are presented in figure 5.

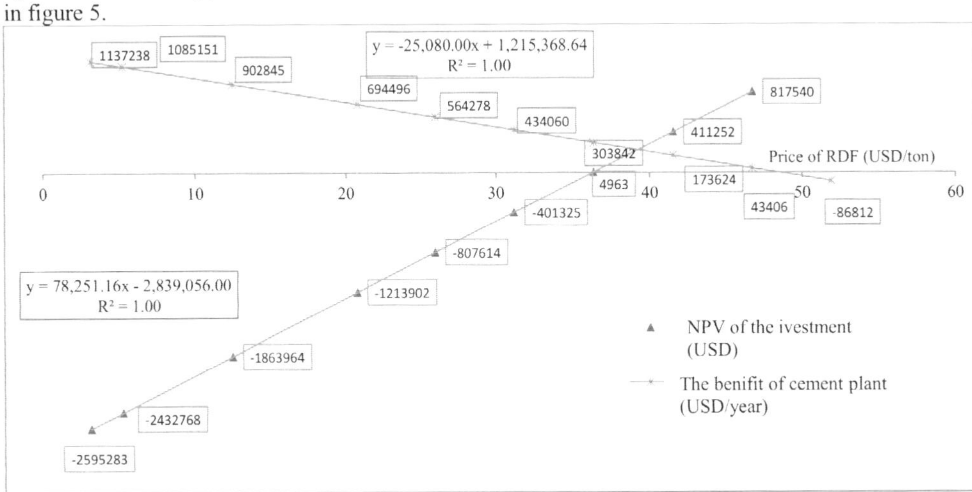

Figure 5. Impact of RDF sales price on NPV and economic gain of the cement plant

Starting at a price of 48.46 USD per ton of RDF, the cement plant neither benefits nor loses anything. Hence, a figure below this price is considered advantageous for the plant. On the other hand, if the price of RDF is higher than 48.46 USD/ton, petroleum coke becomes financially beneficial for the cement plant. The minimum value of RDF sales price for the project to be profitable for the company (scenario 1) is 36.24 USD/ton from the first year of investment, since a figure lower than this price makes the project unprofitable for the company. Other treatment lines can be added to the RDF production unit (composting, metal recovery, RDF-based electricity production unit, etc.). In research, several scenarios were studied in this respect and it is found that RDF production without other treatment lines is economically feasible. In Thailand, a feasibility study was carried out to assess the efficiency of electricity production from 220 tons per day of RDF, and it was suggested that a power plant project with a production capacity of 5-13 MW could be developed [30]. The authors of a study found that the production of RDF for use in cement kilns is economically viable with a net profit of 122.6 USD/ton of RDF [17]. In a cement plant in Turkey, it is found that it can save 9.4 USD/h if pet coke is substituted by 15% RDF (1,760 tons/year of RDF) [4]. In other study of a cement plant in Jordanie, it is found that it can save 486 USD/h with a 15% substitution rate equivalent to 35,424 tons/year of RDF [15]. For the case of the third scenario of this study, the cement plant will save annually an average of 311,410 USD over ten years. For scenario 2, the cement plant will save1,214,290 USD/year because the plant will use RDF directly without purchase. Table 6 shows the benefits of each scenario on the cement plant, the delegate of the landfill and the group of communes.

Table 6. Benefits related to the production and use of RDF in the cement industry

Scenarios	Communes	Cement factory	Delegate of the discharge
Scenario 1	**Economical :** No advantages ; **Environmental:** Reduction of waste buried in the territories of the municipalities; **Social:** Population satisfaction with the quality of service.	**Economical :** If the selling price of RDF is 36.24 USD/ton, the cement plant could make gains from the reduction of PC consumption; **Environmental:** Reduction of gaseous emissions from PCs; **Social:** No social benefit.	**Economical :** Gain made following the sale of RDF at 36.24 USD/ton ; **Environmental:** Increase of the landfill's life span and minimization of the impacts related to leachate and biogas; **Social:** Recruitment of labor.
Scenario 2	**Economical :** No advantages ; **Environmental:** Reduction of waste buried in the territories of the municipalities; **Social:** Satisfaction of the population with the quality of service.	**Economical:** Gain of the cement plant following the substitution of petroleum coke by the RDF it produces ; **Environmental:** Reduction of gaseous emissions from CPs; **Social:** Recruitment of manpower.	**Economical :** no advantage ; **Environmental:** increase of landfill life and minimization of impacts related to leachate and biogas; **Social:** no benefits.
Scenario 3	**Economical:** Gain following the sale of RDF and gain from the minimization of landfilling costs ; **Environmental:** Reduction of the waste buried in the territories of the communes. **Social:** Population satisfaction with quality of service and recruitment of workforce	**Economical :** Gain from purchasing RDF at a price competitive with fossil fuels; **Environmental:** Reduction of gaseous emissions from petroleum coke; **Social:** No social benefit.	**Economical :** no advantage ; **Environmental:** Increase of the landfill's life span and minimization of the impacts related to leachate and biogas; **Social:** No social benefit.

4. Conclusion

This study was designed to investigate the financial feasibility of producing RDF from MSW generated in the Khenifra province and to evaluate their use as an alternative fuel for clinker production. The results of the analysis of the three scenarios show that the investment in a project for RDF production by the group of communes and the cement plant is profitable. On the other hand, the project is not profitable for the delegate of the landfill with a sales price of 36 USD/ton. To make the project profitable for the delegated company, other chains of valorization must be added like composting and bio-methanization The results of this research provide key information on the long-term financial aspects of MSW treatment especially in areas close to cement plants.

Acknowledgements

This work is a part of the doctoral thesis of Ouigmane Abdellah. The authors would like to thank Ms. Charab Said for his help in the data used in this research

5. References

[1] Pieper C, Liedmanna B, Wirtza S, Scherera V, Bodendiekb N and Schaefer S 2020 Interaction of the combustion of refuse derived fuel with the clinker bed in rotary cement kilns: A numerical study Fuel **266**, 117048 DOI:10.1016/j.fuel.2020.117048

[2] IEA, International Energy Agency 2019 Cement Tracking Clean Energy Progress. Disponible at https://www.iea.org/tcep/industry/cement/[Accessed 30 December 2019]

[3]Çankaya S and Pekey BA 2019 comparative life cycle assessment for sustainable cement production in Turkey J Environ Manage **249**, 109362 doi: 10.1016/j.jenvman.2019.109362.

[4] Kara M 2012 Environmental and economic advantages associated with the use of RDF in cement kilns Resour Conserv Recycl**68**, 21-28

[5] APC Association professionnelle des cimentiers 2020 Secteur du ciment au Maroc. Disponible at: www.apc.ma[Accessed 2 Januray 2020].

[6] Fellaoui S and Bounahmidi T 2017 Evaluation of energy efficiency opportunities of a typical Moroccan cement plant: Part I Appl Therm Engdoi : http://dx.doi.org/10.1016/j.applthermaleng.2017.01.010

[7] Arfala Y, Douch J, Assabbane A, Kaaouachi K, Tian H and Hamdani M 2018 Assessment of heavy metals released into the air from the cement kilns co-burning waste: Case of Oujda cement manufacturing (Northeast Morocco) Sustain Environ Res **28**,363-373

[8] Lafarge Hoclim Maroc 2016 augmentation de capital de lafarge holcim marocs.a. au titre de la fusion-absorption de lafargecementoss.a. note d'information lafargeholcimmarocs.a.

[9] World Bank 2018 What a Waste 2.0 : A Global Snapshot of Solid Waste Management to 2050. Available from http://hdl.handle.net/10986/30317

[10] Sabour MR, Alam E and Mostafa Hatami A 2020 Environmental and economic assessment of Enhanced Landfill Mining in Tehran Environ Sci Pollut Res **27**, 34469–34483. https://doi.org/10.1007/s11356-020-09458-w

[11] Gu W, Liu D and Wang C 2021 Energy recovery potential from incineration using municipal solid waste based on multi-scenario analysis in Beijing. Environ Sci Pollut Res. https://doi.org/10.1007/s11356-021-12478-9

[12] Istratea IR, Iribarren D, Galvez-Martosa JL and Dufour J 2020 Review of life-cycle environmental consequences of waste-to-energy solutions on the municipal solid waste management system Resour Conserv Recycl**157**, 104778

[13] Scarlat N, Fahl F and Dallemand JF 2018 Status and Opportunities for Energy Recovery from Municipal Solid Waste in Europe Waste Biomass Valorization. https://doi.org/10.1007/s12649-018-0297-7

[14] Ramachandra TV, Bharath HA, Gouri K and Sun SH 2018 Municipal solid waste: Generation, composition and GHG emissions in Bangalore, India Renew Sust Energ Rev**82**, 1122–1136.

[15] Hemidat S, Saidan M, Al-Zu'bi S, Irshidat M, Nassour A and Nelles M 2019 Potential Utilization of RDF as an Alternative Fuel to be Used in Cement Industry in Jordan.Sustainability **11** : 5819; doi:10.3390/su11205819

[16] Hasib A, Ouigmane A, Boudouch O, El Kacmi R, Bouzaid M and Berkani M 2020 Sustainable Solid Waste Management in Morocco: Co-Incineration of RDF as an Alternative Fuel in Cement Kilns.

[17] Reza B, Soltani A, Ruparathna R, Sadiq R and Hewage K 2013 Environmental and economic aspects of production and utilization of RDF as alternative fuel in cement plants: A case study of MetroVancouver Waste Management Resour Conserv Recycl**81** :105– 114

[18] Shabib A and Abdallah M 2020 Life cycle analysis of waste power plants: systematic framework. *Int J Environ Stud*, DOI: 10.1080/00207233.2019.1708146

[19] Antonio CC and Pacifico MP 2002RDF production plants: I Design and costs Appl Therm Eng **22**, 423–437

[20] Di Maria F and Pavesi G 2005 RDF to energy plant for a central Italian region SUW management system: Energetic and economical analysis Appl Therm Eng **26**, 1291–1300

[21] Usman A, Isbah H and Muhammad N C 2016 Solid waste management practices under public and private sector in Lahore, Pakistan Bulletin environ stud **1**(4) 98-105

[22] HCP, Haut-commissariat au plan 2014 Population légale de la région par province et commune région de Béni Mellal-Khénifra. Accessed at www.hcp.ma

[23] Ouigmane A, Boudouch O, Hasib A, R Elkacmi, M Aadraoui and Berkani M 2021 Characterization of dry fractions contained in municipal solid waste in developing countries: a case study of a region in Morocco J Mat Cycl Waste Manag **23**:1227–1238

[24] Ouigmane A, Boudouch O, Hasib A and Berkani M 2018 Management of Municipal Solid Waste in Morocco: The Size Effect in the Distribution of Combustible Components and Evaluation of the Fuel Fractions Handbook of Environmental Materials Management https://doi.org/10.1007/978-3-319-58538-3_82-1

[25] Sakri A, Aouabed A, Nassour A and Nelles M 2021 Refuse-derived fuel potential production for co-combustion in the cement industry in Algeria Waste Manag Res doi 10.1177/0734242X20982277

[26] CIMAT 2017 Emission d'un emprunt obligataire non cote amortissable sur 7 ans Note d'information ; 2017.

[27] EN 15400 2011 Solid recovered fuels, Determination of calorific value.

[28] Peters MS and Timmerhaus KD 1991 Plant Design and Economics for Chemical Engineers, McGraw Hill.

[29] DGCL 2016 Loi organique relative aux communes au Maroc. Bulletin Officiel N° 6440 du 09 Joumada I 1437

[30] Srisaeng N, Tippayawong N, Korrakot Y and Tippayawong (2017) Energetic and economic feasibility of RDF to energy plant for a local Thai municipality Energy Procedia**110** : 115 – 120

6. Appendix

Table A1. Calculating of the estimated price of RDF

Year of operation	1	2	3	4	5	6	7	8	9	10
Price of petroleum coke PCF (USD/ton)	83	83	83	83	83	83	83	83	83	83
$PCI_{RDF}/PCI_{pet\ coke}$	0.59	0.59	0.59	0.59	0.59	0.59	0.59	0.59	0.59	0.59
Investment costs (USD/year)	526584	80789	80789	80789	80789	80789	39252	39252	39252	39252
Operating costs (USD/year)	1083769	638422	639460	640498	641537	642575	602077	603115	604154	605192
RDF potential in the study area (10^3 tons/year)	25,766	26,044	26,329	26,62	26,917	27,22	27,53	27,847	28,17	28,5
Estimated price of RDF (USD/ton of RDF)	7	25	25	25	25	26	27	28	28	28

Table A2. Estimated capital investment results in USD

Element of cost	Scenario 1	Scenario 2	Scenario 3
Cost of equipment purchased	280,374	280,374	280,374
Installation of equipment	84,112	84,112	84,112
Instrumentation	42,056	42,056	42,056
Piping	28,037	28,037	28,037
Electrical Installations	42,056	42,056	42,056
Buildings	70,093	70,093	70,093
Services and site preparation	84,112	84,112	84,112
Land *	_	14,019	14,019
Total direct cost	630,841	644,860	644,860
Engineering and supervisory manpower	63,084	64,486	64,486
Construction Spending	63,084	64,486	64,486
Overheads	63,084	64,486	64,486
Total Capital Expenditures	207,684	207,684	207,684
Cost of equipment purchased	1,027,778	1,046,002	1,046,002

IOP Conf. Series: Earth and Environmental Science 1090 (2022) 012014 doi:10.1088/1755-1315/1090/1/012014

Table A 3. Estimated operating cost results in USD

Energy consumption	Scenario 1	Scenario 2	Scenario 3
Manpower	12,461	12,461	12,461
Maintenance manpower	74,766	74,766	74,766
Equipment maintenance	7,477	7,477	7,477
Local Tax	15,417	15,690	15,690
Insurance	20,556	20,920	20,920
Cost of treatment	10,278	10,460	10,460
Cost of administration	165,382	168,115	168,115
Distribution and sales	12,187	12,319	12,319
finance charges	51,389	52,300	52,300
Transport of RDF	75,805	75,805	75,805
Disposal of process residues	10,384	10,384	10,384
Annual operating cost	558,878	565,297	565,297

IOP Conf. Series: Earth and Environmental Science 1090 (2022) 012014 doi:10.1088/1755-1315/1090/1/012014

Table A 4.NPV calculation results for scenario 1

Years	0	1	2	3	4	5	6	7	8	9	10
Turnover in USD (ET) *		179 102	638 103	651 018	664 236	677 760	691 596	747 284	761 755	776 550	791 674
Estimated operating costs in USD		557 185	557 632	558 671	559 709	560 748	561 786	562 825	563 863	564 901	565 940
Depreciation and amortization		512 565	80 789	80 789	80 789	80 789	80 789	39 252	39 252	39 252	39 252
Gross results		-890 648	-318	11 558	23 737	36 223	49 021	145 208	158 640	172 397	186 482
CT**		-267 194	-96	3 467	7 121	10 867	14 706	43 562	47 592	51 719	55 945
Net results		-623 454	-223	8 091	16 616	25 356	34 315	101 645	111 048	120 678	130 537
Self-funding capacity	**1 027 778**	-110 889	80 566	88 880	97 405	106 146	115 104	140 898	150 300	159 930	169 790
Increase in WCR***	**155 763**	186 916	207 684								
WCR recovery											550 363
Cash-flow	**-1 183 541**	-297 805	-127 118	88 880	97 405	106 146	115 104	140 898	150 300	159 930	720 153
Discounted Cash-flow	**-1 183 541**	-248 171	-88 276	51 435	46 974	42 658	38 548	39 322	34 955	30 996	116 309
Cumulativeash-flow	**-1 183 541**	-1431712	-1519988	-1 468553	-1421579	-1378922	-1340374	-1301052	-1266097	-1 235 101	-1 118 792

* The turnover was estimated by multiplying the tonnage of RDF by the selling price per ton
** Corporate tax (CT) is taken as 30%
***WCR was estimated by sorting RDF in the first year and second year of operation.

IOP Conf. Series: Earth and Environmental Science 1090 (2022) 012014 doi:10.1088/1755-1315/1090/1/012014

Table A5: NPV calculation results for scenario 2

Years	0	1	2	3	4	5	6	7	8	9	10
Turnover in USD (ET)	-	1163032	1174663	1186293	1197923	1209553	1221184	1232814	1244444	1256075	1267705
Estimated operating costs in USD	-	557185	557632	558671	559709	560748	561786	562825	563863	564901	565940
Depreciation and amortization	-	526584	80789	80789	80789	80789	80789	39252	39252	39252	39252
Gross results	-	79263	536241	546833	557425	568017	578609	630737	641329	651921	662513
CT	-	23779	160872	164050	167227	170405	173583	189221	192399	195576	198754
Net results	-	55484	375369	382783	390197	397612	405026	441516	448930	456345	463759
Self-funding capacity	1046002	582068	456158	463572	470987	478401	485815	480768	488183	495597	503011
Cash-flow	-1046002	582068	456158	463572	470987	478401	485815	480768	488183	495597	503011
Discounted Cash-flow	-1046002	485056	316776	268271	227135	192259	162699	134174	113536	96050	81239
Cumulativeash-flow	-1046002	-560946	-244169	24102	251236	443495	606193	740367	853903	949953	1031192

Table A6. NPV calculation results for scenario 3

Years	0	1	2	3	4	5	6	7	8	9	10
Turnover in USD (ET)	-	821240	1287184	1307193	1327660	1348590	1369989	1433400	1455757	1478603	1501945
Estimated operating costs in USD	-	557185	557632	558671	559709	560748	561786	562825	563863	564901	565940
Depreciation and amortization	-	526584	80789	80789	80789	80789	80789	39252	39252	39252	39252
Gross results	-	-262529	648762	667733	687161	707053	727413	831324	852642	874450	896753
CT	-	-78759	194629	200320	206148	212116	218224	249397	255793	262335	269026
Net results	-	-183771	454133	467413	481013	494937	509189	581926	596850	612115	627727
Self-funding capacity	1046002	342813	534923	548202	561802	575726	589979	621179	636102	651367	666979
Increase in WCR	155763	186916	207684	-	-	-	-	-	-	-	-
WCR recovery	-	-	-	-	-	-	-	-	-	-	550363
Cash-flow	-1201765	155897	327238	548202	561802	575726	589979	621179	636102	651367	1217343
Discounted Cash-flow	-1201765	129914	227249	317247	270931	231371	197583	173360	147937	126239	196608
Cumulativeash-flow	-1201765	-1071851	-844602	-527355	-256425	-25053	172529	345889	493826	620065	816673

Fish waste: Valorisation methods on a local scale

H Essabiri, R Damrani, O Boumalkha, T Hachi, C Laiboud, and E H Abba*

Superior School of Technology of Khenifra (EST-Khenifra), University of Sultan Moulay Slimane, BP 170, Khenifra 54000 Morocco.

*Corresponding author: e.abba@usms.ma

Abstract. Fish waste management is being neglected in most parts of the world. Unfortunately, with the worldwide increase of demand, this has generated many environmental impacts. This negligence can be traced to the ridicule of the value of this resource as well as the many difficulties related to its valorization. This work suggests the most efficient methods to valorize fish waste using biological and ecological means. With minimal investments so that they can easily be adopted by local farmers and landowners. These methods can either use bacterial action such as composting to create bio-fertilizer, ensilage to manufacture liquid silage or lastly, the production of fish feed to minimize the cost and the detrimental effects of fish farming.

Key words: Fish waste; Environmental impacts; Valorisation; Waste management.

1. Introduction

Fishery and the aquaculture industry have developed rapidly in recent years, unfortunately this increase generates loads of fish waste. In fact, according to FAO, among all the quantities of fish caught or farmed in 2018, only 43% is treated as final products fit for human consumption, while 57% is deemed as waste. Fish waste is a highly fermentable product who deteriorates very quickly due to various factors. The spoilage may be caused by the metabolic activity of microorganisms, endogenous enzymatic activity and by the chemical oxidation of lipids [1], the low acidity (PH>6) and the high amount of non-protein nitrogenous compounds found in fish waste also create favorable conditions for the fast growth of the degrading microorganisms.

Traditionally the handling of this category of waste was done by casting it back in aquatic ecosystems, as it originally came from, to be feed for other species without any processing. However, with the increase of the production rate this model of management can no longer be executed, as the large quantities of organic matter rich in microorganisms and enzymes will deteriorate the aquatic ecosystems and change their whole dynamics.

Fortunately, new directives came to light all around the world to regulate the management of food waste and the animal by-products not intended for human consumption, which include fish waste. Generally, the management hierarchy priorities the prevention of waste generation, followed by processing for reuse and recycling, with disposal as the least favored stage of waste management [2]. As such fish waste can be exploited as a resource for agricultural applications, using simple eco-friendly methods that can be easily implemented in local farms and small land businesses, without advanced technologies or important investments.

The methods chosen in this paper are; composting as the fish waste is rich of nutrients such as N, P and Ca[2]. Also, fish silage is an interesting application for fish waste due to its high contents of enzymes that will be catalyzed by strong acids to create a feed supplement for animal breeding. The last method is the production of fish feed from the unused fish scraps using a simple procedure without any machinery or complex process. These methods will encourage small landowners to start their own aquaculture projects to ensure local food security without any impacts on the environment.

2. Fish Compost

Composting is the decomposition of the unused fish parts under oxygen-rich conditions, into carbon dioxide, water, heat, and humus. The process is similar to the composting of organic waste as it goes through the same steps, and can also reduce the volume of fish waste by 70% [3]. The mesophilic phase where the decomposition starts by mesophilic microorganisms in moderate temperatures. Then as it raises above 45°C the mesophilic microorganisms become less competitive and get replaced by thermophilic ones. At 54°C the pathogens start to decay; it is necessary to keep the temperatures under 65°C to avoid killing all the microorganisms as they are necessary to complete the process.

Finally, during the maturation phase the Temperature gradually decreases and the mesophilic fungi and actinomycetes colonize the compost for the breakdown and transformation of humic substances and lignin[4]. Fish wastes are rich of nitrogen phosphorus which makes the final product obtained from the composting more interesting as a soil amendment to improve its texture and fertility and thus becoming a substitute of synthetic fertilizers used in agriculture. Physiochemical parameters such as temperature, moisture content, C/N ratio need to be adjusted and monitored during the whole process. Table 1 [2], shows the composition of fish compost as dry matter.

Table 1. Chemical composition of fish waste compost, in dry matter [2]

Parameter	Unit	Value
Total organic C	g/kg	382.8
Total nitrogen	g/kg	11.4
C/N ratio	-	33.4
Phosphorus	g/kg	2.61
Potassium	g/kg	2.61
Magnesium	g/kg	3.07
Sodium	g/kg	0.31
Calcium	g/kg	3.75
Copper	g/kg	19.3
Cadmium	g/kg	1.52
Chromium	g/kg	5.4
Lead	g/kg	18.8
Nickel	g/kg	9.91
Zinc	g/kg	159.9

To obtain the correct C/N ratio, it is necessary to add a bulking agent such as wood chips or pine bark as used in this case. This will allow the ratio to be in the acceptable norms and also create ventilation circuits needed to oxygenate the mix.

The fish compost fabrication process is very flexible and can be varied to satisfy different needs. As different components can be added to the mix such as seaweeds, showing a strong residual effect of the compost. This mix is suitable as organic soil fertilizer and may be recommended for improving horticultural crop yields [5]. The method of creating fish compost can also be altered to meet different needs, such as the use of different species of worms to enable the vermicompost process and create stable manure as soil amendment easily absorbed by plants [6].

3. Fish silage

Fish silage is a liquid product made from whole fish or parts of fish that are liquefied by the action of enzymes originally found in the fish in the presence of an added acid. The principle behind the manufacturing of fish silages is that the enzymes already present in the fish are spread throughout the entire mass by grinding and mixing done before the start of the process, then the acidity is adjusted to be a catalyzer so as to favor the rapid action of these enzymes and to inhibit bacterial action [7]. Generally, there are two methods for production of fish silage:

- The standard method is as described previously, by adding acid, which lowers the pH sufficiently to prevent microbial spoilage by inhibiting the responsible microorganisms, shown by figure 1. The fish silage becomes liquid because the tissue structures are degraded by enzymes naturally present in the fish. [8]
- The second method employs bacterial fermentation; the process is initiated by mixing minced or chopped fish with a fermentable sugar which favors growth of lactic acid bacteria. These bacteria may be naturally present in the fish. It may be advisable, however, to add the bacteria as a starter culture. The lactic acid bacteria produce acid and antibiotics which together destroy competing spoilage bacteria. Sugar contributes to the preservation effect during the initial stages of fermentation by repressing bacterial production of enzymes which liberate ammonia from amino acids [8].

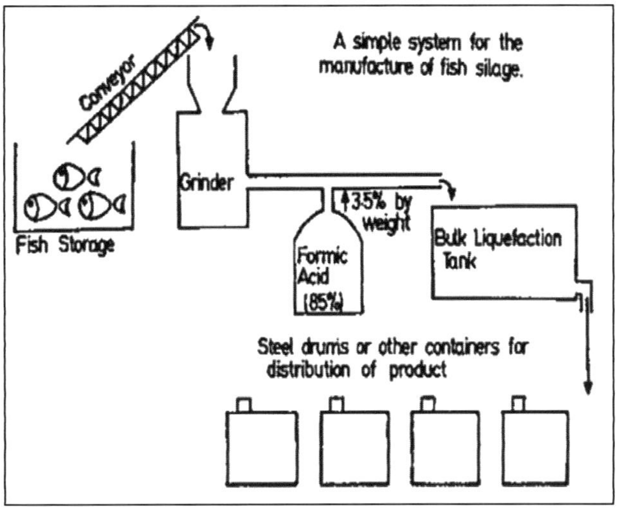

Figure 1. Making of acid preserved fish silage [1].

Hydrochloric acid or sulfuric acid can be used as a substitute for formic acid, as they are reasonably cheap, but a lower pH is required with these mineral acids than with organic ones. This automatically means greater corrosion problems, and in this case the final product has to be neutralized before its use. Therefore, formic acid, considered as an organic acid is a good choice because preservation is achieved at a slightly higher ph, so no neutralization is needed before use. Additionally formic acid has some bacteriostatic action that inhibits spoilage microorganisms' activities and help maintain the silage for longer periods.

Fish silage is used as animal feed supplement, as it is rich in proteins, oils and many other nutrients depending on the species and parts used as raw materials [9]. Table 2 demonstrates the chemical compositions of six types of fish silages using different species and different parts of fish [7].

Table 2. Chemical analysis of six types of fish silage.

	Types of silage					
Analysis	Whole sprat	*Whole herring*	Herring offal	*Whole sand-eels*	White fish offal	*Whole ungutted mackerel*
Crude protein (%) (Nx 6.25)	16.7	15.5	14.5	15.4	15	16.9
Oil content (%)	6.4	4.4	16.3	3.4	0.5	12
Moisture content (%)	74.3	77.7	68.1	77.7	78.9	70.2
Ash content (%)	2.7	2.1	2.6	2.4	4.2	2.1

The preparation of the six silages used the same process with formic acid as catalyzer. The composition of each silage is similar to the one of the raw materials from which it was made of [7].

Fish silage is an all-around animal feed and if correctly made can be safely stored for long periods. As long as the fish waste used as raw material is not rotten, the final product gives a pleasant malty smell during storage and the fabrication process is practically scentless. In appearance fish silage varies between brown gravy and a thin grey, gritty suspension, depending on the fish used for manufacture and the degree of digestion that has taken place[10]. Given that the product is made in large quantities, it is more beneficial that it's used locally or within a short distance of manufacturing site. Ensilage offers a simple and economical way to incorporate most types of unused fish parts and fish waste by converting them into a highly nutritious liquid feed, at a time when the world supply of quality proteins for animal feeding is expected to become both rare and expensive.

4. Fish feed production

Aquaculture is currently the world's fastest-growing food-production sector. For the past 25 years, this industry has experienced remarkable expansion. However, Feed formulation still represents a major restriction to the production rates, as feed alone corresponds to more than 60% of the overall expenses of the industry[11]. Currently fish meal and fish oil prices increased rapidly in the last years, for instance, fish oil prices increased by 115 % in period 2011–2015 from USD 1300 to 2800 per ton [12]. In order to confront this evolution, a low-cost preparation process based on fish waste is crucial. As fish require a high quality, nutritionally balanced diet to grow and maintain healthy populations, the most adequate method is the production of dry feed pellets. Contrary to the silage used in moist feeds, this method makes storage, transport and use easier and more flexible, thus more economical.

The research done previously, suggests that a dry fish feed pellet of high quality can be manufactured by a simple pressing method using a mixture of 50 % fish waste, 30_35 % pregelatinized wheat flour, and 15-20 % soya meal mixed with water (30_35 % of the sum of other ingredients) of 60-65°C for 4-5 min [13].

On a local scale, making fish feed is essentially the same process that is used to make fish balls for human consumption, with the only difference being the use of muscle tissue, bone, skin, cartilage, ligaments, fins, tail and other discarded fish parts. The entirety of these parts needs to be grinded and made into a paste of uniform sized particles. The most commonly used formula is 73% fish processing waste, 2% salt, 25% flour[1]. The pellets are then formed and cooked in hot water at a minimum of 80°C until flotation, then sun-dried to achieve the final product, the production process is demonstrated in the figure 2.

Figure 2. Fish feed formulation steps from waste on a local level [1]

This method as simple and basic as it seems, is capable of preserving the original amino acids and fatty acid profile present in the raw matters and have proved its advantages as a feed formulation technic, employable without any advanced technologies or investments therefore, perfect of local use by small landowners and farmers. The species targeted by this kind of feed formulation must have simple nutritional requirements, and generally be more resistant. Such as Nile tilapia (*Oreochromis niloticus)* the most commercially farmed fish in the world [14].

5. Conclusion

Fish waste should be considered as a very valuable resource with many potential uses. Its valorization doesn't necessary need advanced technologies or important investments, as it can be applied by local farmers or small fish markets. Contributing to reduce environmental impacts and ensuring food security in their communities. The choice of the optimal valorization method is generally based on the nature of fish used as raw material, and the role expected out of it.

6. References

[1] I. S. Boziaris, "Seafood Processing - Technology, Quality and Safety," p. 510 February 2014 ISBN: 978-1-118-34621-1.
[2] M. Radziemska, M. D. Vaverková, D. Adamcová, M. Brtnický, and Z. Mazur, "Valorization of Fish Waste Compost as a Fertilizer for Agricultural Use," Waste Biomass Valor, vol. 10, no. 9, pp. 2537–2545, Sep. 2019, doi: 10.1007/s12649-018-0288-8.
[3] M. E. López-Mosquera, E. Fernández-Lema, R. Villares, R. Corral, B. Alonso, and C. Blanco, "Composting fish waste and seaweed to produce a fertilizer for use in organic agriculture," Procedia Environmental Sciences, vol. 9, pp. 113–117, 2011, doi: 10.1016/j.proenv.2011.11.018.
[4] K. Kazemi, B. Zhang, L. M. Lye, and Z. Zhu, "Evaluation of state and evolution of marine fish waste composting by enzyme activities," Can. J. Civ. Eng., vol. 44, no. 5, pp. 348–357, May 2017, doi: 10.1139/cjce-2016-0377.
[5] M. Illera-Vives, S. Seoane Labandeira, L. M. Brito, A. López-Fabal, and M. E. López-Mosquera, "Evaluation of compost from seaweed and fish waste as a fertilizer for horticultural use," Scientia Horticulturae, vol. 186, pp. 101–107, Apr. 2015, doi: 10.1016/j.scienta.2015.02.008.
[6] H. Chaoui, "Vermicompostage (ou lombricompostage) : Le traitement des déchets organiques par les vers de terre," Nov 2010 revised Jan 2019 p. 8.

[7] I. N. Tatterson and M. L. Windsor, "Fish silage," J. Sci. Food Agric., vol. 25, no. 4, pp. 369–379, Apr. 1974, doi: 10.1002/jsfa.2740250404.

[8] J. Raa, A. Gildberg, and J. N. Olley, "Fish silage: A review," C R C Critical Reviews in Food Science and Nutrition, vol. 16, no. 4, pp. 383–419, Apr. 1982, doi: 10.1080/10408398209527341.

[9] M. Afreen and I. Ucak, "Fish processing wastes used as feed ingredient for animal feed and aquaculture feed," p. 10. July 2019. DOI:10.18331.

[10] I. N. Tatterson, "Fish silage — Preparation, properties and uses," Animal Feed Science and Technology, vol. 7, no. 2, pp. 153–159, Mar. 1982, doi: 10.1016/0377-8401(82)90050-5.

[11] K. Spalvins and D. Blumberga, "Production of Fish Feed and Fish Oil from Waste Biomass Using Microorganisms: Overview of Methods Analyzing Resource Availability," Environmental and Climate Technologies, vol. 22, no. 1, pp. 149–164, Dec. 2018, doi: 10.2478/rtuect-2018-0010.

[12] I. S. Arvanitoyannis and A. Kassaveti, "Fish industry waste: treatments, environmental impacts, current and potential uses," International Journal of Food Science & Technology, vol. 43, no. 4, pp. 726–745, Apr. 2008, doi: 10.1111/j.1365-2621.2006.01513.x.

[13] H. K. Das, M. T. Hattula, O. M. Myllymäki, and Y. Mälkki, "Effects of formulation and processing variables on dry fish feed pellets containing fish waste," J. Sci. Food Agric., vol. 61, no. 2, pp. 181–187, 1993, doi: 10.1002/jsfa.2740610208.

[14] W. Y. Mo, Y. B. Man, and M. H. Wong, "Use of food waste, fish waste and food processing waste for China's aquaculture industry: Needs and challenge," Science of The Total Environment, vol. 613–614, pp. 635–643, Feb. 2018, doi: 10.1016/j.scitotenv.2017.08.321.

Impact of bad odors from the wastewater treatment plant on the well-being of the population of M'rirt (Morocco)

T HACHI[1]*, M HACHI[2], H ESSABIRI[3], D BELGHYTI[1] and E H ABBA[3]

[1] Laboratory of Agro- physiology, Biotechnology, Environment, and Quality, Department of
Biology, Ibn Tofail University, Faculty of Science, Kenitra, Morocco
[2] Laboratory of Botany, Biotechnology and Plant Protection, Department of Biology, Facultyof Science, BP. 133, Ibn Tofail, Kenitra University, Morocco
[3] Higher School of Technology. Sultan Moulay Slimane University B.P. 170, 54000 Khénifra,
Morocco.

* hachitouria03@gmail.com

Abstract. In the city of M'rirt, the odour nuisance caused by the wastewater treatment plant (WWTP) has led to numerous complaints from residents whose quality of life has been affected by these odours. Therefore, the objective of our study is to evaluate, in the field, the "real" impact of bad odours emanating from the WWTP with the help of a questionnaire during a period of 6 months. Our study sample composed of 1473 respondents chosen in a simple random way from the two municipalities surrounding the said odour source. The survey results revealed that among the 1473 respondents 71 (n=1040) confirmed "yes" the odour of the WWTP presents a real olfactory annoyance and 29% said "no" and that several factors are involved in its prediction. The statistical analysis of our study confirmed the negative impact of WWTP odours on the well-being of the city's population, its spatial and temporal distribution with the establishment of a map of the olfactory discomfort in the city.

Key words: STEP of M'rirt, Olfactory pollution, Impact, Survey

1. Introduction

The sense of smell is one of the five senses that connect humans with their environment. It represents more than 3% of the human genetic code, meaning more than 1000 genes according to the work of Buck and Axel, who won the Nobel Prize in Physiology or Medicine in 2004 [1,2], and is manifested in a complex system. It is the oldest and earliest of our senses [3], and one of the most sensitive and evocative: an odor is enough to bring back very old memories and, in many cases, to change our temper [4].

An odor comes from a complex set of molecules present in the air that we breathe in and that the olfactory system perceives, analyzes and decodes [5]. This reflects the complexity of the phenomenon of odor perception, as well as the importance of the individual psychological factor. Therefore, the notion of smell is very subjective, its qualification as well as its quantification are not obviou Disturbances are what is commonly referred to as "environmental pollution" experienced by humans. At the individual level, they translate into a limitation of well-being, the potential or imminent presence of a risk, the lived experience of discomfort associated with one or more physically identifiable environmental components [6]. Any odor becomes a pollution as soon as It is perceived as an excessive disturbance by the population. It is then "olfactory pollution".

Bad odors represent a real problem that concerns all economic sectors. Wastewater, loaded with organic matter particles, nitrogen, Sulphur and phosphorus compounds, can generate unpleasant odors directly (by releasing highly volatile compounds) or indirectly (following a biological process of fermentation in a reducing environment) [7]. Sewage systems (collection system, wastewater treatment plants, by- products) are one of the sources of odor nuisance (Warren Spring Laboratory, USA). Since 1890, the Anglo-Saxon literature was interested in the problems of odor emissions near wastewater treatment plants. However, it is currently becoming more acute [7].

Since its opening in 2003, the M'rirt WWTP has become closely linked to the host community. In fact, its activity generates negative consequences, more likely to deteriorate the quality of their living environment due to the many olfactory nuisances and many diseases. Therefore, the objective of this study is to evaluate the environmental impact of the bad odors emitted by the WWTP on the neighboring population and to map the olfactory nuisance.

2. Material and methods

2.1. Location of the study area

M'rirt region is located in the eastern part of Central Morocco (also called Central Plateau), near the western edge of the Middle Atlas Causse [8]. It is accessible by the national road 08 joining Azrou to Khenifra, as well as by the main road 7032 that connects Meknes to M'rirt (Figure 1). The urban commune of M'rirt had a total population of 42,730 inhabitants in 2014 [9], spread over an area of 17 km 2. With a growth rate of 3.1% [10].

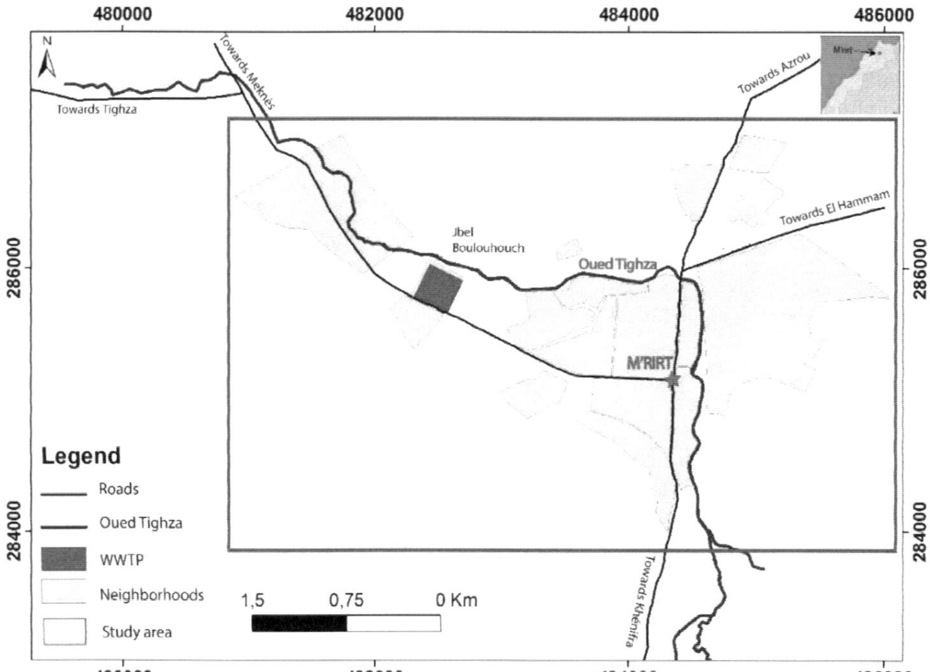

Figure 1. Location map of the study area

The climate of the city is continental with strong variations in temperature: semi-arid with a tendency to subhumid -humid for the rainy seasons or arid for the dry seasons. The average annual temperature of the region is close to 10°C. These temperatures present important daily and annual amplitudes with very hot summers and cold winters. The average interannual rainfall is around 700 mm. The prevailing winds in the region are generally moderately strong with an average speed of 9.3km/h. They are hot in summer and cold in winter and blow from southwest to northeast (gharbi) [10].

The WWTP of M'rirt is a natural lagoon type (4 anaerobic basins followed by 4 optional basins). It is located in the west at 1.8km from the center of M'rirt, 500m from the first settlements of Aït Ami Ali towards the east and less than 100m from Aït Boughzouan towards the south, on the left side of the main road 7032 that connects M'rirt and Meknes (Figure 1). This WWTP has been operational since June 2003, it is the first experience of the ONEE- water branch treatment plants and does not have any specifications. The wastewater of M'rirt transits to the WWTP via a main outfall. According to the sanitation master plan, the WWTP was designed on 9 ha to treat 1800 m $^{3}d^{-1}$ of wastewater but can treat up to 2250 $m^3.d^{-1}$(horizon 2013) and a pollutant load in BOD_5 of about 1200 kg.d-1 and which can reach 1580 kg.d-1 in horizon 2013 [11].

2.2. Methodology

The assessment of the olfactory impact of an already established installation requires an approach oriented to a direct estimate of the odor nuisance by residents because it allows to have the "real" level of the environmental impact [12,13]. In this perspective, several measurement methods were proposed to measure odor nuisance in the receiving environment, including the qualified nose panel, the analysis of complaints, the residents panel and the questionnaire.In order to search for almost all the information on the impact of the WWTP on the population, we adopted the survey method. Using a questionnaire form and a stratified sampling plan. Each stratum corresponds to a district or direction in relation to the WWTP (figure 2). Indeed, the northern direction is not taken into consideration because it is not populated, for the eastern direction (where there is the city of M'rirt) we chose 100 citizens as a sample representing the general population of each neighborhood, while for the southern and western direction, the proportion of the number of people in each sample is almost all the inhabitants who have an age of 16 years and over. The total sample of our study is composed of N=1473 people randomly selected from the population of the city of M'rirt. Thus, a proportional stratified random sampling is adopted (Table 1). Several investigations were carried out. They were extended over a period of 6 months between 2013 and 2014.

The completed surveys are transcribed into a database and processed by SPSS version 21 (System Professional for Social Sciences) software.

Table 1. Distribution of respondents by stratum.

Layer N°	Layer Name	Survey Numbers for each layer
Strate 1	Aït Ami Ali	100
Strate 2	Douar Hssayen	100
Strate 3	Mohamed N'toto	100
Strate 4	Boulouhouch	100
Strate 5	Boulachfare I	100
Strate 6	M'rirt Centre	100
Strate 7	Boulachfare II	100
Strate 8	Nahda	100
Strate 9	Koutoubia	100
Strate 10	Aït Mou	100
Strate 11	Tahjaouit	100
Strate 12	Douar Ghazouani	100
Strate 13	Aït Hajjou	100
Strate 14	Afoud Ikabarn	100
Strate 15	Aït Boughzouan Irchkiken	30
Strate 16	Aït Moussi	43
		N = 1473

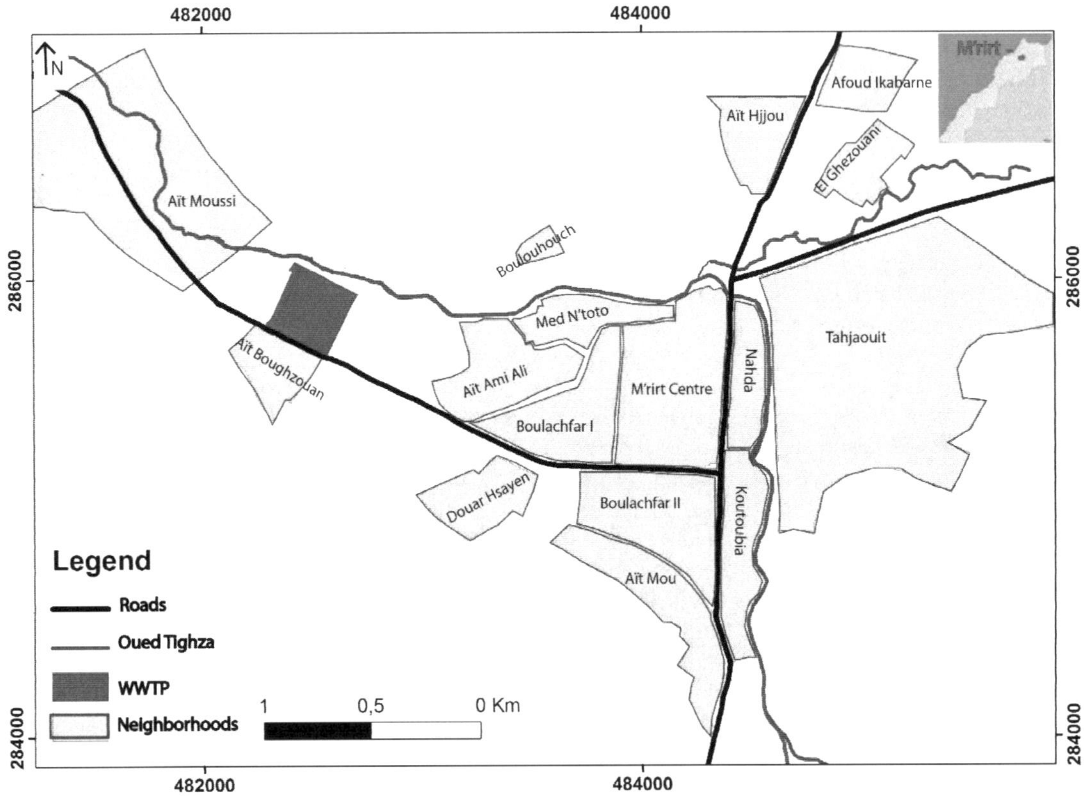

Figure 2. Map of the different neighborhoods surveyed.

3. Results and discussion

3.1. Profile of the informant

3.1.1. Descriptive Analysis of the Sample

The distribution of our sample showed that the average age of the respondents is 38.43± 0.38 years (with a minimum age of 16 years and a maximum age of 90 years), of which 95% are between 38 and 39 years old, this constitutes a large part of our working population. The normality test was highly significant (P-value<0.000), with a skewness coefficient of 0.640± 0.064 and a kurtosis of -0.080±0.127 (Figure 3). On the other hand, 52.3% of the respondents were male (n=770) and 47.7% were female (n=703), the sex ratio was, therefore, balanced (p< 0.085) (Figure 4).

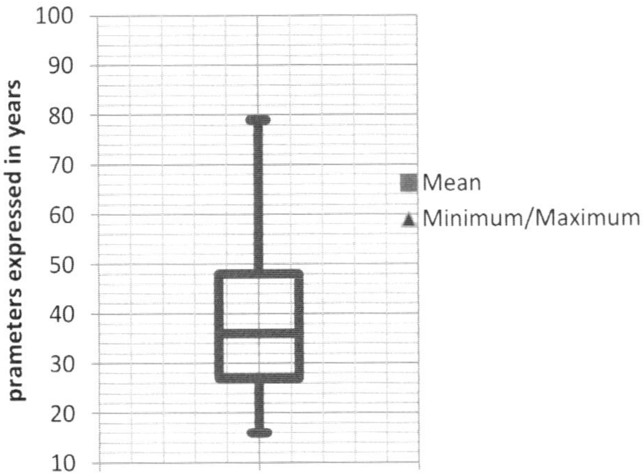

Figure 3. Box plot representation of the age distribution of respondents

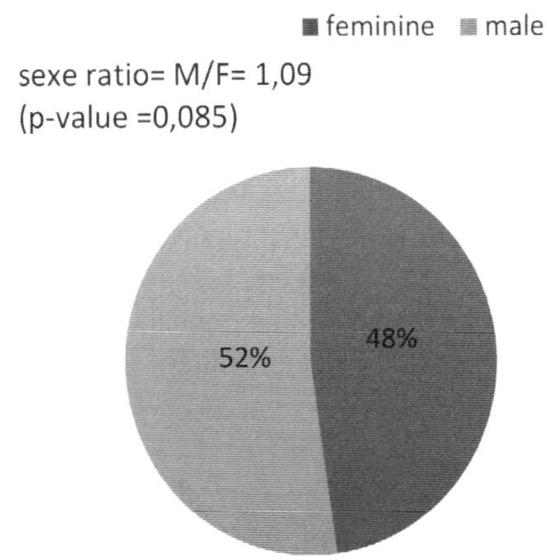

Figure 4. Pie-chart distribution of the sample by sex

3.1.2. Distribution of the sample by orientation and distance from the WWTP neighborhood

It should be noted that 71% (n=1051) of the cases studied declared that they lived in their habitat before 2003 (date of the WWTP's impoundment) and 29% (n=422) returned to it after 2003. Moreover, in addition to the orientation, the distance or not from the treatment plant could be a crucial factor. In fact, we thought of rearranging the neighborhoods into four different groups, based essentially on the distance that separates approximately the centers of the neighborhoods from the anaerobic basins of the WWTP as shown in the following map (Figure 5).

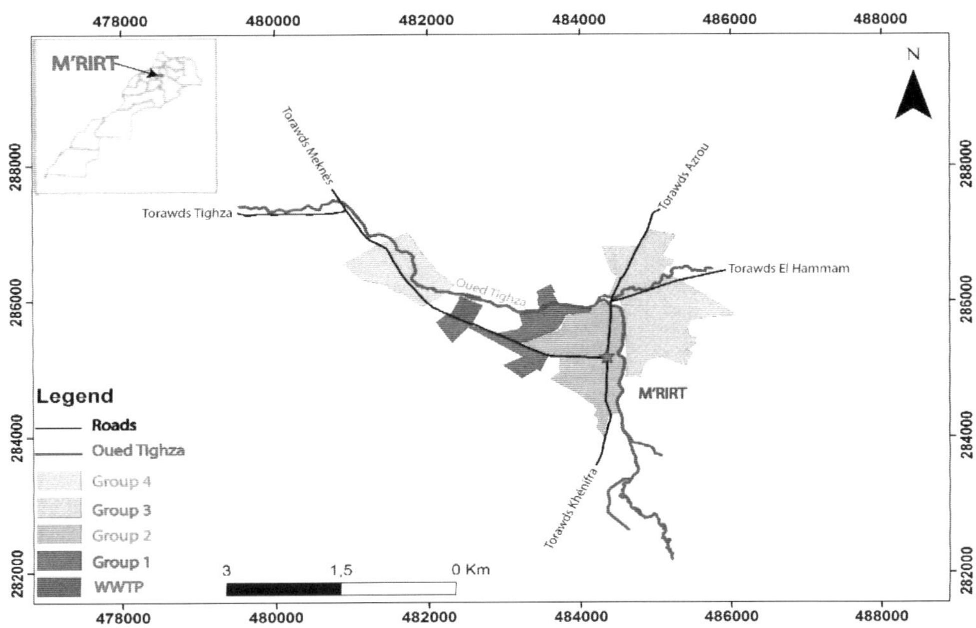

Figure 5. Map of the four neighborhood groups by distance from the WWTP

1. Group 1 composed of the districts located in the East, North-East and South-East directions such as Med N'toto, Ait Ami Ali, Douar Hsayen and Boulouhouch and which are distant of less than

1054m from the source (WWTP), as well as a district Ait Boughzouan Irchkiken located on the South-West side and near (100 m) the station. These districts are all part of the commune of M'rirt.

2. Group 2 includes the districts of Ait Mou, Boulachfar I, Boulachfar II, M'rirt Centre, Hay Nahda and Hay Koutoubia, located in the East, North-East and South-East directions, at a distance ranging from 1054m to 1961m from the WWTP and all belonging to the municipality of M'rirt
3. The group 3 including the districts Tahjaouit, Douar El Ghazouani, Afoud Ikabarn and Ait Hajjou, of the urban commune of M'rirt is distant from the WWTP by more than 1961m in the same direction as the group 1 and
4. Group 4, made up solely of the Aït Moussi district, is 500 m northwest of the treatment plant. This district belongs to the rural commune of El Hammam.

3.1.3. *Descriptive analysis of the samples of the four selected groups.* The distribution of the participants by group is shown in Figure 6. In fact, 41% of them belong to group 2 (distant between 1054 to 1961 meters from the WWTP), while 29% and 27% of the surveyed population belong, respectively, to group 1 (distance closer than 1054 meters) and group 3 (distance farther than 1961 meters from the WWTP). Group 4 alone represents 2.9% of the total population and is 500 meters away from the WWTP.

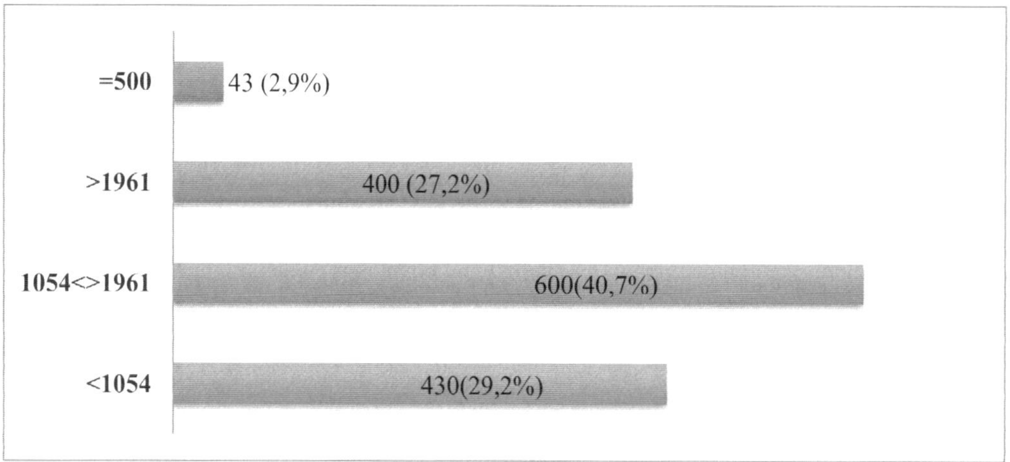

Figure 6. Population distribution in each neighborhood grouping

Furthermore, a descriptive analysis of age and gender, shown in Table 2, shows that 95% of the respondents in groups 1 and 4 are between 42 and 46 years of age and those in groups 2 and 3 are close to 36 years of age. Except for group 4, the distribution of genders in groups 1, 2 and 3 is not balanced ($p<0.05$) (Table 2). The study of the regression between the distance and the age of the respondents showed a strong negative correlation ($p<0.000$), this explains that the age varies in a negative direction to the distance that separates the districts to the WWTP. Similarly, the regression study showed a strong positive relationship between distance and length of residence ($p<0.000$). In the light of these results, it could be said that the citizens in groups closest to the WWTP are the oldest, and the largest percentage of them reside before 2003 in the neighborhood. This may be due to the odors that have limited the urban extension of the city in the direction of the WWTP, as all new residents prefer to move away from it in search of their well- being. Even those who are close to the WWTP want to change their residences but there is no market for real-estate in that zone deemed as a red zone; and even if someone is willing to buy, they give very low prices compared to the prices in the distant and less impacted districts.

Table 2. Descriptive study of age and gender in the four groups.

Distance(m)	Residence before 2003	Age ± Standard error	Male %.	Female %.	Sex ratio
<1054	78,4 (337)	42,83±0.75	44,0 (189)	56,0 (241)	P<0.05 s
1054<>1961	69,5 (417)	36.38±0.53	55.8 (335)	44.2 (265)	P<0.05s
>1961	64,5 (258)	36.21±0.70	56,0(224)	44,0 (176)	P<0.05s
<500	90.7 (039)	43.72±2.70	51.2 (022)	48.8 (021)	P>0,05ns

s: highly significant difference; ns: no difference; (): size;

3.2. *Profile of population well-being*

3.2.1. Spatial study of malodors. On a question asked to the interviewers "is there a perception or not of the bad smell emanating from the wastewater treatment plant? "Out of 1473 respondents, 71% (n=1040) confirmed "yes", the smell represents for them a real olfactory nuisance and 29% (n=433) said "no". The distribution of the respondents to the same question indicated above, according to the neighbourhood is represented in (figure 7). The results of this distribution allowed us to classify the neighborhoods into 4 distinct groups:

A group where all respondents said "yes" for the direct annoyance of the odor emitted by the WWTP such as the districts of Med N'Toto; Ait Ami Ali and Dr Hssayen. Indeed, according to [14] the olfactory nuisance index (ONI) in these neighborhoods is equal to 100.

A second group where few respondents said "no" to the direct effect of the smell such as Boulachfar I and II; Ait Mou; M'rirt Centre; Nahda; Koutoubia and Boulouhouch.

A third group where the number of respondents saying "no" is important, represented by the Tahajouite; Ghazouani; Afoud Ikabarn and Ait Hajjou districts.

The last group representing the districts of Ait Boughazouane and Ait Moussi where almost all the respondents answered "no".

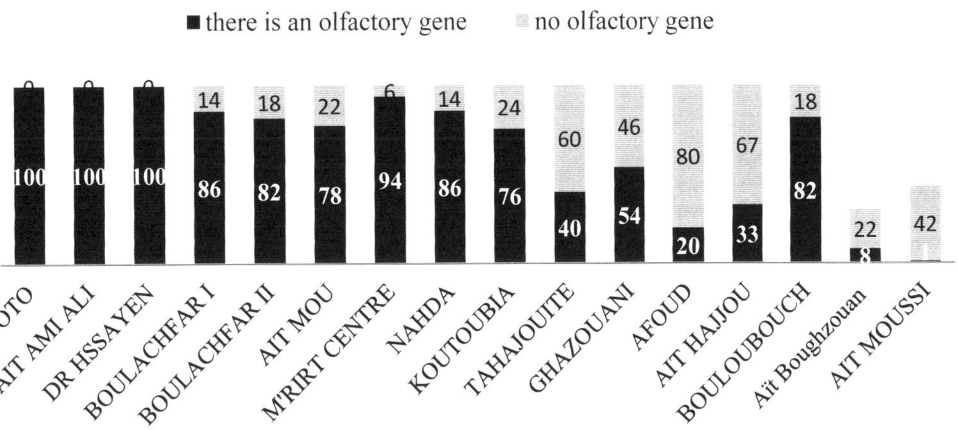

Figure 7. Distribution of responses regarding the perception of odors from the WWTP in each neighborhood.

The chi-square test of independence shows that odor perception and place of residence are strongly related (chi-square=556.091; p<0.000). These results can be explained by the intensity of the odor, which reflects the power of perception felt by the exposed individuals, since the stronger the odors, the more individuals say they are bothered. These results are consistent with the work of [15] and [16]. And the work of [17,18,19], which showed a positive correlation (moderate but not strong) between the concentration of odors emitted by various types of plant and the degree of annoyance as expressed by the people exposed.

3.2.2. Correlation between odor annoyance and distance. In order to investigate a possible link between the distance of the neighborhoods to the sewage treatment plant and the answer to the question "to smell or not to smell", we opted for a chi-square test of independence with a 5% error. The results of this test (Table 3) show a strong relationship (contingency coefficient of 0.484; p<0.000) between odor annoyance and the distance or proximity of the neighborhoods to the WWTP. In fact, the respondents who declared "yes" for the odor annoyance, of the first two groups (<1054 and 1054<>1961) alone represent 85.77%, followed by the neighborhoods of the third group (> 1961) with a percentage of 14.13% and finally the last group (d = 500) where only one respondent declared to smell sometimes. Our results show that the annoyance is felt beyond 1300 m determined as the maximum radius of olfactory nuisance in the theorical impact study carried out by [20] in agreement with the ONEE by modeling the atmospheric dispersion of the odor under ALOHA.

Table 3. Distribution of the perception or not of olfactory annoyance according to the distance from the WWTP and their degree of correlation.

distance	no olfactory discomfort	there is olfactory discomfort	khi-deux	p-value
d <1054	40	390	450,451	0,000 *
1054< d >1961	98	502		
d >1961	253	147		
d =500	42	1		
total	433	1040		

*: highly significant difference at 5% and 3 ddl; d: distance.

However, these results could be, probably, related to several factors including:

- The high intensity in the closest neighborhoods,
- hedonism: the more unpleasant the odors, the greater the level of discomfort felt. Thus, in nearby neighborhoods where the concentration of odors is strong and unpleasant (quality), all respondents report their discomfort, and as they move further away, there is a decrease in the discomfort felt as a result of a dilution of the odors in the air and a decrease in the time and frequency of exposure.
- The olfactory sensitivity in the measure of the importance of the discomfort felt in each district was well highlighted in the districts furthest to the east and especially in the district of Aït Moussi to the west where only one more sensitive respondent declared her discomfort for the bad odors emanating from the WWTP.
- The orientation in relation to the source and the dominant winds which showed that the direction of the dominant winds is from West to East (WE) because the impact was important in the districts which are in the East of the STEP and in the direction where the winds blow contrary to the districts of the West (Aït Moussi) and the South (Aït Boughzouane).
- The distance and the axial position of the neighborhood in relation to the WWTP, as well as its topography, have influenced the dispersion of the odors and the sensation of olfactory discomfort by the inhabitants. This is why, at the level of each group of neighborhoods, a disparity of answers is underlined and sometimes in the same neighborhood the opinions diverge.

3.2.3. *Perception of Odors by Age, Gender and Length of Residence in the Neighborhood.* In order to look for possible relationships between gender, age and length of residence and the level of olfactory annoyance, we used the chi-square test of independence.

- **According to age**

The chi-square test of independence shows that odor perception and age of respondents are strongly related (chi-square=20.601; p<0.001). In addition, in all age categories, more than 60% of respondents expressed discomfort with odors from the WWTP. In fact, people aged between 50 and 60 years are the most affected by odors (more than 80%), whereas young respondents were approximately the least bothered (Table 4). In view of these very high percentages of people expressing total agreement with the odor nuisance and which are well over 50%, this odor is qualified as being of great intensity as pointed out by [5] in the odor management guide.

Table 4. Distribution of olfactory discomfort by age of respondents.

class age	no olfactory discomfort	there is olfactory discomfort	khi-deux	p-value
<20	47(35,07%)	87(64,93%)	20,601	0,001 **
20<>30	117(36,91%)	200(63,09%)		
30<>40	108(27,76%)	281(72,24%)		
40<>50	83(27,39%)	220(72,61%)		
50<>60	36(19,46%)	149(80,54%)		
>60	42(28,97%)	103(71,03%)		
total	433(29,40%)	1040(70,60%)		

**: very highly significant difference at 1%

However, research by [19] has proven the opposite of what was found, young people are more bothered by industrial odors than older people.

- **By gender**

The chi-square test of independence shows no relationship between gender and odor annoyance (chi-square=0.036; $p<0.850$). Indeed, men and women are affected by the annoyance with the same frequency (70%) (Table 5) contrary to the results of [19] who showed that women are more affected than men by the effect of industrial odors. The same result is found by [12] in France.

Table 5: Distribution of olfactory discomfort by sex of respondents

gender	no olfactory discomfort	there is olfactory discomfort	khi-deux	p-value
male	228(29,61%)	542(70,39%)	0,036	0,850
female	205(29,16%)	498(70,84%)		
total	433(29,40%)	1040(70,60%)		

- **According to the length of residence**

The chi-square test of independence shows that odor perception and length of residence are strongly related (chi-square=10.776; $p<0.001$). In fact, people living in the neighborhood after 2003 showed less annoyance than people who lived there before 2003 (Table 6). This behavior was very much in line with the way of living in the neighborhoods of the city of M'rirt, thus preferring urban areas far from the source.

Table 6. Distribution of odor annoyance by length of residence.

Duration of residence	no olfactory discomfort	there is olfactory discomfort	khi-deux	p-value
Before 2003	283(26,93%)	768(73,07%)	10,776	0,001*
After 2003	150(35,55%)	272(64,45%)		
total	433(29,40%)	1040(70,60%)		

*: very highly significant difference at one per thousand

In summary, it can be said that age and residential indicators, such as neighborhood, distance from the WWTP and length of residence in the area, influence the level of odor nuisance experienced by the population.

3.2.4. Temporal study of malodors

For those who responded that they smell the odors (n=1040), they were asked "do they smell the odor during the day, at night, or both day and night, and also in what season of the year?) Since respondents were free to indicate multiple seasons and times of day, there were multiple responses to these two questions.

3.2.5. Daily perception of malodors

The results for daily odor perception are shown in Table 7 and Figure 8. Excluding respondents who answered "no: no annoyance" (n= 433)

	Distance in meters				Chi-square	p-value
	<1054	**1054 1961**	**>1961**	**500**		
day	18	1	12	0	**164,43**	**0,000** *
night	88	268	103	1		
Day/ night	284	233	32	0		
Total	390	502	147	1		

Table 7. Distribution of those who answered "yes"

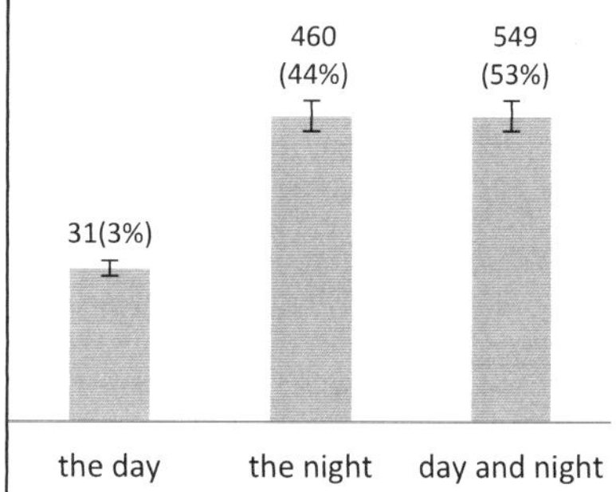

Figure 8. Temporal distribution

*: very highly significant difference at 5% and 6 ddl

According to figure 8, out of the 1040 respondents who declared to smell the odor of the WWTP, 53% prove, seriously, that they smell it day and night of which 94.17% (n= 517) belong to the first two groups (<1054 and 1054<>1961) and out of 44% of the respondents who answered that they smell it only at night, 58.26%, 19.13% and 22.39% belong respectively to the second group (1054 to 1961), the first group (<1054) and the third group (>1961). It should be noted that 58.06%, 3.22% and 38.70% of those who did not smell during the day belong to group 1, group 2 and group 3 respectively.

The chi-square independence test applied to all the neighborhoods in relation to daily odor annoyance shows a strong link between these two factors (chi-square = 164.431; p<0.000), reflected in a contingency coefficient of 0.369 (p<0.000). In fact, according to Figure 9, the evolution of smelling the odor during the day or during the day decreases as a function of the distance separating the neighborhoods from the treatment plant with correlation coefficients of negative sign of rday= -0.64 and rday/night = - 0.96. On the other hand, the evolution of the olfactory annoyance during the night follows a non-linear equation and is strongly linked to the "distance" factor with a coefficient of 0.88. This evolution is marked by a peak in the respondents from the second group of neighborhoods, which could probably be due to the effect of certain factors other than distance, such as atmospheric stability, wind speed and direction during the night and the position and orientation of the neighborhoods in this group. Plus, the temporal variations of the odorous emissions of the site and the meteorological conditions, because during the night and in the absence of solar rays there is no photosynthesis of the algae that produce an important portion of oxygen in the facultative and anaerobic lagoons, so there is an important restoration of the conditions of anaerobiosis that produces the H_2S that is the main cause of these bad odors.

In conclusion, the inhabitants of the neighborhoods in front of the WWTP (East) often smell the odor during the whole day and conversely the neighborhoods further away (distance >1961) are influenced by the odor only at night. However, the neighborhoods in the middle position seem to smell the odor during the day and intensively at night.

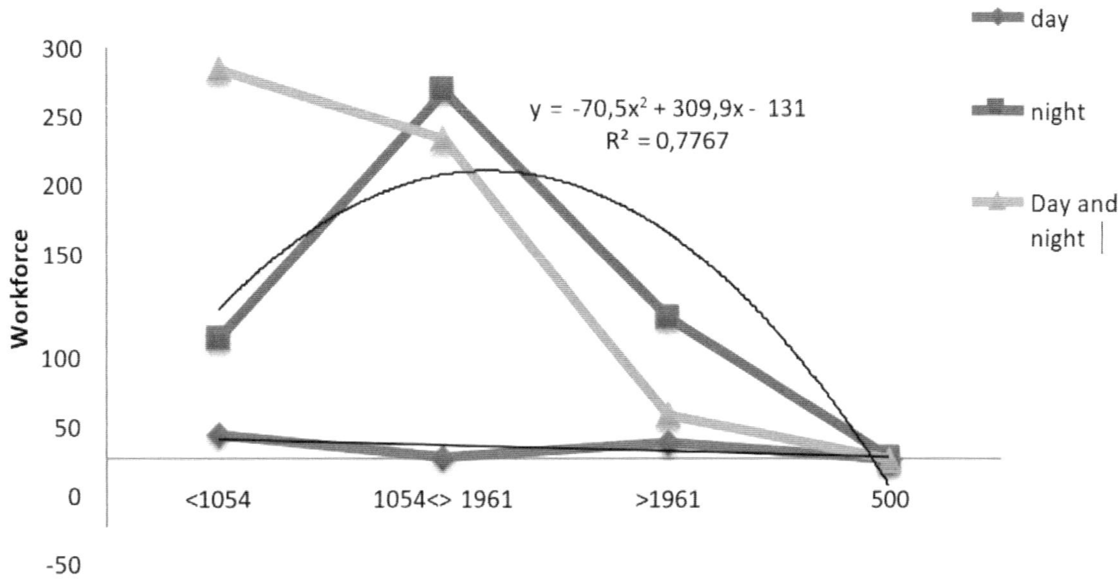

Figure 9. Daily evolution of the odor according to the distance of the group of neighborhoods.

3.2.6. Perception of Odors in Relation to the Four Seasons of the Year

For the question concerning the seasonal olfactory discomfort the results are illustrated in (table 8) and (figure 10).

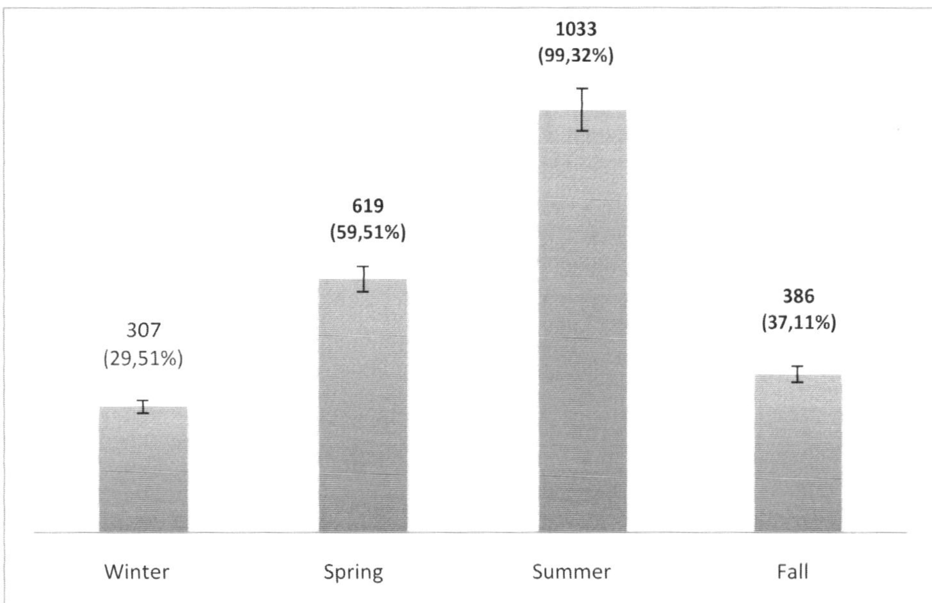

Figure 10. Odor perception by season

Table 8. Distribution of responses to the question regarding the presence of seasonal odors by distance.

	distance between WWTP and neighborhood				khi-deux	p-val
	<1054	1054<>1961	>1961	=500		
Winter	225	64	18	0	238,15	**0,000s**
Spring	349	241	29	0	271,05	**0,000s**
Summer	390	500	142	1	19,58	**0,000s**
Fall	272	108	6	0	299,56	**0,000s**

S: very highly significant difference

The distribution of the respondents who reported smelling the odor is shown in (Figure 10). The results show that 99.32% of the respondents who said they smell the odor during the summer season, of which 37.75%; 48.40%; 13.74% and 0.09% (the only individual who said he smelled the odour in group 4) belonged to group 1 (distance <1050m from the WWTP), group 2 (1054<>1961), group 3 (>1961) and group 4 respectively. Followed by spring with a percentage of 59.51% where 56.38% of the "yes" respondents in this season belong to group 1, 38.93% to group 2 and 4.68% to group 3, yet none of the respondents in group 4 gave up smelling the odor during this season. And out of 37.11% (n=386) of the "yes" respondents smell in autumn 70.46%, 27.97% and 1.55% are respectively from group 1, 2 and 3 (0% for group 4). Winter is the season represented by the low percentage of respondents "yes" smell during winter, which is 29.51% of the total population (n=1040), most of which belong to group 1 with a percentage of 73.28% followed respectively by group 2 and 3 with percentages of 20.84% and 5.86%.

Nevertheless, the chi-square test of independence showed a highly significant difference ($p<0.000$) between the seasons and the distance separating the neighborhoods from the station, so this determines a strong link between these two last variables (Table 8). Indeed, the evolution of the frequency of olfactory discomfort of the respondents according to the season follows a non-linear equation for the first three groups and almost zero for the district of Ait Moussi (Figure 11), with correlation coefficients $r_{<1054}= 0.99$, $r_{1054<>1961} = 0.87$, $r_{>1961}= 0.69$ which show that there is a strong link between the variable season and the variable distance of the districts from the WWTP. Indeed, the maximum of respondents to the above question is marked during the summer, followed by spring and autumn and a minimum of respondents during the winter. This is due to the seasonal difference in atmospheric stability which affects the atmospheric dispersion of the odour plume. However, the distribution of neighbourhoods according to representativeness allows to classify group 1 as the most influenced by WWTP odours, followed by the second group and then the third group.

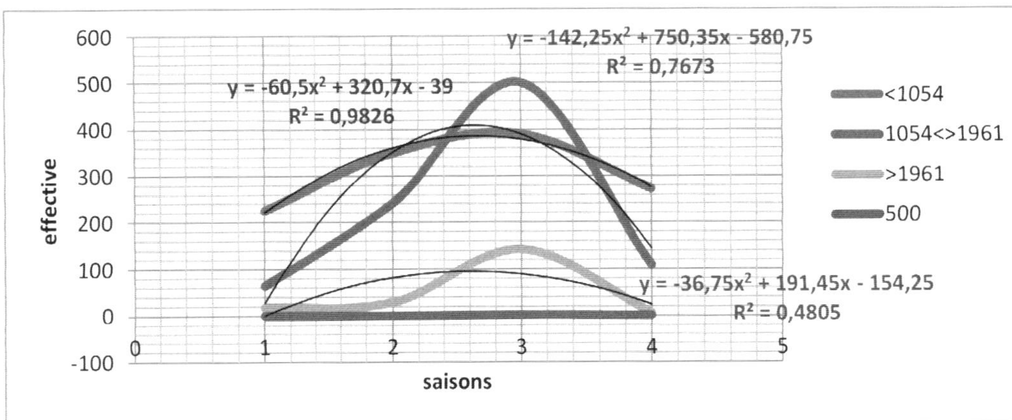

Figure 11. Seasonal evolution of the odor according to the distance from the group of neighborhoods.

In conclusion, the results show that people residing below 1054m smell odors regardless of the season and time of day. However, respondents who are between 1054m and 1961m from the WWTP smell odors day and night throughout the year with a dominance in the summer period and at night. The same pattern of daily and seasonal perception for respondents living above 1961m as for the second group with smaller numbers (Figures 9 and 11). Thus, as one moves away from the treatment plant, the frequency of daily and seasonal exposure decreases. This confirms the fact that the frequent presence of odors in the place of residence leads to a destabilization of the olfactory landscape and causes discomfort, and this is why people who feel that they often smell odors from the WWTP say they are the most bothered. Indeed, this difference in temporal impact is due to the temporal variations of the odorous emissions from the WWTP and to the meteorological conditions which influence the dispersion or the transport and diffusion of the polluting plume in the atmosphere [21] and which can be summarized as follows the horizontal transport of the pollutant cloud, essentially conditioned by the wind speed and direction, it is the advection. Thus, the wind direction is one of the most important parameters, since it determines where the pollutants are directed, in our case towards the residents on the eastern side of the WWTP. As for the wind speed, it is logical to assume that the higher it is, the faster the plume will pass over the receptors and the less time it will have to interact with them. In terms of horizontal transport, therefore, the severity of the impact will be inverse ly proportional to the wind speed. This explains the importance of the impact during the night and in summer the dilution of this cloud by mixing with the ambient air, i.e. the relative displacement of each particle of the cloud with respect to its center of gravity, which moves horizontally: this diffusion process is governed above all by the state of stability of the atmosphere and by the phenomenon of thermal inversion, in other words, by the vertical temperature gradient, which is linked to sunshine and also to wind speed. The impact of the latter is, however, opposite to the one it had on the horizontal transport. Thus, a high wind speed favors the thermal homogeneity of the air layers, and therefore, a greater stability of the air. The more stable the air, the less the plume will be able to dilute and the higher the concentration of pollutants at the receptors. For a source at ground level, in the case of anaerobic ponds, a stable atmosphere produces maximum impact. The most unfavorable conditions for dispersion in this case are therefore :

- low wind speed (about 1m/s)
- persistent wind direction and towards sensitive receptors
- low and persistent mixing height
- covered sky

The interaction between these different phenomena explains the differences in daily and seasonal impacts recorded in our study. It also calls on the specificity of our sense of smell, which has a very rapid response and sensitivity to variations and maxima [22].

3.3. Global analysis of variables by Multiple Correspondence Analysis (MCA)

It is a method of multi-variate analysis that allows the reduction of the number of variables in order to facilitate the reading of the results and to deduce the different links that exist between the variables studied. The multiple correspondence analysis (MCA) of the different variables shows that the two dimensions alone absorb 73% of variation. Cronbach's Alpha coefficient shows that dimension 1 explains 90% of the total variation and dimension 2 explains 75% of the variation (Table 9).

Table 9. Validation of the questionnaire by Cronbach's test, first and second items

Dimension	**Alpha de Cronbach**	**Applied variance**	
		Total	**Inertia**
1	0,909	6,214	0,478
2	0,752	3,268	0,251
Total		9,482	**0,729**
medium	0,855[a]	4,741	0,365

a. La valeur Alpha de Cronbach moyenne est basée sur la valeur propre moyenne.

The analysis of the joint dot plot allows us to classify all the modalities into four distinct groups (Figure 12). The first group on the right at the bottom of the origin shows clearly that the districts of Aït Ami Ali, Med N'toto, Douar Hssayen and Boulouhouch which are part of the group that is far from the WWTP by a distance <1054 meter are the most affected by the bad smell of the WWTP during all the year (because it is the group that represents the answers "yes" for almost all the questions). This group is represented by the two sexes and the age classes between 50 and 60 years and higher than 60 years. The group on the right at the top which includes the districts (from the right to the left) M'rirt Centre, Nahda, Koutoubia, Boulachfare I and II and Aït Mou which are distant from the source between 1054 and 1961 meter this impact has an effect on both sexes aged less than 40 years during the whole year especially in summer and during the night The third group on the left at the top of the origin represented by the districts farthest from the STEP (>1961m) including Douar Ghazouani, Tahjaouit, Afoud Ikabarn and Aït Hajjou where the effect of the olfactory discomfort is less perceptible throughout the year except in summer during the night when the odor is strong and also affects the inhabitants of less than 40 years old men and women. The fourth group represented by the district Ait Boughzouan Irchkiken of the municipality of M'rirt and the tribe Ait Moussi of the municipality of El Hammam where the potential impact of the WWTP is almost zero on both sexes with ages between 40 and 50 years, which emphasizes the importance of the factor orientation of the neighborhood in the measurement of the olfactory discomfort.

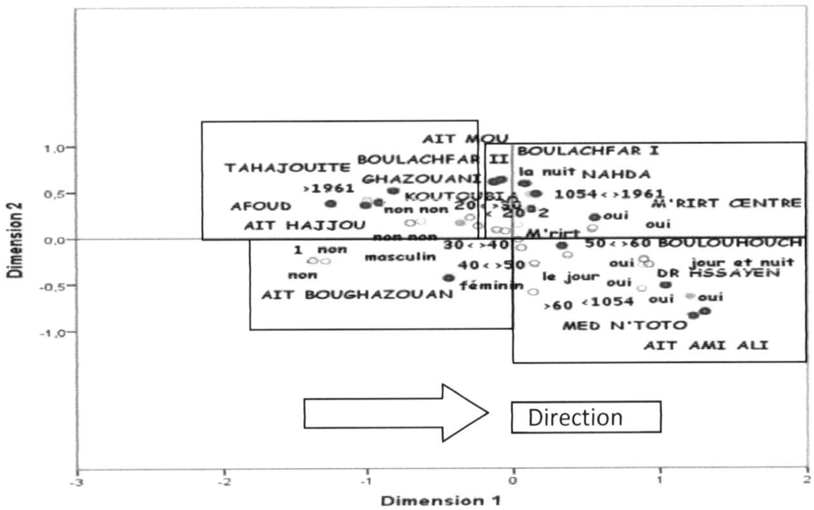

Figure 12. Multiple correspondence analysis (MCA) of the set of variables that make up the data matrix onolfactory annoyance caused by foul odors from the WWTP

3.4. Mapping of the olfactory nuisance in the city of M'rirt.

The map below (Figure 13) shows the degree of odour annoyance experienced in the different neighbourhoods. The percentage of respondents who confirmed that they were bothered by the odours emanating from the WWTP is determined by symbols of different sizes and colours :

- the red colour represents the neighbourhoods where 100% of the respondents are bothered
- the yellow colour represents the neighbourhoods where between 72% and 94% of respondents are bothered
- the brown colour of the symbol groups the neighbourhoods where between 20% and 54% of respondents are bothered
- the green colour of the symbol is attributed to the Aït Moussi district where 2% of respondents declared they were annoyed.

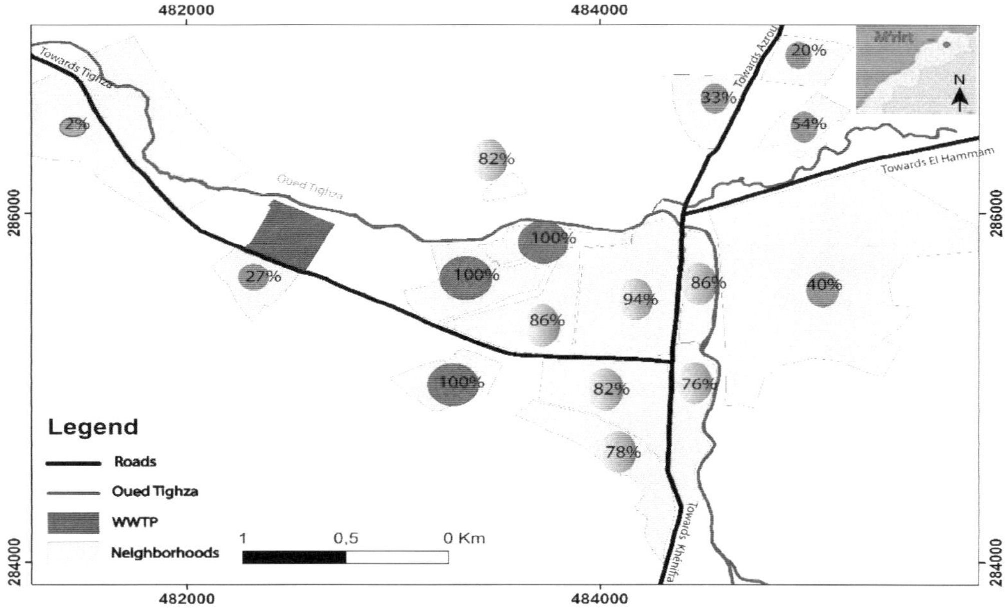

Figure 13. Map of the distribution of odor nuisance in the districts of the city of M'rirt.

The closest neighbourhoods with a distance of less than 1961m from the WWTP in the east, south-east and north-east direction are the most disturbed and the most affected in their well-being. The degree of annoyance decreases with distance in the same direction. However, if we change the direction towards the West and even at 500m from the source, we find that 2% of the respondents are bothered. In the district of Aït Boughzouane Irrchkiken which is 100m south and parallel to the STEP, only 27% of the respondents said they were bothered.

4. Conclusion

Since its commissioning in 2003, the M'rirt WWTP has become a permanent source of foul odors. The continuous exposure to these foul odors can be a risky situation marked by the permanent complaints declared by the residents and by the state of stress and worries about the danger of these odors on the health and well being of the residents. Hence the interest in conducting a study on the impact of these odours on the well-being of the population. This study was carried out by means of a survey using a questionnaire and stratified sampling. Our final sample consisted of 1473 respondents selected in a simple random way.

The answers given by the population surveyed to the question: "Is there a perception or not of the bad smell emanating from the wastewater treatment plant? "Out of 1473 respondents, 71% (n=1040) confirmed "yes", the smell represents for them a real olfactory discomfort (of which 85.77% live in the districts which are at a distance <1054 and $1054< d >1961$ from the WWTP and 14.13% of them live further away ($d > 1961$)) and 29% (n=433) say "no". In the neighborhoods of Mohamed N'Toto; Ait Ami Ali and Douar Hssayen all respondents confirmed their direct discomfort of the smell emitted from the WWTP.

The assessment of odor perception frequency showed a difference in daily and seasonal impacts between neighbourhoods and neighbourhood groups. Indeed, respondents residing within 1054m smell odours regardless of the season or time of day. However, respondents who are between 1054m and 1961m from the WWTP smell odours day and night throughout the year with a dominance in the summer and at night. The same trend of daily and seasonal perception for respondents living above 1961m as those in the second group with lower effectiveness. Thus, as one moves away from the treatment plant, the frequency of daily and seasonal exposure decreases.

The chi-square test of independence shows that there is a strong relationship between odour annoyance and neighbourhood (chi-square=556.091 ; $p<0.000$), distance or proximity of neighbourhoods to the WWTP (chi-square=450.451 ; $p<0.000$), daily and seasonal frequency of malodour perception ($p<0.000$), age ($p<0.000$) and length of residence in the neighbourhood.

The level of discomfort noted by residents using the questionnaire seems to indicate that there is a real problem of odour nuisance and that several factors influence or intervene in its prediction, including: those related to the characteristics of the odours (intensity, concentration, hedonism and frequency), to the conditions of their dispersion in the atmosphere (wind direction, meteorologic a l conditions, topography, etc.), and to individual factors (socio-demographic, sensitivity to nuisances and residential).

The survey showed the impact of bad odours emanating from the WWTP on the well-being of the population of M'rirt, identified the predictors of the olfactory discomfort and confirmed the poor choice of the WWTP site in relation to the direction of the prevailing winds and the need to provide solutions to put an end to these negative impacts on the environment and public health.

5. References

[1]. Buck L.B and Axel R. A., 1991. Novel Multigene Family may Encode Odorant Receptors: a Molecular Basis for Odor Recognition, Celle, vol. 65, 175-187 p. [http://www.cell.com/content/article/abstract?uid=PII009286749190418X] (February 18, 2014).

[2]. The Nobel Prize in Physiology or Medicine, 2004. Nobelprize. Org. [http://www.nobelprize.org/nobel_prizes/medicine/laureates/2004/press.html] (18 Feb. 2014).

[3]. Schaal B., Marlier L., & Soussignan R., 2000. Olfactory function in the human fetus: Evidence from selective neonatal responsiveness to the odor of amniotic fluid. Behavioral Neuroscience, 112(6), 1438-1449 p.

[4]. RECORD., 2006. Odour pollution, sources of odour, regulatory framework, measurement techniques and treatment processes. State of the art, 236p, n°03-0808//0809/1A.

[5]. Anastasakis S., Delisle A., and Sergerie M., 2010. Odour Management Guide. Réseau environnement, Canada, n.d., 88 p.

[6]. Bour C., Romain E., Thouet A., 2006. Évaluations et gestion des risques liés aux situat ions de nuisances olfactives, Ateliers de Santé Environnementale, Ecole Nationale de la Santé Publique. Rennes. 72 p.

[7]. Fonds National pour le Développement des Adductions d'Eau (FNDAE)., 2004. Ministère de l'Agriculture, de l'Alimentation, de la Pêche et des Affaires Rurales, Office International de l'Eau. Lutte contre les odeurs de l'assainissement. Document technique, N° 13, 69 p.

[8]. Combe M, Ferré M & Thauvin J.P., 1977. La Meseta centrale et meseta côtière. In: Ressource en eaux du Maroc: Tome 2. Note et Mémoires du service Géologique du Maroc, n° 231, 162 p.

[9]. HCP High Commission for Planning, 2014. General Census of Population and Housing, Morocco.

[10]. Anonymous, 2010 . Monograph of the urban commune of M'rirt.

[11] ONEP. , 1998. City of M'rirt. Mission I : Establishment of the detailed preliminary designs (D.P.D.) of the sewerage works. Sub-mission I-1 : D.P.A. of the waste water network, pumping station and waste water treatment plant. National Drinking Water Board, Directorate of Studies and Sanitation, Sanitation Division, Vo. II : Pumping station and wastewater treatment plant.

[12]. Pierrette M., 2009. The olfactory environment: from perception to discomfort, the case of an industrial site. Doctoral thesis in Psychology. Paris : University of Paris Ouest Nanterre La Défense, 233p.

[13]. Köster E.P., 1991. Psychophysical assessment methods in environmental studies. In G. Martin & P. Laffort (Eds.), Odours and deodorization in the environment. Paris: Lavoisier.

[14]. Laffort P., 1991. Odour and deodorization in the environment. Paris: Lavoisier.

[15]. Cain W.S., & Moskowitz H.R. 1974. Psychophysical Scaling of Odor. In: Turk A., Johnston J.W., & Moulton D.G., Eds. Human Responses to Environmental Odors. N.Y., Academic Press, 1-32 p.

[16]. Miedema H.M.E., Walpot J.I., Vos H., & Steunenberg C.F., 2000. Exposure annoyance relationships for odour from industrial sources. Atmospheric Environment, 34, 2927-2936 p.

[17]. Winneke G., & Kastka J. 1987. Comparison of odour-annoyance data from different industrial sources: Problems and implications. In H. S. Koelega (Ed.) Environmental annoyance: Characterization, Measurement, and control, Amsterdam, Elsevier. 128-138 p.

[18]. Verschut C., Miedeme H.M.E., Blaauwbroek J., Punter P.H., Majwald K.D., 1991. Koppeling stankconcentratie en stankbeleving. Publicatier-eks Lucht 98. Ministerie van Volkshuisvesting, Ruimtelijke Ordening en Milieubeheer: Leiden, (in Dutch). Cited in Cavalini P.M., 1994. Industrial odorants: the relationship between modeled exposure concentrations and annoyance. Archives of Environmental Health. September/October, vol. 49 (n°.5).

[19]. Cavalini P.M., 1994. Industrial odorants: the relationship between modeled exposure concentrations and annoyance. Archives of Environmental Health. September/October, vol. 49 (No.5): 344-51p.

[20]. Elanaia M., 2012. Odor problem at the M'rirt wastewater treatment plant: Diagnosis and mitigation measures. End of studies project. Université Hassan II Mohammedia - Casablanca Faculty of Sciences and Techniques Mohammedia. 120 p.

[21]. Nicolas J., & Craffe F., 2003. Modelling atmospheric dispersion. Particular case of odours. Fondation Universitaire Luxembourgeoise - Arlon, 40 p.

[22]. Gingras B., Guy C., & Page T., 2003. Odours Environment and Public Health. Fondements et pratiques, s.éd. 515 p.

Improving the physiological responses and stomatal characteristics of *Paspalum vaginatum* during the winter season

D Fetjah[1]*, L F Z Ainlhout[2], Z Idardare[3], L Bouqbis[2]

[1] Laboratory of Biotechnology, Materials, and Environment, Faculty of Sciences, Ibn Zohr University, Agadir, Morocco
[2]Laboratory of Biotechnology, Materials, and Environment, Faculty of Applied Sciences, Ibn Zohr University, Ait Melloul Morocco
[3]Higher Institute of Maritime Fisheries, Agadir, Morocco

*Email: fetjahdounia194@gmail.com

Abstract. Drought is a universal problem that affects plant production and degrades soil. The point of this study was to appraise the physiological responses of *Paspalum vaginatum* to waterstress and to attempt to minimize its impacts using organic amendment in arid and semi-arid climates. In winter 2019, a pot investigation was undertaken for this reason in the south of Morocco on a plastic pot filled with sandy loam soil and various levels of peanut hull biochar and compost on 20% of water holding capacity. The results revealed that adding biochar and compost enhanced photosynthesis, stomatal traits, and water availability in plants cultivated in amended treatments compared to untreated treatments. This study could be a prominent reason for managing luxury golf courses in arid zones.
Keywords: Drought, physiological traits, stomatal traits, water availability.

1. Introduction

Rainfall, poor soil fertility, and water scarcity are indeed the primary factors influencing plant growth [19]. Plants in arid and semi-arid regions have progressed to strengthen and safeguard themselves againstdry spells by sustaining many adaptive responses, such as closing their stomata and tweaking photosynthetic activity [4]. Many research findings have shown that combining biochar and compost with soil improves not only soil quality but also water holding capacity. All of these advantages will help to mitigate the severe effects of the drought [1, 12, 15].

Furthermore, adding biochar to the soil improves its biological, physical, and chemical properties [5]. To recognize the physiological basis of enhanced production and biological consistency, responses of plants once the biochar amendment to sandy areas must be evaluated [17].

Paspalum vaginatum has gained more attention, it is a widespread plant in golf courses and parks due to its resistance to severe drought conditions. Seashore paspalum is a drought-tolerant C4 that can withstand both abiotic and biotic stresses [14]. Recent research has found that banana waste biochar improves the biological characteristics of *P. vaginatum* [8]. In a pot experiment with ten levels of an organic amendment given to a sandy loam poor soil at 20% WHC, this research looked at the effect of peanut hull biochar on *P. vaginatum*. Based on our previous results, we suggested that peanut hull biochar will improve and ensure physiological traits and paspalum growth even under 20% of WHC considered severe drought conditions [7, 8].

2. Materials and methods

2.1. Experimental design

Peanut hull feedstock was pyrolyzed in a Moroccan pyrolysis process heater predicated on Prof Claudia Kammann's (Institute for Plant Ecology, Giessen University, Germany) prototype [3]. The detailed physicochemical properties of peanut hull biochar have previously been reported [7].

The soil for the research was gathered in the south of Morocco, notably in Taroudant, at a depth of 0-20 cm. The sample of soil was filtrated after air-drying and milling to get around 2mm particles. The soil contains 33.39% of coarse silt, 2.05% of fine silt, 1.20 % of clay, 39.96% of coarse sand, and 23.40% of fine sand. For the elemental composition: 0.007% of total nitrogen Nt, 0.201 ‰ of P_2O_5, 2.5% of total organic matter (Mot), 1.45% of total organic carbon (Cot), 0.357 ‰ of K_2O,0.192 ‰ of Na_2O,0.933‰ ofCao, and 0.391 ‰ of MgO. The contents of iron, manganese copper and zinc were, respectively, 0.4, 7.4,0.9, 2.6 ppm [8-11].

Paspalum vaginatum Swartz seeds were sown in plastic trays (33.5 x 50 cm) supplied with saturated peatto provide nutrition for seedlings. Seedlings were transplanted into plastic tubes (13 cm long, 12.5 cm diameter; four plants per tube) supplied either with basic soil or soil supplemented with peanut hull biocharor compost after four weeks. Ten mixes of soil and organic additions (compost and peanut hull biochar) were created, each with four replications: (1) CTR (control with 100% of dry soil), (2) 3B0C containing(97% of dry soil, 3% Biochar and 0% of compost), (3) 3C0B contained only 97% of dry soil and 3% Compost, (4) 3B3C made by 3%Biochar+3%Compost and 94% of dry soil, (5) 3B6C containing3% of Biochar and 6%Compost and 91% of dry soil, (6) 6B0C mixture of 6 % of Biochar + 0% Compostand 94% of dry soil, (7)6B6C mixture of 6 % of Biochar+ 6% Compost and 88% of dry soil, (8) 6C0B mixture of 0% of Biochar + 6% Compost and 94% of soil,(9) 6B3C containing 6% of B and 3%C and 81% of dry soil, and (10)Test contained soil and NPK. In this experiment, the test treatment was used asa supplement to evaluate its effects on those of the peanut hull biochar and compost treatments. Furthermore, all *P. vaginatum* plants were given 20% WHC of sandy loam soil. Different rates of peanuthull biochar were used in this study, with 6 percent B0C corresponding to 137.04 Mg/ ha and 3 percent B0C corresponding to 68.52 Mg/ha [8, 9].

P. vaginatum was watered daily for eight weeks in the pot experiment at the start of the study. The stress treatment lasted three months during the winter season, with a minimum of 20% irrigation water. *P. vaginatum* plants were treated with different rates of peanut hull biochar and compost at 20% of WHC forthe winter season. For the test plant, a data logger UNI-T UT330 was used to take the temperature and humidity of the *P. vaginatum* [8, 9].

2.2 Gas exchange, and chlorophyll pigment content

Using an Infrared Gas Analyzer (IRGA), we analyzed leaf paspalum gas exchange characteristics (LCi-portable photosynthesis System, ADC, Hertfordshire, UK). Fresh leaves were placed within the IRGA, andphysiological features were measured between 9.00 am and 12.00 pm during the day. Three measurementswere taken for each treatment.

2.3. Relative Water Content (RWC)

Weighing 0.5g of a fresh weight determined the relative water content (RWC) of *P. vaginatum* leaves. Afterthat, we soaked the turgid in tap water for 24 hours to determine their weight. The data were made at 6 a.m.and 12 a.m. RWC was computed using the equation after the leaves were oven-dried for 48 hours at 80°C.

$$\mathrm{RWC}\ (\%) = (\mathrm{FW} - \mathrm{DW})/(\mathrm{TW} - \mathrm{DW}) \times 100$$

FW, DW, and TW signify fresh, dry, and turgid weights, respectively. [21, 22].

2.4. Leaf stomatal traits

Stomata imprints were made on the abaxial surface of the leaflets. A thin layer of nail polish was applied tothe surface of the leaf of *P. vaginatum* after it had been cleaned with a gentle brush and left to dry for 10– 15 minutes. After that, the area was taped with transparent tape and carefully removed. The samples were then examined under an optical microscope right away (BX51, Olympus, Tokyo, Japan). [9]. Stomatal number, stomatal density, and size parameters were measured, including **LS**: stomata length, **WS:** stomata width, **Lo:** length of the ostiole, **Wo**: width of the ostiole).

2.5. Statistical analysis

Three replicated measures were used for analyzing data. Tukey test was done to compare the means of the different treatments. All statistical analysis was displayed by Sigma Plot version12.

3. Results and Discussions

3.1. Effect of peanut hull biochar on seashore paspalum growth

P. vaginatum growth was affected by drought. The addition of peanut hull biochar reduces these negative effects on certain treatments (Figure 1). The highest dry weight was noticed in the test treatment which contained soil and NPK fertilizer, in fact, this treatment was done as a supplementary treatment in order to compare its effect with the addition of organic amendment (peanut hull biochar and compost). The 6B6C treatment showed the greatest result in enhancing the weight of seashore paspalum biomass compared to control and succeeded to decrease the negative impact of severe conditions. In line with our results, similar studies showed that adding biochar and compost enhances the plant's growth even in drought [1, 23]. In general, all combined treatments showed an improvement in paspalum growth compared to non-combined treatments. The lowest weight was noticed from the 6C0B treatment which contains 6% of compost and no biochar. This result demonstrates that compost alone is insufficient and that a soil amendment including a combination of biochar and compost is required to extend the duration and improve nutrient uptake [23].

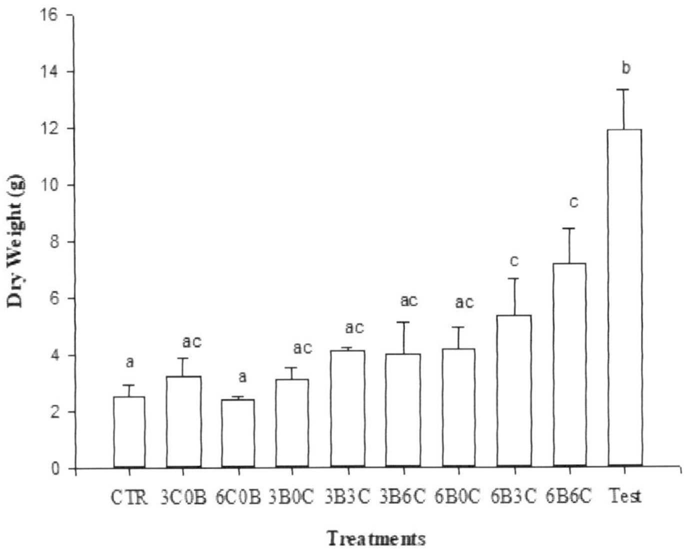

Figure 1. The effect of 20% WHC irrigation on the dry weight of *Paspalum vaginatum*. The valuesrepresent the mean standard deviation (n = 3), and distinct symbols indicate a significant differencebetween treatments at P< 0.001 using the Tukey test.

3.2. Effect of peanut hull biochar on relative water content RWC

The addition of peanut hull biochar had been shown to improve relative water content in certain treatments, especially the combined ones 6B6C, 6B3C compared to non-combined soil amended with peanut hull biochar or compost only 6C0B. Figure 1 showed highly significant differences between the treatments with p-values < 0.001. In fact, as the intensity of drought conditions increased, RWC diminished even more. Figure 2 shows that when biochar and compost were applied to the soil with such a 20% irrigation rate, RWC expanded once, particularly in comparison to biochar or compost as a single addition treatment. For the winter season, plants were treated with varying amounts of peanut hull biochar and compost. Despite the fact that drought has always been one of the factors that reduce relative water content, our peanut hull biochar has been a helpful tool to mitigate its damaging effects. Similar studies have reported that adding organic amendments either biochar or compost had enhanced the relative water content and water status of plants. The study of [13] on which they reported that RWC improved significantly more with the application of a vermicompost-biochar mixture than with the application of biochar alone. Farhangi-Abriz and Torabian,[6] revealed that varying biochar rates greatly raised the relative water content of the bean. [15] explained that biochar might achieve a level of plant water status, signaling a better susceptibility to potential water and salineenvironmental stress, which is consistent with our findings. However, [1] discovered thatadding 2.5 % biochar reduced RWC under drought.

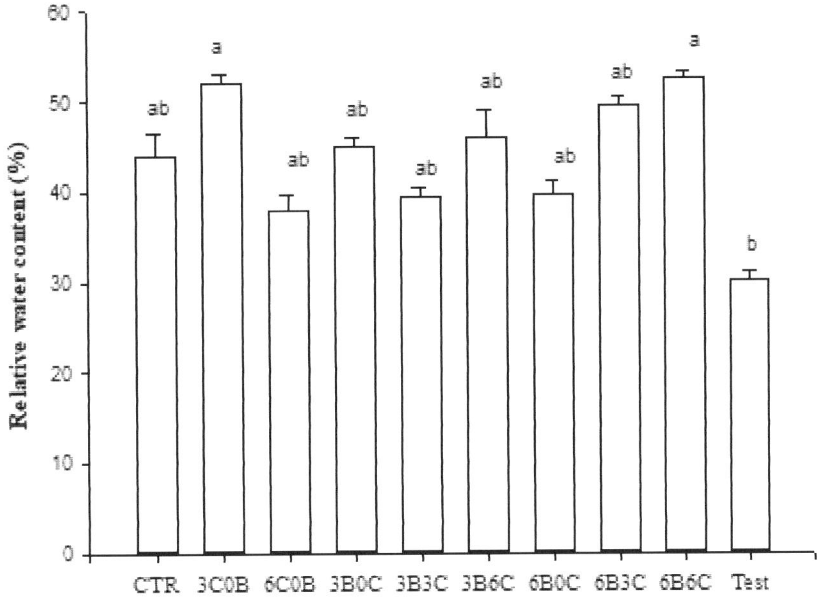

Figure 2. The effect of 20% of WHC irrigation on the relative water content (RWC percent) of the leaves of *Paspalum vaginatum*. The values represent the mean standard deviation (n = 3), and distinct symbols imply a meaningful effect among treatments at P 0.001 using the Tukey test.

3.3. Effect of peanut hull biochar on physiological responses of P. vaginatum

Table 1 provides that peanut hull biochar and compost mixtures diminished transpiration on modified treatments. The treatment 6B3C resulted in a significant decrease. While the biggest value of paspalum leaf transpiration was shown in the control treatment. We observed that the application of peanut hull biochar and compost mixtures had a positive effect on the amended treatments. The values of photosynthesis and stomatal conductance were enhanced with the amended treatments. While, they declined with non-amended treatments. It is known that under drought the physiological responses decrease, in fact, amending soil with biochar and compost had revealed a great result in enhancing photosynthesis and stomatal conductance. The temperature and humidity of the *P. vaginatum* were recorded during the winter season. On average, the temperature of the paspalum was 17.7°C and the humidity mean was 72.5%. In accordance with our findings, similar studies showed that adding biochar improved photosynthesis and stomatal conductance even in harsh conditions [1, 20]. Actually, in the absence of biochar [2] discovered that plants can survive without organic amendments, in addition, adhere to stringent stomatal limitations, which mostly led to a reduction in transpiration E and CO_2 supply for photosynthetic activity escape [16] reported that under drought, seashorepaspalum limited the gas exchange process by maintaining its stomatal conductance and closing its stomata. As an adjustment to tough conditions, *P. vaginatum* tended to close its stomata as an adjustment mode to avoid the water escape.

Table 1. Paspalum leaf gas exchange parameters (A, net photosynthesis; gs, stomatal conductance; Ci, intercellular concentration of CO2; E, transpiration rate; WUE, water use efficiency (A/E) of seashorepaspalum leaf.

Treatments	E(mmol·m⁻²s⁻¹)	A(µmol·m⁻²s⁻¹)	gs(mol·m⁻²s⁻¹)	WUE (mmol·m⁻²s⁻¹/mmol·m⁻²s⁻¹)	Ci
CTR	3,74±0,05a	6,98±0,25ab	0,176±0,02a	1,86±0,04a	193±2,49a
3C0B	3,16±0,02ab	6,90±0,33ab	0,0916±0,01a	2,19±0,12ab	105±1,89de
6C0B	2,50±0,21ab	8,54±0,33ab	0,056±0,01a	3,48±0,43ab	96±0,33def
3B0C	2,94±0,41ab	7,06±0,25ab	0,119±0,01a	2,48±0,30ab	154±2,84bc
3B3C	3,17±0,02ab	8,24±0,58ab	0,0597±0,02a	2,60±0,20ab	105±2,96d
3B6C	2,13±0,12ab	7,39±0,74ab	0,105±0,01a	3,48±0,35ab	203±4,17a
6B0C	1,44±0,34ab	6,32±0,7ab	0,196±0,24a	4,89±1,23ab	113±2,51d
6B3C	0,87±0,04b	11,17±0,54ab	0,292±0,22a	12,75±0,13b	88±6,06e
6B6C	1,23±0,05ab	12,26±0,08a	0,215±0,10a	10,04±0,49ab	78±5,68ef
Test	2,16±0,05ab	6,37±0,01b	0,0774±0,01a	2,95±0,18ab	157±5,78cd

3.4. Effect of peanut hull biochar on stomatal traits of P. vaginatum
Under stress levels of 20% of WHC, we observed that stomatal density (SD) decreased even with compostand biochar combination addition (Table 2), except for the two treatments 3B3C and 3COB. Stomatal density decreased in ginger, which is consistent with our findings [25]. Furthermore, Xu and Zhou,[24] discovered that under moderate drought, stomatal density steadily improved while overall water was reduced, whereas under severe drought, stomatal density reduced, implying that grass can respond to environmental changes through leaf plasticity. In general, water stress increased length and width of stomataof amended treatments. In comparison with our results, [18] found that drought caused an increasein stomata length and a decline in stomata width in the ear organs.

Table 2. Effect of drought on stomatal traits of *P. vaginatum* treated with different rates of peanut hull biochar and compost at 20% of WHC for winter season

Treatments	SD (number mm⁻²)	LS (µm)	WS (µm)	Lo (µm)	Wo (µm)
CTR	93,51±0,78ac	31,33±0,72a	21,00±0,57ac	23,17±0,92ac	5,6±0,11ab
3C0B	101,10±0,67c	32,35±0,23a	21,50±0,86ace	24,00±0,57abc	5,53±0,21ab
6C0B	75,33±0,88d	31,67±1,66a	23,83±0,72c	20,67±2,18abc	5,4±0,23ab
3B0C	65,58±1,96e	33,44±0,29a	18,10±0,58de	26,10±1,40a	5,1±0,05ab
3B3C	124,42±1,26b	32,23±0,39a	22,00±1,15ac	25,93±0,93ab	4,22±0,04b
3B6C	74,372±0,80d	36,13±1,04a	32,43±1,23b	19,17±0,55bc	7,53±0,04a
6B0C	73,30±2,03de	37,43±0,26a	31,90±0,28b	20,20±0,20abc	5,11±0,07ab
6B3C	73,25±2,76d	32,50±0,28a	20,50±0,28ac	25,00±1,44ac	5,33±0,33ab
6B6C	74,75±1,28d	32,67±0,16a	22,50±0,28ac	22,00±2,08ac	5,80±0,11ab
Test	75,03±2,01d	31,30±0,20a	16,00±0,57d	18,23±1,75c	3,33±0,08b

4. Conclusion
In sum, applying peanut hull biochar with 20% of WHC improved the physiological responses of *Paspalum vaginatum*. The amended treatments had an incredible response of the photosynthetic activity and developed biomass weight under drought, which is a quality to resistance. Furthermore, the seashore

paspalum tended to positively enhance relative water content for combined amended treatments compared to unamended treatments. Therefore, peanut hull biochar and compost addition had succeeded in improving plant growth in sandy loam soil. The amendment of organic amendments had led to increased biomass production of seashore paspalum by improving plant water status, and photosynthesis. These physiological parameters respond positively to biochar and compost mixtures application, leading to enhanced biomass production and conserve water.

5. References

[1] Abideen Z, Koyro H W, Huchzermeyer B, Ansari R, Zulfiqar F, Gul B 2020 Plant Biology 22 (2) 259-266.
[2] Akhtar S S, Andersen M N, Liu F 2015. *Agricultural Water Management* 158 61-68.
[3] Bouqbis L, Daoud S, Koyro H W, Kammann C I, Ainlhout L F Z, Harrouni M C 2016 *InternationalJournal of Recycling Organic Waste in Agriculture (IJROWA)*5 361-365.
[4] Condon A G, Richards R A, Rebetzke G J, Farquhar G D 2004 *Journal of Experimental Botany* 55 (407)2447-2460.
[5] Cornelissen G, Martinsen V, Shitumbanuma V, Alling V, Breedveld G D, Rutherford D W, Sparrevik M,Hale S E, Obia A, Mulder J 2013 *Agronomy* 3 (2) 256-274.
[6] Farhangi-Abriz S, Torabian S 2018 *Environmental Science and Pollution Research* 25 (12) 11556-11564.
[7] Fetjah D, Ainlhout L F E, Ihssane B, Houari A, Idardare Z, Bouqbis L 2021 *Journal of EcologicalEngineering* 22 (4)36-46.
[8] Fetjah D, Ainlhout L F E, Ihssane B, Bouqbis L 2021 *Journal of Ecological Engineering* 22 (8) 1-10.
[9] Fetjah D, Ihssane B, Ainlhout L F E, Idardare Z, Bouqbis L 2022 *Journal of Ecological Engineering* 23 (1)216-227.
[10] Fetjah D, Ainlhout L F E, Idardare Z, Ihssane B, Bouqbis L 2022 *Sustainability* 14 (3)1-13.
[11] Fetjah D, Idardare Z, Ihssane B, Ainlhout L F E, Bouqbis L 2022 *Biology* 11 (4)1-14.
[12] Glaser B, Lehmann J, Zech W 2002 *Biology and Fertility of soils* 15 219-230.
[13] Hafez E M, Omara A E D, Alhumaydhi FA, El-Esawi M A 2021 *Physiologia Plantarum* 172 587- 602.
[14] Ibitayo O, JD B, MJ B 1981 *HortScience* 16 (5) 683- 684.
[15] Kammann C I, Linsel S, Gößling J W, Koyro H-W, 2011 *Plant and soil* 345 195-210.
[16] Katuwal K B, Tishchenko V, Jespersen D 2021 *Crop Science* 61 2121-2134.
[17] Keshavarz Afshar R, Hashemi M, DaCosta M, Spargo J, Sadeghpour A 2016 *Communications in SoilScience and Plant Analysis* 47 (6) 743- 752.
[18] Li Yuping, Li H, Li Yuanyuan, Zhang S 2017 *The Crop Journal* 5 (3) 231-239
[19] Liu F, Chen Y, Lu H, Shao H 2017 *Science of the Total Environment* 578 67-73.
[20] Nadeem S M, Imran M, Naveed M, Khan M Y, Ahmad M, Zahir Z A, Crowley D E 2017 *Journal of theScience of Food and Agriculture* 97 (15) 5139-5141.
[21] Nikpour-Rashidabad N, Tavasolee A, Torabian S, Farhangi-Abriz S 2019 *Archives of Biological Sciences* 71 (2) 321-327.
[22] Sairam R K, Rao K, Srivastava G C 2002 *Plant Science* 163 1037-1046.
[23] Vaughn S F, Dinelli F D, Jackson M A, Vaughan M M, Peterson S C 2018 *Industrial Crops and Products* 111 667-672.
[24] Xu Z, Zhou G 2008 *Journal of Experimental Botany* 59 (12) 3317-3325.
[25] Xu K, Zou G, Zhao Y 2003 *Ying Yong Sheng Tai Xue Bao* 14 (10) 1645-1648.

Environmental Impact Assessment of the quarries on Grou River: Khenifra region- Morocco

A Boutsougame[1,*], M Khaffou[1], H Aassine[2], H Ouazzani[3], M Alaoui[2]

[1]: Higher School of Technology- Khénifra-, P.B: 170-54000, Khénifra, Morocco.
[2]: Sciences and Technologies Faculty of Sultan Moulay Slimane University, Beni Mellal, Morocco.
[3]: Faculty of Sciences Meknes, Department of Geology, B.P11201 Zitoune, Meknes, Morocco.
[*] E-mail: a.boutsougame@usms.ma

Abstract:
Khenifra region is rich in geological resources, which attracted a lot of investment, especially the quarrying. The Grou River watershed contains a lot of quarries of alluvial material extraction which causes negative impacts on its environment such as landscape modification, water quality degradation, dust emission and, soil erosion. The evaluation of water quality by measuring physicochemical parameters showed a deterioration of its quality in the downstream area compared to the upstream area. To respond to this situation, we suggest a set of mitigation measure and propose the respect of monitoring and surveillance plans.

Keywords : Environment, Impact, Quarries, Quality, Grou River, Khenifra.

1. Introduction:

The evaluation as a catalyst for the development, before the project is undertaken, is known as the Environmental Impact Assessment (EIA), which has become a compulsory requirement in Morocco and before implementing the decision of the project [1].The general mode of urbanization several Moroccan cities including Khénifra, in addition to the recent implementation of major structuring projects, are one of the main causes leading to the increase in the demand for construction materials, to meet the needs of this fabulous market. This will generate consequences for the ecosystem of this region and /or the socio-economic activity of the populations [2] and subsequently will raise the concern about the environmental and landscape impacts of the industries of extracting alluvial materials.

This environmental impact study concerns the quarries of Sidi Lamine located West of Khenifra. The study presented consists, through the geographic information system (GIS), of locating the quarries, the water sampling stations on Oued Grou (a tributary of Oued Bouregreg) upstream and downstream of the active quarries. In addition to the study based on physico-chemical parameters of water, the synthesis of other parameters resulting from point observations is adopted, such as dust

IOP Conf. Series: Earth and Environmental Science 1090 (2022) 012018 doi:10.1088/1755-1315/1090/1/012018

emissions, space consumption, noise and vibrations, impacts on soil quality and agriculture, impacts on the local flora and fauna. The EIA law (12-03) in Morocco benefits the conservation of biodiversity and the protection of the environment has been introduced as compulsory in Morocco since 2003.

From this perspective, this work aims is to identify and assess the environmental and socio-economic impacts of the exploitation of quarries by the extraction of alluvial materials, the pressure that this activity exerts on the environment, and finally the responses that stakeholders must provide to preserve the environment.

The environmental impact study has the following objectives:

- Assess in advance the possible direct and indirect effects of quarries on Oued Grou ecosystem,
- Indentify temporary and permanent projects that potentially would have a negative impact on the environment.
- Suggest actions to eliminate mitigate and compensate the negative effects of the project.
- High light and improve the positive impacts of the project on the environment.

2. Material and methods

Following the location of the quarries using the GIS, we took water samples at different stations in Oued Grou (Fig. 1). The sampling was carried out in March 2021 after a period of precipitation at one station upstream and two stations downstream from the confluence of Oued and the sloping streams of the active quarries.

The methodology presented consists of the study is based on quantifying physico-chemical parameters of water (pH, temperature, dissolved oxygen, electrical conductivity and turbidity). The analysis of other variables resulting from observation in the field was combined, such as dust emissions, land use, noises and vibrations, impacts on fauna, and flora, impacts on soil quality, and agriculture.

Table 1: stations monitored along the Oued Grou river with geographical coordinates and values obtained for each parameter.

Monitored stations		Geographic coordinates	Parameters				
			T (°C)	**pH**	**Dissolved oxygen (Mg/l)**	**Electrical conductivity (Us/cm)**	**Turbidity (NTU)**
Station Upstream	Sam.1	32°56'24.8"N 5°55'31.7"W	17,2	8,1	8,2	1780,4	15
	Sam.2		16	8	8	1775,2	13
	Sam.3		18,4	8,4	8,5	1785,6	14
Stations Downstream 1	Sam.1	32°56'00.5"N 5°57'33.0"W	18,7	8,33	7,9	2773,53	36
	Sam.2		17,6	8,2	7,8	2766,2	32
	Sam.3		19,8	8,82	8,1	2765,33	34
Stations Downstream 2	Sam.1	32°55'55.1"N 5°57'41.3"W	18,4	8,28	8	2752,5	31
	Sam.2		17,5	8,1	7,9	2756,21	29
	Sam.3		19,3	8,68	8,2	2755,65	27

2.1. Presentation of the study area

The study was carried out on three stations at the Oued Grou river near the village of Sidi Lamine in the province of Khenifra (Figure 1).

2.1.1. Pluviométrie. According to data obtained from the Center of Agricultural Works in the city of Khenifra, the rainy period extends from December to April, with an annual average of 516 mm calculated over 31 years (1971-2002, [3]). In general, the study area is characterized by a semi-arid climate with cool winters, excessive summer heat and torrential rainfall.

2.1.2. Geology. Formations are made up of rocks that range from the Precambrian to the Devonian [4]. It correspond to shales, limestones, marbles, thin polygenic breccias, clays in addition to sandstones.

The essentially impermeable nature of the bedrock, the relief often steeply sloping and the intensity of the precipitation induces a predominance of runoff which impedes pedogenetic processes.

This entrains the fine materials which are quickly evacuated by the flows of the hydrographic network (high content of suspended elements). Coarse materials can also be eroded during heavy downpours, giving off concentrated flows that incise surface formations and the bedrock itself. Thus, where runoff dominates, the soils are little or no evolved [5]. The soils are mainly lithosols, derived from a schistose source rock permanently stripped by erosion [6].

In the case of the Khenifra marbles, the rock is white, gray to black in color, and sometimes cherry or nodular in appearance.

2.2. Establishment of alluvial material extraction quarries

The production of building materials in the region of Khenifra is provided by several quarries dealing with the extraction of alluvial materials covering the demand of the local and regional market in this area. The number of these quarries is in remarkable evolution with a strong positive correlation with the demand for building materials which depends on the downstream activity of building and public works.

This is the reason why we have tried to model these environmental data and using System Geographic Information (GIS), QGIS in particular in order to:

- Manage environmental information;
- Develop support tools for decision making;
- Continuously monitor the state of the environment.

To this end, and using this modelling the environmental information collected will allow us to contribute to the development of a systemic and prospective approach to the state of the environment by pooling the ecological systems of the region, carrying out analyzes of the state of the environment and producing results.

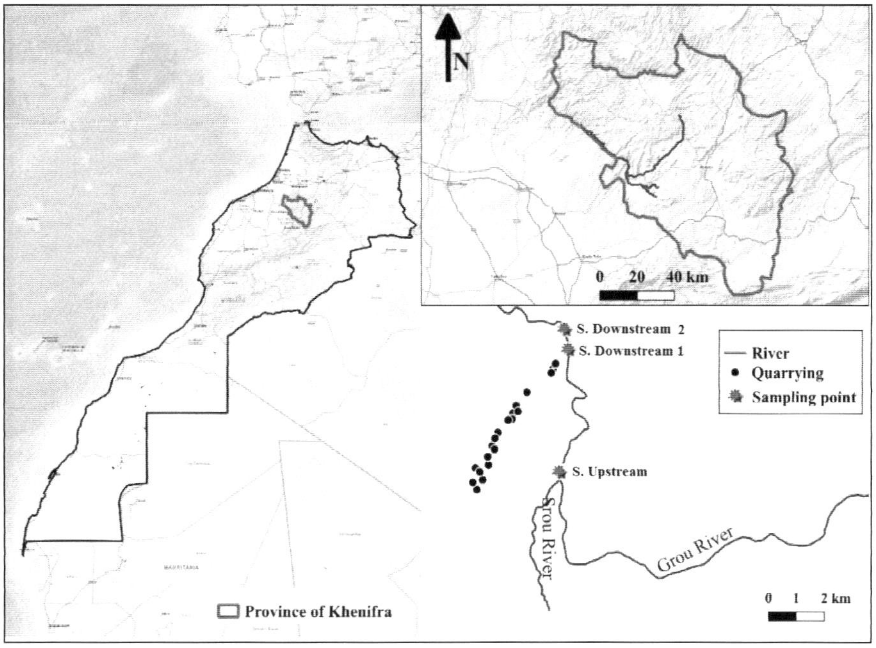

Figure 1. Geographical location of the study area, the studied quarries and the sampling stations.

3. Results and discussion

3.1. Impact on water quality

To assess the impact of the activity of the studied quarries on water quality, physico-chemical parameters were examined in this study.

Sampling was carried out at three stations on the Oued Grou river in March 2020 following a period of flooding, 1 station upstream (S.Up), and 2 downstream (S.Do1 and S.Do2).

3.1.1. Temperature. Water temperature is an important factor in organic production. It acts on the physical and chemical properties of water: in particular, its density, its viscosity, the solubility of its gases (in particular that of oxygen O_2) and the speed of chemical and biochemical reactions. In fact, aquatic life depends on this parameter indirectly too [7]. The results obtained in the area of our study indicate that this parameter does not show large variations from one station to another, with a minimum of 16 °C and a maximum of 19.8 °C.

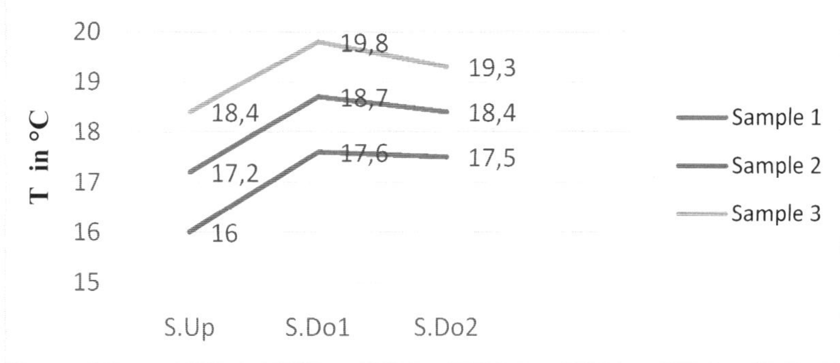

Figure 2: Temperature variation upstream and downstream

IOP Conf. Series: Earth and Environmental Science 1090 (2022) 012018 doi:10.1088/1755-1315/1090/1/012018

3.1.2. pH. The pH values recorded in our study area (upstream-downstream of the Khénifra quarrymen) vary between 8 and 8.82. In this zone there is a slight increase in pH which has an alkaline character justified by the type of limestone soil in the region, passing from upstream to downstream.

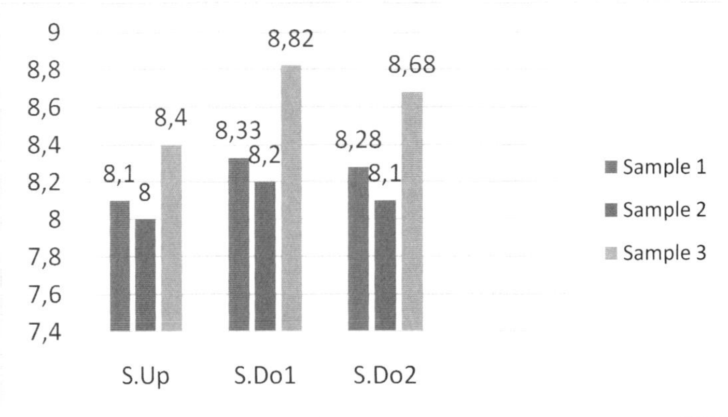

Figure 3: Measurement of the pH

3.1.3. Dissolved oxygen. In the study area, the value of 8.2 mg/l of dissolved oxygen is recorded at the first station (upstream), then there is a decrease in this value to 8 mg/l (downstream). This slight decrease is probably explained by the use of O_2 by aerobic micro-organisms to degrade organic matter.

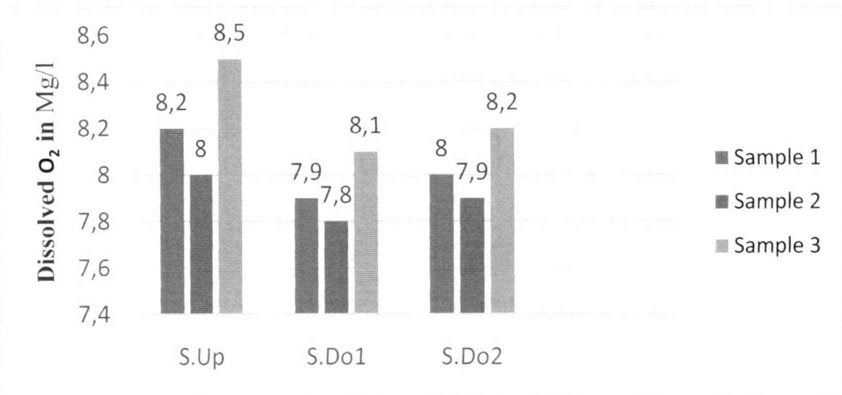

Figure 4: The variation in dissolved O_2 (mg / l) upstream and downstream

3.1.4. Electrical conductivity. The important values of the conductivity recorded are 2856,2 µs/cm due to the type of clay and limestone soil crossed by the water. The increase in these values downstream is caused by the mineralization of water from quarrying activities.

IOP Conf. Series: Earth and Environmental Science 1090 (2022) 012018 doi:10.1088/1755-1315/1090/1/012018

Figure 5: Change in electrical conductivity (us/cm)

3.1.5. Turbidity. Measuring turbidity allows us to collect visual information about the water. It reflects the presence of particles in suspension in the water: organic debris, clays, and microscopic organisms. In this study area, there is a remarkable difference between the first and the third stations explained by the runoff of rainwater and the wind, which carries several particles warning quarrymen.

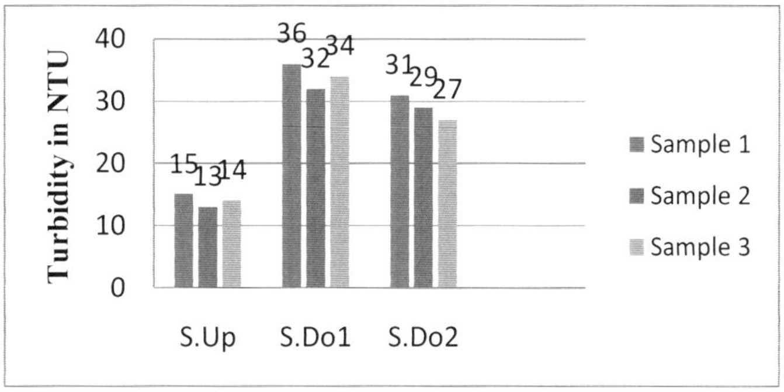

Figure 6: Turbidity change (NTU)

The digging of land and the extraction of materials (figure 7. A) generate a disturbance of the water quality in the quarry which lead to eutrophication indicating an advanced state of pollution locally and therefore algae bloom (figure 7. B).

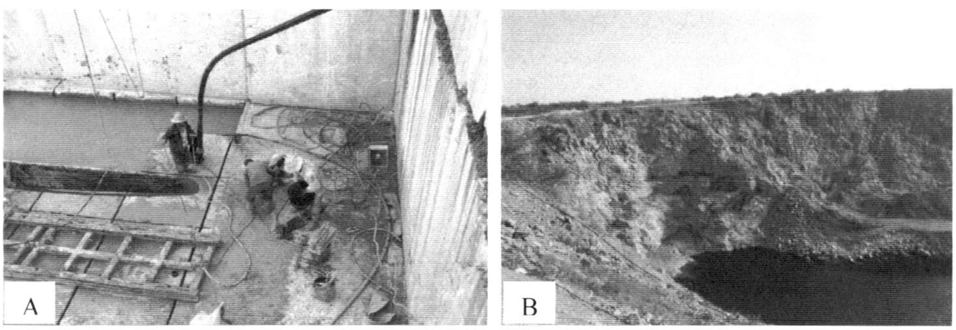

Figure 7: Pumping units installed in the outcrops of the aquifer (A) and eutrophication (B).

3.1.6. Dust emissions. On the Sidi Lamine quarry, dust emissions noticed with the naked eye result from the cutting of materials during blasting, their treatment, and storage, as well as the movement of machines on unpaved tracks (figure 8). These emissions are higher since the materials extracted and processed are dry.

Figure 8: Dust emissions resulting from the extraction of materials (A) and the movement of machinery on unpaved tracks (B).

3.2. Noise and vibrations

The weekly explosives fire in massive rock quarries and the noise of heavy machinery, in addition to crushing and screening equipment, induce noise and vibration emissions.

The vibration affects the stability of the infrastructures and the houses of the people living in Sidi Lamine in proximity to the operations of large open-air works.

Figure 9: Open-air work of heavy machinery near Sidi Lamine.

3.3. Impacts on flora and fauna

The location of the quarry activities (extraction, trampling, earthworks, etc) has a strong negative influence on the local ecosystem which causes the removal of native vegetation. Moreover, due to dust accumulation on vegetation leaves during machinery transportation, the photosynthesis process should be affected leading locally to an outbreak in the local flora composition and habitat loss and fragmentation. Consequently, the native fauna is seriously affected [8].

3.4. Impacts on soil quality

The negative impacts that can affect the soil are physical and chemical:

- Physical impacts, by soil compaction by the circulation of heavy machinery on undeveloped tracks;

IOP Conf. Series: Earth and Environmental Science 1090 (2022) 012018 doi:10.1088/1755-1315/1090/1/012018

- Chemical impacts, by the degradation of the biological quality of the soil by the discharge of solid and liquid wastes (oils, greases, etc.). This destroys the microorganisms in the soil and makes it unsuitable for agriculture.

3.5. Impact on social value

Creation of jobs through the exploitation of quarries can cause the decline in agricultural activities leading to an increase in rural migration [9].

3.6. Space consumption

The thickness of the alluvial deposit of quarries in the Khenifra region can reach 35m [5], the consumption of space linked to this method of extraction is much greater than that of massive rock deposits (height of limestone deposits about 15 to 80 m, eruptive> 100 m) for similar productions.

If landscape integration measures are not taken, massive rock quarries on the hillside, as well as the sprawl of the landscape by bodies of water can constitute an attack on the quality of natural sites.

3.7. Road transport

In areas where quarry workers are located, there is a real impact caused by the deterioration of the roadways linked to mud deposits and the circulation of bulky and heavily loaded vehicles on public roads.

3.8. Air impacts

Gases from explosives used in blasting and exhaust operations from equipment used in puncturing, loading and transporting materials, as well as Dusts from finds~~,~~ and other operations in open-cast quarries and dust from crushing and separating materials causes pollution of the ambient air.

3.9. Positive impacts

The quarries are generally responsible of many socio-economic changes:

- Settlement of the local population,
- Taxes benefits to the rural municipalities
- Considerable quantity of materials for the benefit of the building and civil engineering sectors.
- Improving the region's brand image by attracting new investors and promoters.

3.9.1. Some mitigation measures

Means make it possible to reduce the noise impact of quarries, such as: choosing of less noisy equipment, using of insulating materials and devices, installation of internal traffic routes and treatment areas. As for vibrations, to prevent the deterioration of constructions, their characteristics can be controlled by the use of suitable mining techniques (delayed initiation)

3.10. Recommendations for environmental protection

- ✓ Construct buildings to house a workshop for maintaining machinery;
- ✓ Storage of hydrocarbons and emptying of oils;
- ✓ The stripped vegetable soil and the sterile resulting from the exploitation must be accumulated on a non-exploitable plot and protected against erosion by runoff water;
- ✓ Respect the loading level of trucks and their tarpaulins during transport;
- ✓ Regularly water the traffic lanes and the extraction perimeter when necessary (dry period) to stabilize the dust clouds.

4. Conclusion

The evaluation of the impacts of the quarries of Sidi Lamine, shows that these quarries have more negative than positive impacts, knowing that, if we add the cumulative fact of other quarries throughout the territory of the Khenifra region, the results will surely be more important.

The identification of these negative impacts by phase shows that local actors and promoters are required to take, among other things, the environmental protection measures mentioned above in order to enhance these quarries which present significant socio-economic potential and participate in local development as well.

5. References

[1] Loi n° 12-03 relative aux études d'impact sur l'environnement. Dahir n° 1-03-60 du 10 rabiï I 1424 (12 mai 2003). Bulletin officiel n° 5118 du 19 juin 2003.
[2] Sanae A 2018 Politique des grands projets et gouvernance urbaine à Casablanca, URL : http://journals.openedition.org/emam/1446 ; DOI : https://doi.org/10.4000/emam.1446
[3] Driouech F, Déqué M & Mokssit A 2008 Numerical simulation of the probability distribution function of precipitation over Morocco. Climate Dyn. Springer. DOl 10.1007/s00382-008-04306.
[4] Verset Y 1988 Mémoire explicatif de la carte géologique (feuille de Kasbat-Tadla). Notes et Mém. Serv. Géol. Maroc, n°340bis, 132p.
[5] Cherrad B 1997 Le bassin versant de Oued Grou (plateau central marocain) étude hydro-climatologique thèse de Doctorat de l'université de Metz (Mention Géographie. 272 p.
[6] Beaudet G 1969 Le plateau central marocain et ses bordures: étude géomorphologique. Thèse Lettres, Paris, 478p.
[7] Kumar MP A 2012 A review of permissible limits of drinking water. Indian Journal of Occupational and Environmental Medicine 16: 40-44.
[8] Mature V 2004 Liste des conventions relatives à l'Environnement signées et ratifiées par le Maroc -24 pages.
[9] Julie C, Renaud J, Elise B, Fabrice B, Thierry D 2018 Impact of quarry exploitation and disuse on pedogenesis. CATENA, Elsevier, 2018, 160 (160), pp. 354-365. ff10.1016/j.catena.2017.09.012ff. ffhal-01790691f

Evaluation of cadmium levels in bivalve molluscs from Dakhla Bay in south of Morocco from 2012 to 2020.

M Dellal[1,2*], S Benbrahim[3], EM Ait Chatou[1], A Errhaif[2], A Benhra[4], M Anhichem[5], F Bouthir[5], R Chfiri[5], M Jayed[5] and M Malki[2].

(1) Marine Environment Surveillance and Monitoring Laboratory, INRH Regional Center in Dakhla- Morocco
(2) Environmental Health Laboratory, Ain Chock Faculty of Sciences, Hassan 2 University, Casablanca- Morocco
(3) Marine Environment Surveillance Department, INRH, Casablanca- Morocco
(4) URD of Chemical contaminants, INRH, Casablanca- Morocco
(5) Chemistry Laboratory, INRH, Casablanca-Morocco
- INRH (National Institute of Fisheries Research)

*E-mail: dellal@inrh.ma

Abstract: The present study illustrates the monitoring results of the assessment of the contamination of 5 species bivalve molluscs by cadmium in shellfish production areas of Dakhla Bay during the period from 2012 to 2020. The analysis of the samples is carried out by mass spectrometry using a mass spectrometer of the kind (ICP-MS). The results of this assessment show a difference in the accumulation of cadmium by species of bivalve molluscs of the two groups, in particular group 2 (burrowing bivalve molluscs) and group 3 (non-burrowing bivalve molluscs). Also, a slight increase in cadmium is noted in the different species of bivalve molluscs filter during the period 2017-2019, in all the shellfish production areas. However, the cadmium levels recorded remain below the toxicity thresholds. A decrease in these levels is clearly detected from the year 2020 in bivalve molluscs from all shellfish production areas of Dakhla bay.

Keywords: bivalve molluscs (LBM), cadmium, shellfish production areas, Dakhla bay.

1. Introduction

The bay of Dakhla, is one of the most important paralic environments in Morocco, it represents one of the regions of the richest coastal areas in fish; and is also one of the highest sites with aquaculture potential. Fishing and aquaculture are a vital socio-economic activity. The development of this activity required the construction of a new port 5 km from the bay pass, whose activities were launched during 2001. This makes the latter an attractive pole for the realization of various types of investment related to fisheries, aquaculture, tourism and recreation. On the other hand, these different activities represent a strong pressure attacking this paralic environment. This requires special attention and the elaboration of a program to protect and preserve this ecosystem [1].

Thus, to resolve the problems of the various contaminants, in particular biological and chemical that may condemn this paralic environment, studies to assess the quality and healthiness of this environment have been carried out continuously since 2001 by the INRH (National Institute of Fisheries Research), the objective of which is to provide elements of support for sustainable management for local actors. It also aims at providing information on the quality of this environment and its products as well as the protection of the consumer of fishery products [2]

Heavy metals are among the chemical contaminants monitored. Their uses are multiple and very diversified. They can vary from lead additives in fuels to silver salts in the photographic industry, to nickel or cadmium in accumulator batteries, zinc from gutters or chromium from stainless steels, copper from the electrical industry or arsenic from phytosanitary products. On the other hand, they are also sources of contamination. In fact, during all the development, use and / or recycling phases of these products, metals are released into the environment, either directly into continental or marine waters or into the atmosphere transported by winds. They are associated with aerosols before being deposited by dry or wet means on the surface of the land or the ocean. They are thus present in all the compartments of the environment, both because they are naturally present as natural sources or as an anthropogenic source, given that certain human activities favor their dispersion in the air in waters (sea and fresh). Thus, they present a potential danger for the consumer of the fishery product due to their possibility of concentration in marine species, their difficult elimination and their wide distribution in the aquatic environment [3].

The present study is based on an evaluation of the contamination for the environment and of filter-feeding and burrowing bivalve molluscs (group 2 and 3), by cadmium in the shellfish production areas of Dakhla bay during the period from 2012 to 2020. It also relies on a comparison of the cadmium accumulation capacity by the LBM species of the two groups and by the LBM species within each group.

2. Research Methods

2.1. Study zone

The bay of Dakhla, located on the Moroccan Saharan coast, is 37 km long and 10 to 12 km wide, its surface area is 400 km^2 and has a maximum depth of 20 m. It is oriented NE-SW. It is relatively narrow and open to the south on the Atlantic Ocean and it is composed of sandy and sandy-muddy sediment. It is subdivided into two parts, a southern part characterized by an alternation of furrows and shoals and a northern part represented by a vast basin whose average depth is about 15 meters [3]. It is characterized by high biological productivity, mainly at the bottom of the bay where the dissolved oxygen content — the result of photosynthesis— is high. This high ecosystem productivity is associated with the upwelling recirculation cell at the level of the continental shelf and the Canary Islands current. However, fishing is strictly prohibited in the bay (AMP / ZMP), as it is the spawning ground for many species of fish, crustaceans and molluscs, which ensures their sustainability [4].

However, the phenomena of upwelling cold water, rich in nutrients and mineral salts, can be a factor in the enrichment of the waters of the area by metallic traces, particular in the bay of Dakhla. By means of the force of the currents, these ocean waters can also disperse anthropogenic inputs of wastewater and industrial water and move them from downstream to upstream of the bay. Shellfish production areas located further from sources of pollution can be impacted [5].

2.2. Sampling areas and sampling frequency

The areas studied are the parts of the bay that are the richest in natural shellfish beds as well as the parts reserved for the culture of cupped oysters and clams. The assessment of the contamination of LBMs by cadmium in natural or farmed deposits focused on burrowing and filtering bivalve molluscs used as an integrating support for water contamination (figure1).

IOP Conf. Series: Earth and Environmental Science 1090 (2022) 012019 doi:10.1088/1755-1315/1090/1/012019

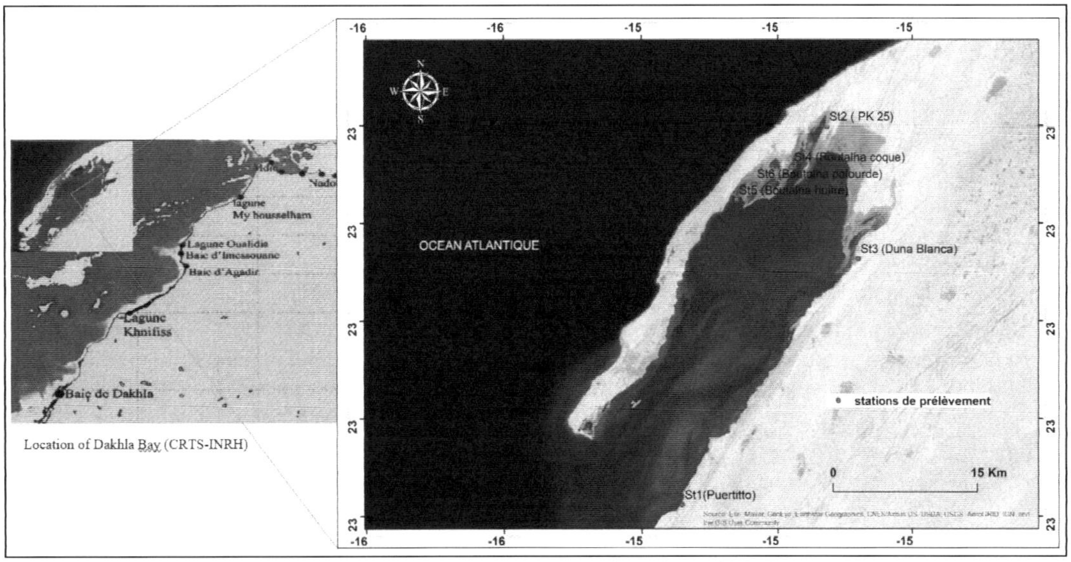

Figure 1. Location of sampling stations in Dakhla bay.

The species sampled are those exploited in the following areas, namely the mussel (*Perna perna*) in the Puertitto area, the razor clam (*Solen marginatus*) at PK 25 and the common shell (*Cerastoderma edule*) at Boutalha (Table 1).

In the farming areas, the sampled species are the ones exploited, namely the cupped oyster (*Crassostrea gigas*), in the Boutalha and Dunablanca areas, and the clam (*Rudetapes dessecatus*) in Boutalha (Table 1). The frequency adopted for the evaluation of the contamination of LBMs by Cadmium is six-monthly for a study-period from 2012 to 2020.

Table 1. Location of sea bivalve molluscs sampling stations in Dakhla bay.

Station / area	Name	Geographical coordinates	MBV species
St 1	Puertitto (Natural Mussel site)	23°34'16 N 15°54'13 W	*Perna perna*
St 2	PK25 (Knife natural site)	23°54'45 N 15°46'04 W	*Solen marginatus*
St3	Dunablanca (Oyster breeding site)	23°47'36 N 15°44'30 W	*Crassostrea gigas*
St4	Boutalha (Common shell natural site)	23°52'26 N 15°48'01 W	*Cerastoderma edule*
St5	Boutalha(Clam Farm site)	23°51'36 N 15°50'11 W	*Rudetapes dessecatus*
St6	Boutalha (Oyster breeding site)	23°50'49 N 15°51'09 W	*Crassostrea gigas*

2.3. Analyzes

The sampling strategy permitted the collection of bivalve molluscs. Specimens of each bivalve molluscs species were depurated in the seawater from the local sampling sites for about 24h hours in order to remove the digestive contents so as not to interfere with the content of cadmium in the shellfish tissue. The soft part removed from shells and then homogenized by mixing in rotating glass and stored in sealed plastic bottles in vacuum desiccators. All precautions are taken in order to avoid contamination during the steps of preparation and Measurements and then vessels were tested individually for blanks and screened before use. Ultrapure water and acid of high purity cleaned by sub boiling distillation are used.

The digestion procedure is carried out according to method for the pressure digestion of foodstuffs "NM EN 13805: 2015". Determination of Cd is performed by inductively coupled plasma mass spectrometry (ICP-MS) for direct analysis of liquid samples (iCAP-Q series mass spectrometry Thermo Fisher Scientific), according to "NM EN 15763: 2012". The expected values of Cd in the samples are expressed in mg/kg of fresh weight.

The criteria for evaluating the chemical quality of shellfish production areas is based on the contamination thresholds set by the Circular of the Ministry of Agriculture and Maritime Fisheries 1508-12 relating to the production and the market for live bivalve molluscs (15 August 2012), the Directive of the Minister of Agriculture, Maritime Fisheries, Rural Development and Water and Forests n ° 1950-17 of 07 August 2017 relating to the health classification of maritime shellfish production areas and regulation (EC) n ° 854/2004 of the European Parliament and of the Council of 29 April 2004 laying down specific rules for the organization of official controls on products of animal origin intended for human consumption.

3. Results and discussions

The results of the analyzes of cadmium in LBMs indicate the following different cases:

The assessment of the chemical contamination of cadmium in the LBMs of shellfish production areas in Dakhla Bay during the period from 2012 to 2020, does not reveal any contamination. Indeed, the cadmium contents oscillate between a minimum of 0.1 mg / kg PF noted in burrowing species collected during the year 2012 and 1.55 mg / kg PF noted in oysters collected during the winter season 2018 in the Boutalha area. On the other hand, the maximum level noted does not exceed the national regulatory thresholds (Circular 1508-12 (threshold = 1 mg / kg PF) and the Decree 1950-17 (threshold = 2 mg / kg PF) (figure 2).

By comparing the accumulation of cadmium in burrowing and filtering LBM, we see that the cadmium levels detected in burrowing LBM are lower than the cadmium contents accumulated in filtering LBM. On the other hand, these levels are lower than the national and international regulatory thresholds (EC 854/2004 (threshold = 1 mg / kg PF)), which confirms the significant capacity of cadmium accumulation in filter feeders by input to burrowers. Very generally, burrowing shellfish (clams, cockles, tellins) are much less contaminated than non-burrowing bivalves [6].

Also, these results reveal the possibility of the significant presence of cadmium, in the water column as in the sediment. Indeed, the results of the evaluation of the sediment quality of the bay of Dakhla reveal that the latter was assessed through the determination of the contents of the main metallic trace elements (cadmium, lead, mercury, chromium, copper and zinc) [7]. An increase in cadmium levels in LBM, especially in filter feeders (Hollow oysters), is recorded to have exceeded European regulatory thresholds from the end of 2017 to the winter 2018 (figure 2).

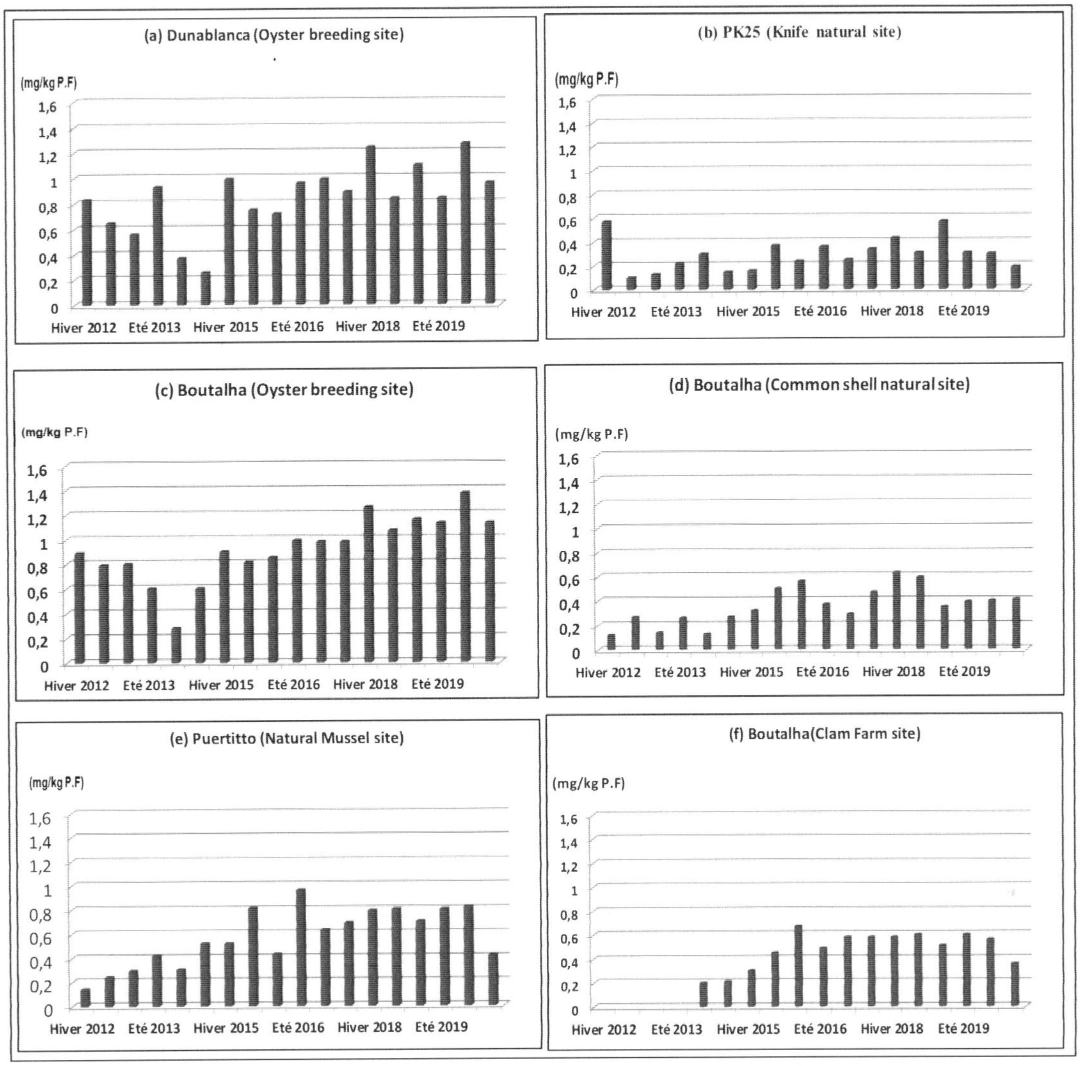

Figure 2. Evaluation of cadmium levels in bivalve molluscs from Dakhla Bay (2012-2020).

The results of the statistical analysis show a very significant difference between the two species of filter-feeding LBM ($p < 0.05$) (Table 2). Indeed, the cadmium contents accumulated in the Hollow Oysters (*Crassostrea gigas*) in particular of the two zones of Boutalha and Dunablanca of the bay of Dakhla are higher than those accumulated by the mussels (*perna perna*) of the zone of Puertitto of the bay of Dakhla throughout the study period (2012-2020).

Table 2: Comparison of LBM Filters (Anova test 1).

ANALYSIS OF VARIANCE : ONE FACTOR						
DETAILED REPORT						
Groups	Number of samples	Sum	Mean	Variance		
Puertitto (Natural Mussel site)	18	10,28	0,571111111	0,05741046		
Boutalha (Oyster breeding site)	18	16,59	0,921666667	0,06837941		
Dunablanca (Oyster breeding site)	18	15,15	0,841666667	0,07007353		
ANALYSIS OF VARIANCE						
Source of variations	sum of squares	Degree of freedom	Mean of squares	F	Probability	Critical value for F
Between groups	1,214937037	2	0,607468519	9,30447224	0,00035899	3,178799292
Inside the groups	3,329677778	51	0,0652878			
Total	4,544614815	53				

These results confirm that there is a difference between these two species of filter feeders, indicating that the Oyster species is much more sensitive than the mussel species (figure 3). Indeed, previous research studies have shown that there are important differences within group 3 of non-burrowing bivalve molluscs, such as that which exists in several metals (cadmium, zinc) in particular much more accumulated by oysters than by mussels [6].

The expected results also show that the levels of cadmium accumulated by the mussels have never exceeded national and international regulatory thresholds (figure 3).

A decrease in the cadmium contents accumulated by the two species of filtering LBM is detected from the summer of 2020 (figure 3).

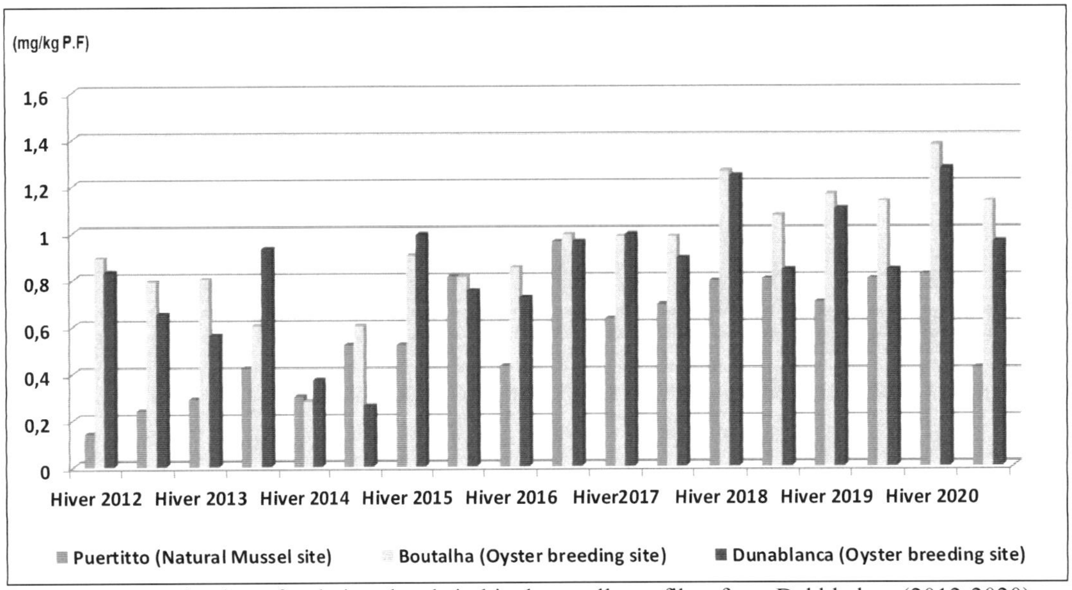

Figure 3. Evaluation of cadmium levels in bivalve molluscs filter from Dakhla bay (2012-2020).

The results of the statistical analysis show that the difference in the capacity of cadmium accumulation is not significant between the three species of burrowing LBM ($p > 0.05$) (table 3). In fact, the cadmium contents accumulated by the three burrowing species, in particular the common shell (*Cersoderma edule*), the clam (*Rudetapes dessecatus*) and the razor clam (*Solen marginatus*) taken from shellfish-growing areas, Boutalha and PK25 are similar and oscillate between a minimum of 0.1 mg / kg PF and a maximum of 0.67 mg / kg PF (figure 4).

Table 3: Comparison of LBM Burrowing (Anova test 2).

ANALYSIS OF VARIANCE : ONE FACTOR

RAPPORT DÉTAILLÉ

Groups	*Number of samples*	*Sum*	*Mean*	*Variance*
PK25 (GKnife natural site)	18	5,3	0,294444444	0,01812026
Boutalha (Common shell natural site)	18	6,47	0,359444444	0,02319379
Boutalha (Clam Farm site)	18	6,69	0,371666667	0,05950882

ANALYSIS OF VARIANCE

Source of variations	*sum of squares*	*Degree of freedom*	*Mean of squares*	*F*	*Probability*	*Critical value for F*
Between groups	0,062025926	2	0,031012963	0,92279543	0,40393935	3,178799292
Inside the groups	1,713988889	51	0,033607625			
Total	1,776014815	53				

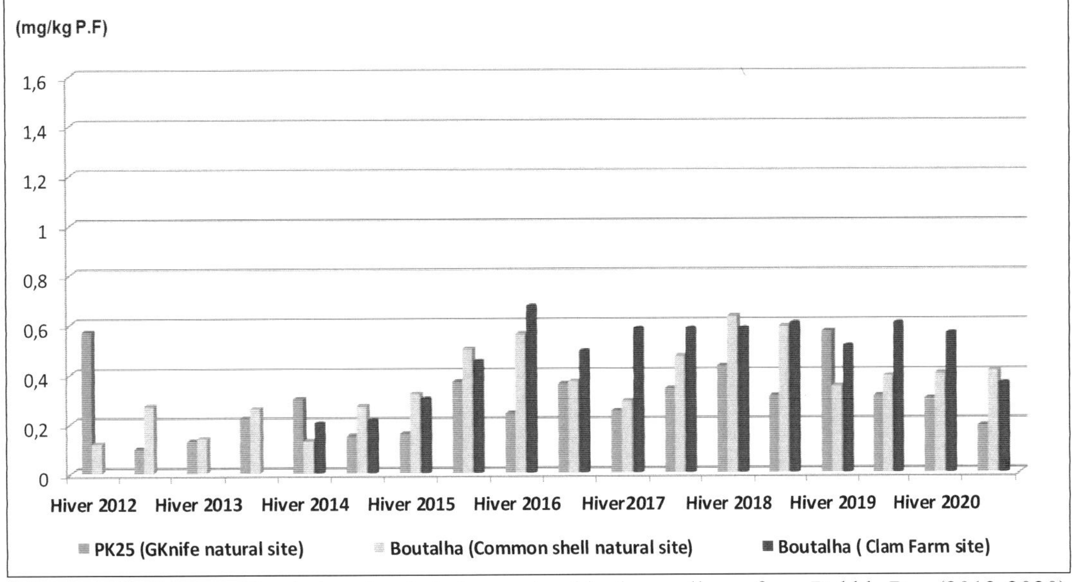

Figure 4. Evaluation of cadmium levels in burrowing bivalve molluscs from Dakhla Bay (2012-2020)

4. Conclusion

The results of the evolution of Cadmium in the LBM of the shellfish production areas of the bay of Dakhla for the period from 2012 to 2020, indicate no contamination of LBM was detected in shellfish production areas of Dakhla Bay by cadmium. Also the cadmium contents noted in the LBM are below the contamination thresholds set by national and international regulations for burrowing LBM and for the mussel species of filtering LBM.

A significant difference in the capacity of cadmium accumulation between the two species of filter feeders is revealed. The species of Hollow Oyster is much more sensitive and accumulates much more cadmium than the species of Mussel. On the other hand, the capacity to accumulate cadmium is high in filter-feeding LBM and are much more sensitive than burrowing LBM.

The cadmium accumulation capacity by burrowing LBM is similar for the three burrowing species, in particular the common shell (*Cersoderma edule*), the clam (*Rudetapes dessecatus*) and the razor clam (*Solen marginatus*);

An increase in cadmium levels in the two groups of LBM is noted to have exceed the international regulatory threshold for oysters from year 2018. However, a decrease in these levels is noted in both groups of LBM from the summer season 2020.

5. References

[1] Orbi A, Dafir JM, Berraho A 1995 Etude pluridisciplinaire de la baie de Dakhla. Travaux et document N°86 de l'INRH, Maroc. Rapport interne 26p.
[2] INRH 2015 Plan stratégique.
[3] Stellio C Juin 2005 Modélisation de la bioaccumulation de métaux traces (Hg, Cd, Pb, Cu et Zn) chez la moule (Mytilus galloprovincialis) en milieu méditerranéen. 314-p11.
[4] Agence du sud Janvier 2014 Etude de la vulnérabilité de la baie d'oued Eddahab- Elaboration du plan de gestion et du plan d'action et la charte environnementale. Mission II, Version définitive 131-N374-12b.
[5] INRH Octobre 2015 Etude sanitaire de la zone de Boutalha Nord, Région de Dakhla- p36.
[6] Grouhel A, Knoery J, Marchand P Février 2019 Programme de surveillance chimique sanitaire et technique définitif - convention relative aux actions de surveillance sanitaire année 2018. Réf: Ifremer n° 17/2216082. 5-36.
[7] Anhichem M, Dellal M, Chfiri R, Yahyaoui A, Benbrahim S Etude de la qualité des eaux et des sédiments de la Baie de Dakhla au Maroc. Qualité physicochimique et contamination Métallique, 2017- p17.

IOP Conf. Series: Earth and Environmental Science 1090 (2022) 012020 doi:10.1088/1755-1315/1090/1/012020

Olive and green tea leaves extract in *Pleurotus ostreatus* var.florida culture media: Effect on mycelial linear growth rate,diameter and growth induction index

W Werghemmi[1*], S Abou Fayssal[2,3], H Mazouz[4], H Hajjaj[4] and L Hajji[1]

[1] Laboratory of Bioactives and Environmental Health, Faculty of Sciences, Universityof Moulay Ismail, Meknes, Morocco
[2] Department of Agronomy, Faculty of Agronomy, University of Forestry, 10 KlimentOhridski Blvd. 1797, Sofia, Bulgaria
[3] Department of Plant Production, Faculty of Agriculture, Lebanese University, Dekwaneh, Beirut, Lebanon
[4] Laboratory of Plant Biotechnology and Valorization of Bio-resources, Faculty ofSciences, University of Moulay Ismail, Meknes, Morocco

* wafa.werghemmi0709@gmail.com

Abstract. The exponential disposal of agro-industrial wastes onto the environment has endangered all forms of life. The implementation of these wastes in mushroom production is an eco-friendly and promising solution. The effect of olive and tea leaves extracts represented in culture media treatments: potato dextrose agar (PDA) 80% + tea extract (TE) 20% (T2), PDA 80% + olive extract (OE) 20% (T3), PDA 80% + OE 10% + TE 10% (T4), PDA 70% + OE20% + TE 10% (T5) on mycelial growth diameter (MGD) and linear growth rate (MLGR) of *Pleurotus ostreatus* var. florida was compared to the one of PDA used as control (T1) at 22, 25 and 28°C inoculation temperatures. Optimum MGD was observed at 28°C in T1 (day 2), while it was significantly increased by 1.1–1.4 folds (days 4, 6, 8) in culture media containing plant extracts nevertheless the inoculation temperature. MLGR was improved by 102%–145% in olive/tea culture media compared to PDA nevertheless the inoculation temperature. Optimum growth induction index (GII) was observed in T3 (22.2%) at 22°C, T5 (21.9%) at 25°C and T4 (18.2%) at 28°C. These findings suggest the combination of olive and tea leaves extracts in the production of *Pleurotus ostreatus*.

1. Introduction

Nowadays, the world's population continues to arise in parallel to increased human poverty and tremendous environmental pollution. The hazardous burning and disposal of agro-industrial wastes in different worldwide regions are basically affecting and encountering different life forms. Several initiatives were launched in order to find solutions for these emerging issues [1–4].

Morocco is a traditional olive producer since the ancient ages and ranks six in terms of worldwide olive oil production [5]. The Moroccan government lead a project increasing the national olive production area from 773.000 ha in 2009 to 1.220.000 ha in 2020 according to the International Olive Council [6]. This has naturally increased the amount of olive by-products wastes that need to be wisely treated before their discharge into the environment as each litre of olive oil results in 6.2 kg of pruning residues (leaves and branches) [7]. Burning these wastes resulted in an air pollution represented by the generation of around 750g of CO_2 per litre of produced olive oil [7].

Pleurotus ostreatus (*P. ostreatus*, oyster mushroom) is a lignocellulosic fungus that grows on lignocellulosic material through a process of bioconversion. It is ranked as the second most cultivated mushroom in the world [8] and is well known for its high nutritional value (carbohydrates, protein, minerals) and medicinal properties (vitamins and antioxidants) [9]. It is mainly grown on wheat straw substrate [10,11] beside many other substitutes like spent mushroom substrate (SMS) [12,13], spent coffee grounds [14], maize wastewater [15], tea leaves [16] and so on. It is noteworthy that the type of substrate affects majorly the yield and composition of produced mushrooms [17,18].

Olive by-products were found to be recycled via bioremediation, a treatment process consisting on growing mushrooms on agro-industrial wastes. In this context, it was reported that the incorporation of olive mill wastes in *Pleurotus ostreatus* substrate resulted in increased yields and β-glucans [19]. Researchers acknowledged the decreased amounts of sugars, fat, saturated fatty acids (SFA) and increased amounts of polyunsaturated fatty acids (PUFA) as well as monounsaturated ones (MUFA) in mushrooms grown on substrates containing olive pruning residues; thus, increased health and nutritional attributes [9]. In a similar vein, Abou 10] considered the mixture of olive pruning residues and spent coffee grounds a promising solution for wheat straw scarcity, resulting in high yields and higher MUFA essentially needed for LDL level's regulation. Other researchers used olive press cake mixed with several agro-industrial wastes to produce a cellulose medium for growing *Pleurotus ostreatus* [20]. They reported the successful colonization of this medium by oyster mushroom between 15 and 30°C associated with decreased linear growth rate in comparison with potato dextrose agar (PDA) (used as control). These authors also noted that the daily linear growth rateof *P. ostreatus* on olive press cake alone was two folds lower than on PDA. The use of green tea extract as growing medium for *P. ostreatus* was investigated by 21]; they acknowledged anincreased enzymatic activity of the mushroom with no mention of its impact on linear growth rate.

Olive leaves hold valuable concentrations of cellulose, hemicellulose and lignin [22] essential for mushroom growth and production. Moreover, olive leaf extract is well known to enclose therapeutic effects in food and medical applications [23] due to their richness in phenols, flavones and flavonols. To our knowledge, this is the first time that fresh olive leaves extract was used combined or not with dried green tea leaves extract in *P. ostreatus* var. florida in vitro production. Therefore, the current study aims the investigation of fresh olive leaves and dried tea leaves extracts' effect on the mycelial growth diameter and linear growth rate of oyster mushroom in comparison with PDA.

2. Materials and methods

2.1. *Mushroom culture*

Pleurotus ostreatus var. florida, sub-cultured on PDA (BAM Media M127), was procured from "Champignons du Maroc" (Rabat, Morocco) and initially imported from Belgium. It was stored at in the refrigerator at 4°C until the beginning of the experiment.

2.2. *Plant sample collection*

Fresh olive leaves, pruned form the orchards of the "Moulay Ismail Meknes" university (33.856292° N, 35.574372° E) on March 2020, were used to prepare a leaf olive extract. Dried green tea "Sultan", purchased form a local Moroccan store, was adopted for green tea extract in order to push mycelial growth in in-vitro conditions.

2.3. *Aqueous extraction from olive and tea leaves*

2.3.1. Preparation of olive leaves extract. Fresh olive leaves were washed several times in order to get rid of all dust until perfect cleaning. Then, they were a were placed in a clean and well aerated areato drain completely and sterilized using an autoclave (WOF–155 WiseVen, DAIHAN Scientific Co., Ltd., Korea) at a temperature and pressure of 121°C and 15 psi respectively for15 min. A precise digital balance (PS 450.3Y, Radwag, Poland) was used to weight 60 g of fresh olive leaves and 300 mL of distilled water were then added. The extract was placed on a hotplate equipped with magnetic stirrer (MR Hei-Tec, Heidolph, Germany). After 10 minutes of boiling, the extract was left for cooling, then filtered for 24 hours using a filter paper. Subsequently, the obtained solution was centrifuged for 20 minutes at a speed of 5000 rpm (Sigma D-37520, Sigma Laborzentrifugen GmbH, Germany). Then, the needed part for mediapreparation was frozen at –20°C for optimum preservation.

2.3.2. Preparation of tea leaves extract. Same procedure for olive leaves extract was adopted for the preparation of tea leaves extract with some difference (figure 1). Dried tea leaves were already pasteurized from its original source and placed in packages; therefore, no initial washing was performed.

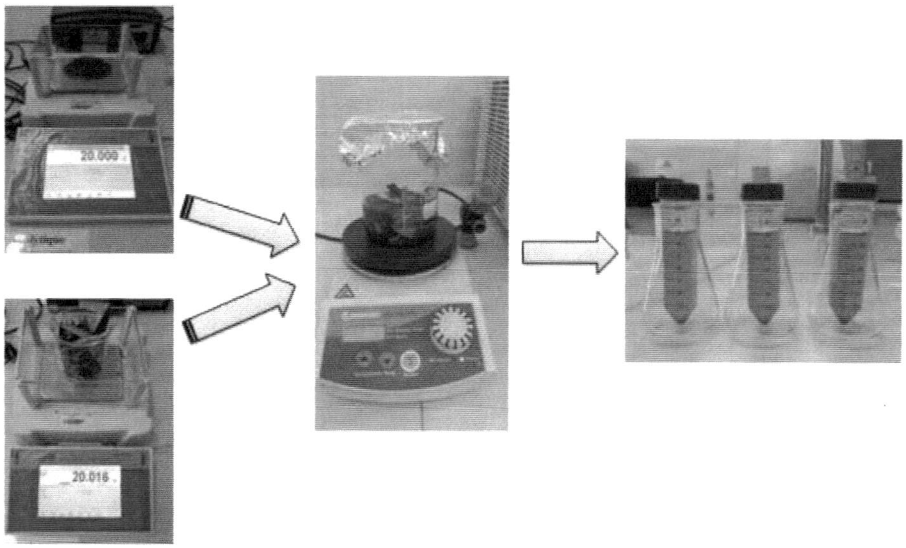

Figure 1. Plant leaves aqueous extraction's steps

2.4. Preparation of the culture media

Culture media were prepared following the instructions on the commercial box. 40 g of powdered medium were suspended in 1L of distilled water. PDA medium, composed of 4 g potatoes, 20 g dextrose, 20 g agar powder, was used as control (T1). Four culture media types (treatments) were prepared by replacing a part of distilled water with prepared aqueous solutions (extracts) of olive and tea leaves. Four flasks were prepared separately as follow: media contained 8 g of PDA powder, 160 mL of distilled water and 40 mL of olive/tea extract (T2 and T3). 160 mL of distilled water, 20 mL of tea extract and 20 mL of olive extract were added to 8 g of PDA powder (T4). 140 mL of distilled water, 40 mL of olive extract and 20 mL of tea extract were added to 8 g of powdered PDA (T5) (table1). pH of all culture media was adjusted to 6.5 using 0.1 mol of potassium hydroxide (KOH).

Table 1. Culture media composition basedon olive and tea extracts.

Aqueous solution	T2	T3	T4	T5
Olive extract	0%	20%	10%	10%
Tea extract	20%	0%	10%	20%
Distilled water	80%	80%	80%	70%

2.4.1. Sterilization of culture media. All prepared media were subjected to a sterilization process carried out in a conventional autoclave for 15 min at 121°C and a pressure of 15 psi.

2.4.2. Inoculation and incubation of culture media. After autoclaving, all culture media were placed ina sterile host for few minutes to cool down. Then, around 25 mL of each culture medium are placed in 85 mm sterile petri dishes and left to solidify. For each medium, 9 replicates were inoculated with a piece of mycelium of around 5 mm × 5 mm in the centre. Each 3 replicates were placed into incubator at corresponding temperatures of 22°C, 25°C and 28°C, respectively.

2.5. Calculation of studied parameters

2.5.1. Mycelial growth diameter (MGD). The mycelial growth diameter (MGD), expressed in mm, was calculated every 48 hours using the following equation [24]:

$$MGD\ (\text{mm}) = (D1 + D2)/2 \qquad (1)$$

The following figure shows clearly the mentioned methodology of calculation (figure 2).

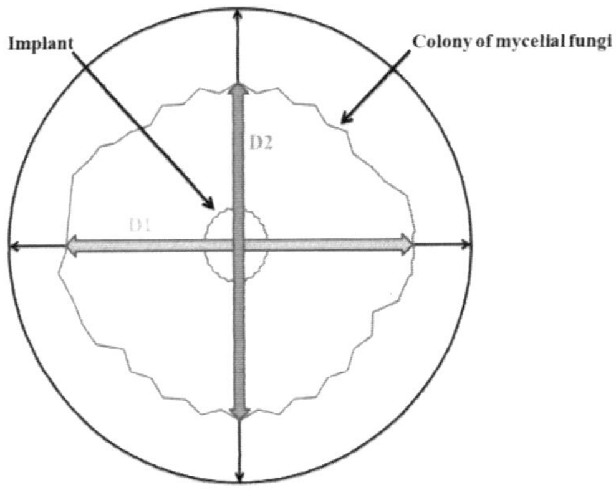

Figure 2. Mycelial growth diameter calculation

2.5.2. Mycelial linear growth rate (MLGR). The mycelial linear growth rate (MLGR), expressed inmm/day, was calculated following the equation [24]:

$$MLGR\ (\frac{mm}{day}) = \frac{Diameter\ average\ n' - Diameter\ average\ n}{n - n'} \qquad (2)$$

2.5.3. Growth induction index (GII). The growth induction index (GII), expressed in percentage (%),was calculated after six days of incubation following the equation [24]:

$$GII\ (\%) = \left(\frac{Diameter\ of\ the\ control\ PDA - Diameter\ of\ the\ treatment}{Diameter\ of\ the\ control\ PDA}\right) \times 100 \qquad (3)$$

2.6. Spawn preparation for different grains culture media

2.6.1. Spawn preparation steps. Three different types of grains (wheat, sorghum and maize) were tested for their ability to produce white mycelium. The first step consisted on the removal of any dirt and impurities form the seeds. Then, all seeds were rinsed several times under running water for more approval of the cleaning process. For optimal water absorption, seeds were immersed in distilled water, where the latter's volume was double of the former's one. They were then boiled and left to drain for one hour. As the nature of each grain type varies, the boiling time was adjusted accordingly: 15 min for wheat, 10 min for sorghum and 30 min for corn. For each 100 g of seeds, 1% dextrose, 1.5% yeast extract, 2% calcium carbonate, 4% calcium sulphate were weighted and mixed with the seeds. The latter were then autoclaved at 121°C for one hour. After cooling in clean environment, seeds were inoculated with the mycelium and then incubated until complete bleaching at a temperatureof 25°C (figure 3).

Figure 3. Seeds preparation

2.6.2. Estimation of spawn grain MGD and MLGR. In order to estimate the MGD and MLGR, petri dishes (90 mm) were filled with a 3–4 mm layer of sterilized seeds inoculated with a 1 cm × 1 cm piece of mycelium. Studied parameters were calculated following the equations mention in section 2.5.

2.7. Statistical analysis. One-Way ANOVA and Duncan tests were applied for statistical analysis of findings using SPSS 25®.

3. Results and discussion

3.1. Effect of inoculation temperature on mycelial growth diameter of treatments

Results showed that at 22°C (day 2), the mycelial growth diameter (MGD) was significantly higher by 2.2 mm when 20% of olive extract (OE) were added to potato dextrose agar (PDA) in comparison withcontrol (T1: PDA) (table 2, figures 4–6). Except for T4 (PDA 80% + OE 10% + TE 10%), all culture media containing olive and/or tea (TE) extracts showed a significantly increased MGD by 3.0–8.8 mmin comparison with T1 at a temperature of inoculation of 25°C. Increasing the latter to 28°C resulted in a significantly lower MGD of the culture medium containing OE and/or TE in comparison with T1 (lower by 0.8–2.5 mm).

Table 2. Mycelial growth diameter (MGD) (mm) of different treatments at Day 2.

Treatment	MGD (Temperature 22°C)	MGD (Temperature 25°C)	MGD (Temperature 28°C)
T1	21.3±0.1d	18.2±0.1a	31.3±0.1d
T2	17.3±0.1b	21.2±0.1b	28.8±0.1a
T3	23.5±0.1e	27.0±0.1c	29.0±0.1a
T4	15.8±0.1a	18.3±0.1a	30.5±0.1c
T5	18.3±0.1c	24.8±0.1c	29.8±0.1b
p_{value}	<0.01	<0.01	<0.01

T1 : PDA, T2 : PDA 80% + TE 20%, T3 : PDA 80% + OE 20%, T4 : PDA 80% + OE 10% + TE 10%, T5 : D PDA 70% + OE 20% + TE 10%, MGD: mycelial growth diameter. Values are means ± SD; Means within the same column followed by the same letters are not significantly different at p<0.05 according to Duncan's multiple range test.

Except for T2 (PDA 80% + TE 20%), all treatments containing OE and/or TE inoculated at 22°C (day 4) had a significantly higher MGD (higher by 1.2–6.0 mm) compared to PDA (T1) (table 3). At highertemperatures (25 and 28°C), all extracts containing OE and/or TE had significantly higher MGDs than T1 by 3.5–14.3 mm and 0.5–4.5 mm, respectively.

Table 3. Mycelial growth diameter (MGD) (mm) of different treatments at Day 4.

Treatment	MGD (Temperature 22°C)	MGD (Temperature 25°C)	MGD (Temperature 28°C)
T1	33.3±0.2b	35.7±0.2a	51.3±0.2a
T2	31.3±0.2a	40.2±0.2c	53.7±0.2c
T3	39.3±0.2e	47.2±0.2d	51.8±0.2b
T4	34.5±0.2c	39.2±0.2b	55.8±0.2e
T5	37.0±0.2d	50.0±0.2e	54.3±0.2d
p_{value}	<0.01	<0.01	<0.01

T1 : PDA, T2 : PDA 80% + TE 20%, T3 : PDA 80% + OE 20%, T4 : PDA 80% + OE 10% + TE 10%, T5 : PDA 70% + OE 20% + TE 10%, MGD: mycelial growth diameter. Values are means ± SD; Means within the same column followed by the same letters are not significantly different at p<0.05 according to Duncan's multiple range test.

In a similar trend to day 4, same observation was clear at day 6 at different inoculation temperatures (table 4).

Table 4. Mycelial growth diameter (MGD) (mm) of different treatments at Day 6.

Treatment	MGD (Temperature 22°C)	MGD (Temperature 25°C)	MGD (Temperature 28°C)
T1	48.7±0.1a	49.3±0.1a	69.5±0.1a
T2	48.8±0.1a	53.8±0.1c	83.0±0.1b
T3	59.5±0.1d	58.8±0.1d	84.2±0.1d
T4	53.8±0.1b	53.2±0.1b	85.0±0.0e
T5	56.5±0.1c	63.2±0.1e	83.8±0.1c
p_{value}	<0.01	<0.01	<0.01

T1 : PDA, T2 : PDA 80% + TE 20%, T3 : PDA 80% + OE 20%, T4 : PDA 80% + OE 10% + TE 10%, T5 : PDA 70% + OE 20% + TE 10%, MGD: mycelial growth diameter. Values are means ± SD; Means within the same column followed by the same letters are not significantly different at p<0.05 according to Duncan's multiple range test.

At day 8, all treatments containing OE and/or TE showed significantly higher MGD (5.1–10.5 mm/day) in comparison with PDA (73.2 mm) when they were all incubated at 22°C (table 5). At an incubation temperature of 25°C, a similar trend was observed with a significant increase by 1.5 mm (full colonization of the culture medium). At 28°C, all treatments had fully colonized the culture media (85.0 mm).

Table 5. Mycelial growth diameter (MGD) (mm) of different treatments at Day 8.

Treatment	MGD (Temperature 22°C)	MGD (Temperature 25°C)	MGD (Temperature 28°C)
T1	73.2±0.3a	83.5±0.3a	85.0±0.0a
T2	78.3±0.3b	85.0±0.0b	85.0±0.0a
T3	83.7±0.3d	85.0±0.0b	85.0±0.0a
T4	82.0±0.3c	85.0±0.0b	85.0±0.0a
T5	82.0±0.3c	85.0±0.0b	85.0±0.0a
p_{value}	<0.01	<0.01	1.00

T1 : PDA 100%, T2 : PDA 80% + TE 20%, T3 : PDA 80% + OE 20%, T4 : PDA 80% + OE 10% + TE 10%, T5 : PDA 70% + OE 20% + TE 10%, MGD : mycelial growth diameter. Values are means ± SD; Means within the same column followed by the same letters are not significantly different at $p<0.05$ according to Duncan's multiple range test.

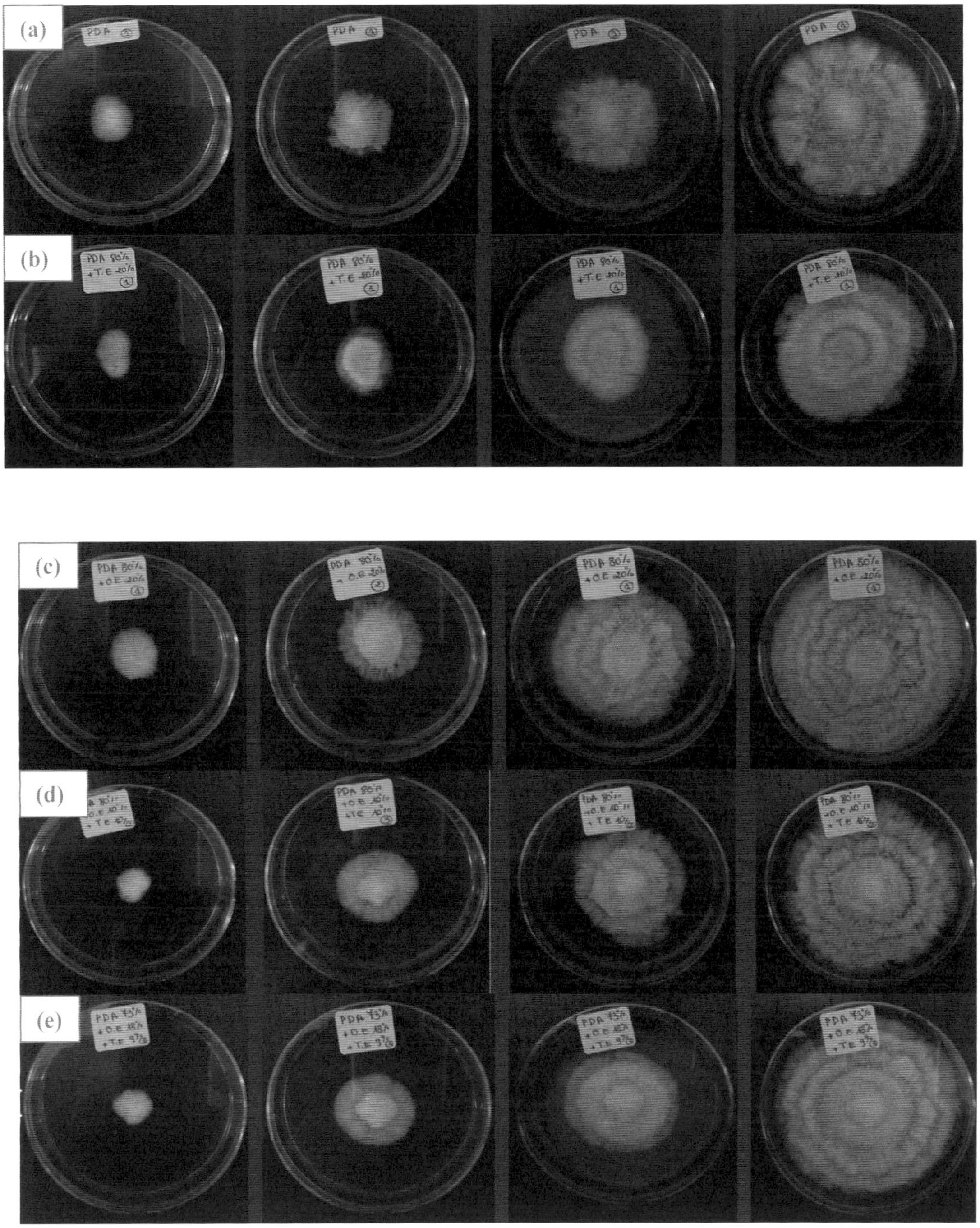

Figure 4. Mycelial growth diameter at 22°C; (a) T1: PDA, (b) T2: PDA 80% + TE 20%, (c) T3: PDA80% + OE 20%, (d) T4: PDA 80% + OE 10% + TE 10%, (e) T5: PDA 70% + OE 20% + TE 10%

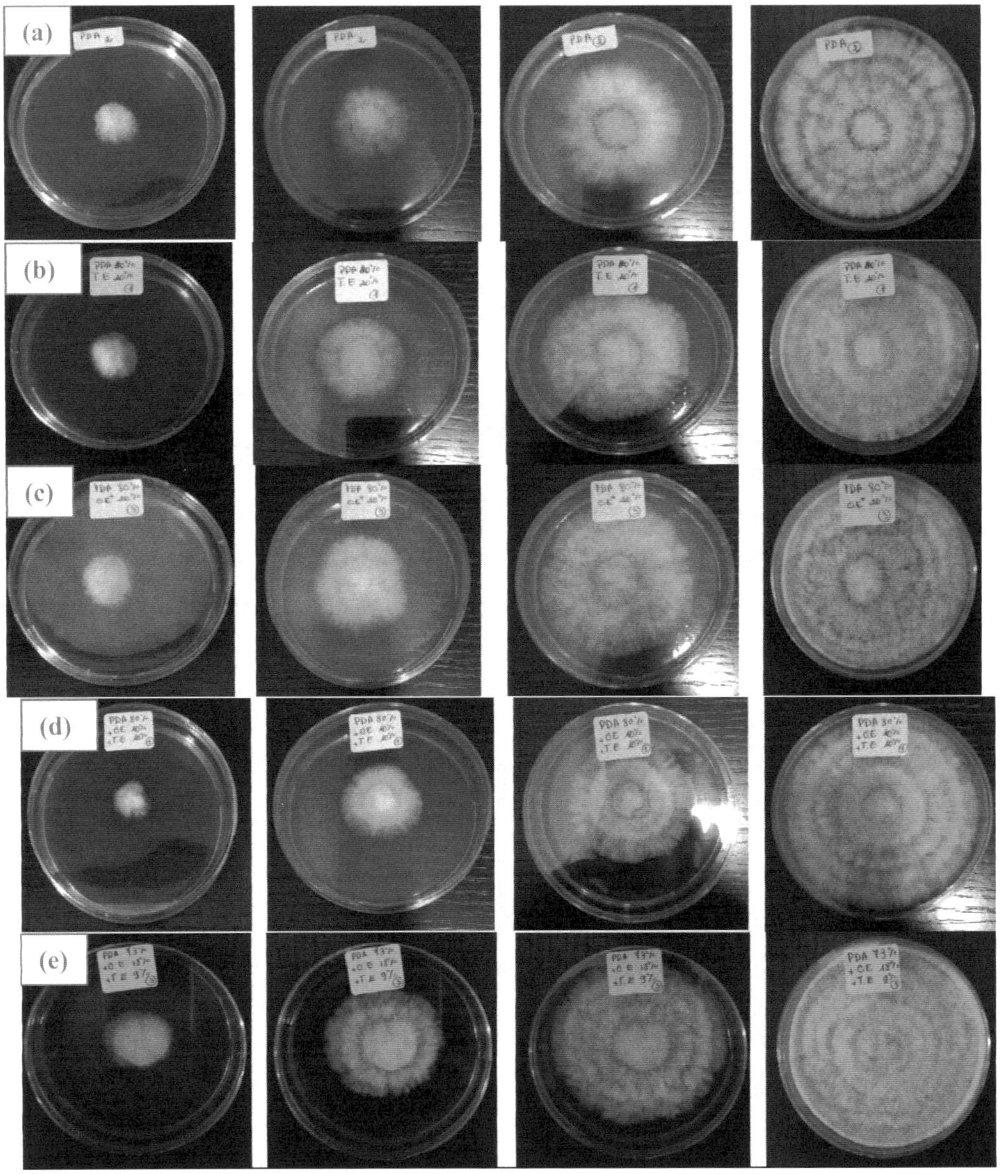

Figure 5. Mycelial growth diameter at 25°C; (a) T1: PDA, (b) T2: PDA 80% + TE 20%, (c) T3: PDA80% + OE 20%, (d) T4: PDA 80% + OE 10% + TE 10%, (e) T5: PDA 70% + OE 20% + TE 10%

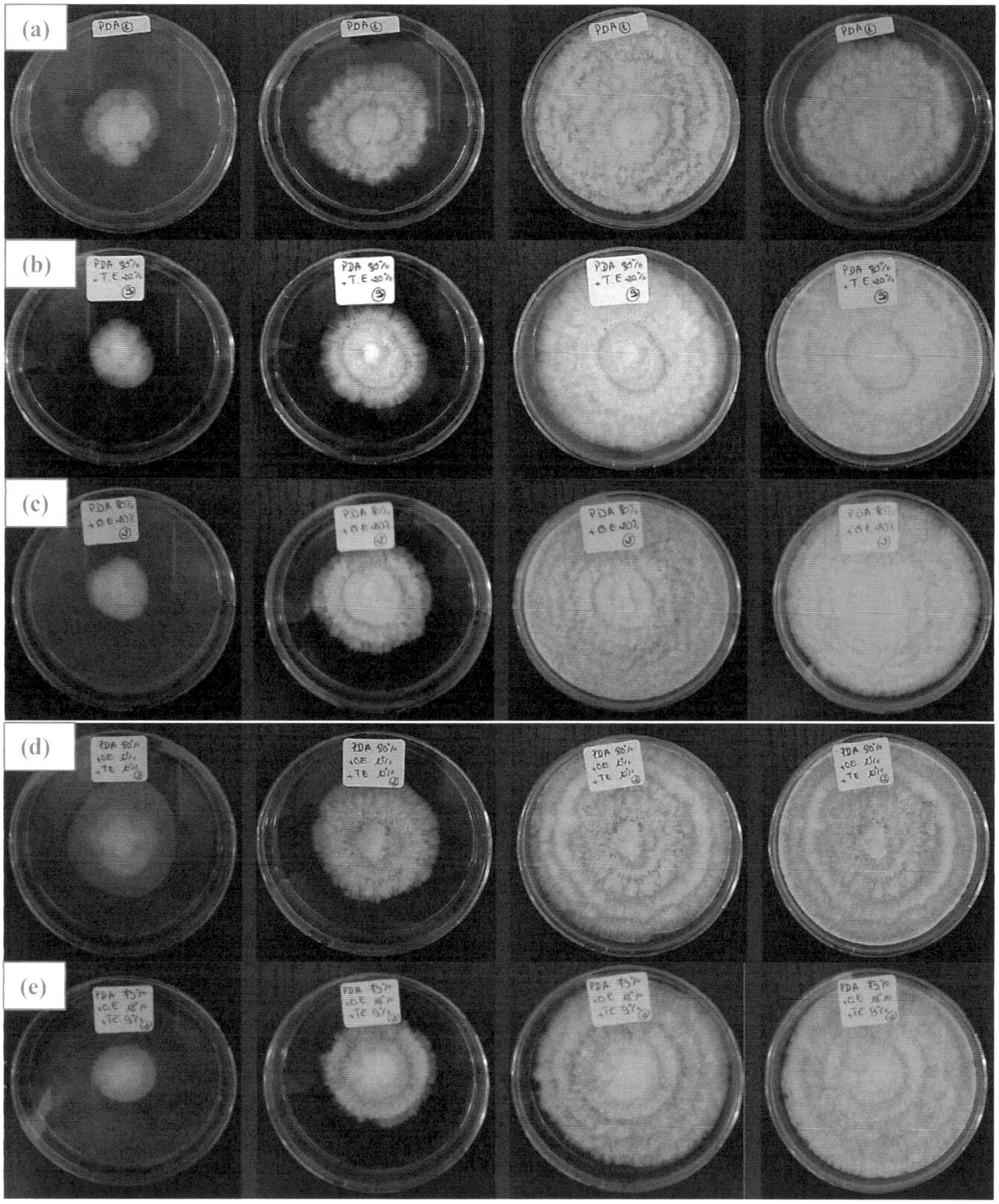

Figure 6. Mycelial growth diameter at 28°C; (a) T1: PDA, (b) T2: PDA 80% + TE 20%, (c) T3: PDA80% + OE 20%, (d) T4: PDA 80% + OE 10% + TE 10%, (e) T5: PDA 70% + OE 20% + TE 10%

It was found that *Pleurotus ostreatus* grown in-vitro on PDA had a mycelial growth diameter of around 35 mm after four days of incubation [24]. The findings of the current study showed comparable MGR on PDA culture medium with Owaid et al. [24] at a temperature of 25°C and higher MGR by 6 mm at a temperature of 28°C. This may be related to the strength of the spawn used. Other researchers reported the colonization of 100 mm on PDA by *P. ostreatus* in 17 days

[21]; which means the colonization of 85 mm in around 14 days. The role of olive extract in the culture medium in increasing MGR to around 47.0–49.0 mm at day 4 was pointed out by [24]. In the current study, adding 20% OE or 20% OE + 10% TE to the culture media resulted in comparable and higher MGR by 3 mm at 25°C as well as higher by 4.8–5.3 mm at 28°C compared to the same aforementioned study. It was previously noted that a full colonization of media containing olive extracts occurred at day 8 [24]. These findings are consistent with our ones at 25°C and 28°C. However, mixing 10% OE with 10% TE in the culture medium resulted in full colonization of the latter in six days at 28°C. This may point the role of tea extract when combined with olive leaves one in the hastening of mycelial growth. Green tea extract contains valuable enzymes [26] that may have been liberated and thus helped in the decomposition of holocellulose compounds (found in olive and tea extracts) and the provision of needed energy for *P. ostreatus* mycelium growth [9,10].

3.2. Effect of inoculation temperature on growth induction index of treatments
At 22°C, the highest growth induction index (GII) was observed when 20% OE were added to PDA (22.2%) (table 6). It was significantly higher by 21.9% compared to T2 (PDA 80% + TE 20%) and by 6.2–11.8% in comparison with extracts inoculums containing both OE and TE (T4 and T5). At 25°C, the culture medium containing 20% OE and 10% TE (T5) showed a considerable induction index significantly higher by 5.7–14.7% in comparison with the remaining inoculums. The culture medium containing equal proportions of OE and TE (10%) (T4) showed a significantly higher GII (higher by 0.8–1.9%) at 28°C compared to the remaining treatments. Only the GII of T5 at 25°C (21.9%) showed comparable value to [24] while the remaining treatments had lower values nevertheless the incubation temperature.

Table 6. Average growth induction index (GII) (%) of different treatments.

Treatment	GII (Temperature 22°C)	GII (Temperature 25°C)	GII (Temperature 28°C)
T2	0.3±0.03a	8.4±0.02b	17.4±0.01c
T3	22.2±0.03d	16.2±0.02c	17.0±0.01b
T4	10.4±0.03b	7.2±0.02a	18.2±0.01d
T5	16.0±0.03c	21.9±0.02d	16.3±0.01a
p_{value}	<0.01	<0.01	<0.01

T2 : PDA 80% + TE 20%, T3 : PDA 80% + OE 20%, T4 : PDA 80% + OE 10% + TE 10%, T5 : PDA 70% + OE 20% + TE 10%, GII: growth induction index. Values are means ± SD; Means within the same column followed by the same letters are not significantly different at p<0.05 according to Duncan's multiple range test.

3.3. Effect of inoculation temperature on mycelial linear growth rate on different types of grains
Nevertheless the temperature of inoculation, all inoculums containing OE and/or TE showed significantly higher mycelial linear growth rate (MLGR) by 0.5–1.4 mm/day (22°C), 0.1–0.6 mm/day (25°C) and 4.0–4.3 mm/day (28°C) in comparison with PDA culture (table 7). It was reported that a linear growth rate of around 11 mm/day was noted when *P. ostreatus* was cultivated on olive-based extracts [24]. In a same vein, others mentioned that olive press cake-based medium lead to a linear growth rate of 3.4 mm/day for *P. ostreatus* mycelium [20]. The current study showed linear growth rates higher by 2.5–2.8 mm/day compared to Owaid et al. [24] and much more higher (by 10.1–10.5 mm/day) than Zervakis et al. [20] at an inoculation temperature of 28°C. The latter detected a linear growth rate of 3.0 mm/day on PDA medium being lower than the values in the current study nevertheless the inoculation temperature. The high linear growth rates observed in media containing both olive and tea extracts may point out the high valuable compounds of Moroccan-origin olive and tea by-products and their hidden beneficial effect on mushroom production.

Table 7. Mycelial linear growth rate (MLGR) (mm/day) of different treatments.

Treatment	MLGR (Temperature 22°C)	MLGR (Temperature 25°C)	MLGR (Temperature 28°C)
T1	3.4±0.02a	5.2±0.03a	9.5±0.02a
T2	3.9±0.02b	5.3±0.03b	13.6±0.02c
T3	4.5±0.02c	5.4±0.03c	13.8±0.02e

T4	4.7±0.02d	5.8±0.03d	13.7±0.02d
T5	4.8±0.02d	5.8±0.03d	13.5±0.02b
p_{value}	<0.01	<0.01	<0.01

T1 : PDA, T2 : PDA 80% + TE 20%, T3 : PDA 80% + OE 20%, T4 : PDA 80% + OE 10% + TE 10%, T5 : PDA 70% + OE 20% + TE 10%, MLGR: mycelial linear growth rate. Values are means ± SD; Means within the same column followed by the same letters are not significantly different at p<0.05 according to Duncan's multiple range test.

3.4. *Effect of spawn grain type on MGD and MLGR*

Since day 2 until day 6, sorghum grain-based culture medium showed a significantly higher MGD compared to wheat and maize grains-based culture media (higher by 1.6, 17.9 and 27.8 mm, and 1.6, 17.0 and 22.8 mm for wheat and maize, respectively) (table 8, figure 7). The earliest full mycelial colonization of the culture media was detected at day 8 when using sorghum grains (T2). The least full mycelial colonization of the culture media was observed for wheat grains at day 12. While comparing between wheat and maize grains, it is observed that except at day 2, the mycelium colonized significantly faster the latter rather than the former.

Table 8. Mycelial growth diameter (MGD) (mm) of different treatments.

Treatment	T1	T2	T3	p_{value}
MGD (Day 2)	20.7±0.2a	22.3±0.2b	20.7±0.2a	<0.01
MGD (Day 4)	35.3±0.1a	53.2±0.1c	36.2±0.1b	<0.01
MGD (Day 6)	52.2±0.1a	80.0±0.1c	57.2±0.1b	<0.01
MGD (Day 8)	72.7±0.3a	90.0±0.0c	82.2±0.3b	<0.01
MGD (Day 10)	81.3±0.2a	90.0±0.0b	90.0±0.0b	<0.01
MGD (Day 12)	90.0±0.0	90.0±0.0	90.0±0.0	-

T1 : Wheat grain-based culture medium, T2 : Sorghum grain-based culture medium, T3 : Maize grain-based culture medium, MGD: mycelial growth diameter. Values are means ± SD; Means within the same row followed by the same letters are not significantly different at p<0.05 according to Duncan's multiple range test.

In concurrence, the MLGR was significantly higher in T2 (culture media based on sorghum grains) in comparison with wheat and maize-based cultures (T1 and T3) (higher by 2.6 mm/day and 1.0 mm/day,respectively) (table 9). An earlier study found that the linear growth rate of *P. ostreatus* on wheat grains was around 3.5–4.0 mm/day at an inoculation temperature of 25°C [26] being lower by 40–46%lower than in the current study. An MGR of 31.5–38.6 mm was acknowledged after 8 days of inoculation on white and red sorghum grains [27] being much lower than in the current study (90 mm). A range of 3.5–5 mm/days of linear growth rate was previously reported on maize [28] being less than 0.5 fold compared to the current study.

Table 9. Mycelial growth rate (mm/day) of different treatments.

Treatment	T1	T2	T3	p_{value}
MLGR (Temperature 25°C)	8.7±0.2a	11.3±0.1c	10.3±0.1b	<0.01

T1 : Wheat grain-based culture medium, T2 : Sorghum grain-based culture medium, T3 : Maize grain-based culture medium, MLGR: mycelial linear growth rate. Values are means ± SD; Means within the same row followed by the same letters are not significantly different at p<0.05 according to Duncan's multiple range test.

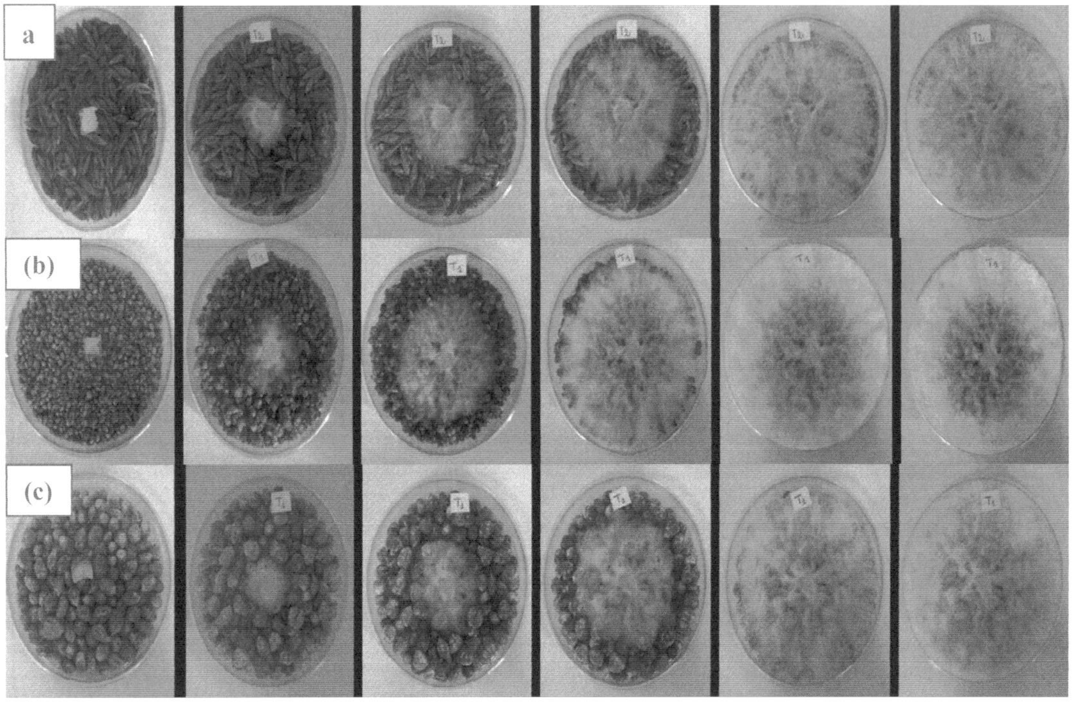

Figure 7. Mycelial growth diameter at 25°C; (a) T1: Wheat grain-based medium, (b) T2: Sorghumgrain-based medium, (c) T3: Maize grain-based medium

4. Conclusion

The incorporation of Moroccan fresh olive leaves and dried green tea leaves alone or combined in extracts for the production of *P. ostreatus* var. florida was investigated. Results showed increased mycelial growth diameter especially in days 6 and 8 in comparison with control (PDA). Moreover, linear growth rate showed significant increases in result to 20% OE, 20% TE, 10% OE + 10% TE and 20% OE + 10% TE incorporations in the culture media in comparison with PDA-based medium nevertheless the inoculation temperature. The growth induction index of *P. ostreatus* in olive/tea leaves extracts-based media was improved by more than 50 folds at 22°C, around 3 folds at 25°C and 1.1 folds at 28°C in comparison with PDA. These results suggest the combination or not of olive/tea leaves extracts as promising culture media for the growth of *Pleurotus ostreatus*. Further research are suggested on the chemical and antioxidants composition of these fully colonized media. Currently undergoing researches aim, for the first time, the valorization of underrated olive and tea by-products in the production of mushroom carpophores on the territory of Morocco Kingdom. This will reduce waste disposal and enrich the Moroccan cuisine with highly nutritious products at relatively low costs.

5. References

[1] Kumar V, Valadez-Blanco R, Kumar P, Singh J and Kumar P 2020 Effects of treated sugar mill effluent and rice straw on substrate properties under milky mushroom (*Calocybe indica* P&C) production: Nutrient utilization and growth kinetics studies. *Env. Tech. Inn.* 101041.
[2] Kumar V, Kumar P, Singh J and Kumar P 2021a Kinetics of nutrients remediation from sugar industry effluent-treated substrate using *Agaricus bisporus*: mushroom yield and biochemical potentials. *3 Biotech* 11(4).
[3] Kumar P, Kumar V, Goala M, Singh J and Kumar P 2021b Integrated use of treated dairy

wastewater and agro-residue for *Agaricus bisporus* mushroom cultivation: Experimental and kinetics studies. *Biocat. Agric. Biotech.* 32 101940.

[4] Kumar P, Kumar V, Adelodun B, Bedekovíc D, Kos I, Širíc I, Alamri S A M, Alrumman S A, Eid E M, Abou Fayssal S *et al* 2022 Sustainable Use of Sewage Sludge as a Casing Material for Button Mushroom (*Agaricus bisporus*) Cultivation: Experimental and Prediction Modeling Studies for Uptake of Metal Elements. *J. Fungi* 8 112.

[5] International Olive Council (IOC) 2013 *General description of olive growing in Morocco* 10.

[6] International Olive Council (IOC) 2019 *The olive grove in Morocco* 0255–996X.

[7] Avraamides M and Fatta D 2008 Resource consumption and emissions from olive oil production: a life cycle inventory case study in Cyprus. *J. Clean. Pro.* 16 809–821.

[8] Royse D J, Baars J and Tan Q 2017 Current overview of mushroom production in the world *Edible and medicinal mushrooms: Technology and applications* eds D C Zied and A Pardo Giménez (New York: John Wiley and Sons) pp 5–12.

[9] Abou Fayssal S, Alsanad M A, El Sebaaly Z, Ismail A I H and Sassine Y N 2020 Valorization of Olive Pruning Residues through Bioconversion into Edible Mushroom *Pleurotus ostreatus* (Jacq. Ex Fr.) P. Kumm. (1871) of Improved Nutritional Value *Scientifica* 2020 3950357.

[10] Abou Fayssal S, El Sebaaly Z, Alsanad M A, Najjar R, Bohme M, Yordanova M H and Sassine Y N 2021a Combined effect of olive pruning residues and spent coffee grounds on *Pleurotusostreatus* production, composition, and nutritional value *PLoS ONE* 16 e0255794.

[11] Naim L, El Sebaaly Z, Abou Fayssal S, Alsanad M A, Najjar R, Yordanova M H and Sassine Y N 2021 The use of nitrogen-rich nano-supplements affects substrate temperature, delays the production cycle, and increases yield of *Pleurotus ostreatus*. *Acta Hortic.* 1327 831–840.

[12] Naim L, Alsanad M A, El Sebaaly Z, Shaban N, Abou Fayssal S and Sassine Y N 2020a Variation of *Pleurotus ostreatus* (Jacq. Ex Fr.) P. Kumm. (1871) performance subjected to different doses and timings of nano-urea *Saudi J. Biol. Sci.* 27 1573–1579.

[13] Naim L, Alsanad M A, Shaban N, El Sebaaly Z, Abou Fayssal S and Sassine Y N 2020b Production and composition of *Pleurotus ostreatus* cultivated on Lithovit®-Amino25 supplemented spent substrate *AMB Express* 10 188.

[14] Alsanad M A, Sassine Y N, El Sebaaly Z and Abou Fayssal S 2021 Spent coffee grounds influence on *Pleurotus ostreatus* production, composition, fatty acid profile, and lignocellulose biodegradation capacity *CyTA – J. Food* 19 11–20.

[15] Loss E, Royer A, Rodrigues M and Barana A C 2009 Use of maize wastewater for the cultivation of the *Pleurotus* spp. mushroom and optimization of its biological efficiency *J. Haz. Mat.* 166(23) 1522–1525.

[16] Yang D, Liang J, Wang Y, Sun F, Tao H, Xu Q and Wan X 2016 Tea waste: an effective and economic substrate for oyster mushroom cultivation *J. Sci. Food Agric.* 96(2) 680–684.

[17] Abou Fayssal S, Alsanad M A, Yordanova M H, El Sebaaly Z, Najjar R and Sassine Y N 2021b Effect of olive pruning residues on substrate temperature and production of oyster mushroom*(Pleurotus ostreatus)* *Acta Hortic.* 1327 245–252.

[18] Sassine Y N, Naim L, El Sebaaly Z, Abou Fayssal S, Alsanad M A and Yordanova M H 2021 Nano urea effects on *Pleurotus ostreatus* nutritional value depending on the dose and timing of application *Sci. Rep.* 11 5588.

[19] Koutrotsios G, Kalogeropoulos N, Kaliora A C and Zervakis G I 2018 Toward an increased functionality in oyster *(Pleurotus)* mushrooms produced on grape marc or olive mill wastes serving as sources of bioactive compounds *J. Agric. Food Chem.* 66(24) 5971–5983.

[20] Zervakis G, Philippoussis A, Ioannidou S and Diamantopoulou P 2001 Mycelium growth kinetics and optimal temperature conditions for the cultivation of edible mushroom species on lignocellulosic substrates *Folia Microbiol.* 46(3) 231–234.

[21] Illuri R, Kumar M, Eyini M, Veeramanikandan V, Almaary K S, Elbadawi Y B, Biraqdar M A and Balaji P 2021 Production, partial purification and characterization of ligninolytic enzymes

from selected basidiomycetes mushroom fungi *Saudi J. Biol. Sci.*

[22] Espeso J, Isaza A, Lee J Y, Sörensen P M, Jurado P, Avena-Bustillos Rd J, Olaizola M and Arboleya J C 2021 Olive Leaf Waste Management *Front. Sustain. Food Syst.* 5 660582.

[23] Romero-García J M, Lama-Muñoz A, Rodríguez-Gutiérrez G, Moya M, Ruiz E, Fernández-Bolaños J *et al.* 2016 Obtaining sugars and natural antioxidants from olive leaves by steam-explosion *Food Chem.* 210, 457–465.

[24] Owaid M N, Jaloot A S and Ahmed D M 2018 Influence of *Ficus carica* and *Olea europaea* leaves extracts on the mycelial growth of mushrooms in vitro *Acta Ecol.Sin.*

[25] Kochman J, Jakubczyk K, Antoniewicz J, Mruk H and Janda K 2021 Health Benefits and Chemical Composition of Matcha Green Tea: A Review *Molecules* 26 85.

[26] Zárate-Salazar J R, Santos M N, Caballero E N M *et al.* 2020 Use of lignocellulosic corn and rice wastes as substrates for oyster mushroom (*Pleurotus ostreatus* Jacq.) cultivation. *SN Appl. Sci.* 2 1904.

[27] Jiskani M M, Bhatti M I, Wagan K H, Pathan M A and Bhatti A G 2007 Determination of sorghum grains for spawn growth of oyster mushroom *Pleurotus ostreatus* (Jacq. Ex. Fr) Kummer *Pak. J. Bot.* 39(7) 2681–2684.

[28] Mkhize S S, Cloete J, Basson A K and Zharare G E 2016 Performance of *Pleurotus ostreatus* mushroom grown on maize stalk residues supplemented with various levels of maize flour and wheat bran *Food Sci. Technol* 36(4) 598–605.

Ethnobotany and diversity of medicinal and food plants used in the region Chtouka Ait Baha and Tiznit (Western Anti-Atlas), Morocco

M Barkaoui [1,2 *], F Msanda [2], H Boubaker [3] , O Elasri [3] , R El Boullani [2] , H Benchelha[4] and M Chahboune[1]

[1] Hassan First University of Settat, Higher Institute of Health Sciences, Laboratory of Health Sciences and Technologies. Settat.
[2] University of Ibn Zohr, Faculty of Sciences, Laboratory of Biotechnology and Natural Resources Valorization, B.P. 8106, Agadir, Morocco.
[3] University of Ibn Zohr, Faculty of Sciences, Laboratory of Microbial Biotechnology and Plant Protection, B.P. 8106, Agadir, Morocco.
[4] Université Ibn Tofaïl, Faculté des Sciences, Laboratoire de Biologie et Santé, Kenitra, Maroc

*Corresponding author: mohamed.barkaoui@uhp.ac.ma

Abstract. Western Anti-Atlas of Morocco represents an important area with interesting knowledge of medicinal and food plants, particularly the use of medicinal plants for human healthcare. Moreover, the geographic location of this region makes this site a remarkable area of incredible biodiversity and the highest levels of endemism. This work aims to make an inventory of plant species used in Chtouka Ait Baha and Tiznit provinces in folk medicine to manage different diseases and, at the same time, used in food. The survey was carried out using semi-structured questionnaires. A total of 450 interviews were conducted with traditional health practitioners and knowledgeable villagers. The data were analyzed using the use-value (UV) and the relative frequency citation (RFC). Eighty-four species belonging to 37 families were inventoried in this survey, with Lamiaceae as the dominating family. Two species that showed a significant value of UV (UV≥1.50) and RFC (RFC≥0.50) are *Thymus leptobotrys* Murb. and *Thymus saturejoides* Coss. with the highest value obtained for *Thymus leptobotrys* Murb. (UV=2.60). The digestive system was the highest foremost treated disease in the region (22% of treated infections). The obtained results also showed that seeds and leaves were the most cited plant part used, and the decoction is the preferred mode of preparation. The method of consumption the medicinal plants as food is fresh or dried with a percentage of 36%.

Keywords: Ethnobotanical survey; Western Anti-Atlas; medicinal food plants; ethnobotanical indices; Morocco

1. Introduction

For a long time ago, with a sense of curiosity, the innate principle of the doctrines, and sometimes the hazard, our ancestors used the plants as a food source to cure their ills. This simple knowledge is transmitted to the future generation and influenced by gender, age, and ethnicity [1]. The plants were always the first source of drugs for human healthcare, and they continued to provide them with new and original remedies [2]. Indeed, the biological factors constitute a part of local knowledge culture; some species used are not randomly selected, especially the food plants but fairly chosen based on conscious and unconscious factors [3].

Today, throughout the world, more than 20000 to 25000 plants are used in the pharmacopeia, and more than 50% of the pharmaceutical products available on the market are the natural origin [4]. In other words, the World Health Organization (WHO) estimates that approximately 65–80% of the world's population meets their primary health care needs through traditional medicine [5]. In addition, more than 80% of the population uses traditional medicine and medicinal plants for primary health care [6]. Recently, the knowledge and consumption of medicinal plants embedded in indigenous cultures have been influenced by the "contemporary way of life," making the new generations not perpetuate the way to identify and process the native species [7]. For this, many legislation and regulations have been recommended that appropriate strategies enhance sustainable utilization and management of medicinal plants, focus on local approaches involving traditional medicinal knowledge [8]. Furthermore, ethnobotanical studies are the most reliable approaches for discovering new drugs and several active compounds. They have been found from plants based on ethnobotanical information. Indeed, the World Health Organization (WHO) has provided evidence that ethnomedical details can lead to valuable drug discovery [9].

Although there are numerous reports, published works, thesis, dissertations, books, inventories, media reports, and monographs on the diversity of medicinal plants within the Mediterranean area. Morocco is one of the specific countries in Africa and Mediterranean countries with diverse climatic conditions and biotopes that almost any plant can grow. Generally, they have more than 4200 spontaneous species, and some 1500 introduced species have been cataloged [10], for the 500 species were identified as having potential in the phytotherapy [11]. In addition, Morocco has a long medical tradition and ancestral knowledge of phytotherapy along with botanical diversity. However, local ethnobotanical expertise from many areas of Morocco, especially in the south, remains poorly known [12-14].

The study area (Western Anti-Atlas) is located in a semiarid region between the High Atlas, and the Mediterranean to the north, the Atlantic Ocean to the west, and the Sahara to the south. Given the orographic and climatic variations, it is considered among the most remarkable regions of North Africa as a region of incredible biodiversity and the highest levels of plant endemism [15]. Moreover, the local people's adaptation to the arid and semiarid conditions in the Western Anti-Atlas region. The ancestral knowledge is fundamental in using and managing natural resources in the first instance to encounter their nutritional needs and treat their illnesses. Therefore, this study aimed to identify and document local knowledge of medicinal food plants commonly used to treat different diseases by the people of the Western Anti-Atlas region.

2. Material and methods

2.1. Study area

The study was carried out in 18 communes (16 rural communes and two urban communes) who are situated in the Chtouka Ait Baha and Tiznit province (Moroccan central-western); between the latitudes 29°25' and 30°32' with a surface of 2600 km^2 (Figure 1). Geographically, the area corresponds to the western side of the mountains of the Anti-Atlas; UNESCO classified it in 1998 as Arganeary Reserve Biosphere. The Mediterranean's climate and precipitation levels vary between 200 and 500 mm according to the location and the altitude.

The region is characterized by various geographic profiles, including littoral, Atlas Mountains and Sahara. It is a region of incredible biodiversity constituted by plant formations *Argania spinosa* (L.) Skeels and shrublands dominated by *Cistus creticus* L., *Lavandula dentata*, *Artemisia herba-alba* Asso, and *Thymus saturejoides* Coss [15].

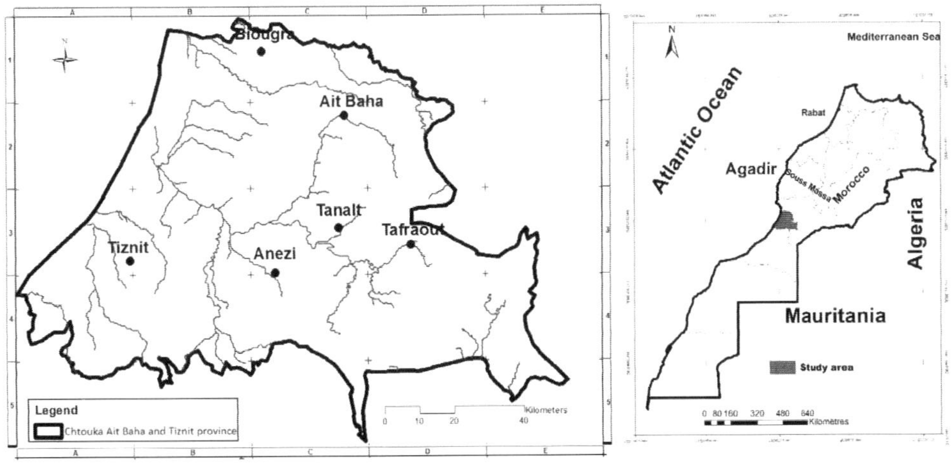

Figure 1. Study area

2.2. Data collection and analysis

The present ethnobotanical survey was carried out from January 2014 to December 2018, and we conducted 450 interviews with 18 communes of Chtouka Ait Baha and Tiznit province. All data has been collected from conversations with local people in the *Tashelhit* dialect (local Berber language). In addition, a questionnaire was processed and modeled according to different ethnobotanical surveys [16-19].

The data on traditional medicinal plants knowledge used by the local people were collected through semi-structured questionnaires [20]. A medicinal property was accepted as valid only if at least five independent interviewees mentioned it. The collected plants were identified by using "Flora of Morocco" [21-23]. The scientific names and endemic of plant species were determined according to the Plant List (http://www.theplantlist.org).

2.3. Quantitative analysis of data

We interpret the results of our ethnobotanical survey; the descriptive data during fieldwork was analyzed by quantitative ethnobotanical indexes like Use Value (UV) and Relative citation frequency (RFC).

2.3.1 Use value

The use-value is a quantitative method that demonstrates the relative importance of species known locally, was also calculated (Rossato, De LeitãO-Filho, and Begossi 1999; Silva, Albuquerque, and Nascimento 2004; Phillips and Gentry 1993):

$$UV = \Sigma U_i/n$$

Where: U_i = the number of uses mentioned by each informant for a given species, n = the total number of informants.

2.3.2 Relative Frequency of Citation

Relative Frequency of Citation (RFC), obtained by dividing the number of informants who mention the use of the species, also known as the frequency of citation (FC), by the number of informants participating in the survey (N), was also calculated (Tardío, Pardo-de-Santayana, and Morales 2006):

$$RFC = FC \backslash N \quad (0 < RFC < 1)$$

3. Results and discussion

3.1. Ethno-demography of study participants

In the present study, 450 local informants were interviewed and classified into different demographic categories. 234 females (52%) and 216 males (48%) have traditional knowledge of medicinal food plants. The study showed that therapeutic food plant use frequency increased with age. The people older than 50 years has a frequency of use of medicinal food plants by 46%, followed by age categories [41 to 50], [31 to 40], [21 to 30], and finally type under the age of 20 with 22%, 20.89%, 9.33%, and 1.78%, respectively (Table 1). The knowledge of medicinal food plants in the study area was obtained from older informants (about 50% were 50 years old). Furthermore, these age groups have more knowledge of medicinal plants than the other age groups; it is obtained after a long-accumulated experience and received from their older family members. On the other hand, younger people usually move to urban areas to look for paid employment and are poorly linked to traditional medicine [18, 24]. The study showed that women take the most knowledge because women learn mainly from their mothers through routine observations, and they take care of small children when they are sick [12, 25]. The study area is a rural region (90%) which has a lower level of schooling (59%), particularly in women's population, which explains 70% of the woman interviewed are illiterate (Table 1); this result confirmed the Moroccan General Census of Population and Housing [26].

Table 1. Socio-demographic characteristics of interviews

Socio-demographic variables				Percentage (%)
Gender	Male			48%
	Female			52%
Family situation	Married			67%
	Single			14%
	Divorce			3%
	Widower			16%
Age (years)	[17 – 20]	Male	72%	1.78%
		Female	28%	
	[21 – 30]	Male	63%	9.33%
		Female	37%	
	[31 – 40]	Male	45%	20.89%
		Female	55%	
	[41 – 50]	Male	44%	22%
		Female	56%	
	[51 – 87]	Male	42%	46%
		Female	58%	
Education	Illiterate	Male	29%	59%
		Female	71%	
	Primary	Male	66%	24%
		Female	34%	
	Secondary	Male	89%	15%
		Female	11%	
	university	Male	78%	2%
		Female	22%	

3.2. Floristic analysis

We have identified eighty-four medicinal food plants belonging to 37 plant families. The first three families well used in traditional medicine represented in the study region were Lamiaceae with 16 species, Apiaceae with ten species, and Leguminosae with 07 species (Table 2). These three families (Lamiaceae, Apiaceae, and Leguminosae) are the most used in the studies conducted in neighboring semiarid and arid regions, like in Agadir Ida Ou Tanane [25, 27] and Tata provinces [12]. Generally, most studies conducted in the different areas of Morocco prove that the most significant number of medicinal species used in phytotherapy belongs to the Lamiaceae botanical family [28, 29].

3.3. Therapeutic uses and ethnobotanical data analysis
In this ethnobotanical survey, we recorded information on 84 medicinal food plants species. The plants were listed in alphabetical order by families and species. Information such as the local name of plant species used parts, modes of consumption, method of preparation, and ethnobotanical indices (UV and RFC) are provided for each species. Vernacular names were included as Arabic and Roman alphabets referred to them. In this study, we compared medicinal food plants and their uses in our study area, based on ethnobotanical studies realized in Morocco. The large number of plant species used in the study area indicates dependence on a great diversity of ethnomedicinally plant species. It represents a good indicator of the profound knowledge of herbal plants held by the local people living in the Western Anti-Atlas semiarid region (Table 2).

Table 2: List of medicinal food plants used in Chtouka Ait Baha and Tiznit areas

Voucher	Scientific Name and Family	Local Name	Modes of consumption	RFC	UV	Therapeutic uses or treated diseases	Part used	Preparation	Administration
	Amaryllidaceae								
CAB T003	*Allium cepa* L.	Azalim	Bulb boiled and eaten as a vegetable or raw in salads	0.067	1.37	Ear Pains, Hypotensive, Anti-Diabetic, Headache	Bulb	Boiled in oil, Decoction, Crude, Fumigation	Ear drops, Oral Inhalation
CAB T004	*Allium sativum* L.	Tiskhert/ Touma	Bulb boiled and eaten as a vegetable or raw in salads	0.220	1.63	Aphrodisiac, Emmenagogue, Ear Pains, Hypotensive, Stomachic, Laxative, Cough, Expectorant, Eye pain, Antiseptic, Wound Healing, Anti-Diabetic, Dermatosis, Haircare	Bulb	Crude, Boiled in oil, Decoction, Cataplasm, Friction	Oral Ear drops, Applied Externally
	Apiaceae								
CAB T008	*Anethum foeniculoides* Maire & Wilczek	Amssa / Besbas	Bulb boiled and eaten as a vegetable or raw in salads; Seeds used as a food seasoning	0.040	1.17	Diuretic, Antispasmodic, Carminative, Laxative	Seed, Stem, and leaf	Decoction, Maceration, Crude	Oral
CAB T009	*Apium graveolens* L.	Krafess	Stem and leaf cooked as a vegetable in the soup; Seeds powder used as a condiment	0.022	1.40	Aphrodisiac, Diuretic, Anxiety, Antispasmodic, Carminative	Seed, Stem, and leaf	Decoction, Powder	Oral
CAB T027	*Carum carvi* L.	Lkarweia	Seeds are used as a soup seasoning or condiment	0.291	1.31	Hypotensive, Antidiarrheal, Hepatitis, Antispasmodic, Antiemetic, Anti-Diabetic, Vertigo	Seed, Stem, and leaf	Crude, Decoction, Infusion, Maceration, Powder	Oral
CAB T044	*Coriandrum sativum* L.	Kasbour	Stem and leaf cooked as a vegetable or raw in salads; Seeds powder used as a condiment	0.044	1.25	Diuretic, Emmenagogue, Colds, Laxative, Stomachic, Cough	Seed, Stem, and leaf	Infusion, Decoction	Oral
CAB T050	*Cuminum cyminum* L.	Lkmmoun	Seeds powder is used as a condiment	0.131	1.25	Emmenagogue, Antidiarrheal	Seed	Decoction, Powder, Maceration	Oral
CAB T055	*Daucus carota* L.	Khizzou	The root is boiled and eaten as a vegetable or raw in salads	0.036	1.38	Antidiarrheal, Eye pain	Seed, Root	Decoction, Cataplasm	Oral, Applied Externally

CAB T073	*Foeniculum vulgare* Mill.	Nafaa	Seeds are used as a food seasoning	0.111	1.36	Diuretic, Emmenagogue, Colds, Antispasmodic, Carminative, Catarrh, cough, Anti-Diabetic	Seed	Decoction, Infusion, Decoction, Infusion, Crude	Oral
CAB T127	*Pastinaca sativa* L.	Left Lmahfour	The root is boiled and eaten as a vegetable	0.036	1.13	Carminative, Cholagogue, Anti-Diabetic	Root	Crude	Oral
CAB T132	*Petroselinum crispum*(Mill.) Fuss	Lmaadnouss	Stem and leaf cooked as a vegetable or raw in salads; Seeds powder used as a condiment	0.053	1.29	Kidney Stone, Colds, Carminative, Stomachic, Skincare	Seed, Stem, and leaf	Decoction, Powder, Infusion	Oral, Applied Externally
CAB T135	*Pimpinella anisum* L.	Habbat Hlawa	Seeds are used as a food seasoning	0.022	1.40	Diuretic, Antispasmodic, Carminative, Galactagogue, Anti-Diabetic	Seed	Decoction, Infusion	Oral
	Asteraceae								
CAB T014	*Artemisia absinthium* L.	Chiba	Leaf used as an aromatic infusion for tea	0.084	1.45	Diuretic, Ear Pains, Colds Antispasmodic, Febrifuge	Leaf	Decoction, Infusion, Boiled in oil	Oral Ear drops
CAB T053	*Cynara cardunculus* L.	Lkherchouf	Stem boiled and eaten as a vegetable	0.024	1.09	Hepatitis, Anaemia, Anti-Diabetic	Flower, Stem and leaf	Decoction	Oral
CAB T092	*Podospermumlaciniatum*subsp.*decumbens*(Guss.)	Aghrim	Leaf and stem used raw in salads	0.024	1.18	Cholagogue, Anti-Diabetic Dermatosis	Stem and leaf	Crude, Cataplasm	Oral, Applied Externally
	Brassicaceae								
CAB T020	*Brassica rapa* L.	Left	The root is boiled and eaten as a vegetable	0.060	1.11	Ear Pains, Cough, Expectorant, Anti-Diabetic	Root	Crude, Decoction	Ear drops, Oral
CAB T101	*Lepidium sativum* L.	Hebb Rchad	Seeds are used as a food seasoning	0.204	1.25	Phytoestrogen, Colds, Antidiarrheal, Cough	Seed	Infusion, Crude, Decoction, Powder	Oral
CAB T145	*Raphanusraphanistrum*subsp.*sativus*(L.) Domin	Leafjel	Root has eaten raw in salads	0.011	1.20	Choleretic, Anti-Diabetic	Root	Crude	Oral
	Cactaceae								

CAB T124	*Opuntiaficus-indica* (L.) Mill.	Lhndia / Aknari	Fruit is eaten fresh or dried	0.191	1.20	Diuretic, Antidiarrheal, Anti-Diabetic, Haircare, Cancer	Stem, Flower, Fruit, Fruit Oil	Decoction, Infusion, Powder, Crude	Oral, Applied Externally
	Capparaceae								
CAB T024	*Capparis spinosa* L.	Tailoulout / Lkbar	Flower buds are boiled and eaten as a vegetable	0.082	1.24	Antirheumatic, Hepatitis, Cholagogue, Pancreatitis, Wound Healing, Nervousness, Anxiety	Leaf, Flower buds, Fruit	Decoction, Decoction, Powder, Cataplasm, Fumigation	Oral, Applied Externally Inhalation
	Cucurbitaceae								
CAB T039	*Citrulluslanatus*(Thunb.) Matsum. & Nakai	Delah	Fruit eaten fresh	0.009	1.00	Aphrodisiac, Diuretic, Hypertensive Eye pain Anti-Diabetic Vermifuge	Fruit	Crude	Oral
CAB T048	*Cucumis sativus* L.	Lkhiar/ Tadamst	Fruit used raw in salads	0.040	1.06		Fruit	Crude, Cataplasm	Oral, Applied Externally
CAB T049	*Cucurbita maxima* Duchesne	Lgraa/ Takhsayt	Fruit boiled and eaten as a vegetable	0.013	1.00		Seed	Crude	Oral
	Leguminosae								
CAB T030	*Ceratonia siliqua* L.	Tikida/ Lkharoub	Fruit is eaten dried	0.200	1.30	Kidney Stone, Antidiarrheal, Antispasmodic, Antiemetic, Astringent, Wound Healing, Anti-Diabetic	Leaf, Fruit	Decoction, Powder, Cataplasm	Oral, Applied Externally
CAB T033	*Cicer arietinum* L.	Lhmesse	Seeds are boiled and eaten as a food	0.016	1.14	Aphrodisiac, Antidiarrheal, Aperitif	Seed	Maceration, Decoction, Powder	Oral
CAB T076	*Glycine max (*L.) Merr.	Soja	Seeds are boiled and eaten as a food	0.020	1.44	Aperitif, Anti-Diabetic, Anaemia	Seed	Maceration, Crude	Oral
CAB T100	*Lens culinaris* Medik.	Laadess	Seeds are boiled and eaten as a food	0.011	1.00	Anemia	Seed	Decoction	Oral
CAB T109	*Medicago sativa* L.	lFessa	Leaf boiled and eaten as a vegetable in couscous	0.022	1.30	Phytoestrogen	Seed, Stem, and leaf	Decoction, Crude	Oral

CAB T170	*Trigonella foenum-graecum* L.	Lhelba / Tifidas	Seeds are boiled and eaten as a food	0.280	1.25	Diuretic, Hypotensive, Hepatitis, Aperitif, Carminative, Tuberculosis, Cough, Anti-Diabetic, Epileptic	Seed	Decoction, Infusion, Maceration, Crude, Powder	Oral
CAB T173	*Vicia faba* L.	Ibaoin	Seed pods are boiled and eaten as a vegetable; Seeds are cooked and eaten as a food	0.011	1.00	Cough, Expectorant	Seed	Powder	Oral
	Fagaceae								
CAB T144	*Quercus ilex* L.	Tasaft / Lblout	Almond eaten raw	0.027	1.17	Stomachic, Cough	Leaf, Seed	Decoction, Crude, Powder	Oral
	Geraniaceae								
CAB T129	*Pelargonium graveolens* L'Hér.	Laatercha	Leaf used as an aromatic infusion for tea	0.073	1.12	Diuretic, Colds Febrifuge	Leaf	Decoction, Infusion	Oral
	Iridaceae								
CAB T047	*Crocus sativus* L.	Zaafran Lhor	Stamens are used as a condiment or aromatic infusion for tea	0.176	1.16	Diuretic, Emmenagogue, Antirheumatic, Antispasmodic, Laxative, Eye pain, Headache, Dermatosis, Skincare	Flower	Decoction, Infusion, Maceration, Friction	Eye drops, Oral, Applied Externally
	Juclandaceae								
CAB T086	*Juglans regia* L.	Douj / Gergaa	Almond eaten raw	0.067	1.07	Diuretic, Emmenagogue, Toothache, Gingivitis	Bark, Leaf, Root	Decoction, Friction	Oral, Applied Externally
	Juncaceae								
CAB T087	*Juncus rigidus* Desf.	Smar	Seeds boiled and eaten in the soup	0.009	1.00	Diuretic	Seed	Decoction	Oral
	Lamiaceae								

CAB T094	*Lavandula dentata* L.	Timzeria	Leaf used as an aromatic infusion for tea or coffee	0.440	1.15	Diuretic, Emmenagogue, Ear Pains, Colds, Antidiarrheal, Antioxidant, Antispasmodic, Catarrh, Cough, Anti-Diabetic, Dermatosis, Skincare	Leaf, Stem and leaf, Flower	Cataplasm, Decoction, Fumigation, Crude, Friction	Oral, Fumigation, Ear drops, Applied Externally, Inhalation
CAB T095	*Lavandula maroccana* Murb.	Igazioien	Leaf used as an aromatic infusion for tea or coffee	0.047	1.33	Diuretic, Emmenagogue, Catarrh, Cough, Wound Healing, Anti-Diabetic, Febrifuge	Stem and leaf	Decoction, Infusion	Oral, Applied Externally
CAB T096	*Lavandula multifida* L.	Igiz	Leaf used as an aromatic infusion for tea or coffee	0.227	1.29	Diuretic, Emmenagogue, Stomatitis, Catarrh, Cough, Wound Healing, Anti-Diabetic, Febrifuge, Dermatosis, Skincare	Leaf, Stem and leaf, Flower	Decoction, Infusion, Cataplasm, Friction	Oral, Applied Externally
CAB T097	*Lavandula officinalis* Chaix	Lkhzama	Leaf used as an aromatic infusion for tea or coffee	0.058	1.38	Diuretic, Emmenagogue, Antirheumatic, Antispasmodic, Carminative, Febrifuge, Anxiety	Leaf	Decoction, Infusion, Cataplasm	Oral
CAB T098	*Lavandula stoechas* L. subsp. *atlantica* Br. - Bl.	Imzeria	Leaf used as an aromatic infusion for tea or coffee	0.067	1.27	Diuretic, Emmenagogue, Hypertensive, Astringent, Wound Healing, Anti-Diabetic	Leaf	Decoction, Infusion, Fumigation	Oral, Inhalation Applied Externally
CAB T111	*Mentha pulegium* L.	Flyou	Leaf and stem used as an aromatic infusion for tea	0.056	1.16	Colds, Catarrh, Cough, Febrifuge	Leaf, Stem, and leaf	Decoction, Infusion	Oral
CAB T112	*Mentha suaveolens* Ehrh.	Timija / Timersad	Leaf used as an aromatic infusion for tea	0.224	1.24	Diuretic, Emmenagogue, Colds, Hepatitis, Stomachic, Catarrh, Vertigo	Leaf, Stem, and leaf	Decoction, Infusion, Powder	Oral
CAB T113	*Mentha viridis* (L.) L.	Naânaâ	Leaf and stem used as an aromatic infusion for tea	0.018	1.50	Anxiety, Skincare, Haircare	Leaf, Stem, and leaf	Decoction, Infusion, Cataplasm	Oral, Applied Externally
CAB T120	*Ocimum basilicum* L.	Lahbaq	Leaf used raw in salads or food seasoning	0.104	1.30	Diuretic, Carminative, Stomachic, Cough, Expectorant, Insect Bite, Anxiety, Dermatosis	Leaf, Stem, and leaf	Decoction, Infusion Cataplasm, Friction	Oral, Applied Externally

IOP Conf. Series: Earth and Environmental Science 1090 (2022) 012021 doi:10.1088/1755-1315/1090/1/012021

CAB T125	*Origanum majorana* L.	Merdeddouch	Leaf used as an aromatic infusion for tea	0.058	1.12	Emmenagogue, Colds Cough, Expectorant	Leaf, Stem and leaf	Infusion	Oral
CAB T152	*Rosmarinus officinalis* L.	Azir	Leaf used as an aromatic infusion for tea or coffee	0.324	1.16	Emmenagogue, Hypotensive, Antidiarrheal, Antispasmodic, Carminative, Muscle tear, Cough, Expectorant, Anti-Diabetic, Febrifuge, Skincare	Leaf, Stem, and leaf	Decoction, Infusion, Maceration, Cataplasm, Friction	Oral, Applied Externally
CAB T157	*Salvia officinalis* L.	Salmia	Leaf and stem used as an aromatic infusion for tea	0.113	1.35	Diuretic, Emmenagogue, Hypotensive, Anti-Diabetic, Febrifuge	Leaf	Decoction, Infusion	Oral
CAB T168	*Thymus leptobotrys* Murb.	Azougni	Leaf used as an aromatic infusion for tea or coffee	0.882	2.04	Diuretic, Colds Hepatitis, Toothache, Antispasmodic, Cholagogue, Gingivitis, Vermifuge, Expectorant, Cough, Wound Healing Febrifuge, Anaemia, Headache	Leaf, Stem, and leaf	Cataplasm, Decoction, Friction, Infusion, Maceration, Crude, Powder, Fumigation	Oral, Applied Externally , Inhalation
CAB T169	*Thymus saturejoides* Coss.	Assherkna	Leaf used as an aromatic infusion for tea or coffee	0.640	1.99	Emmenagogue, Antirheumatic, Antioxidant, Antispasmodic, Carminative, Cough, Expectorant, Anti-Diabetic, Headache, Antiseptic, Vermifuge	Leaf, Stem, and leaf	Cataplasm, Decoction, Friction, Infusion, Maceration, Crude, Powder, Fumigation	Oral, Inhalation
CAB T035	*Cinnamomum verum* J.Presl	Lquerfa	Bark powder is used as a condiment	0.040	1.33	Emmenagogue, Aphrodisiac, Colds	Bark	Powder	Oral
CAB T093	*Laurus nobilis* L.	Taselt/Oiaraqt Saiedna Moussa	Leaf used as a condiment	0.049	1.32	Diuretic, Colds Febrifuge Vertigo, Headache	Leaf	Decoction, Infusion	Oral
	Linaceae								
CAB T103	*Linum usitatissimum* L.	Zarriaat Al Kettane	Seeds are used as a food seasoning	0.096	1.28	Phytoestrogen, Colds, Laxative, Catarrh, Cough	Seed	Decoction, Infusion, Fumigation, Crude, Powder	Oral, Inhalation
	Malvaceae								

CAB T107	*Malva sylvestris* L.	Oabgir/Khbiza	Leaf used raw in salads	0.102	1.37	Diuretic, Emmenagogue, Colds	Leaf, Stem, and leaf	Decoction, Crude	Oral
	Moraceae								
CAB T072	*Ficus carica* L.	Milsoud/Lkar mous	Fruit is eaten fresh or dried	0.038	1.35	Hypotensive, Laxative, Wound, Healing, Dermatosis	Latex, Stem and leaf, Fruit	Decoction, Crude Friction	Oral, Applied Externally
	Musaceae								
CAB T115	*Musa* spp.	Lbannan	Fruit eat en fresh	0.011	1.00	Dermatosis	pericarp	Friction	Applied Externally
	Myristicacea								
CAB T116	*Myristica fragrans* Houtt.	Lgouza	Seeds powder is used as a condiment	0.042	1.16	Emmenagogue, Colds, Aperitif, Cough, Anti-Diabetic	Seed	Powder	Oral
CAB T067	*Syzygium aromaticum* (L.) Merr. & L.M.Perry	Lkrenfel	Flower buds used as a condiment	0.136	1.23	Diuretic, Emmenagogue, Antirheumatic, Colds, Analgesic, Toothache, Anti-Inflammatory, Lumbago	Flower	Decoction, Infusion, Powder Cataplasm	Oral, Applied Externally
	Oleaceae								
CAB T122	*Olea europaea* L.	Zitoun	Fruit oil used as a food	0.217	1.20	Ear Pains, Anti-Hemorrhoid, Toothache, Antispasmodic, Astringent, Wound Healing, Anti-Diabetic, Headache, Dermatosis	Leaf, Fruit oil	Decoction, Cataplasm, Friction, Maceration, Crude, Infusion	Ear drops, Applied Externally , Oral
	Palmaceae								
CAB T134	*Phoenix dactylifera* L.	Tignie / Tmer	Fruit eaten fresh or dried	0.078	1.14	Phytoestrogen, Anxiety, Carminative, Stomachic, Eye pain	Fruit, Seed, Flower	Crude, Powder Friction	Oral, Applied Externally
	Piperaceae								
CAB T136	*Piper nigrum* L.	Lbzar	Seeds powder is used as a condiment	0.029	1.15	Colds, Ossification, Catarrh	Seed	Powder	Oral
CAB T082	*Hordeum vulgare* L.	Timzine	Seed powder is used as a food	0.016	1.00	Aperitif, Carminative	Seed	Decoction	Oral

CAB T130	*Pennisetum glaucum* (L.) R.Br.	Illan	Seed powder is used as a food	0.082	1.43	Antidiarrheal, Ossification	Seed	Powder	Oral
CAB T178	*Zea mays* L.	Asngar /Drat	Seed powder is used as a food	0.013	1.00	Kidney Stone, Diuretic	Flower	Decoction	Oral
	Polygonaceae								
CAB T154	*Rumex vesicarius* L.	Basmom	Leaf and stem used raw in salads	0.013	1.00	Aperitif, Cholagogue	Stem and leaf	Crude	Oral
	Portulacaceae								
CAB T140	*Portulaca oleracea* L.	Rajla	Leaf and stem used raw in salads	0.011	1.00	Laxative	Stem and leaf	Decoction, Crude	Oral
	Punicaceae								
CAB T143	*Punica granatum* L.	Rman	Fruit eat en fresh	0.129	1.22	Antidiarrheal, Hepatitis, Cholagogue, Gingivitis, Anti-Diabetic, Anaemia	pericarp	Maceration, Crude, Powder, Decoction, Infusion	Oral
CAB T119	*Nigella sativa* L.	Sanouj	Seeds are used as a condiment	0.258	1.23	Diuretic, Emmenagogue, Colds, Hepatitis, Antispasmodic, Catarrh, Cough, Anti-Measles, Alzheimer	Seed	Decoction, Fumigation, Crude, Powder	Oral, Inhalation
	Rhamnaceae								
CAB T180	*Ziziphus lotus* (L.) Lam.	Azouggar	Fruit is eaten dried	0.096	1.37	Kidney Stone, Colds, Hepatitis, Anti-Diabetic, Anaemia, Anti-sorcery, evil eye	Leaf, Fruit	Decoction, Infusion, Powder	Oral
	Rosaceae								
CAB T106	*Malus domestica* Borkh.	Tefah	Fruit eat en fresh	0.013	1.00	Carminative, Stomachic	Fruit	Crude	Oral
CAB T141	*Prunus domestica* L.	Tabrkekcht	Fruit eat en fresh or dried	0.011	1.00	Laxative	Fruit	Crude	Oral
CAB T142	*Prunus dulcis* (Mill.) D. A. Webb	Louz /Imrzig	Almond eaten raw	0.053	1.29	Diuretic, Choleretic, Anti-Diabetic	Seed	Infusion, Crude	Oral
CAB T150	*Rosa canina* L.	Tighfert	Fruit eaten fresh	0.016	1.00	Laxative	Flower	Crude	Oral
	Rubiaceae								

IOP Conf. Series: Earth and Environmental Science 1090 (2022) 012021 doi:10.1088/1755-1315/1090/1/012021

CAB T043	*Coffea arabica* L.	Qahoia	Seeds powder boiled and drunk as a beverage	0.013	1.33	Catarrh, Anti-Diabetic	Seed	Decoction	Oral
	Rutacea								
CAB T040	*Citrus limon* (L.) Obsbek	Lhamed	Fruit used as a food	0.104	1.60	Kidney Stone, Gingivitis, Catarrh, Cough, Expectorant, Febrifuge, Dermatosis	Fruit	Crude, Cataplasm, Friction	Applied Externally , Oral
CAB T041	*Citrus aurantium*L.	Limone	Fruit eat en raw	0.031	1.36	Diuretic, Colds Laxative, Catarrh, Expectorant	Fruit, Flower	Infusion, Crude	Oral
	Sapotaceae								
CAB T011	*Argania spinosa* (L.) Skeels	Argane/ Alhorjan	Fruit oil used as a food	0.222	1.94	Ear Pains, Antirheumatic, Laxative, Stomachic, Muscle Tear, Eye pain Antiseptic, Wound Healing, Anti-Diabetic, Dermatosis, Skincare, Haircare	Fruit oil, Seed	Crude, Friction Cataplasm	Ear drops, Oral, Applied Externally
	Solanaceae								
CAB T025	*Capsicumannuum*L.	Tifelfelt	Fruit boiled and eaten as a vegetable or raw in salads	0.076	1.09	Colds, Aperitif, Carminative, Gingivitis	Fruit	Crude, Powder, Decoction	Oral
	Theaceae								
CAB T023	*Camelliasinensis*(L.) Kuntze	Attai	Leaf boiled and drunk as a beverage	0.016	1.29	Antispasmodic, Anti-Diabetic, Dermatosis	Leaf	Crude, Infusion Cataplasm	Oral, Applied Externally
	Urticaceae								
CAB T171	*Urtica dioica* L.	Taznagt / Tigzenin	Leaf used raw in salads	0.084	1.63	Kidney Stone, Antirheumatic, Antispasmodic, Astringent, Carminative, Arthritis, Asthma, Anti-Diabetic, Anaemia	Stem and leaf	Infusion, Decoction, Crude, Cataplasm	Oral, Applied Externally
	Verbenaceae								
CAB T005	*Aloysia citriodora* Palau	Lwiza	Leaf boiled and drunk as a beverage or aromatic infusion for tea	0.196	1.20	Diuretic, Cholagogue, Laxative, Sedative, Stomachic, Catarrh, Febrifuge, Anaemia, Anxiety, Headache	Leaf	Decoction, Infusion	Oral
	Vitaceae								

CAB T175	*Vitis vinifera* L.	Adile/ Laineb	Fruit eaten fresh or dried	0.004	1.00	Febrifuge	Leaf	Cataplasm	Applied Externally
	Zingiberaceae								
CAB T006	*Alpinia officinarum* Hance	Khedenjal	The root powder is used as a condiment	0.016	1.29	Emmenagogue, Cough	Root	Decoction, Powder	Oral
CAB T061	*Elettaria cardamomum* (L.) Maton	Quaii Quola	Seeds powder used as a condiment	0.018	1.38	Colds, Carminative	Seed	Powder	Oral
CAB T179	*Zingiber officinale* Roscoe	Skinjbir	The root powder is used as a condiment	0.111	1.50	Diuretic, Emmenagogue, Colds, Analgesic, Antispasmodic, Cough, Catarrh, Expectorant, Headache	Root	Decoction, Powder	Oral

3.4. Use Value (UV) and Relative Frequency of Citation (RFC)
For this study, the UV values of collected plants are arranged between 1.00 and 2.04 (Table 2). Out of 84 collected plants, 06 species showed significant UV (UVs ≥ 1.52), *Thymus leptobotrys* Murb. (2.04), *Thymus saturejoides* Coss. (1.99), *Argania spinosa* (L.), Skeels (1.94), *Allium sativum* L. (1.63), *Urtica dioica* L. (1.63), and *Citrus limon* (L.) Burm. f. (1.60). On the other hand, RFC values of collected plants are arranged between [0.004; 0.882], and two plant species were identified with RFC greater than 0.50; *Thymus leptobotrys* Murb. (0,882) and *Thymus saturejoides* Coss. (0,640). Similarly, these two plants have significant value for UV, and they are an endemic species to Morocco.

Indeed, Thyme has an important place in traditional herbal medicine everywhere in Morocco and Algeria [12, 25]. In addition, it is widely used in Moroccan cuisine as a condiment or an aromatic infusion for tea or coffee. So, these plants are in danger of extinction due to their overexploitation and are widely exported as herbs or essential oils [18].

3.5. Type of Diseases treated
In this study, the majority of medicinal species are used primarily against diseases of the digestive system, with a percentage of 22%, followed by endocrine and genito-urinary diseases (16%), cardiovascular diseases (11%), the rest of the conditions are under 10% (Figure. 2). In ethnobotanical studies carried out nationally [16] and internationally [35, 36], digestive system diseases are the most commonly treated among all the disorders treated by medicinal food plants because these plants are used for their double functioning as food and as medicine.

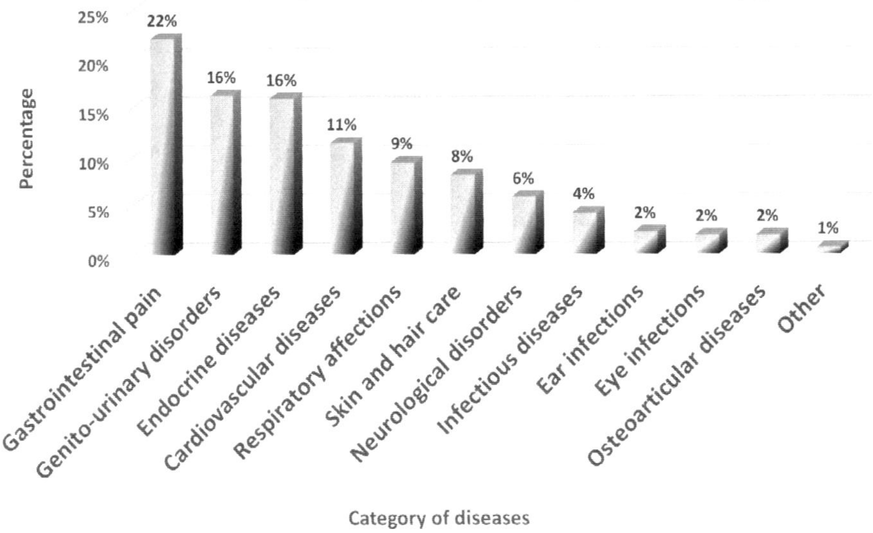

Figure 2. Type of diseases treated

3.6. Plant part used, route of administration, and method of preparation
The plant used in medicinal preparations in our study is principally the seeds and leaves. It is 46% of herbal preparation, followed by the leafy stems, which include 18%, fruit 11%, other plants part (bulb, Fruit, stem, roots, pericarp, latex, fruit oil, stem, bark) are less than 10% (Figure 3).

This study showed that plant parts (seeds and leaves) play an essential role in herbal medicine preparation in Chtouka Ait Baha and Teznit regions. These plant parts used are similar to those brought back by other authors [12, 16, 25, 35].

Benefit from active compounds of medicinal plants, diverse methods are integrated (decoction, infusion, powder, etc.). The results shown in the graph in Figure 4 are as follows: Decoction is the most common method of preparation with a percentage of 27%, followed by the crude, which constituted 18%, infusion at 16%, powder, cataplasm, friction, maceration and other (fumigation and boiled in oil) 13%, 9%, 6%, 5% respectively, used for the preparation of the medicines food plants. However, the practitioners administer their remedies prepared by oral (65%), followed by the applied externally (22%), inhalation (6%), ear drops (5%), and other forms (eye drops, vagina) with 2% (Figure 5).

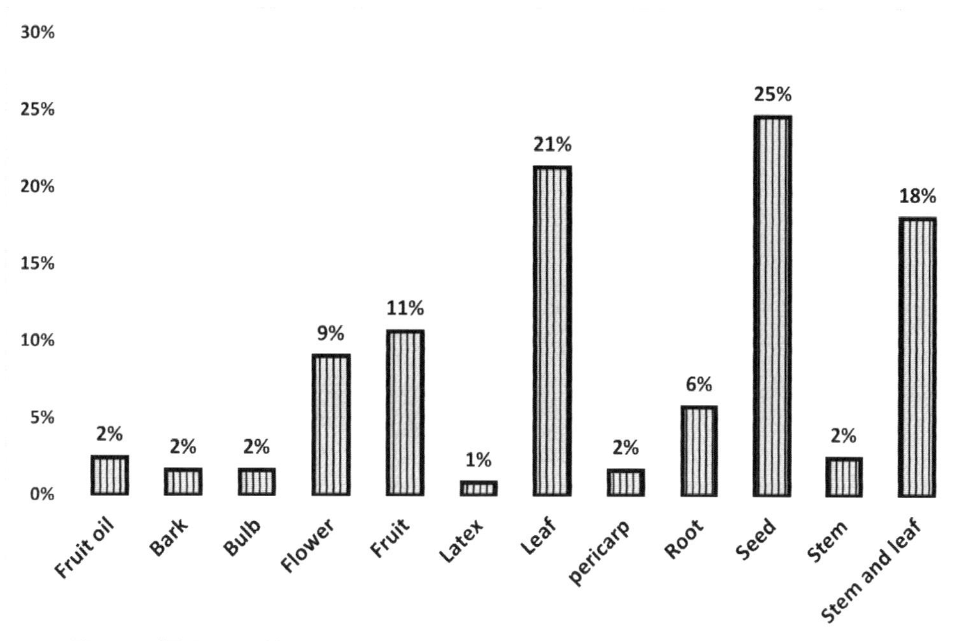

Figure 3. Plant parts used

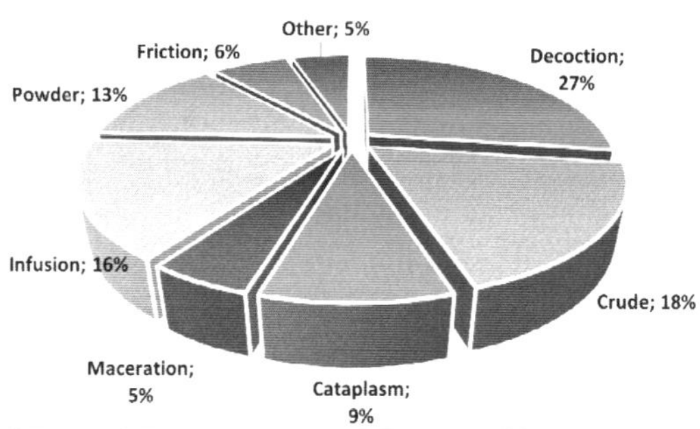

Figure 4. Preparation methods

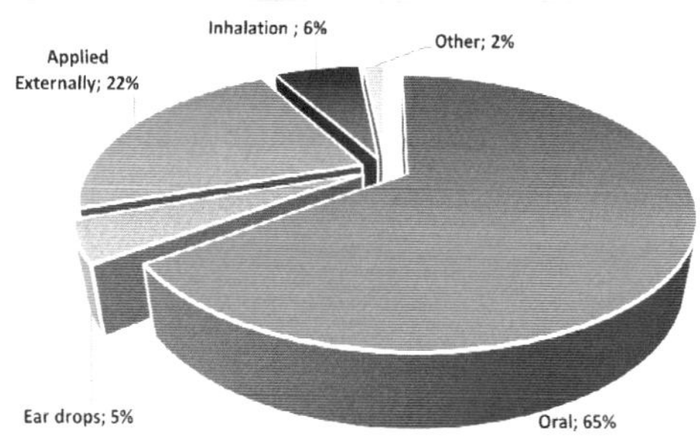

Figure 5. Routes of administration

3.7. Modes of consumption of medicinal plants as food

The mode of consumption of medicinal plants as food mostly consumed in our study was fresh or dried with a percentage of 36%, followed by the specie as boiled and eaten as food which constituted 25%, specie used as a condiment 21%, the rest of specie used as an aromatic infusion for tea or coffee with 18% (figure 6). The result indicated that the local population always chooses the simplest method to consume the plants; moreover, most Moroccan consummation is fruits or salads.

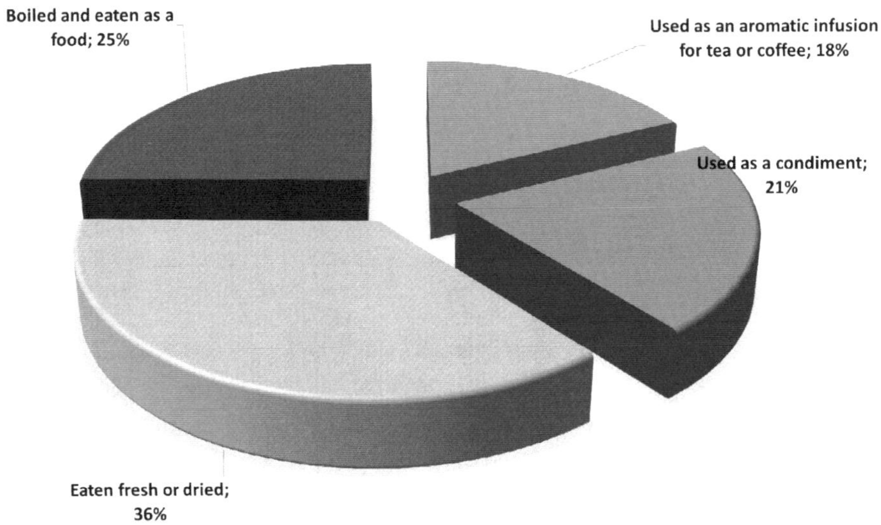

Figure 6. Modes of consumption of medicinal plants as food

4. Conclusion

Our study region contained many medicinal food plants (84 plant species), plants with high UV, and RFC (*Thymus leptobotrys* Murb., *Thymus saturejoides* Coss.). It was the most preferred species in the study area. However, these plants need more attention and require additional research. Diseases of the digestive system are the most treated among all the conditions treated. The plant part mainly used in medicinal preparations in our study is the seeds and leaves; the decoctions were the most common practice method, and most medicines are administered orally.

Despite the region's arid conditions, the area has a rich diversity of medicinal plant species. The tradition of using plants for medicinal and food is still alive in the local community. Still, recently this tradition has been gradually declining. Therefore, awareness is needed among the local people on ecological use and conservation of local flora.

5. References

[1] Ferreira Júnior, W.S., F.R. Santoro, A.B. Nascimento, A.H. Ladio, and U.P. Albuquerque, *The role of individuals in the resilience of local medical systems based on the use of medicinal plants – a hypothesis.* Ethnobiology and Conservation, 2013. 2(0).

[2] Leduc, C., J. Coonishish, P. Haddad, and A. Cuerrier, *Plants used by the Cree Nation of Eeyou Istchee (Quebec, Canada) for the treatment of diabetes: A novel approach in quantitative ethnobotany.* J Ethnopharmacol, 2006. 105(1-2): p. 55-63.

[3] Hora, J.S.L., I.S. Feitosa, U.P. Albuquerque, M.A. Ramos, and P.M. Medeiros, *Drivers of species' use for fuelwood purposes: A case study in the Brazilian semiarid region.* Journal of Arid Environments, 2021. 185: p. 104324.

[4] Hamilton, A., P. Shengji, J. Kessy, A.A. Khan, S. Lagos-Witte, and Z.K. Shinwari, *The purposes and teaching of applied ethnobotany.* 2003: United Nations Educational, Scientific and Cultural Organization (UNESCO).

[5] Calixto, J.B., *Twenty-five years of research on medicinal plants in Latin America: a personal view.* J Ethnopharmacol, 2005. 100(1-2): p. 131-4.
[6] WHO, *(World Health Organization) traditional medicine strategy 2002-2005.* 2002, Geneva: World Health Organization.
[7] Wanjohi, B.K., V. Sudoi, E.W. Njenga, and W.K. Kipkore, *An Ethnobotanical Study of Traditional Knowledge and Uses of Medicinal Wild Plants among the Marakwet Community in Kenya.* Evidence-Based Complementary and Alternative Medicine, 2020. 2020: p. 3208634.
[8] Pieroni, A., A. Nedelcheva, and Y. Dogan, *Local knowledge of medicinal plants and wild food plants among Tatars and Romanians in Dobruja (South-East Romania).* Genetic Resources and Crop Evolution, 2015. 62(4): p. 605-620.
[9] Ajibesin, K.K., B.A. Ekpo, D.N. Bala, E.E. Essien, and S.A. Adesanya, *Ethnobotanical survey of Akwa Ibom State of Nigeria.* Journal of Ethnopharmacology, 2008. 115(3): p. 387-408.
[10] Jahandiez, E. and R. Maire, *Catalogue des plantes du Maroc:(Spermatophytes et Ptéridophytes). Tome premier. Ptéridophytes, Gymnospermes et Monocotylédones*. 1931: Minerva.
[11] Sijelmassi, A., *Les plantes médicinales du Maroc*. 1996: Editions Le Fennec.
[12] Abouri, M., A. El Mousadik, F. Msanda, H. Boubaker, B. Saadi, and K. Cherifi, *An ethnobotanical survey of medicinal plants used in the Tata Province, Morocco.* J Int Med Plants Res, 2012. 1(7): p. 99-123.
[13] Bellakhdar, J., *La pharmacopée marocaine traditionnelle.* 1997: París-Casablanca: Ibis Press-Éditions Le Fennec.
[14] Idm'hand, E., F. Msanda, and K. Cherifi, *Ethnobotanical study and biodiversity of medicinal plants used in the Tarfaya Province, Morocco.* Acta Ecologica Sinica, 2020. 40(2): p. 134-144.
[15] Msanda, F., A. El Aboudi, and J.-P. Peltier, *Biodiversité et biogéographie de l'arganeraie marocaine.* J Cahiers Agricultures, 2005. 14(4): p. 357-364 (1).
[16] Zouhri, A. and L. Aarab, *Ethnobotanical Survey of Medicinal and Aromatic Plants Used by the People of Targuist in the North of Morocco.* Der Pharma Chemica, 2018. 10(5)g: p. 11-18.
[17] Eddouks, M., M. Ajebli, and M. Hebi, *Ethnopharmacological survey of medicinal plants used in Daraa-Tafilalet region (Province of Errachidia), Morocco.* J Ethnopharmacol, 2017. 198: p. 516-530.
[18] Barkaoui, M., A. Katiri, H. Boubaker, and F. Msanda, *Ethnobotanical survey of medicinal plants used in the traditional treatment of diabetes in Chtouka Ait Baha and Tiznit (Western Anti-Atlas), Morocco.* J Ethnopharmacol, 2017. 198: p. 338-350.
[19] Skalli, S., R. Hassikou, and M. Arahou, *An ethnobotanical survey of medicinal plants used for diabetes treatment in Rabat, Morocco.* Heliyon, 2019. 5(3): p. e01421.
[20] Cotton, C.M. and P. Wilkie, *Ethnobotany: principles and applications*. 1996: John Wiley & Sons Chichester.
[21] Fennane, M., M. Ibn Tattou, J. Mathez, A. Ouyahya, and J. Oualidi, *Flore Pratique du Maroc, Vol. 1: Pteridophyta, Gymnospermae, Angiospermae (Lauraceae-Neuradaceae): Manuel de Détermination.* Travaux de l'Institut Scientifique, série botanique, 1999: p. 36.
[22] Fennane, M., M. Ibn Tattou, J. Mathez, A. Ouyahya, and J. Oualidi, *Flore pratique du Maroc. Vol. 2. Angiospermae (Leguminosae-Lentibulariaceae).* Travaux de l'Institut Scientifique, série botanique, 2007.
[23] Fennane, M., M. Ibn Tattou, J. Mathez, A. Ouyahya, and J. Oualidi, *El oualidi J. Flore pratique du Maroc, Manuel de détermination des plantes vasculaires.* Vol. III. Rabat. Travaux de l'Institut Scientifique, 2014: p. 450.
[24] Zougagh, S., A. Belghiti, T. Rochd, I. Zerdani, and J. Mouslim, *Medicinal and Aromatic Plants Used in Traditional Treatment of the Oral Pathology: The Ethnobotanical Survey in the Economic Capital Casablanca, Morocco (North Africa).* Nat Prod Bioprospect, 2019. 9(1): p. 35-48.
[25] Ouhaddou, H., H. Boubaker, F. Msanda, and A. El Mousadik, *An ethnobotanical study of medicinal plants of the Agadir Ida Ou Tanane province (southwest Morocco).* Journal of Applied Biosciences, 2015. 84(1).
[26] HCP, *Haut commissariat au plan: General census of population and housing. Government of Morocco. Online: https://www.hcp.ma/downloads/RGPH-2014_t17441.html. Accessed 13 Mai 2019.* 2014.

[27] Saadi, B., F. Msanda, and H. Boubaker, *Contributions of folk medicine knowledge in Southwestern Morocco: The case of rural communities of Imouzzer Ida Outanane Region.* J of Medicinal Plant Research, 2013: p. 135-145.
[28] Mechchate, H., I. Es-safi, F.z. Jawhari, A. Bari, A. Grafov, and D. Bousta, *Ethnobotanical survey about the management of diabetes with medicinal plants used by diabetic patient in Region of Fez-Meknes, Morocco.* Ethnobotany Research and Applications; Vol 19 (2020), 2020.
[29] Fouad, Z., E.G.L. El Azzouzi Fatiha, and Z. Lahcen, *Ethnobotanical survey of medicinal plants used in the traditional treatment of diabetes and gout in the north of Morocco (Tangier, Tetouan and Chefchaouen cities).* J Plant Archives, 2019. 19(2): p. 2731-2737.
[30] Lee, C., S.Y. Kim, S. Eum, J.H. Paik, T.T. Bach, A.M. Darshetkar, R.K. Choudhary, D.V. Hai, B.H. Quang, N.T. Thanh, and S. Choi, *Ethnobotanical study on medicinal plants used by local Van Kieu ethnic people of Bac Huong Hoa nature reserve, Vietnam.* J Ethnopharmacol, 2019. 231: p. 283-294.
[31] Ouelbani, R., S. Bensari, T.N. Mouas, and D. Khelifi, *Ethnobotanical investigations on plants used in folk medicine in the regions of Constantine and Mila (North-East of Algeria).* J Ethnopharmacol, 2016. 194: p. 196-218.

IOP Conf. Series: Earth and Environmental Science 1090 (2022) 012022 doi:10.1088/1755-1315/1090/1/012022

Chemical composition, Antibacterial and Antioxidant Activities of the Essential Oil of *Lavandula Pedunculata* from Khenifra Morocco

M Chroho[1], J El Karkouri[2], N Hadi[2], B Elmoumen[2], T Zair[2] and L Bouissane[1]

[1] Laboratory of Molecular Chemistry, Materials and Catalysis, Faculty of Sciences and Technology Beni Mellal (FSTBM), Sultan Moulay Sliman University, Morocco.
[2] Research team Chemistry of Bioactive Molecules and Environment, Laboratory of Innovative Materials and Biotechnologies of Natural Resources, Faculty of Sciences, Moulay Ismaïl University, Meknes, Morocco.

chroho_mounia@yahoo.fr

Abstract. Essential oils (EO) extracted from medicinal and aromatic plants (MAP) have proven to be good sources of bioactive compounds with antibacterial and antioxidant properties. This work presents the study of the chemical composition of Essential Oil extracted from *Lavandula Pedunculata* from the middle atlas of Morocco and its antioxidant and antibacterial power. The EO was obtained by hydrodistillation and analyzed by gas chromatography coupled to mass spectrometry. The antibacterial power of EO was estimated by the disk diffusion method against six bacterial species stemming from the hospital. Antioxidant activity was tested by two methods: measuring the trapping power of DPPH radical and measuring the Ferric Reducing/Antioxidant Potential (FRAP). The chromatographic analysis of *Lavandula Pedunculata* essential oil allowed the identification of 55 compounds representing 98.9% of the essential oil, of which the major compounds are *camphor* (44.16%), *1 –epi-cubenol* (8.61%), *fenchone* (7.48%), *camphene* (6.45%) and *α-pinene* (4.86%). Antibacterial power tests showed that *L. Pedunculata* EO effect on all the tested bacteria is lower than that of the most powerful Antibiotic and even powerless on certain strains. The obtained results from antioxidant power tests by both methods proved the antioxidant power of *Lavandula Pedunculata* EO but was less effective than ascorbic acid. The obtained results indicate that the plant EO's antibacterial and antioxidant activities are not powerful compared to antibiotics and ascorbic acid respectively, but enough to explain the traditional therapeutic use of the plant.

1. Introduction

Medicinal plants have been a source of medicines and health care for humans since ancient times, they represent natural sources of important new active ingredients. They can act as natural antibiotics and antioxidants and help overcome two major human health problems: antibiotic resistance and oxidative stress. In this context, several studies have been conducted to evaluate the antimicrobial and antioxidant activities of Essential Oils (EOs) extracted from certain medicinal plants. Thus, several EOs have shown remarkable antibacterial and/or antioxidant effects.

Lavenders are aromatic and medicinal plants belonging to the family *Lamiaceae*, they exist under more than 20 species and are widely spread in the Mediterranean region [1,2]. Lavenders have very high commercial and industrial values and are used in the food industry, perfume shops, and pharmaceutical preparations [3,4]. Recently, it has been referred that they are also used in agriculture mainly as crop candidates for arid lands and as efficient bioinsecticides [5].

Essential Oil (EO) of *Lavandula* species has shown an antibacterial power [6–9], antispasmodic and sedative properties [10] as well as acaricidal [11], antifungal [2,12] and antioxidant activities [13–15], there biological activities against the growth of a wide range of microorganisms were also distinguished [16], it was reported also that *Lavandula Pedunculata* EO presents a good efficiency of the inhibition of the corrosion of the soft steel and can be considered as promising green inhibitors of corrosion [17].

Lavenders are a part of the rich and diversified Moroccan flora. Eleven species were quoted among the aromatic and medicinal plants of Morocco (*L. augustifolia* [18], *L. multifida* [9], *L. dentata*, *L. Pedunculata*, *L. abrialis* [19], *L. stoechas* [20], *L. atlantica*, *L. maroccana* [21], *L. rejdalii* [22], *L. officinalis* [23] and *L. tenuisecta* [24]), they are used by the population in traditional medicine for the treatment of diverse affections such cough, intestinal pains and diseases related to the urinary system, as well as to calm the renal and rheumatic pains [21,25].

Lavandula Pedunculata is one of the species of lavender growing in Morocco, It can reach up to 70 cm tall where only the long-stalked spikes can reach 24 cm and are full of lilac flowers and covered by different kinds of glandular trichomes [16]. Ethnobotanical studies listed this specie in the region of Khenifra in the Middle Atlas of Morocco [21,26], its vernacular name in these regions is " Lhalhal " [21] in moroccan dialect or "izr" [26] in berber of the Moroccan Middle Atlas.

The aim of this work is the valorization of *Lavandula Pedunculata*, native from Khenifra region located in the Middle Atlas of Morocco, by studying the chemical composition of its essential oil and the evaluation of its antimicrobial and antioxidant powers.

2. Material and methods

2.1. Plant material

The plant was collected in full blooming, in May 2019, from the site Assif Bou Zekour (on the sides of a waterway) where it grows wildly. The site is located in the southeast of Khenifra city at a height of 850 m in the geographic localization (32°54'45.1"N 5°39'33.8"W) in the Middle Atlas of Morocco. Identification of species was confirmed in the Scientific Institute of Rabat.

The collected samples were dried for ten days at room temperature in the shade.

2.2. Phytochemical screening

It aims to highlight the important families of secondary metabolites contained in the plant. This is a qualitative study based on coloring and/or precipitation reactions from different plant extracts. These extracts are obtained by decoction, infusion, or maceration by solvents. Phytochemical screening is also based on the use of several reagents [27,28]. Table 1 lists the reagents used in the research of different families.

Table 1. Reagents used in the search for secondary metabolites families

Family sought	Reagents used
Alcaloïds	Dragendorff reagent
Catchic Tanins	Isoamyl alcohol and hydrochloric acid
Gallic Tanins	Stiasny reagent, sodium acetate, and iron chloride
Sterols and Triterpenes	Acetic anhydride and concentrated sulphuric acid
Flavonoïds	Dilute hydrochloric alcohol, magnesium shavings, and iso-amylic
Quinonic substances	Chloroform, dilute ammonia, and hydrochloric acid

IOP Conf. Series: Earth and Environmental Science 1090 (2022) 012022 doi:10.1088/1755-1315/1090/1/012022

2.3. Essential Oil extraction and analysis

The extraction of essential oil was made by hydrodistillation in Clevenger device. Three distillations were realized by boiling during three hours 100 g of plant material with one liter of water. The obtained essential oil was stored in darkness at 4°C.

The analysis of the EO chemical composition was performed by Gas chromatography (GC) and mass spectrometry (MS). Gas chromatograph used is Thermo Electron type : Trace GC Ultra equipped with DB-5 capillary column (5% *phenyl-methyl-siloxane*) (inner diameter 30m x0. 25mm, thickness: 0.25µm). Detection is provided by a flame ionization detector (FID) fed by a mixture of H2/Air gases. The carrier gas used is nitrogen with a flow rate of 1 ml/min. The injection mode is split (splitting ratio: 1/50, flow rate: 66 ml/min. Temperature programming goes from 50 to 200°C for 5min, with a gradient of 4°C/min. Mass spectroscopy was performed by a Thermo Electron Trace MS system (Thermo Electron: Trace GC Ultra; Polaris Q MS) mass spectrometer. Fragmentation is performed by electronic impact under a 70eV field.

The EO chemical composition was identified by comparing the kovat indices (KI) of compounds with those of known reference products in the literature. Kovat indices are calculated by comparing the retention time of a compound with those of a series of linear alkanes (C7-C40) [29]. The determination of the chemical composition is complemented by a comparison of mass indices and spectra, obtained by gas chromatography coupled with mass spectrometry (GC/MS), with different references [30].

2.4. Antibacterial activity

2.4.1. Choice and preparation of bacterial strains. To study the antibacterial activity we chose pathogenic bacterial strains stemming from the hospital center of Khenifra. Six bacterial species were tested; *Pseudomonas aeruginosa* (Pseudo), *Escherichia Coli* (EC1, EC2, EC3), *Staphylococcus aureus* (Staph A1, Staph A2), *Staphylococcus blanc* (Staph B), *Proteus* (Proteus) and *Klebsiella pneumonia* (Kleb).

Bacterial strains are sowed in Petri dishes containing the environment of culture Meuler Hinton. The inoculum is then prepared by taking a sterile swab of a colony of bacterium and decanting it in a tube containing 2.5 ml of sterile physiological water and then we shake strongly. For every bacterium, four boxes are sowed by the same bacterial suspension (same inoculum) in order to have the same concentration. After drying of boxes, one will serve for the realization of the antibiogram and the three other ones to estimate the antibacterial activity of the EO. The average of the inhibition zones measurements is noted.

2.4.2. Susceptibility tests of bacteria to antibiotics (Antibiograms). The antibiotic susceptibility profiles of the strains are performed according to recommendations of the French Microbiological society *[31]* and the European Committee on Antimicrobial Susceptibility Testing (EUCAST) *[32]*.

The antibiogram is realized by employing a disks distributor; Nine disks of different antibiotics (ATB) were tested on one of the two prepared dishes (*Doxycycline, Imipénème, Amoxicilline, Triméthoprime, Ceftriaxone, Cefalexine, Ciprofloxacin, Ticracilline, Ofloxacin*). Eight antibiotics are arranged in regular intervals all around the dish and one in the center. We incubate the dishes in the steam room at 37°C for 24 hours. Then we measure the diameter of every zone of inhibition.

2.4.3. Disk diffusion tests. On the second dish sowed by the same inoculum, 4µl of the studied EO is placed, then incubated in the steam room at 37°C for 24 hours. The diameter of every zone of inhibition is measured and compared with the realized antibiograms.

2.5. *Antioxidant activity*

The antioxidant activity was estimated by two methods: DPPH and FRAP. The first method measures the trapping power of DPPH radical (*1,1-diphényl-di-picrylhydrazyl*), and the second method FRAP measures the antioxidant power of reducing ferric iron (Fe^{3+}) to ferrous iron (Fe^{2+}).

2.5.1. Trapping power of DPPH radical. For evaluating the power of scavenging the free radical DPPH (1,1-diphényl-di-picrylhydrazyl). An ethanolic solution of DPPH is prepared by solubilizing 2.4 mg of DPPH in 100 ml of ethanol.

To 200 µl (184 mg) of Essential Oil, 800 µl of ethanol are added to prepare a mother solution of concentration 200 µl of EO/ml (184mg /ml). From this mother solution, a series of dilutions are prepared to obtain concentrations ranging from 1mg/ml to 250 mg/ml. The tests are made by mixing 200 µl of every solution of EO at various concentrations with 2.8 ml of the ethanolic solution of DPPH. After 30 minutes in darkness, the absorbance is read at 517nm. The inhibition of the free radical DPPH by Ascorbic acid was also analyzed for comparison.

The antioxidant power is estimated by two parameters:

- The percentage of inhibition of DPPH (I %) was calculated from the absorbance of the sample in the following way:

$$I\% = (A\ blanc - A\ sample)\ X\ 100\ /\ A\ blanc$$

A blanc is the absorbance of the solution of DPPH in ethanol.

- The concentration of antioxidant required for reducing the initial concentration by 50 % (IC_{50}), It is inversely related to the antioxidant capacity, the more its value is low, the more the antioxidant activity of a compound is important. This parameter is graphically defined from the curve of the absorbencies according to the concentrations.

2.5.2. Ferric Reducing/ Antioxidant Potential (FRAP). It is a simple and not expensive procedure that estimates the antioxidant level of a sample. It is based on the reducing potential of antioxidants in the EO that react with ferric ions (Fe^{3+}) provided by potassium ferricyanide $K_3Fe(CN)_6$ and reduce them to ferrous ions (Fe^{2+}). The method used is that described by Zovko Koncic in 2010 *[33]*.

EO dilutions with concentrations ranging from 0 to 70 mg/ml are prepared. 0.5 ml of each solution is mixed with 2.5 ml of a phosphate buffer solution (0.2 M, pH 6.6) and 2.5 ml of a potassium ferricyanide solution $K_3Fe(CN)_6$ (1%). The mixtures are incubated in a water bath at 50°C for 20 minutes. Afterward, 2.5 ml of trichloracetic acid (10%) is added to stop the reaction. The mixtures are then centrifuged at 3000 turns for 10 minutes. In the end, 2.5 ml of the supernatant of each concentration is mixed with 2.5 ml of distilled water and 0.5 ml of an aqueous solution of $FeCl_3$ at 0.1%. The absorbance is measured at 700 nm. The increase in absorbance in the reaction medium indicates the increase in the reducing power of the sample. Ascorbic acid is used as a positive control, its absorbance is measured under the same conditions as the sample.

The antioxidant capacity is expressed by the determination of the Effective Concentration (EC_{50}) which corresponds to an absorbance equal to 0.5. This parameter is used to compare the reducing activity of the sample and the control.

3. Results and discussion

The phytochemical characterization tests carried out on the extracts of *Lavandula Pedunculata* flowers showed the presence of the following chemical families: catechic and gallic tannins, flavonoids, sterols, and triterpenes. On the other hand, the absence of alcaloïds and anthocyanins is noted.

This qualitative chemical analysis of the plant differs from study to study. This is due to the effect of the biotope (climate and soil) and the vegetative stage and may affect biological activities. However, it is noted that the absence of alkaloids has been frequently cited [34–36].

3.1. Essential Oil yield

The yield of Essential Oil extracted of the dry flowers of *Lavandula Pedunculata* is 2.39 %. This yield obtained with our samples collected in full blooming in the site Bou Zekour - Khenifra at a height of 850 m is upper than that found in other studies 1.97 % [19] and 1.83 % [34] for *L. Pedunculata* of Morocco. This result is also very important than that found in a study on the EO of *L. Pedunculata* collected from Cunda Ayvak island on the West of Türkiye and which is 1.33 % [37].

3.2. Chemical composition of the Essential Oil

Chromatographic analysis of the essential oil of *Lavandula Pedunculata* allowed the identification of 55 compounds representing 98.90% of total compounds of the EO (Table 2), and of which the majority compounds are *camphor* (44.16%), *1-epicubenol* (8.61%), *fenchone* (7.48 %), c*amphene* (6.45%) and *α-pinene* (4.86%).

Table 2. Chemical composition of *Lavandula Pedunculata* Essential Oil

KI	Compound	Formule	%	KI	Compound	Formule	%
926	Tricyclene	$C_{10}H_{16}$	0.60	1235	Ocimene (E)	$C_{10}H_{14}O$	0.22
939	**α-pinene**	$C_{10}H_{16}$	**4.86**	**1285**	**Isobornylacetate**	$C_{12}H_{20}O_2$	**0.74**
954	**Camphene**	$C_{10}H_{16}$	**6.45**	1290	Thymol	$C_{10}H_{14}O$	1.50
960	Thuja 2.4 (10) diene	$C_{10}H_{14}$	0.23	1299	Carvacrol	$C_{10}H_{14}O$	0.43
967	Verbenene	$C_{10}H_{14}$	0.14	1408	Caryophyllene (z)	$C_{15}H_{24}$	0.10
1011	S-Carene	$C_{10}H_{16}$	0.45	1454	Clavene	$C_{15}H_{21}$	0.34
1024	P-Cymene	$C_{10}H_{14}$	1.72	1529	Calamenene (cis)	$C_{15}H_{22}$	0.53
1026	O-Cymene	$C_{10}H_{14}$	0.74	1538	α-Cadinene	$C_{15}H_{24}$	0.09
1059	γ-terpinene	$C_{10}H_{16}$	0.43	1545	α-Calacorene	$C_{15}H_{20}$	0.30
1090	Dehydro-Linalool	$C_{10}H_{16}O$	0.38	1574	γ-Asarone	$C_{12}H_{16}O_3$	0.12
1076	Vertocitral (cis)	$C_9H_{14}O$	0.22	1579	Sesquisobinen Hydraté	$C_{15}H_{26}O$	0.16
1086	**Fenchone**	$C_{10}H_{16}O$	**7.48**	1608	Humuleneepoxide	$C_{15}H_{24}O$	0.30
1113	Camphenol (6)	$C_{10}H_{16}O$	0.17	1619	Janenol	$C_{15}H_{26}O$	0.12
1120	Sabina KetoneDehydro	$C_9H_{12}O$	1.53	**1628**	**Cubenol (1-epi)**	$C_{15}H_{26}O$	**8.61**
1122	Myccenol	$C_{10}H_{18}O$	0.34	1651	Cedr-8(15)-en-9-ol	$C_{15}H_{24}O$	0.15
1126	Campholenal (2)	$C_{10}H_{16}O$	0.71	1641	Cedrenal (1.7.diepiα)	$C_{15}H_{22}O$	0.16
1144	Verbenol (trans)	$C_{10}H_{16}O$	0.05	1644	Muurolol	$C_{15}H_{26}O$	0.63
1146	**Camphor**	$C_{10}H_{16}O$	**44.16**	1654	α-Codinol	$C_{15}H_{26}O$	1.32
1150	Eucarvone	$C_{10}H_{14}O$	0.39	1660	Calamenen-10-ol (cis)	$C_{15}H_{22}O$	0.11
1160	Isoborneol	$C_{10}H_{18}O$	3.51	1669	Calamenen-10-ol (trans)	$C_{15}H_{22}O$	0.11
1164	Isocitrol (2)	$C_{10}H_{16}O$	0.11	1656	Exalatacin	$C_{12}H_{14}O$	0.52
1177	Terpinene-4-ol	$C_{10}H_{18}O$	0.45	1676	Codalene	$C_{15}H_{18}$	0.27
1182	P-Cymen-8-ol	$C_{10}H_{14}O$	1.07	1689	Muurol-5-en-one (cis)	$C_{14}H_{22}O$	1.10
1184	Thuj-3-en-10-ol	$C_{10}H_{14}O$	0.22	1678	Apiole	$C_{12}H_{14}O_4$	0.54
1195	Myrtenal	$C_{10}H_{14}O$	0.31	1681	Ishwarone	$C_{15}H_{22}O$	0.10
1195	Myrtenol	$C_{10}H_{16}O$	0.24	1713	Calamenene (cis)	$C_{15}H_{22}O$	0.53
1205	Verbenone	$C_{10}H_{14}O$	2.35	1714	Nuciferal (2)	$C_{15}H_{20}O$	0.09
1216	Carveol (trans)	$C_{10}H_{16}O$	0.41	**Total (%)**			**98.90**

Monoterpènes oxygénés	**Monoterpènes hydrogénés**	**Sesquiterpènes oxygénés**	**Sesquiterpènes hydrogénés**
64.12%	16%	12.39%	1.63%

The total compounds are divided into oxygenated monoterpenes (64.12%), hydrogenated monoterpenes (16%), followed by oxygenated sesquiterpènes (12.39%), and finally hydrogenated sesquiterpènes (1.63%).

Some studies carried out on *Lavandula Pedunculata* from Morocco have also reported this chemotype Camphor. Indeed, in one study the composition *Camphor* (46.36 %), *Fenchone* (13.19%), and *α-pinene* (10.74%) has been found [19]. In another work that studied *L. Pedunculata* from Tafraout in the Southwest Morocco, the essential oil has been characterized by a high amount of *camphor* (53.1%) and the other major components were *camphene* (6.1%) and *1.8-Cineole* (6.5%) [17].

Other studies have presented other chemotypes. The type *fenchone/camphor* characterized by close majority percentages in *fenchone* and *camphor* was quoted by *Harborne and Williams* and they found 24% of *camphor* and 20% of *fenchone* [38], also a composition of 40.6% of *camphor* and 38 % of *fenchone* for *L. Pedunculata* harvested in the South of Portugal have been reported [39]. Three other types: *1,8 cineole/camphor*, *1,8 cineol,* and *fenchone* were reported for *L. Pedunculata* from the center and the north of Portugal [4,12,40].

The chemical composition of the Essential Oil of *L. Pedunculata* consists of three major components (*fenchone*, *1,8-cineol,* and *camphor*) [12]. Some studies even specified that it is generally dominated by *fenchone* and *camphor* and that the sum of their percentages varies between 47 and 83% [40]. This is applied in our study where *fenchone* and *camphor* represent together 51.64%.

The same chemotypes are also present for *L. stoechas*. The composition 48% *camphor* and 22 % *fenchone* for *L. stoechas* from Australia was reported [8] and the analysis of the EO of *Lavandula stoechas* from Tunisia has allowed the identification of *fenchone* followed by *camphor* as major components [41]. Indeed *L. Pedunculata* and *L. stoechas* contain the same major compound and the same chemical polymorphism even if they are morphologically very different [12].

3.3. Antibacterial activity

3.3.1. Susceptibility of bacteria to antibiotics. The susceptibility tests of bacteria to antibiotics (antibiogram) (Table 3) show that bacterial strains EC3 and Pseudo are the most resistant to antibiotics. Only three antibiotics over nine presented inhibition diameters different from zero. On the other hand, bacterial strains Kleb and Staph A1showed no resistance to all nine tested antibiotics.

Table 3. Inhibition zones measurements of antibiotics in mm.

ATB \ Bacteria	Staph B	EC1	EC2	Proteus	EC3	Pseudo	Kleb	Staph A1	Staph A2
DO 30	2	10	7	17	10	0	20	36	18
IPM 10	15	**26**	26	0	**22**	24	**30**	**38**	**42**
AMC 30	0	8	0,5	0	14	0	14	34	22
SXT 25	8		0	19	0	0	22	26	0
CRO 30	1	8	22	22	0	0	28	12	12
CN 15	8	12	12	14	10	8	14	20	14
CIP 5	12	0	**28**	**34**	0	**30**	26	28	20
TIM 85	0	0	0	14	0	0	16	20	12
OFX 5	11	0	26	26	0	20	24	30	24

DO : Doxycycline, IPM: Imipénème, AMC: Amoxicilline, SXT: Triméthoprime, CRO: Ceftriaxone, CN : Cefalexine, CIP : Ciprofloxacine, TIM : Ticracilline, OFX : Ofloxacine, [shaded] : Highest diameter of inhibition.

Ticarcillin (TIM 85) is the least effective antibiotic and five bacterial strains were resistant to it, while Cefalexine (CN 15) was effective on all the bacteria and pronounced diameters of inhibition going from 8 to 20 mm.

The biggest diameters of inhibition were accomplished by the effect of only two antibiotics IPM 10 and CIP 5, Imipénème (IPM 10) showed superior diameters of inhibition on strains Staph B, EC1, EC3, Kleb, Staph A1, and Staph A2, and Ciprofloxacin (CIP 5) had the big effect of inhibition on strains EC2, Proteus and Pseudo.

3.3.2. Disk diffusion tests. The obtained results concerning the antibacterial activity of EO of *L. Pedunculata* by the Disk diffusion method (Table 4) show in the first place that the antimicrobial effect of EO of *L. Pedunculata* for all the tested bacteria is lower than that of the most powerful antibiotic. Three strains (EC 1, EC 2, and Pseudo) were very resistant to *L. Pedunculata* EO and that was clear by the zero inhibition diameters they presented, this EO on the other strains wasn't actionless but its activity was very weak compared with that of the witness antibiotic.

Table 4. Inhibition zones measurements of *L. Pedunculata* EO in mm.

Bacteria / EO &ATB+	Staph B	EC1	EC2	Proteus	EC3	Pseudo	Kleb	Staph A1	Staph A2
Lavandula Pedunculata	6	0	0	6	12	0	14	10	8
ATB+	15 IPM 10	26 IPM 10	28 CIP 5	34 CIP 5	22 IPM 10	30 CIP 5	30 IPM 10	38 IPM 10	42 IPM 10

ATB+: Powerful antibiotic

Similar results were obtained [19] where an antibacterial power of *L. Pedunculata* EO against *Escherichia coli* and *Klebsiella pneumonia* was recorded while it was hopeless against *Pseudomonas aeruginosa*. The same effect was cited for other species of Lavender: *L. stoechas* [42] and *L. dentata* [43]. Similar results have been also presented for a set of lavenders from Australia Essential Oil [8], which did not lead to a remarkable antibacterial activity and were quite inactive on Pseudomonas aeruginosa. Hydrosols and extracts had no antibacterial activity also.

Besides, other studies reported a significant antibacterial power of lavenders, the essential oil of both species of lavender (*Pedenculata* and *dendata*) was active toward bacteria (*E. coli*, *K. pneumonia*, *S. aureus,* and *P.mirabilis*). It has been noticed that *L. Pedunculata* turns out to be less active than *L. dendata* [34].

Lavender where Camphor is the majority compound was also cited to have antimicrobial activity against *Shigellaflexneri*, *Staphylococcus aureus*, *E. Coli,* and *Salmonella typhimurium* [44]. This proves once again that the major compound is not inevitably responsible for the antibacterial activity of the EOs. This conclusion was deducted also by [45].

It thus turns out that not all the lavenders have remarkable antibacterial power, it is not enough to know the majority compound and to attribute the activity to it, but all the compounds of EO act in synergy some with the others. Some studies showed that EO has a biological activity upper than that of the combination of the majority compound isolated, the same studies indicated that even the minority compound plays a critical role in the biological activity of the EO [12,46].

The antibacterial mode of action of essential oils is complex and still under deep investigation and study. The fact that essential oils are more powerful on Gram-positive bacteria than Gram-negative bacteria is explained. In Gram-negative bacteria, the outer membrane creates a barrier toward macromolecules and hydrophobic compounds because it contains hydrophilic lipopolysaccharides (LPS), which justify the higher tolerance of Gram-negative bacteria toward hydrophobic antimicrobial compounds present in essential oils [47].

3.4. Antioxidant activity

Trapping power of DPPH radical. The results obtained from the measurement tests of inhibition percentage (I %) of the DPPH radical are registered in Figure 1. It is clear that Lavandula Pedunculata Essential oil does present an antioxidant effect and the percentage of inhibition of the free radical increases with the increase of concentration and reaches its maximal value at the concentration of 100 mg/ml. The maximal value of the antioxidant power is about 85 %. The value of the parameter IC_{50}, the quantity of antioxidant required to decrease by 50 % the concentration of the free radical, for Lavandula Pedunculata essential oil, is in the order of 21.14mg/ml (Figure 3). Figure 2 presents the inhibition percentage of DPPH according to ascorbic acid concentrations, the percentage of inhibition of the free radical approaches its maximal values since the concentration is 10mg/ml and the parameter IC_{50} is in the order of 3.54 mg/ml (Figure 3).

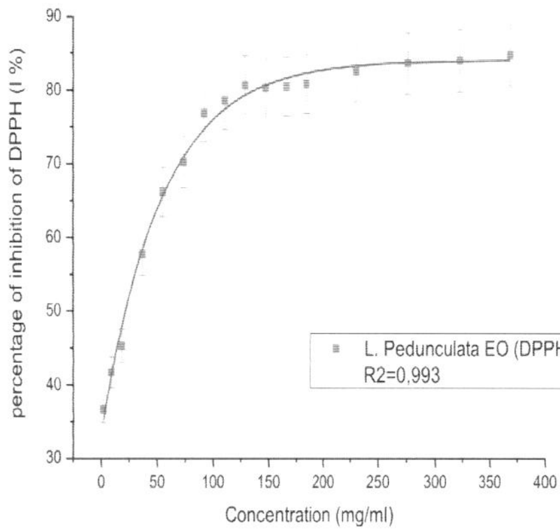

Figure 1. Antioxidant activity *of Lavandula Pedunculata* EO by DPPH

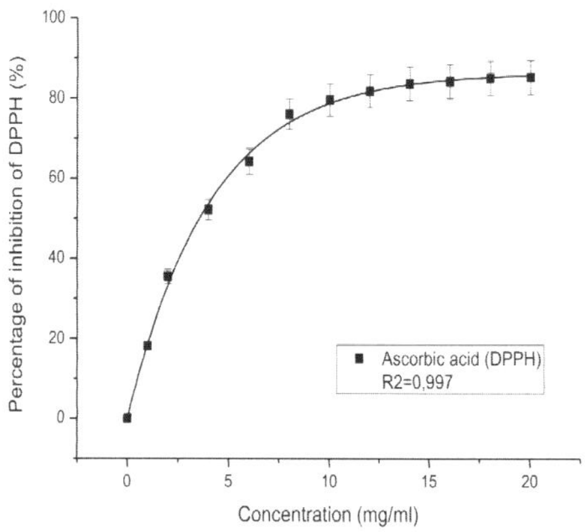

Figure 2. Antioxidant activity of ascorbic acid by DPPH

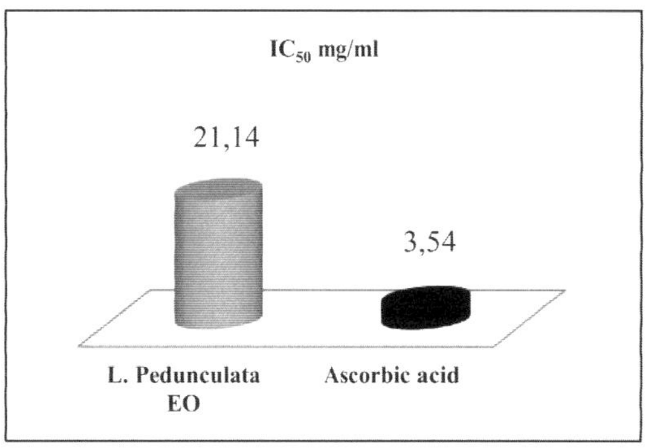

Figure 3. IC_{50} values for *Lavandula Pedunculata* Essential Oil and Ascorbic acid

Ferric Reducing/ Antioxidant Potential (FRAP). The antioxidant power of Lavandula Pedunculata Essential Oil has also been tested by the FRAP method (Figure 4), we note the existence of an antioxidant effect that increases with the increase of concentration. The EC_{50} parameter required concentration for absorbance of 0.5 is of the order of 8.65mg/ml. The antioxidant power of *Lavandula Pedunculata* Essential Oil is compared to that of ascorbic acid also evaluated by FRAP (Figure 5). The latter shows a much stronger antioxidant power than that of *Lavandula Pedunculata* EO (Figure 6).

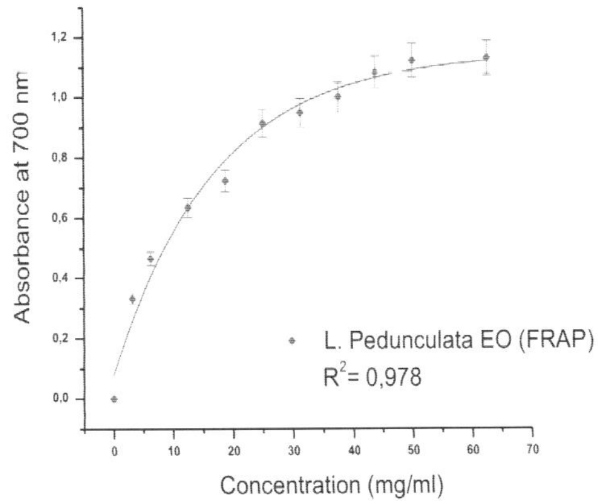

Figure 4. Antioxydant activity of *Lavandula Pedunculata* EO by FRAP

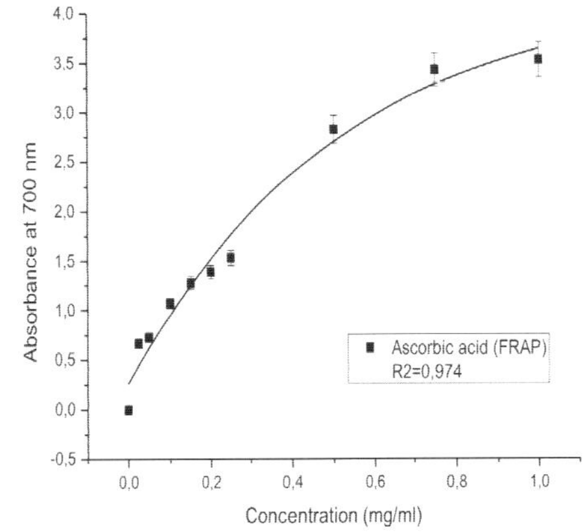

Figure 5. Antioxydant activity of ascorbic acide by FRAP

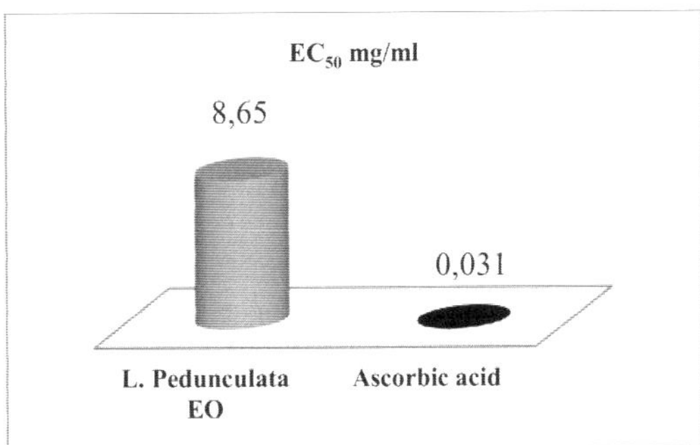

Figure 6. EC_{50} values for *Lavandula Pedunculata* Essential Oil and ascorbic acid

The comparison of the antioxidant power of *Lavandula Pedunculata* EO, tested by two methods (DPPH and FRAP), with the ascorbic acid shows that the antioxidant activity of the studied EO is lower than ascorbic acid antioxidant activity. For the latter one, the percentage of inhibition of the free radical reaches its maximal values (near 85 %) since concentrations are much lower than those of the studied EO. In addition, the values of IC_{50}, for the lavender essential oil and ascorbic acid, show a wide gap in favor of the ascorbic acid. This one presents the power to reduce 50 % of the free radical by only the concentration of 3.54 mg/ml while the studied EO doesn't reach this percentage of inhibition until 21.14 mg/ml.

Similar conclusions were quoted for other lavenders, *Lavandula officinalis* [34,48] and *Lavandula stoechas* [49], which presented a low antioxidant activity compared with the most powerful antioxidant in an aqueous medium; the ascorbic acid. Despite this weak power, the use of these EO as antioxidant complements has been proposed thanks to their liposolubility [49].

On the other hand [50] compared the inhibition of linoleic acid peroxidation by lavender essential oil and by ascorbic acid, and they found that the antioxidant activity of lavender essential oil against the peroxidation of lipids is stronger than ascorbic acid.

4. Conclusion

In this paper, we studied the chemical composition of the essential oil of *Lavandula Pedunculata* from Khenifra region in the Middle Atlas of Morocco and we estimated its antibacterial and antioxidant power. The obtained results showed that the EO studied consists essentially of *camphor* (44.16%), 1-*epi cubenol* (8.61%), *fenchone* (7.48%), *camphene* (6.45%), and *α-pinene* (4.86%). The antimicrobial effect of *L. Pedunculata* EO tested on six bacterial species is lower than that of the powerful Antibiotic in every case. This shows that lavenders do not present the same antibacterial properties. These properties are not due only to the majority compound but to the synergy of all the constituents. The tests of antioxidant activity proved the existence of antioxidant power but it was less effective than that of the ascorbic acid. Despite that, this essential oil can be used as an antioxidant complement thanks to its liposolubility.

References

[1] Hazzit M, Baaliouamer A, Veríssimo A R, Faleiro M L and Miguel M G 2009 Chemical composition and biological activities of Algerian Thymus oils *Food Chemistry* **116** 714–21

[2] Angioni A, Barra A, Coroneo V, Dessi S and Cabras P 2006 Chemical Composition, Seasonal Variability, and Antifungal Activity of *Lavandula stoechas* L. ssp. *stoechas* Essential Oils from Stem/Leaves and Flowers *Journal of Agricultural and Food Chemistry* **54** 4364–70

[3] Boelens M H 1995 Chemical and Sensory Evaluation of Lavandula Oils *Perfumer and flavorist* 23–51

[4] Zuzarte M R, Dinis A M, Cavaleiro C, Salgueiro L R and Canhoto J M 2010 Trichomes, essential oils and in vitro propagation of Lavandula pedunculata (Lamiaceae) *Industrial Crops and Products* **32** 580–7

[5] González-Coloma A, Martín-Benito D, Mohamed N, García-Vallejo M C and Soria A C 2006 Antifeedant effects and chemical composition of essential oils from different populations of Lavandula luisieri L. *Biochemical Systematics and Ecology* **34** 609–16

[6] Dadalioglu I and Evrendilek G A 2004 Chemical compositions and antibacterial effects of essential oils of Turkish oregano (Origanum minutiflorum), bay laurel (Laurus nobilis), Spanish lavender (Lavandula stoechas L.), and fennel (Foeniculum vulgare) on common foodborne pathogens *J. Agric. Food Chem.* **52** 8255–60

[7] Djenane D, Aïder M, Yangüela J, Idir L, Gómez D and Roncalés P 2012 Antioxidant and antibacterial effects of Lavandula and Mentha essential oils in minced beef inoculated with E. coli O157:H7 and S. aureus during storage at abuse refrigeration temperature *Meat Science* **92** 667–74

[8] Moon T, Wilkinson J M and Cavanagh H M A 2006 Antibacterial activity of essential oils, hydrosols and plant extracts from Australian grown Lavandula spp. *International Journal of Aromatherapy* **16** 9–14

[9] Soro N K, Majdouli K, Moutaouakil K, El hilali F and Khabbal Y 2014 Chemical composition and Antibacterial power of Lavandula multifida L. essential oil against multiresistant strains of Escherichia coli, Pseudomonas aeruginosa and Klebsiella pneumoniae isolated in hospital *International Journal of Innovation and Scientific Research* **9** 464–72

[10] Cavanagh H M A and Wilkinson J M 2002 Biological activities of Lavender essential oil *Phytotherapy Research* **16** 301–8

[11] Perrucci S, Macchioni G, Cioni P C, Flamini G, Morelli I and Taccini F 1996 The activity of volatile compounds from Lavandula angustifolia against Psoroptes cuniculi *Phytotherapy Research* **10** 5–8

[12] Zuzarte M, Gonçalves M J, Cavaleiro C, Dinis A M, Canhoto J M and Salgueiro L R 2009 Chemical Composition and Antifungal Activity of the Essential Oils of *Lavandula pedunculata* (Miller) Cav. *Chemistry & Biodiversity* **6** 1283–92

[13] Matos F, Miguel M G, Duarte J, Venâncio F, Moiteiro C, Correia A I D, Figueiredo A C, Barroso J G and Pedro L G 2009 Antioxidant Capacity of the Essential Oils From Lavandula luisieri, L. stoechas subsp. lusitanica, L. stoechas subsp. lusitanica x L. luisieri and L. viridis Grown in Algarve (Portugal) *Journal of Essential Oil Research* **21** 327–36

[14] Benabdelkader T, Zitouni A, Guitton Y, Jullien F, Maitre D, Casabianca H, Legendre L and Kameli A 2011 Essential oils from wild populations of Algerian Lavandula stoechas L.: composition, chemical variability, and in vitro biological properties *Chem. Biodivers.* **8** 937–53

[15] Benabdelkader T 2012 Biodiversité, bioactivité et biosynthèse des composés terpéniques volatils des lavandes ailées, Lavandula stoechas sensu lato, un complexe d'espèces méditerranéennes d'intérêt pharmacologique *Université Jean Monnet - Saint-Etienne* 282

[16] Baptista R, Madureira A M, Jorge R, Adão R, Duarte A, Duarte N, Lopes M M and Teixeira G 2015 Antioxidant and Antimycotic Activities of Two Native Lavandula Species from Portugal *Evidence-Based Complementary and Alternative Medicine*

[17] Bouazama S, Costa J, Desjobere J M, BenAli A, Guenbour A, Bellaouchou A, Zarrouk A and Tabyaoui M 2017 Use of Lavandula Pedunculata Essential Oil as a Green Corrosion Inhibitor for Mild Steel in 1 M Hydrochloric Acid Solution *Journal of Materials and Environmental Sciences* 10

[18] El haras K, Daagare A, Mesfioui A and Ouhssine M 2013 Activité antibactérienne de l'huile essentielle des inflorescences de Laurus Nobilis et Lavandula Angustifolia *Afrique SCIENCE* **9** 134–41

[19] Soro N K, Majdouli K, Khabbal Y and Zair T 2014 Chemical composition and antibacterial activity of Lavandula species L.dentata L. , L. pedunculata Mill and Lavandula abrialis essential oils from Morocco against food- borne and nosocomial pathogens *International Journal of Innovation and Applied Studies* **7** 774–81

[20] Aafi A, Taleb M S and Fechtal M 2002 *Espèces remarquables de la flore du Maroc* (Édition MCEF)

[21] Bachiri L, Labazi N, Daoudi A, Ibijbijien J, Nassiri L, Echchegadda G and Mokhtari F 2015 Etude ethnobotanique de quelques lavandes marocaines spontanées *International Journal of Biological and Chemical Sciences* **9** 1308

[22] Gharby S, Asdadi A, Ibourki M, Hamdouch A, Ainane T and Hassani L A I 2020 Chemical Characterization of The Essential Oil From Aerial Parts of Lavandula rejdalii Upson & Jury, a Medicinal Plant Endemic to Morocco *Journal of Essential Oil Bearing Plants* **23** 1422–7

[23] Alnamer R, Alaoui K, Bouidida E H, Benjouad A and Cherrah Y 2011 Sedative and Hypnotic Activities of the Methanolic and Aqueous Extracts of Lavandula officinalis from Morocco *Advances in Pharmacological Sciences* **2012** 270824

[24] Sayout A, Ouarhach A, Dilagui I, Soraa N and Romane A 2020 Antibacterial activity and chemical composition of essential oil from Lavandula tenuisecta Coss.ex Ball. an endemic species from Morocco *European Journal of Integrative Medicine* **33** 101017

[25] Mounia C, Zakaria K, Touriya Z and Fatima E H 2016 Ethnobotanical Survey of Local Medicinal Plants in the Urban District of Khenifra – Morocco *IJIRSET* **5** 11

[26] Nassiri L, Zarkani S, Daoudi A, Bammou M, Bouiamrine E H and Ibijbijen J 2016 Contribution à l'élaboration d'un catalogue ethnobotanique de la commune rurale d'Aguelmous (Province de Khénifra, Maroc)/[Contribution to the establishment of ethno botanical catalog of Aguelmos (Khenifra, Morocco] *International Journal of Innovation and Applied Studies* **17** 373–87

[27] Bruneton J 2009 *Pharmacognosie : Phytochimie, Plantes médicinales* (Cachan: Tec & Doc Lavoisier)

[28] N'Guessan K, Kadja B, Zirihi G, Traoré D and Aké-Assi L 2009 Screening phytochimique de quelques plantes médicinales ivoiriennes utilisées en pays Krobou (Agboville, Côte-d'Ivoire) *Sciences & Nature* **6**

[29] Kovats E. 1965 Gas chromatographic characterization of organic substances in the retention index system *Advances in Chromatography* 229–47

[30] Adams R P 2007 *Identification of Essential Oil Components By Gas Chromatography/Mass Spectrometry* (Carol Stream, Ill: Allured Pub Corp)

[31] R. Bonnet 2013 Comité de l'antibiogramme de la société française de microbiologie: Recommandations 2013

[32] Leclercq R, Cantón R, Brown D F J, Giske C G, Heisig P, MacGowan A P, Mouton J W, Nordmann P, Rodloff A C, Rossolini G M, Soussy C-J, Steinbakk M, Winstanley T G and Kahlmeter G 2013 EUCAST expert rules in antimicrobial susceptibility testing *Clin. Microbiol. Infect.* **19** 141–60

[33] Zovko Končić M, Kremer D, Karlović K and Kosalec I 2010 Evaluation of antioxidant activities and phenolic content of Berberis vulgaris L. and Berberis croatica Horvat *Food and Chemical Toxicology* **48** 2176–80

[34] Bachiri L, Bammou M, Echchegadda G, Ibijbijen J, Rhaffari L E, Haloui Z and Nassiri L 2017 Composition Chimique Et Activité Antimicrobienne Des Huiles Essentielles De Deux Espèces De Lavande : Lavandula Dentata Spp. Dentata Et Lavandula Peduncultata Spp. Pedunculata *European Scientific Journal* **13** 19

[35] Teixeira G, Correia A I, Vasconcelos T, Feijão D and Madureira A M 2013 Lavandula stoechas subsp. luisieri and L. pedunculata - phytochemical study, micromorphology and histochemistry *Revista de Ciências Agrárias* **36** 220–8

[36] Boufellous M, Lrhorfi L A, Berrani A, Haoud H E, Zaher A, Bouhaddioui B and Bengueddour R 2017 Phytochemical screening of a medicinal plant: Lavandula stoechas (Lamiaceae) **6** 56–62

[37] Topçu G and Gören A C 2007 Biological Activity of Diterpenoids Isolated from Anatolian Lamiaceae Plants *Rec. Nat. Prod.* 16

[38] Harborne J B and Williams C A 2003 Phytochemistry of the genus Lavandula *Lavender The Genus Lavandula* (Maria Lis-Balchin)

[39] Costa P, Gonçalves S, Valentão P, Andrade P B, Almeida C, Nogueira J M F and Romano A 2013 Metabolic profile and biological activities of Lavandula pedunculata subsp. lusitanica (Chaytor) Franco: Studies on the essential oil and polar extracts *Food Chemistry* **141** 2501–6

[40] D. Feijão M, Teixeira G, Vasconcelos M, Rodrigues L, Correia A, Sanches J, Pedro L, Barroso J and Figueiredo A 2011 Essential oil variability and trichomes morphology from Lavandula pedunculata (Mill.) Cav. grown at Mata Experimental do Escaroupim (Portugal) *Planta Medica* **77** 1297–1297

[41] Bouzouita N, Kachouri F, Hamdi M, Chaabouni M M, Aissa R B, Zgoulli S, Thonart P, Carlier A, Marlier M and Lognay G C 2005 Volatile Constituents and Antimicrobial Activity of *Lavandula stoechas* L. Oil from Tunisia *Journal of Essential Oil Research* **17** 584–6

[42] Mohammedi Z and Atik F 2011 Antibacterial activity of essential oils from Cistus ladaniferus L.and Lavandula stoechas L. *International Journal of PharmTech Research* 484–7

[43] Imelouane B, Elbachiri A, Ankit M, Benzeid H and Khedid K 2009 Physico-Chemical Compositions and Antimicrobial Activity of Essential Oil of Eastern Moroccan Lavandula dentata *Int. J. Agric. Biol.* **11** 6

[44] Jianu C, Pop G, Gruia A T and Horhat F G 2013 Chemical Composition and Antimicrobial Activity of Essential Oils of Lavender (Lavandula angustifolia) and Lavandin (Lavandula x intermedia) Grown in Western Romania *Int. J. Agric. Biol.* **15** 5

[45] Ghalem B R and Mohamed B 2008 Antibacterial activity of leaf essential oils of Eucalyptus globulus and Eucalyptus camaldulensis *African Journal of Pharmacy and Pharmacology* **2** 211–5

[46] Berger R G 2007 *Flavours and Fragrances: Chemistry, Bioprocessing and Sustainability* (Berlin Heidelberg: Springer-Verlag)

[47] Hyldgaard M, Mygind T and Meyer R 2012 Essential Oils in Food Preservation: Mode of Action, Synergies, and Interactions with Food Matrix Components *Frontiers in Microbiology* **3**

[48] Imène L 2012 Etude des activités antioxydante et antifongique de l'huile essentielle des fleurs sèches de Lavandula officinalis : application aux moisissures des légumes secs *Nature & Technologie* 44–52

[49] Zohra M and Fawzia A 2012 Pouvoir antifongique et antioxydant de l'huile essentielle de Lavandula stoechas L. *Nature & Technologie* 34–9

[50] Hui L, He L, Huan L, XiaoLan L and AiGuo Z 2010 Chemical composition of lavender essential oil and its antioxidant activity and inhibition against rhinitis- related bacteria *African Journal of Microbiology Research* **4** 5

Fencing land impacts on plant biodiversity and argan trees dynamic in the Ida-Ou-Tanane (central western of Morocco)

M Ait Haddou[1], M Ben El Caid[2], A Aydda[1], Y Bouchriti[1], A Wanaim[1], H Gougueni[1] and S Ezaidi[1]

[1]Geosciences, Environment and Geomatic Laboratory (GEG), Department of Geology, Faculty of Sciences, Ibn Zohr University, Agadir 8106, Morocco
[2]Laboratory of Biology and health, Department of Biology, Abdelmalek Essaâdi University, PO Box 2121, 93002 Tétouan, Morocco
E-mail: mohamed.aithaddou@edu.uiz.ac.ma

Abstract. The argan trees are subject to strong anthropgenic pressure that prevents its natural regeneration. This study aims to evaluate the effect of lands fencing on the redynamization of the argan trees through analyzing and comparing the specific richness in two closer parcels. The first parcel have been fenced since 2004, whereas the second one is an open land. In this study, the floristic diversity indices were investigated and compared based on field inventory from both parcels. The results obtained show the beneficial effect of the fence on the dynamics of the vegetation. The plant specific richness is higher for the fenced parcel than unfenced parcel (37 and 21 species, respectively). For the fenced parcel, the maximum diversity index is about 3.61, while it is about 3.04 for the unfenced parcel. The Shannon-Weaver index (H') and the evenness (E) are also higher for the fenced parcel where the cover rate is 57% compared to 32% for the unfenced parcel. The average of the computed argan trees, in the fenced parcel, is 116 trees/ha of size ranging between 2 to 5 m. For the unfenced parcel, the average is 83 trees/ha of size of 4 m. Additionally, inside the fenced parcel, the natural regeneration by seedling is well marked, where considerable number of argan trees towere identified during field survey. These results highlight the importance to use fencing in the generation of the endemic plants.

1. Introduction

The centre-west of Morocco is the exclusive territory of a vast forest based on the argan trees (*Argania spinosa* (L.) Skeels), recently incorporated into the genus Sideroxylon [1, 2]. It is an endemic, thermophilic and typical forest species of the Moroccan infra-Mediterranean domain. In terms of plant biodiversity, the argan trees reserve is the most important in the south. It is characterised by its flora which is mainly made up of Mediterranean taxa and others adapted to an arid climate [3]. It was declared as UNESCO Biosphere Reserve in 1998. It is a national and universal heritage with a great wealth of endemic or rare plants of multiple use (aromatic, medicinal, melliferous and pastoral) [4, 3], and to preserve the soils against water erosion [5]. This reflects a strong relationship between humans and this endemic resource [6].

The argan trees represent the backbone of the traditional rural economy, and has become, in recent decades, the main engine of development for the local population. However, this reserve is more or less marked by the impact of notably climate change and anthropic effects [7, 8]. Is likely confronted with enhanced climate variability, climate change and socio-economic constraints in the future [9]. The current exploitation methods only aggravate the ecological situation of the natural resources in this reserve, where the argan trees are the main component threatened by the overexploitation of plant resources, the clearing of soils, and the absence of natural regeneration [10]. Confronted with the problem of the degradation of the vegetation cover, the Provincial Directorate of Water and Forests and the Fight against Desertification of Agadir (DPEFLCD) has launched a programme for the regeneration and development of the argan trees plantation through permanently fencing of some places of the argan trees reserve.

The current study aims to elucidate the effect of fencing on the preservation of the argan trees and on the dynamics of plant species in the Ida-Ou-Tanane (central western of Morocco). The study compared the specific richness and floristic diversity in two parcels (fenced and unfenced).

2. Study area

The study was conducted in the Asserssif perimeter at the foothills of the Ida-Ou-Tanane massif subject to the "Soussien" currents of the south-west [11]. With an area of about 50 hectares, the parcel was fenced in 2004 and is located 18 km north-east of the city of Agadir (Figure 1). From the ecological point of view, the study area is essentially consist of argan trees and is under a worm upper arid atmosphere, marked by 300 days of insolation annually with an monthly measurements in Agadir station indicate an increase in temperature over the last 50 years and seasonal irregularities a downward trend in rainfall [12, 13], and an average rainfall of 242 mm (Figure 2 and 3). The dry season extending 8 months from Mars to October and rainy season for the rest 4 months (Figure 4). Furthermore, the water resources vulnerability in Agadir Ida -Ou-Tanane have reached very high levels [14].

The study area is located in a geographical area classified as an Argan Biosphere Reserve (ABR) and is an important grazing area for goats and camels [15]. The substratum is mostly made up of hard and compact limestone, topped by a soil that is not very developed. It is highly influenced by the Atlantic Ocean; the Agadir station is characterized by moderate temperatures (18.5 °C annual average). The cold Canary currents soften the temperature by the sea breeze and increase the humidity of the air, causing the fog. The maximum is in August or September [16, 17]. The oceanic façade plays a very important role in the distribution of vegetation.

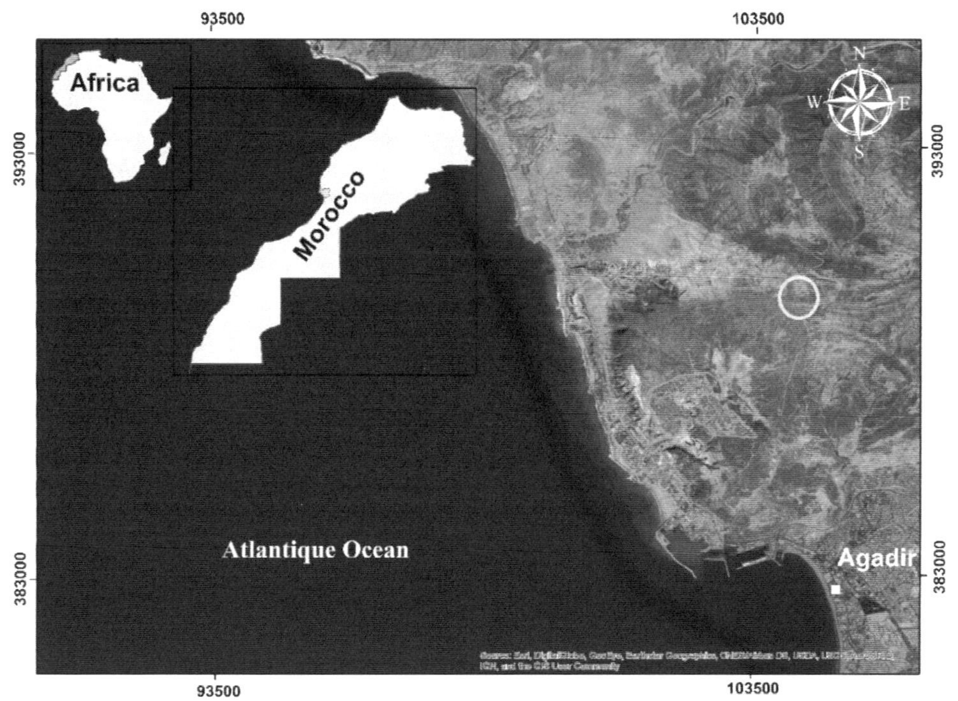

Figure 1. Location map of the study area of Ida-Ou-Tanane showing the sampling parcels (yellow circle).

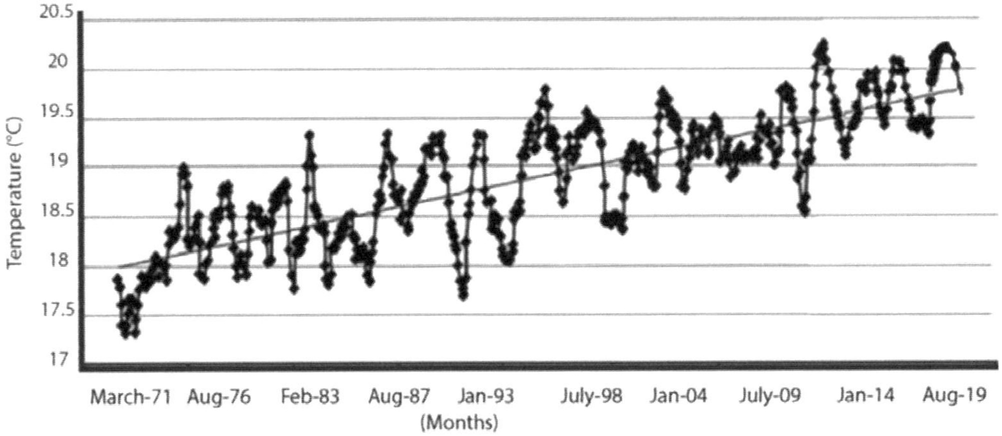

Figure 2. Seasonal variation of mean monthly temperature in Agadir city [14].

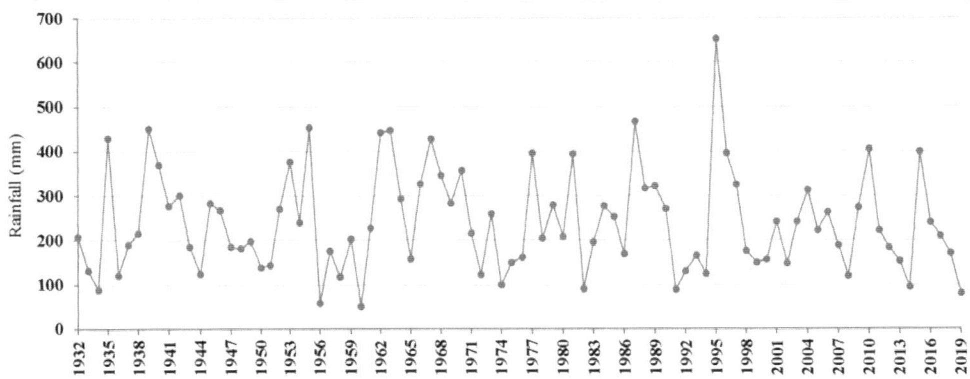

Figure 3. Mean annual rainfall during the period 1932-2019 measured at the Agadir city station

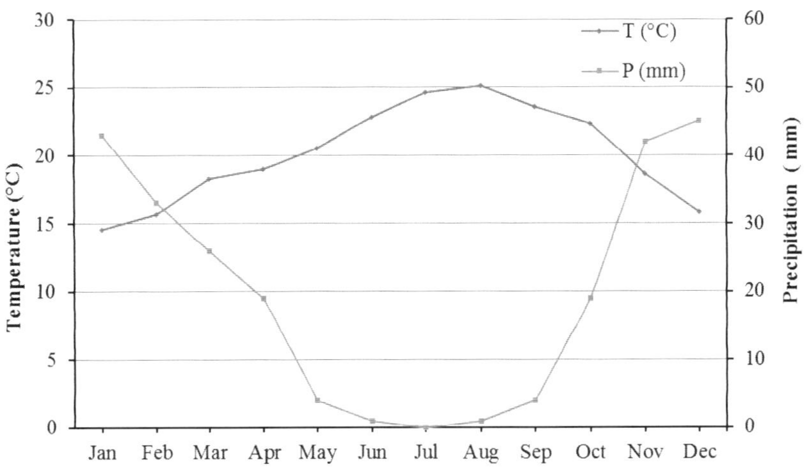

Figure 4. Bagnoul and Gaussen diagram of the Agadir city

3. Materials and methods

3.1. Vegetation sampling
The study field was sampled using the phytosociological methodology of quadrats along transects spaced 200-250 m apart [18]. Three transects per plot were established. Each transect consists of ten surveys with a minimum area for each square plot of about 100 m² (10m X 10m) and spaced 200 m apart, which considered sufficiently representative in Mediterranean Formations [19, 20]. The surveys were carried out in such a way as to ensure exhaustive coverage of the two closer parcels at a rate of 30 vegetation surveys per parcel. The abundance and dominance of the floristic species record were estimated using the scale (1-5) of Braun-Blanquet and Géhu & Rivas-Martínez [21, 22]. The identification of species was done using the national floras [23, 24]. For plant groups, we used the reference of [25]. After the inventory step, several indices were evaluated using the equations and methodologies described in following lines.

Climatic data covering the study area for the period 1932-2019 was collected from Agency of Hydraulic Basin of Souss-Massa (ABHSM). The collected data consists of monthly precipitation and monthly temperature data in the Agadir station.

3.2. Diversity Indices
The floristic diversity can be estimated using several indices and equations (1-5) (Table 1).

Table 1. Floristic diversity indices used in this study.

Diversity Index	Index formula	Description
Maximum Diversity (H_{max}) [26]	$H_{max} = Log_2(S)$ (1)	H_{max} is Shannon's maximum diversity index S: Total number of species present
Shannon–Weaver Index (H') [27, 28]	$H' = \sum fi * Log_2 fi$ (2) $fi = ni/\sum ni$ (3)	f_i: Specific contribution of species i n_i: Absolute frequency of species i
Evenness index (E) [29]	$R = H'/H_{max}$ (4)	Comparison of the measured diversity with the maximum theoretical diversity
Beta diversity or Jaccard similarity index (Cj) [30, 31]	$Cj = a/(a+b+c) * 100$ (5)	A: Number of species common in both parcels B: Number of species in parcel A not in parcel B C: Number of species in parcel B not in parcel A

3.3. Global vegetation cover
The average global cover was evaluated on the basis of the soil cover of the vegetation strata (tree, shrub and herbaceous).

3.4. Life form and disturbance index
The relative representation of life form (biological types) provides precious indications on the structure, physiognomy and adaptive strategies of the flora and vegetation to the environmental conditions [32]. In this study, we used the biological type's classification of Raunkiaer [33], which is the most widely used one Romane [34]. This classification consists to categorize the plants based on the position of the survival organs (growth meristems). There are five main classes that can be defined: Therophytes or annual plants (Th), Geophytes or Cryptophytes (Cr), Hemicryptophytes (Hem), chamephytes (Ch), and Phanerophytes (Ph).

The disturbance index defined by [35] makes it possible to quantify the therophytisation of the territory (equation 6).

$$\boldsymbol{Id} = \textbf{N umber of chamephytes} + \textbf{Number of therophytes/Total number of species} \quad (6)$$

IOP Conf. Series: Earth and Environmental Science 1090 (2022) 012023 doi:10.1088/1755-1315/1090/1/012023

3.5. Density of argan trees and natural regeneration
The density of argan trees was estimated by averaging the number of trees computed in each plot in the fenced and unfenced parcels per hectare

The natural regeneration of the argan trees is defined as all the spontaneous natural processes of restoration of the seedlings, favouring the return of the argan trees without any direct human intervention. Practically, we computed the number of argan seedlings under the same argan trees.

4. Results and discussion

The field study was realized during mid-March to the end of May 2012 period. Thirty floristic records were realized. The species richness, floristic diversity indices, recovery, density of argan trees and its natural regeneration, were determined in the two parcels (fenced and unfenced). All species recorded during this study are listed in table 2.

Tableau 2. List of vascular plants recorded in the two plots studied (+) : species present, (-) : species absent, FP : fenced parcel, UP : unfenced parcel, L. form : Life form, Ph : Phanerophytes, Nph : Nanophanerophytes, Ch : Chamephytes, Th : Therophytes, Hem : Hemicryptophytes, Cr : Cryptophytes).

Plant Species	FP	UP	L. form	Family
Sideroxylon spinosum L.	+	+	Ph	Sapotaceae
Launaea arborescens (Batt.) Maire	+	+	Nph	Asteraceae
Cladanthus arabicus (L.) Cass.	+	-	Th	
*Artemisia reptans (*Chr.) Smith.	+	-	Ch	
Kleinia anteuphorbium (L.) Haw.	+	+	NPh	
Scolymus hispanicus L.	+	-	Th	
Carthamus sp	+	-	Ch	
Echinops spinosus L.	-	+	Hem	
Atractylis cancellate L.	+	-	Th	
Chamaecytisus mollis (Cav.) G.& B.	+	+	Nph	Fabaceae
Ononis natrix L.	+	+	Ch	
Genista ifniensis (Caballero, A.)	+	-	Nph	
Genista tricuspidata Desf.	+	+	Nph	
Vachellia gummifera (Willd.) K. & B.	-	+	Ph	
Scorpiurus muricatus L.	+	-	Th	
Cenchrus ciliaris L.	+	-	Cr	Poaceae
Stipellula capensis (Thunb.) R. & H.	+	+	Th	
Ballota cinerea (Desr.) Briq.	+	-	Ch	Lamiaceae
Teucrium polium L.	+	-	Ch	
Thymus satureioides Cosson	+	-	Ch	
Lavandula maroccana Murb.	+	+	Ch	
Lavandula dentata L.	+	+	NPh	
Asphodelus ramosus L.	+	+	Cr	Asphodelaceae
Drimia maritima (L.) Stearn	-	+	Cr	Asparagaceae
Diplotaxis sp	+	-	Th	Brassicaceae
Lycium intricatum Boiss.	+	-	Nph	Solanaceae
Withania frutescens (L.) Pauquy	+	-	Nph	
Searsia tripartita (Ucria) Moffett	+	+	Nph	Anacardiaceae
Searsia pentaphylla (Jacq.) F.A. Barkley	+	-	Nph	
Pistacia lentiscus L.	+	+	Nph	
Euphorbia officinarum (L.) subsp. officinarum	-	+	Nph	Euphorbiaceae
Ziziphus lotus L.	-	+	Nph	Rhamnaceae

Paronychia argentea Lam.	+	-	Hem	Caryophyllaceae
Apteranthes europaea (Guss.) Plowes	+	-	Ch	Apocynaceae
Periploca angustifolia Labill.	+	+	Nph	
Ephedra altissima Desf.	+	+	Nph	Ephedraceae
Olea europaea L.	+	+	Ph	Oleaceae
Limonium sinuatum (L.) Miller	+	-	Th	Plumbaginaceae
Kickxia heterophylla (Schousb.) Dandy	+	-	Ch	Plantaginaceae
Echium horridum Batt.	+	-	Th	Boraginaceae
Atriplex halimus L.	+	-	Nph	Amaranthaceae
Eryngium ilicifolium Lam.	+	+	Th	Apiaceae
Number of species	37	21		

4.1. Specific richness

The floristic surveys realized in the two parcels allowed us to highlight a global floristic list of 42 species (Figure 5). The fenced parcel shows a specific richness of 37 species grouped in 34 genres and 18 families. The best represented family is the Asteraceae (7 species), followed by the Fabaceae and Lamiaceae with five species for each of the two families. These four families alone represent more than 54% of the species richness in this parcel, while Poaceae, Solanaceae and Apocynaceae are represented by only two species, the remaining 11 families are "mono-specific".

The species richness in the unfenced parcel identified 21 plant species organized in 14 families and 20 different genres. The Fabaceae family is the best represented with four species, followed by the Asteraceae with three species; the Anacardiaceae and Lamiaceae are each represented by two species; these four families alone represent more than 52% of the species richness in this parcel. The remaining ten families are present with only one species each.

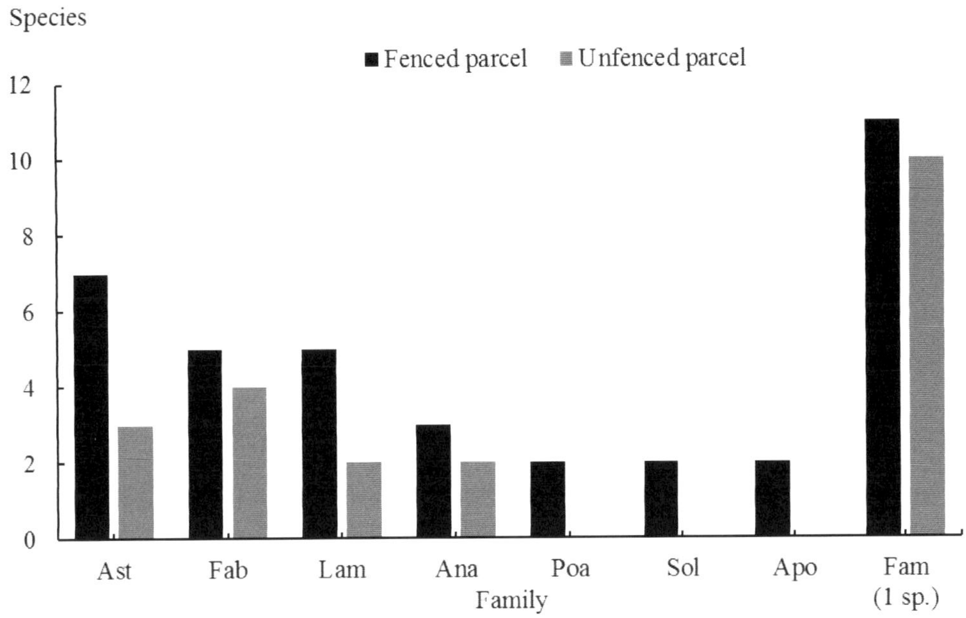

Figure 5. Number of species (sp.) per family in the two parcels. (Ast: Asteraceae, Fab: Fabaceae, Lam: Lamiaceae, Ana: Anacardiaceae, Poa :Poaceae, Sol :Solanaceae, Apo: Apocynaceae).

The fenced parcel appears to be the most diversified, hosting almost all species (37 species), or 88% of the flora recorded in the study area, while the unfenced parcel has only 21 species, or 50% of the total flora (Figure 6). The common flora between the two parcels includes 16 species. Twenty-one species (*Thymus satureioides, Genista ifniensis, Artemisia reptans, Kleinia anteuphorbium, Searsia pentaphylla...*) develop only in the fenced parcel and are maintained due to the defense. These species are experiencing a significant decline in their range and *Apteranthes europaea* represents a species at high risk of extinction in the study area.

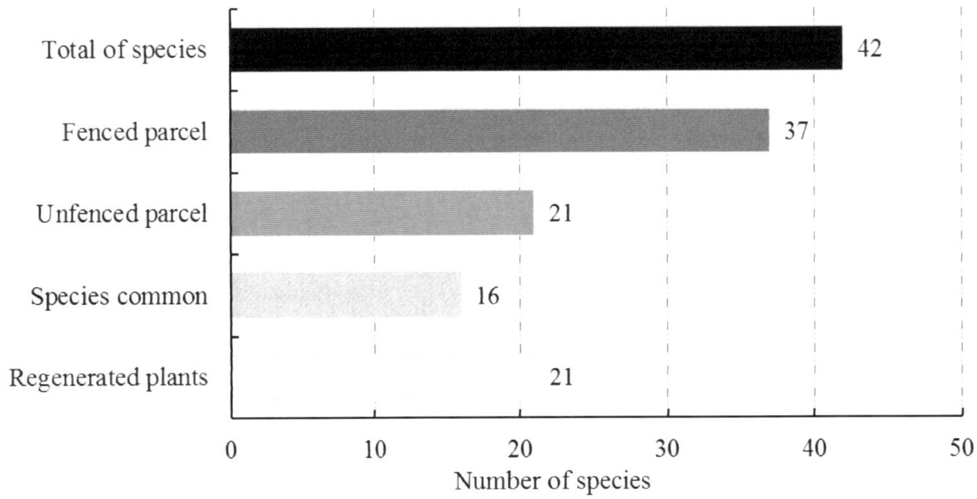

Figure 6. Species richness variation in the two parcels.

The floristic poverty of the parcel except for defending seems to be due to overgrazing and human destruction. These degradation factors lead to a modification of the floristic composition in this parcel. Among the 42 species observed, it should be noted the presence of a certain number of endemic species namely: *Sideroxylon spinosum, Genista ifniensis, Chamaecytisus mollis, Euphorbia officinarum (L.) subsp. officinarum, Vachellia gummifera and Lavandula maroccana*, which is equivalent about 14% as a count of endemism. In addition to its endemic plants, this area accommodate a number of forage species, such as: *Chamaecytisus mollis*, *Genista tricuspidata*, *Genista ifniensis*, *Scorpiurus muricatus*, *Cenchrus ciliaris*, *Stipellula capensis, etc.* and species of medicinal and aromatic interest, like *Lavandula dentata*, *Artemisia reptans*, *Ballota cinerea*, *Teucrium polium* and *Thymus satureioides*, etc. Of note that the number of species recorded in the two parcels can be slightly revised upward, taken into account the factor of dryness that has reduced this year the germination about ten/annual species of argan trees. In light of this, the maintenance of a greater specific richness in the fenced parcel could be result of its setting in defense.

4.2. Floristic diversity

Floristic diversity indices are essential criteria for estimating the biodiversity of a plant community. The Table 3 listed a clear difference between the floristic diversity indices for both parcels. The maximum diversity index value is 3.61 and 3.04, respectively, for the fenced and unfenced parcels. The Shannon-Weaver index (H') and regularity (E) are also higher in the fenced parcel. Which mean that these indices are positively influenced by fencing.

Table 3. Species diversity indices and Jaccard's index in the area studies.

Plot	Maximum Diversity Index (H_{max})	Shannon and Weaver Index (H')	Regularity Index (E)	Jaccard Index (Cj)
Fenced	3.61	3.46	0.95	38.1 %
Unfenced	3.04	2.73	0.89	

The moderately high diversity in the fenced parcel is explained by a slight equality of individual contributions to the vegetation cover, as the more species, the Shannon index is higher. On the other hand, outside the protected area, there are a regular decrease in the specific diversity index (H') and the equitability (E). This decrease can be explained by a very irregular distribution of cover between species because of the disturbances that occur in the environment, which favors the relative abundance of certain species over others. The presence of Asphodel, Jujube, Ononis and Echinops in this floristic set indicates beginning of degradation, indicating increase modification of the argan trees landscapes and erosion of the plant biodiversity. The degree of similarity (Jaccard index) between the fenced parcel and the unfenced parcel tends to 40 %, reflecting low similarity of flora between parcels and demonstrating tendency towards floral heterogeneity. These results are in accordance with those of [36] who show the benificent effect of the setting of defences in the improvement of the floristic list and the return of certain species of the original floristic procession, with pastoral vocation at the level of three steppe formations.

4.3. Global vegetation cover (GC)

The variation area covered by vegetation is illustrated in Figure 7.
For the fenced parcel, the vegetation cover is about 57% while the bare land represents only 43%. More than half of this recently protected area is covered by vegetation. In contrast, for the unfenced parcel, the vegetation cover and bare land are about 32% and 68%, respectively.

The vegetation is much degraded in the unfenced parcel cause of the intensive grazing that contributes to the regression of many fodder species appreciated by the animals such as: *Genista ifniensis, Chamaecytisus mollis Genista tricuspidata, Stipellula capensis* ... which are part of the floral procession of the argan grove, and promotes the sustainability of species less appetizing, such as *Launaeae arborescens, Ononis natrix and Echinops spinosus*. Other studies confirmed the importance to use fencing to increase vegetation cover [18, 37], phytomass and organic matter in the soil [38].

It should be noted that even in the fenced parcel and because dryness recorded during this year, many herds (ex. camels) entered the fenced parcel and caused extensive vegetation damage.

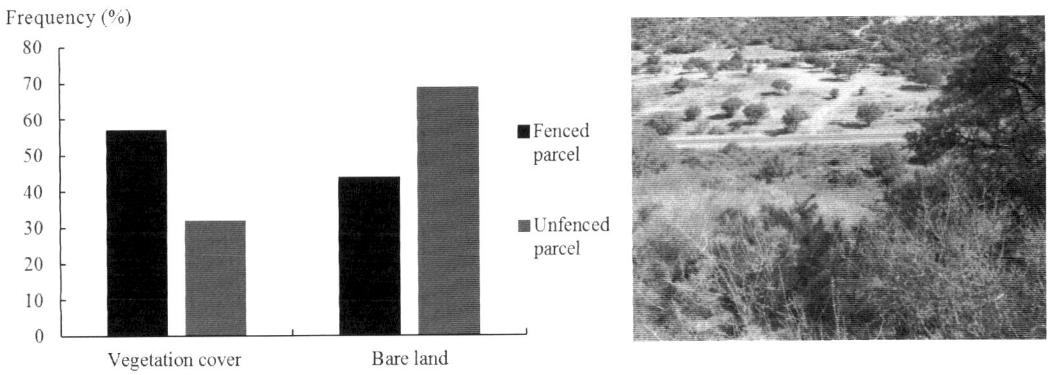

(a) (b)

Figure 7. (a) Comparisons of percentages of vegetation cover and bare land in the zone studied, (b) Environments with distinctly different vegetation density and biodiversity.

4.4. Life form and disturbance index

The biological spectrum of the plants inventoried inside the fenced parcel shows that Phanerophytes and Nanophanerophytes ranked first (40%), followed by Chamephytes (27%), Therophytes (24%), Hemicryptophytes (3%) and Crytophytes (6%) (Figure 8). For the unfenced parcel, the biological spectrum indicates that Phanerophytes-Nanophanerophytes, Chamephytes, Therophytes, Cryptophytes and Hemicryptophytes cover 66%, 10%, 9%, 10% and 5% of the territory (Figure 8).

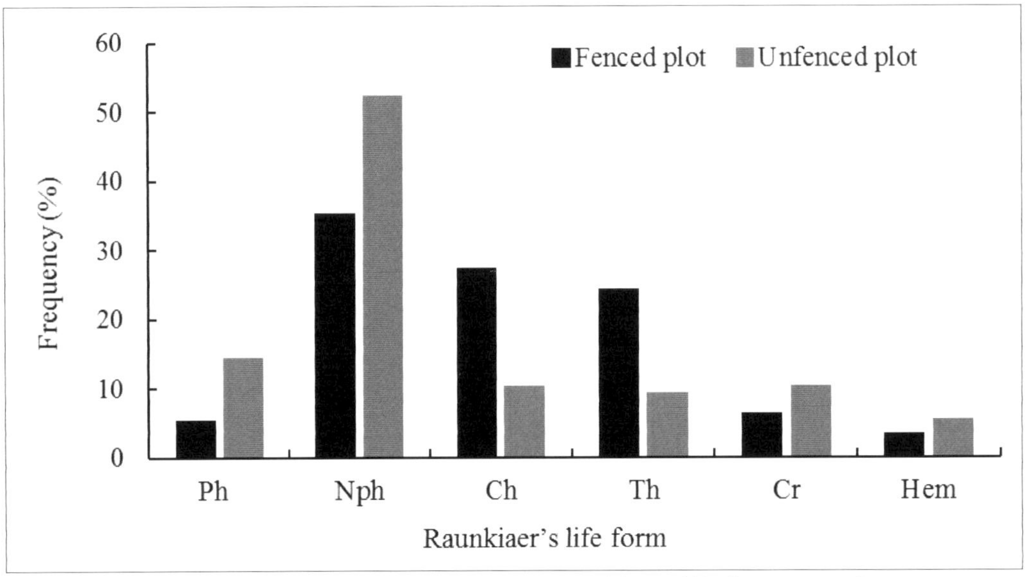

Figure 8. Frequency of plant species (Raunkiaer's life-form percent).

In general, the spectrum composition reveals predominance of Phanerophytes and Nanophanerophytes compared to the other biological classes. For both parcels, the pattern of the biological classes is: Pha and Nph > Ch > Th > Cr > Hem.

The Nanophanerophytes occupy the first place and represent more than half of the biological spectrum in the open environment (unfenced parcel). The abundance of Nanophanerophytes and Phanerophytes in this environment, it may be explained by the capacity of the taxa occurrences (*Sideroxylon spinosum*, *Zizyphus lotus*, *Searsia tripartita*, *Vachellia gummifera*, *Launea arborescens*, *Euphorbia officinarum (L.) subsp. officinarum,* etc.) to adapt the xenological conditions and support grazing, which are more xerophilic and may herds not liked it [17]. According to [38], the overgrazing prevents the flowering and fruiting of palatable annuals.

The high spectrum frequency of the Chamephytes and Therophytes in the close environment (fenced parcel) compared to the open environment may be explained by the beneficial effect of the defenses. In fact, the Chamephytes of the open environment are aromatic and medicinal plants (*Thymus satureioides*, *Artemisia reptans*, *Teucium polium*, etc.), which are harvested and uprooted by the local population and thier herds (goats, sheep and camels). The Therophytes or annual and biennial plants are generally fodder plants much liked to the cattle (*Stipellula capensis*, *Scorpiurus muricatus*). In addition, the Hemicryptophytes and Cryptophytes are of lower spectrum frequency in both parcels, which may be explained by water deficit and soil poverty (less developed soil) in the study area [39, 40]. The Cryptophytes are represented by three species: *Cenchrus ciliaris, Asphodelus ramosus and Drimia maritima*. Furthermore, there are transition from vegetation dominated by Nanophanerophytes, Chamephytes and Therophytes in the fenced parcel, to vegetation with Nanophanerophytic in the unfenced parcel. Which indicating positive effect of the fencing on perennially of Chamephytes and Therophytes and distribution of different biological classes.

The disturbance index is about 51% for the fenced plot, indicating strong degradation of the argan trees of the studied area.

4.5. Density of argan trees and natural regeneration

4.5.1 Density of argan trees

Without taken into account the number of shoots of woody trees, the average of the computed argan trees, in the fenced parcel, is 116 trees/ha of size ranging between 2 to 5 m. For the unfenced parcel, the average is 83 trees/ha of size of 4 m. Whatever the average computed for the fenced parcel, it is lower and may be explained by the impact of anthropogenic practices, such as overgrazing and clearing, were exercised by the local community before the installation of the fence. In addition, the lower average computed for the unfenced parcel (83 trees/ha), it may be explained by the landscape of this parcel that is characterized by low vegetation density (spare vegetation) caused by the anthropogenic practices (Figure 9), as confirmed by long-term analysis (2007–2018) [41]. These results indicate the positive effect of defense setting (fence) in the preservation of the overmentioned plants. [18] also confirmed the beneficial effect of the setting in defend on the density, the vigour and the productivity of the trees and also the regeneration and the development of the argan trees.

Figure 9. Photo shows spare vegetation in the unfenced parcel (UP) and fenced parcel (FP).

4.5.2 Natural regeneration

Inside the fenced parcel, the natural regeneration by seedling is well marked, where considerable number of argan trees up to 30 seedlings of height exceeds 50 cm were identified during field survey (Figure 10). Also, in this close environment, the young plants are always sheltered in the clumps of argan trees and in the clumps of broom, cytisus, pistachio, Tizgha, ensuring their protection and development. These observations confirm the positive effect of fencing on fruits stay and seeds germination. In contrast, outside the fenced parcel, the natural regeneration is rare and not well marked, which may explained by anthropogenic practices. [42] reported that the violation of the enclosure by herds in the open environments is one of the main factors causing forest regeneration failure.

Of note that during the field survey we observed presence of Cepes resulting from argan trees stumps growth, indicating development of rejection and burgeon located at the base near the roots.

In addition, the rare cases of natural regeneration of the argan trees in the open environment it may be explained by several factors as:

- The intensive collection of argan fruits by local population;
- The spare vegetation cover, where the dense vegetation cover ensures protection of young seedlings from herds teeth;
- The climate impact, where the consecutive drought may be caused low vegetation recovery;
- The degradation of soil quality due to overgrazing, which causes enormous damage to the vegetation that participates in matter organic formation and soil improvement.

Figure 10. Photos show natural regeneration of argan trees by sowing inside the fenced parcel.

5. Conclusion and recommendations
Referring to all indices and criteria used (specific richness, floristic diversity, recovery rate, biological spectrum, density and regeneration of the argan trees), this study reported a significant difference of vegetation in the fenced and unfenced parcels. Where all are moderately high in the case of the fenced parcel.

Considering that the factors limiting the productivity and sustainability of the argan trees, such as climatic constraints (frequency and severity of drought years) and anthropogenic practices (overgrazing, the collection of wood and plants for commercial purposes or for combustible) that causing deterioration of vegetation and soil degradation, we conclude that fencing strategy is the simplest one to maintain plant biodiversity and fight against desertification. It is an essential approach for better preservation and conservation of the argan trees, and it is indeed the only way to promote the recovery of the ecosystem to argan trees and to advance the stands of this valuable species supporting the traditional economy.

Finally, facing the worst climate conditions and anthropogenic pressure, the reconstitution of the argan trees may be difficult, even impossible in the long term without adopting a strategy to adapt this ecosystem to the expectations of the indigenous population through:

- Awareness, education, and mobilization of the local population to share a vision for a sustainable management of argan resources.
- Organization of the users of the pastoral resources in collective courses in order to ensure a sustainable management of the argan trees.
- Preservation (fencing the other open lands) and application of an appropriate forestry (sustainable and periodically updated) to maintain the multifunctional services of the forest.
- Taken into account climate change factor and difficulty of natural regeneration in reforestation programs and rejuvenation of the argan trees. Also, to search, in a perspective linked to climate change, potentially favourable areas for the future introduction of *Sideroxylon spinosum*
- Contractual compensation of local communities for the extension of the period of fencing.

6. Acknowledgements
The authors would like to thank Pr. Fouad Msanda and Pr. Belkacem Kabbachi (Ibn Zohr University - Agadir) for their judicious advice. We also would like to thank ABHSM for making climate data available for this work.

7. References
[1] Stride G, Nylinder S & Swenson U 2014 *Australian Systematic Botany*, 27(2), 104–118. https://doi.org/10.1071/SB14010

[2] Khayi S, Gaboun F, Pirro S, Tatusova T, El Mousadik A, Ghazal H, & Mentag R 2020 *Plants*, 9(10), 1354. https://doi.org/10.3390/plants9101354

[3] Msanda F, Mayad E.H and Furze J.N 2021 *Environ Sci. Pollut. Res.* 28, 64156–64165. https://doi.org/10.1007/s11356-020-11936-0

[4] Benabid A and Melhaoui Y 2011 *Écosystèmes naturels à arganier (Argania spinosa), patrimoine national et universel : Bio-écologie, phytosociologie, phytodynamique et ethnobotanique ; Restauration et certification.* Actes du 1er congrès internaional de l'Arganier (pages 39-47). Agadir, Maroc.

[5] Ait Haddou M, Kabbachi B, Aydda A, Bouchriti Y, Gougueni H, En-Naciry M & Aichi A 2022 *E3S Web of Conferences*. Vol. 337, p. 02002. https://doi.org/10.1051/e3sconf/202233702002

[6] El Wahidi F 2013 *Dynamiques des dégradations de l'arganeraie et modélisation spatiale de l'évolution forestière : scénarios d'aménagement pour une gouvernance locale.* Thèse Doct. En Sciences Agronomique et Ingénierie Biologique. Université Catholique de Louvain, 264 p.

[7] Moukrim S, Lahssini S, Rhazi M, Alaoui H.M, Benabou A, Wahby I, ... & Rhazi L 2019 *Agroforest Syst* 93, 1209–1219. https://doi.org/10.1007/s10457-018-0232-8

[8] Zhao X, Dupontet L Cheddadi R, Kölling M, Reddad H, Groeneveld J, Ain-Lhout F.Z, Bouimetarhan I 2019 *Quaternary Science Reviews,* Volume 221, 105889. https://doi.org/10.1016/j.quascirev.2019.105889

[9] Baqloul A, Schefuß E, Kölling M, Dupont L, Groeneveld J, Zhao X, ... & Bouimetarhan I 2021 *The Holocene*, 31(5), 699-708. https://doi.org/10.1177/0959683620988053

[10] Sebbar B, Moumni A, Lahrouni A, Chehbouni A, Belghazi T & Maksoudi B 2021 *Journal of Sustainable Forestry*, https://doi.org/10.1080/10549811.2021.1897018

[11] Haddou M.A., Kabbachi B, Aydda A, Gougueni H, & Bouchriti Y 2020 *E3S Web of Conferences*. Vol. 183, p. 02003. https://doi.org/10.1051/e3sconf/202018302003

[12] Bouchaou L, Tagma T, Boutaleb S, Hssaisoune M, & El Morjani Z.E.A 2011 *Climate Change and its Impacts on Groundwater Resources in Morocco: The Case of the Souss-Massa basin.* In: Treidel, H., Martin-Bordes, J.L., Jason, J.G editors, Climate change effects on groundwater resources: a global synthesis of findings and recommendations. London, 129.

[13] MDCE 2016 *La Troisième Communication Nationale à la Conventioncadre des Nations Unies sur les Changements Climatiques.* Rabat, Maroc, ministère de l'Environnement, 132 p. http://www.un-gsp.org/sites/default/files/documents/3069145_cite_divoire-nc3-1 cote_divoire_-_third_national_communication.pdf

[14] Mansir I, Bouchaou L, Chebli B, Ait Brahim Y, and Choukr-Allah R 2021 *Hydrological Sciences Journal*, 66:7, 1151-1168. https://doi.org/10.1080/02626667.2021.1924379

[15] El-Ghazouani F, El-Ouahmani N, Teixidor-Toneu I, Yacoubi B, Zekhnini A 2021 *European Journal of Integrative Medicine*, Volume 42, 101284. https://doi.org/10.1016/j.eujim.2021.101284

[16] Msansda F 2004 *Végétation de l'Anti-Atlas occidental et de sa retombée saharienne (Maroc)* : essai de synthèse. Thèse Doctorat ès-Sciences, Univ. Ibn Zohr Fac. Sciences Agadir 114p.

[17] Msanda F, El Aboudi A, et Peltier J.P 2005 *Cahiers Agricultures*, 14(4), 357-364.

[18] Achour A, Aroui A, Defaa C, El Mousadik A et Msanda F 2011 *Effet de la Mise en Défens sur la Richesse Floristique et la Densité dans deux Arganeraies de Plaine.* Actes du Premier Congrès International de l'Arganier, Agadir 15 - 17 Décembre 2011.

[19] Fennane M 2003 *Ecol. Mediterr*, 29 (1) : 87-106. https://doi.org/10.3406/ecmed.2003.1531

[20] Benabdellah M.A, Amrani S.M, Elhaitoum A et Ollier S 2010 *Ecol. Mediterr*, tome 36 n°2,. pp. 97-120. https://doi.org/10.3406/ecmed.2010.1369

[21] Braun-Blanquet J, and Lalucat Jo J 1979 *Fitosociología. Bases para el Estudio de las Comunidades Vegetales*; Blume: Madrid, Spain, 820p.

[22] Géhu J.M and Rivas-Martínez S 1981 *Notions Fondamentales de Phytosociologie.* In Berichte der Internationalen Symposien der Internationalen Vereinigung für Vegetationskunde; Dierschk, H., Ed.; Cramer: Vaduz, Liechtenstein, pp. 5–33

[23] Fennane M, Ibn Tattou M, Ouyahya A and El Oulaidi J 1999 *Flore pratique du Maroc*. Vol 2, Travaux Institut Scientifique, Sér. Botanique, (36).

[24] Valdés B, Rejdali M, Achhal El Kadmiri A, Jury S.L and Montserrat J.M 2002 *Catalogue des Plantes Vasculaires du Nord du Maroc Incluant des Clés d'Identification*, 1 et 2. Madrid, 1002 p. Editorial CSIC-CSIC Press.
[25] Benabid A 2000 *Flore et Ecosystèmes du Maroc. Evaluation et Préservation de la Biodiversité.* Paris: Ibis Press, 359p.
[26] Guo Y, Gong P, and Amundson R 2003 *Geoderma*, vol. 117, no. 1-2, pp. 99–115. https://doi.org/10.1016/S0016-7061(03)00137-X
[27] Kent M and Coker P 1992 *Vegetation Description and Analysis: A Practical Approach*, John Wiley & Sons, Chichester, UK.
[28] Rakotoarimanana V, Gondard H, Ranaivoarivelo N and Carriere S 2008 *Sécheresse*; 19(1) : 39-46.
[29] Augusto L, Dupouey J and Ranger J 2003 *For. Sci.*, 60, pp. 823-831. https://doi.org/10.1051/forest:2003077
[30] Magurran A.E 2003 *Measuring biological diversity*. John Wiley & Sons. 256 p.
[31] Oluyinka Christopher A 2020 *Int. J. For. Res*, vol. 2020. https://doi.org/10.1155/2020/4865845
[32] Messaoudéne M, Laribi M & Derridj A 2007 *Bois Forêts Trop*. 291 : 75-81. https://doi.org/10.19182/bft2007.291.a20359
[33] Raunkiaer Ch 1934 *The Life-Forms of Plants and Plant Geography*. Clarendon Press, Oxford, 2, 104 p.
[34] Romane F 1987 *Efficacité de la Distribution des Formes de Croissance des Végétaux pour l'Analyse de la Végétation à l'Echelle Régionale. Cas de quelques taillis de chêne vert du Languedoc*. Thèse de Doctorat, Faculté des Sciences et Techniques de Saint Jérôme, Université de Droit d'Economie et des Sciences d'Aix Marseille, 153 p.
[35] Gomila H, 1993 *Incidences du Débroussaillement sur la Flore, la Végétation et le Sol, dans le Sud-Est 179 de la France*. Thèse en sciences. Faculté des Sciences et Techniques de St Jérôme, Aix-Marseille III : 195 pp.
[36] Tbib A, Chaieb M 2004 *La mise en défense des parcours en zones arides : avantages écologiques et obstacles socio-économiques*. In : Ferchichi A. (collab.). *Réhabilitation des pâturages et des parcours en milieux méditerranéens*. Zaragoza : CIHEAM, 2004. p. 473-476 (Cahier Options Méditerranéennes ; n. 62)
[37] Benarchid K, Khatori M, Hilali S 2020 Afr J Ecol. 2020;00:1–11. https://doi.org/10.1111/aje.12726
[38] Khalid F, Benabdeli K, & Morsli B 2015. *Revue d'écologie (La Terre Et La Vie)*, 70, 16–31. http://hdl.handle.net/2042/56307
[39] Van Rooyen M.W., Theron G.K., Grobbelaar N 1990 *Journal of Arid Environments*, 19(2), 133–145. https://doi.org/10.1016/S0140-1963(18)30812-7
[40] Carvalho da Costa R, Soares de Araújo F, Wilson Lima-Verde L 2007 *Journal of Arid Environments*, 68(2), 237-247. https://doi.org/10.1016/j.jaridenv.2006.06.003
[41] Ezaidi S, Aydda A, Kabbachi B, Althuwaynee O.F, Ezaidi A, Haddou M.A, Idoumskine I, Thorpe J, Park H.J, and Kim S.W 2022 *Remote Sensing Applications: Society and Environment*, 27, 100800. https://doi.org/10.1016/j.rsase.2022.100800
[42] Defaa C, Achour A, Hossayni A, Bellefontaine R, El Mousadik A and Msanda F 2013 *Analyse de l'Itinéraire Technique d'un Périmètre Exceptionnellement Réussi de Régénération d'arganier*, in Actes du 1er Congrès International sur l'Arganier, Institut National de la Recherche Agronomique, Agadir, Morocco, pp. 83–92.

Biodiversity of intestinal parasites in urban waters of the Ouislane River in central Morocco

K Ouarrak [1*], H El Omari[1], and A Chahlaoui[1]

[1] Natural Resources Management and Development Team, Laboratory of Health and Environment, Faculty of Sciences, Moulay Ismail University, Meknes, Morocco

*Corresponding author: kouarrak@yahoo.com

Abstract: The objective of the present study is to shed light on the parasitic load characterization of wastewater samples collected from four selected stations (O1, O2, O3 and O4 selected from upstream to downstream, on the bed of the Ouislane River in central Morocco. Parasite load flux (quantitative and qualitative) was determined monthly for one year, from January through December 2017, using direct spawning examination, the modified Bailenger concentration technique and the Faust flotation technique. Parasitological examination allowed the identification of helminth eggs and protozoan cysts, on the one hand, and the evaluation of the variation in the concentration of the parasitic load of this wastewater, on the other hand. The results of the present work allow the identification of 17 taxa: *Ascaris lumbricoïdes, Trichuris trichiura, Capillaria sp, Ancylostoma diudinal, Enterobius vermicularis, Toxocara canis, Strongyloïdes stercolaris, Hymenolepis nana, Hymenolepis diminuta, Taenia saginata, Fasciola hépatica, Giardia lamblia, Entamoeba coli, Entamoeba histolytica, Blastocystis hominis, Balantidium coli* and *Cryptosporidium parvum.* with a predominance of protozoan cysts over helminths eggs. The evolution of the parasitic load over time indicates the ubiquity of these parasitic species in raw wastewater at variable concentrations, which are closely related to the periods of human activities and according to the seasons. However, the high load of these intestinal parasites in the urban waters of Ouislane wadi will have a potential impact on human and livestock health. Hence the interest to prohibit their use commonly observed in the field and their practice in agriculture.

Keywords: Ouislane wadi, Parasitic Charge, Meknes City, Central Morocco.

1. Introduction

Wastewater discharged directly into waterways without any prior treatment is one of the most worrying environmental problems [1]. Indeed, this domestic and industrial wastewater carries numerous pathogenic micro-organisms that have a sanitary impact on human and animal populations [2, 3, 4]. Urban effluents are a source of life for frequently identified parasitic species [5]. On the other hand, parasites have a greater power of resistance in the environment than bacteria and viruses. In contrast, protozoan cysts can survive in wastewater for up to one month, while helminth eggs can survive for up to twelve months [6]. Transmission of resistant forms of parasites occurs through ingestion of eggs, cysts, and oocysts via the fecal-oral route, contact between people and between people and other animals, consumption of water contaminated with host feces (humans and other nonhuman animals), recreational activities, and contaminated surface water, in addition to consumption of contaminated food [7]. However, these discharges of raw sewage loaded with fecal pathogens into the natural environment are used in urban and peri-urban agriculture activities, including the watering of sports fields. The World Health Organization estimates that a single helminthic egg or protozoan cyst can act as an etiological agent of a particular parasitic disease [8, 9, and 10]. The objective of this work is to assess the degree of contamination of surface waters of the Ouislane wadi by identifying and quantifying eggs and cysts of parasitic species and to alert the authorities to the risks associated with pollution and reuse of urban wastewater in its raw state.

2. Materials and Methods

2.1. *Study area.*

The prefecture of Meknes is a predominantly urban subdivision of the Moroccan region of Fez-Meknes. Due to its geographical location, it is subject to a temperate to semi-arid Mediterranean climate, with an aridity index of 19.23 mm/°C, and a rainfall characterized by a great interannual variability. The city is crossed by three wadis: Ouislane wadi, Boufekran wadi and Bouishak wadi (Figure 1). This work was conducted in the Ouislane commune in Meknes located in central Morocco. According to the 2014 census, the study area is characterized by a significant urban and demographic expansion: encloses 87910 inhabitants [11]. Agricultural activity and animal production in this locality are important due to their proximity to Ouislane wadi. The stations chosen for this study are four: O1, O2, O3, and O4 located on the bed of the Ouislane wadi, they are upstream to downstream. These stations were chosen taking into account, among other things, their length and the anthropic pressure to which they are subjected.

2.2. *Sampling and analysis technique of collected waters*

During the year 2017, monthly sampling campaigns had been conducted from upstream to downstream. A total of 48 wastewater samples covering all stations (O1, O2, O3 and O4), were taken at 11am (this is the time when domestic wastewater discharges are significant [12].

IOP Conf. Series: Earth and Environmental Science 1090 (2022) 012024 doi:10.1088/1755-1315/1090/1/012024

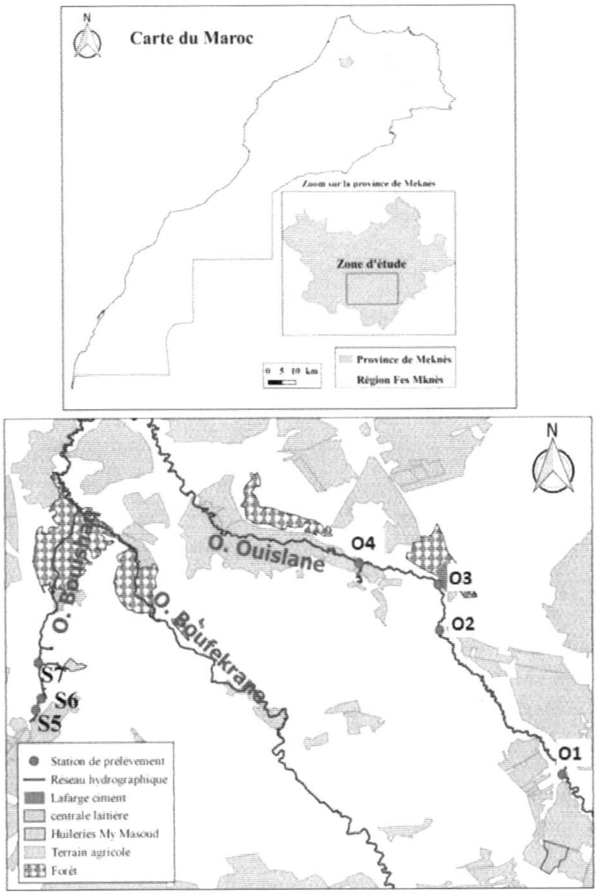

Figure 1. Presentation of the study area and sampling stations

At each station, 1L of wastewater was collected in sterile polyethylene bottles and fixed with 10% formalin. The identification of the parasites in the Ouislane wadi effluents was conducted in two ways: a fresh examination which consisted in observing the vegetative and larval forms and an examination after concentration of the resistant forms by the modified Bailenger method [13] and by the Faust technique. The counting of the different taxa was done by the Mac Master cell of 0.3 ml capacity. In addition to these routine examinations, other techniques were performed: M.I.F. (Merthiolate-iodine-formalin) staining. The different species of parasites were identified on the basis of the outline, appearance, color, thickness and content of the parasite eggs and cysts [14, 15]. The number of parasite species contained in our wastewater samples was determined using the following formula:

$$N = \frac{A * X}{P * V}$$

With:
N: number of eggs or cysts per liter of sample,
A: average number of eggs or cysts counted on three slides,
X: volume of the final product (ml),
P: slide capacity (0.15 ml),
V: volume of the initial sample (in liter).
The results obtained from the parasitological analysis of urban waters of Ouislane wadi were statistically processed using Excel software.

3. Results and discussion

On a local scale, the wastewater of Ouislane wadi harbors 17 taxa: *Ascaris lumbricoïdes, Trichuris trichiura, Capillaria sp, Ancylostoma diudinal, Enterobius vermicularis, Toxocara canis, Strongyloïdes stercolaris, Hymenolepis nana, Hymenolepis diminuta, Taenia saginata, Fasciola hépatica, Giardia lamblia, Entamoeba coli, Entamoeba histolytica, Blastocystis hominis, Balantidium coli, Cryptosporidium parvum* (Table 1).

Table 1. Taxa counted in the effluents of Ouislane wadi and their percentage of occurrence (% Oc), (+: presence; -: absence) during the year 2017

Espèces / Stations	O1	O2	O3	O4	% Oc
Ascaris lumbricoïdes	+	+	+	+	100
Trichuris trichiura	+	+	+	+	100
Capillaria sp	+	+	+	+	100
Ancylostoma diudinal	+	+	+	+	100
Enterobius vermicularis	+	+	+	+	100
Toxocara canis	+	+	+	+	100
Strongyloïdes stercolaris	+	+	+	+	100
Hymenolepis nana	+	+	+	+	100
Hymenolepis diminuta	+	+	+	+	100
Taenia saginata	+	+	+	+	100
Fasciola hépatica	-	+	+	+	75
Giardia lamblia (10^3)	+	+	+	+	100
Entamoeba coli (10^3)	+	+	+	+	100
Entamoeba histolytica (10^3)	+	+	+	+	100
Blastocystis hominis (10^3)	+	+	+	+	100
Balantidium coli ($10^{3)}$	-	+	+	+	75
Cryptosporidium parvum (10^3)	-	+	+	-	50
Nombre de taxons	14	17	17	16	

Overall, the number of taxa found in our study is higher than that found by [16] in Tunisia, who identified 11 parasite species, which is the highest number of parasite species detected in a single study compared to other studies. This is followed by the studies conducted by Salama [17] in Morocco and Okojokwu [18] in Nigeria where 10 parasite species were identified in each study. It should be noted that the analysis of Table 1, allows us to conclude that the majority of the taxa identified are highlighted by other work done in Morocco [19, 20, 21, 22]. In terms of occurrence, *Ascaris lumbricoïdes, Trichuris trichiura, Capillaria sp, Ancylostoma diudinal, Enterobius vermicularis, Hymenolepis nana, Giardia lamblia, Entamoeba coli, Entamoeba histolytica* and *Blastocystis hominis* are the most common species.

The distribution of taxa by station (Table 1) shows an increase in species at station O2, which could be explained by the fact that this station is a collector that discharges industrial effluents directly into the Ouislane wadi stream without any treatment. This station seems to have a higher raw wastewater production than the other stations. This makes the station a source of pollution for the receiving environment. The quantitative analysis of parasitic species in the wastewater collected in the different selected stations contain forms of parasitic resistance, in protozoan cysts and helminth eggs in different proportions (Figure 2). The higher mode revealed through the study is that of the protozoa, represented by six species with a frequency equivalent to 41.2%, followed by nematodes with a frequency of 35.30% (seven species identified). The class of cestodes has a frequency of 17.6% (contains three species) and the frequency of trematodes is only 5.9% (one species identified). The predominance of protozoa may be due to their greater prevalence. This is confirmed by local epidemiological data presented in some countries.

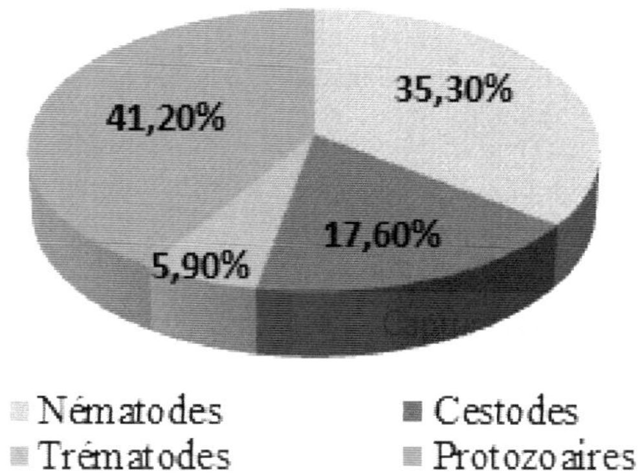

Figure 2. Frequency of occurrence of groups of parasites identified in the wastewater of Ouislane wadi during the year 2017

As for the quantitative analysis of helminths, it was possible to distinguish three helminthic groups: nematodes, cestodes and trematodes, with a predominance of the former on the latter. Such outcome corroborates the findings of other authors [23] who reported the predominance of these parasitic species. Parasite eggs concentration varied among the stations studied (Table 2). At station O1, the number of parasite eggs is 780 eggs/liter of which 72.8% (n=568) are Nematodes and 18% (n=140) are cestodes, no trematodes were detected. Station O2 contained 2130 eggs/liter of which 69.4% (n=1479) were nematodes, 32.2% (n=709) were cestodes and 0.05% (n=12) were trematodes. Station O3 is characterized by egg contents of 80.7% (n=857), 19.8% (n=210) and 0.02% which are nematodes, cestodes and trematodes respectively. The downstream station O4 has 81.3% (n=694) of nematodes, 18% (n=159) of cestodes and 0.02% (n=2) of trematodes. The dominance of the nematode class could be explained by the high resistance of these parasites to environmental stress.

The eggs of *Ascaris lumbricoides* obtained in all stations could be interpreted by their ecology, biology, as well as the high resistance of these parasites in these waters.

This variability of concentrations would find its explanation in the typology of wastewater drained to this site and the density of the population connected to the site surveyed. The reduction of the parasitic load at station O3 located downstream of station O2 could be explained by the dilution of this load by the important flow of the Ouislan wadi. The analysis of the content of protozoa identified in the urban waters of Ouislane wadi, shows that the class of amoebae is more important than the class of flagellates and ciliates. It should even be noted that the station with the highest number of parasitic species is station O2 (Table 2). A comparison of the persistent protozoan species in the different selected stations shows that *Giardia lamblia* is the most abundant species with 31%, 39%, 34% and 38% at O1, O2, O3 and O4 respectively.

Table 2 . Specific richness of parasites collected in urban waters of the selected stations

	Classes / Taxons	Stations O1	O2	O3	O4
Helminths	NEMATHODES				
	Ascaris lumbricoïdes	175	561	276	208
	Trichuris trichiura	91	215	170	134
	Capillaria sp	50	166	91	82
	Ancylostoma diudinal	103	202	117	108
	Enterobius vermicularis	82	135	112	105
	Toxocara canis	22	109	32	20
	Strongyloïdes stercolaris,	45	91	59	37
	CESTODES				
	Hymenolepis nana	63	356	97	59
	Hymenolepis diminuta	76	340	108	95
	Taenia saginata	1	13	5	3
	TREMATODES				
	Fasciola hépatica	0	12	3	2
Total number of helminths		780	2130	1061	853
PROTOZOARS	FLAGELLES				
	Giardia lamblia (10^3)	1,35	4,47	2,24	1,86
	AMIBES				
	Entamoeba coli (10^3)	0,87	2,01	1,08	0,95
	Entamoeba histolytica (10^3)	0,76	1,33	0,91	0,89
	Blastocystis hominis (10^3)	1,36	2,83	1,75	1,04
	CILIES				
	Balantidium coli (10^3)	0	0,79	0,23	0,11
	COCCIDES				
	Cryptosporidium parvum (10^3)	0	0,12	0,3	0
Total number of protozoa (10^3)		4,34	11,55	6,51	4,85

Our findings show that 81.8% of parasites in common with other works [25], followed by 63.6% of parasites in common with other results [20, 21]. Similarly, 45.5% of parasites in common were found by other researchers [22, 25]. The ubiquitous suite of parasitic species in all urban waters in our study consists of *Ascaris lumbricoides, Trichuris trichiura, Capillaria sp, Ancylostoma diudinal, Enterobius vermicularis, Toxocara canis, Hymenolepis nana, Hymenolepis diminuta and Taenia saginata, Giardia lamblia, Entamoeba coli, Entamoeba histolytica* and *Blastocystis hominis*. In addition, *Cryptosporidium parvum* and *Giardia intestinalis*, are the two most common protozoa associated with waterborne epidemics [26]. Gastroenteritis is a common disease caused by these waterborne protozoa.

4. Conclusion

The obtained results have allowed us to deduce that our watercourse is much polluted. In fact, this urban wastewater does not meet the WHO recommendations and the Moroccan standards for surface water used for irrigation. It is essential to make the population aware of the health risks related to the use of untreated surface water.

5. References

[1] Akpo Y, Sawadogo GJ, et Degnon R.G 2013. Évaluation de la contamination parasitologique des eaux usées domestiques à la station d'épuration de Cambérène (Dakar). Journal of Applied Biosciences, 69: 5449 – 5455.

[2] Okoh A. I, Odjadjare E. E, Igbinosa E. O, and Osode A. N 2007. Wastewater treatment plants as a source of microbial pathogens in receiving watersheds, African J. Biotechnol., vol. 6, no. 25, pp. 2932–2944.

[3] Kazi T. G, Arain M. B, Jamali M. K, Jalbani, N, Afridi H. I, Sarfraz R. A, Baig J. A, Shah A. Q 2009. Assessment of water quality of polluted lake using multivariate statistical techniques: A case study. Ecotoxicology and Environmental Safety 72, 301-309.

[4] Zurita F and White J. R. 2014. Comparative Study of Three Two-Stage Hybrid Ecological Wastewater Treatment Systems for Producing High Nutrient, Reclaimed Water for Irrigation Reuse in Developing Countries," Water, vol. 6, pp. 213–228.

[5] Hatam-Nahavandi K, Mahvi A. H, Mohebali M. , Keshavarz H. , Mobedi I. and Rezaeian M. 2015. Detection of parasitic particles in domestic and urban wastewater and assessment of removal efficiency of treatment plants in Tehran, Iran, J. Environ. Heal. Sci. Eng., vol. 13, no. 1

[6] Stott R., 2003. Fate and behavior of parasites in wastewater treatment systems," in Handbook of Water and Wastewater Microbiology, D. Mara and N. Horan, Eds. London: Academic Press, pp. 492–521

[7] Ryan u.n.a, Fayer R; Xiao I. 2014. *Cryptosporidium* species in humans and animals: current understanding and research needs. *Parasitology*, v. 141, n. 13, p. 1667-1685. http://doi.org/10.1017/s0031182014001085 https://doi.org/http://doi.org/10.1017/s0031182014001085

[8] OMS : Organisation mondiale de la santé. L'utilisation des eaux usées en agriculture et en aquaculture : recommandations à visées sanitaires, série de rapports techniques n° 778, Genève, 1989.

[9] Jimenez B, 2007. Helminth ova removal from wastewater for agriculture and aquaculture reuse," Water Sci. Techno-, vol. 1, no. 2, pp. 485–493.

[10] Dungeni M, and Momba M. N. B, 2010. The abundance of Cryptosporidium and Giardia spp. in treated effluents produced by four wastewater treatment plants in the Gauteng Province of South Africa, Water SA, vol. 36, no. 4, pp. 425–431.

[11] RDPH : Recensement Général de la Population et de l'Habitat 2014

[12] Ouarrak K, Chahlaoui A., and Taha I. 2017. Monitoring of the parasitic load wastewater of oued Ouislane and oued Bouishak from the city of Meknes, Morocco. IJISR Vol. 7, pp. 1063-1067.

[13] Bailenger J.Coprologie parasitaire et fonctionnelle. Bordeaux, France, Imprimerie E. Drouillar, 1973

[14] Golvan J., 1985. Les nouvelles techniques de parasitologie. Ed. Flammarion MédecineSciences, 67p

[15] Bouhoum K. and Schwartzbrod J., 1989.Quantification of helminth eggs in wastewater. Zentralblatt für Hygiene und Umweltmedizin, 188: 322-330

[16] Sabbahi S., Trad M., Ben Ayed L., and Marzougui N. 2018. Occurrence of intestinal parasites in sewage samples and efficiency of wastewater treatment systems in Tunisia S. Water Qual. Res. J., vol. In press, pp. 1–15

[17] Salama Y., Chennaoui M., El amraoui M., Mountadar M., Rihani M., and Assobhei O. 2016. Evaluation of the qualitative and quantitative of parasitic load and Environmental Risks of Raw Sewage Rejected on the Coast of the City of El Jadida (Morocco), J. Mater. Environ., vol. 7, no. 5, pp. 1614–1623.

[18] Okojokwu O., Inabo H., and Yakubu S. 2014. Parasitological Profile of Raw Wastewater and the Efficacy of Biosand Filter in Reduction of Parasite Ova and Cysts, J. Appl. Sci. Environ. Manag., vol. 18, no. 1, pp. 5–9.

[19] Khallaayoune K, 1996. Réutilisation des eaux usées en agriculture : risque de transmission de parasites à l'homme et à l'animal. Homme, Terre et Eaux, 26 (103): 14-19.

[20] Dssouli K., Kharboua M., Khallaayoune K., Haloui B., El Halouani H. 2001. Etude de la contaminationparasitologique des cultures irriguées par les eaux usées dans le Maroc Oriental. Actes Inst. Agron. Vet., 21 (4): 215-225

[21] Belghyti D., El Kharrim K., Bachikh J. & Gabrion C. 2007. Caractérisation parasitologique des eaux usées du Lac Fouarat (Kénitra-Maroc) et évaluation du niveau épidémiologique dans une population rurale en contact avec ces eaux. Actes Inst. Agron. Vet., 21 (4) : 155-177
[22] Raweh S., El Kharrim K., Cissé M., El Guamri Y., Abchir Y. & belghyti d, 2010. Aspect parasitologique des eaux usées rejetées dans le bas Sebou (Kenitra, Maroc). World J. Biol. Res., 3 (1) : 18-29.
[23] EL Ouali Lalami A, zanibou A, Bekhti K, Zerrouq F. et Merzouki M. 2014 Contrôle de la qualité microbiologique des eaux usées domestiques et industrielles de la ville de Fès au Maroc. Journal Material Environmental Science, 5 (1) 2325 -2332
[24] Sylla, I. and Belghyti, D. 2008. Analyse parasitologique des eaux usées brutes de la ville de Sidi Yahia du Gharb (Maroc). Revue Mondiale de la Recherche Biologique. Vol 001, Issue 2, 1-10.
[25] Talouizte H., Merzouki M., El Ouali Lalami A., Bennani L., Benlemlih M. 2007. Evolution de la charge microbienne de la laitue irriguée avec les eaux usées urbaines de la ville de Fès au Maroc, Tribune de l'eau. 642
[26] Baldursson S.; karanis P. 2011. Waterborne transmission of protozoan parasites: review of worldwide outbreaks - an update 2004-2010. Water Research, v. 45, n. 20, p. 6603-6614. https://doi.org/10.1016/j.watres.2011.10.013.

Floristic analysis of the medicinal flora used by the population of the Central Middle Atlas (Morocco)

M Hachi[1], T Hachi[2]*, H Essabiri[3], T Ainane[3], A El Yaakoubi[3] and L Zidane[1]

[1] Laboratoire of Botany, Biotechnology and Plant Protection, Department of Biology, Faculty of Science, BP. 133, University Ibn Tofail, Kenitra, Morocco

[2] Laboratory of Biotechnology, Environment and Quality, Department of Biology, Faculty of Science, PO Box 133, University Ibn Tofail, Kenitra, Morocco

[3] Heigher School of Technology. Sultan Moulay Slimane University B.P. 170, 54000 Khénifra, Morocco.

*Corresponding author: hachitouria03@gmail.com

Abstract. Within the scope of the exploitation of natural resources, an ethnobotanical and floristic study was carried out in the Central Middle Atlas to identify the medicinal plants used in traditional medicine and to collect as much information as possible about the local knowledge held by the local population. Ethnobotanical field surveys were conducted during two campaigns (2013 and 2014) using a questionnaire sheet. The location of the different ethnobotanical and floristic surveying environments was identified by the stratified sampling method. The survey was carried out among 1560 inhabitants randomly selected from the population of the study area with a sample of 40 persons per stratum. The results of these surveys have enabled us to draw up a catalog of 267 medicinal species divided into 211 genera and 79 families, among which: 6 are the most represented, including Asteraceae, Lamiaceae, Apiaceae, Rosaceae, Fabaceae and Poaceae. Oregano (Origanum compactum Benth.) is the most used species. The elaborated catalog is characterized by the presence of endemic species of Morocco and also by the presence of rare or threatened taxa with a percentage of 6, 36 % of the listed species. The study contributes to improving knowledge about traditional medicine and safeguarding folk knowledge. but abusive gathering can put wild medicinal species at risk, many populations of local medicinal species suffer from overexploitation. Thus, it is necessary to provide legislative measures capable of protecting them and also to create specialized nurseries in order to curb the abuse of use and direct it towards a more rational and scientific use.

Key words: Floristic, Medicinal plants, Central Middle Atlas, Morocco.

1. Introduction

Plants populated the planet long before humans and were first used to feed them through gathering and cultivation [1]. Their use has rapidly evolved by finding their therapeutic properties to treat injuries and diseases. The use of aromas was also known to ancient civilizations for religious, cosmetic as well as therapeutic uses [2].

According to 2003 World Health Organization (WHO) statistics, 80% of the world's population uses traditional medicine to meet primary health care needs.

The high cost of medicines and the inability of many developing countries to provide medicines has forced local communities to turn to medicinal plants that are proven to be effective, safe, inexpensive, and culturally acceptable [3].

Medicinal plants also establish inestimable resources for the pharmaceutical industry. According to a WHO estimate, they represent about 70% of the basic materials of modern pharmaceuticals. About 25% of the medicines produced and marketed worldwide come from plants, and at least 35 000 plant species are used for medical purposes [4]. Africa is characterized by an important diversity of medicinal plants [5].

Moreover, the geographical and climatic diversity, which characterizes Morocco, provides it a wide range of species including 600 medicinal and aromatic plants [6]. Indeed, Morocco is a major producer and exporter of fresh plants and essential oils.

The Middle Atlas, has a diversity of fauna and flora very important; this diversity is supported, among other things, by the existence of three levels of vegetation, the mesomediterranean, the upper Mediterranean and the Mediterranean-mountain [7]. It harbors the most important forest potential of the country, endowed with major, ecological, socio-economic and cultural roles and offering services and uses of exceptional importance. We find there one-tenth of the world's area of the Atlas cedar, one-quarter of the world's population of monkeys, and several wetlands classified as Ramsar sites [8].

The sector of aromatic and medicinal plants in Morocco represents a non-negligible value for the Moroccan economy. Existing activities in this field have allowed to export an average of about 970 tons of essential oils and various extracts, obtained from spontaneous or cultivated plants, For a value of about 150 million Dirham [9].

But the high demand and uncontrolled collection of medicinal plants is a real threat to plant resources. In addition to this, there are other equally anthropogenic actions such as the cutting of firewood, bush fires, shifting cultivation, charcoal making, overgrazing and overexploitation. According to Mehdioui and Kahouadji [10], excessive exploitation of plants can become harmful if it exceeds the tolerable threshold of renewal and regeneration of the natural resources used.

The Moroccan population has a rich and ancient tradition in the field of phytotherapy [11]. Indeed local knowledge is not always distributed uniformly in a social group, nor is it known to all members of the group. This is true for knowledge about medicinal plants, where generally health specialists (traditional healers) have more in-depth knowledge than ordinary people [12,13].

From this situation, the present study aims to inventory medicinal plants and traditional practices exploited by the population of the Central Middle Atlas region and to build a database in order to preserve an ancestral knowledge based essentially on an oral tradition. This knowledge undoubtedly contributes to a good knowledge of plant resources for a rational and sustainable exploitation.

2. Material and methods

2.1. Study area

The Central Middle Atlas region represents the ethnobotanical and floristic study area of the medicinal plants used by its population (figure 1).

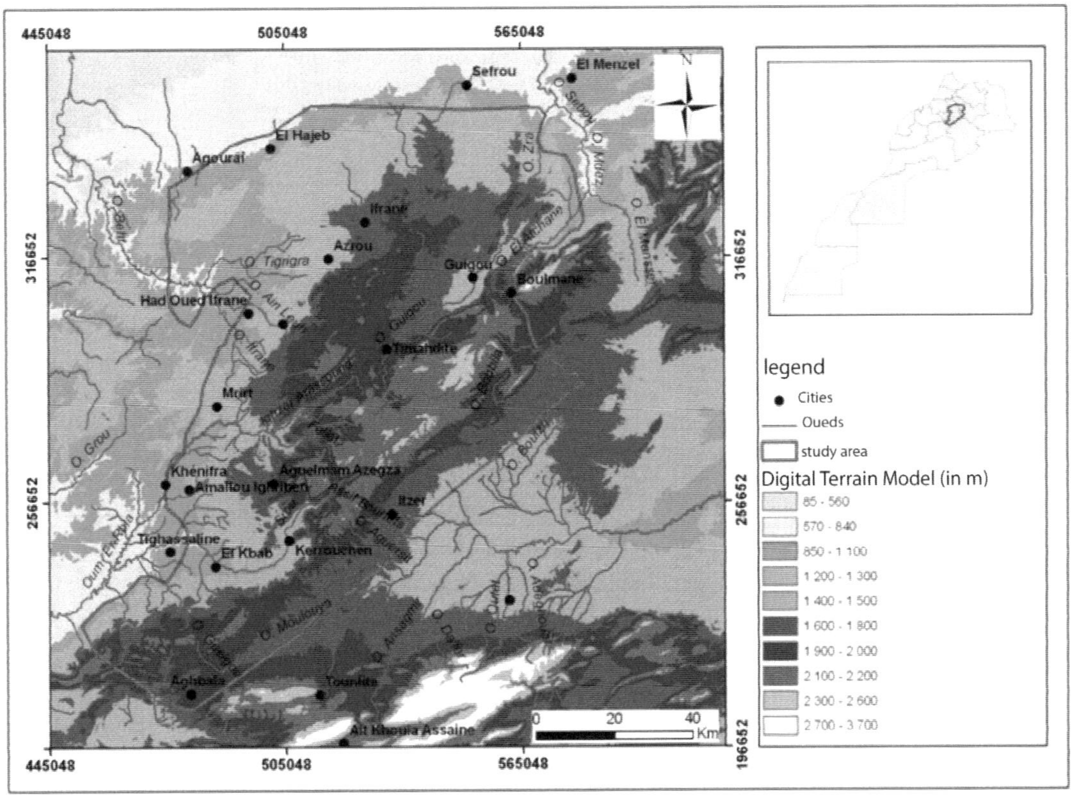

Figure 1. Location of the Central Middle Atlas region.

The Middle Atlas is part of the Atlas chains [14]. It occupies an intermediate position between the Rif to the north, and the High Atlas to the south. Along an EW section, the Middle Atlas is bordered by the high plateaus of the Oran- Moroccan meseta to the east and the coastal meseta to the west. It's bounded on the north by the plain of Sais (Neogene sediment) and the front of the rifain layer, on the northeast by the basin of Guercif and on south and southeast by the depression of the Moulouya [15].

Globally, the Middle Atlas structure corresponds, on one hand, to large synclinal basins with axes that are parallel to the chain and, on the other hand, to narrow anticlinal ridges, sometimes, intruded with gabbroic rocks [16]. It consists of two structural assemblies that are separated by the North Middle Atlas fault [15]: The tabular Middle Atlas or Middle Atlas cause to NW and the folded Middle Atlas to SE. The first is mainly formed by a Paleozoic basement covered with a large Mesozoic cover, all of it is dotted with scattered volcanic effusions of the quaternary age [17,18]. The folded Middle Atlas corresponds to the eastern part of the chain, it is distinguished from the sub-tabular plateau by the presence of ridges that form reliefs, and those reliefs are pointing toward the chain direction. These anticline ridges, framing the synclinal depressions, continued to function, during the Middle Jurassic, as depocentres and quite thick calcareous marls [16].

The pedological context of the studied area according to the (figure 2) shows the existence of several types of soils:

1/ podzolized mountain soils; chernozem type soils and eroded soils that develop on Mesozoic, calcareous rocks;

2/ Mountain forest soils, podzolized; podzolized soils with no apparent podzolization (red soils and yellow soils) these two varieties of soil are stony, also there are red soils that develop on cracks of Mesozoic limestone rocks and in the clearings we find humus carbonate soils;

3/ Podzolic soils and podzolized red soils, brown soils and soils of chernozem type. These types of soils have developed on the alteration products of Paleozoic and eruptive rocks;

4/ brown soils of the forests; tirso-chernozem soils and dark chestnut soils of the plateaus; red soils of hamris type;

5/ chestnut or chernozem soils, frequently stony;

6/ podzolic soils, red soils; podzolized brown soils, red soils, brown soils and humus carbonate soils;

7/ light chestnut soils of the high plateaus of eastern Morocco, developed mainly on sedimentary rocks of the upper Tertiary.

It is also noted that the distribution of these soil types in the study area, and dominated by the soil type 2 (figure 2):

- Podzolized mountain forest soils that bear holm oak and other shrubby species. - Red and yellow podzolized soils that generally support cedar forests. These soils are located in the central and eastern part of the study area.

Soil types 1 and 3 (figure 2) cover almost the same area of the study area. Soil type 1 occupies the northern part, soil type 2 occupies the western part. Both of these soil types contain evergreen forests.

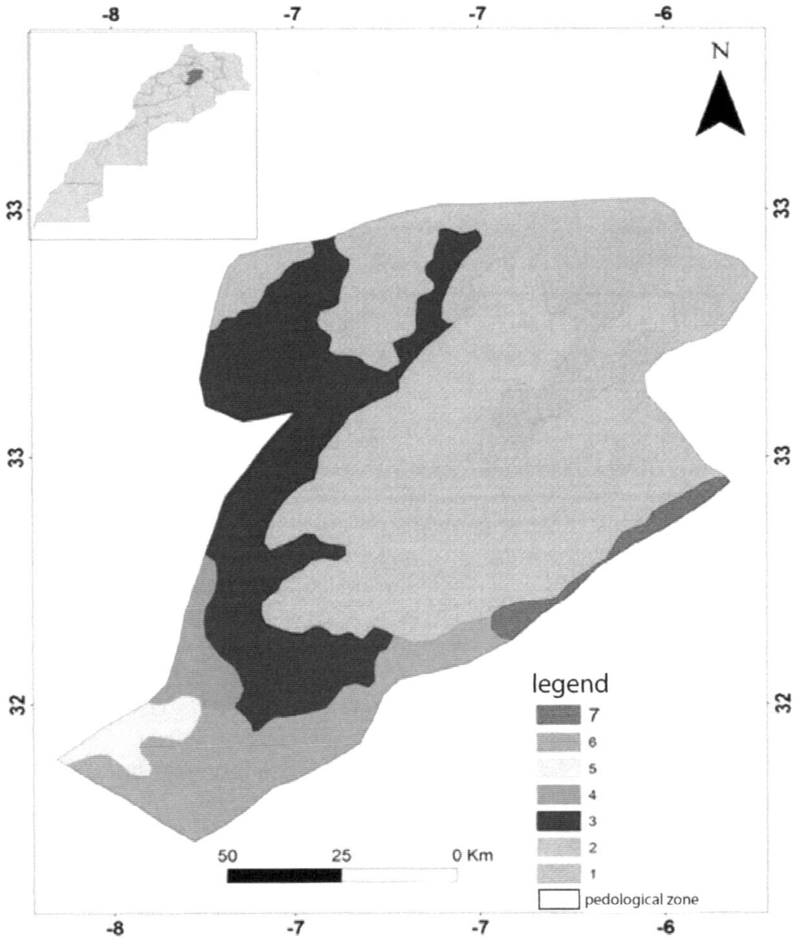

Figure 2. Different types of soils in the study area extracted from the preliminary sketch of the soil map of Morocco, 1/1.500.000[19] . 1. Podzolized mountain soils. 2. Mountain forest soils. 3. Podzolic soils and podzolized red soils. 4. Humus-Carbonate soils. 5. Chestnut soils. 6. Podzolic soils. 7. Light brown soils.

IOP Conf. Series: Earth and Environmental Science 1090 (2022) 012025 doi:10.1088/1755-1315/1090/1/012025

The climate in the Middle Atlas is of a mountain Mediterranean type, it is characterized by a wet and cold climate. This particular climate is due to its altitudinal position, its geographical location and its exposure to marine influences [20].The differences in average temperature recorded in the Middle Atlas are mainly the result of altitudinal variation [20]. This average varies between 3 and 16°C for stations above 1000 m altitude (Azrou, Ifrane) while it is higher for stations at lower altitudes (Khenifra: 17.8°C).

The Middle Atlas, known as the key water tower of Morocco, is the best watered and richest range of wetlands, including natural lakes, rivers and fresh sources, offering varied ecological habitats and fostering a great biodiversity [21].

The Middle Atlas is generally very rich in plant species, because of the great diversity of the ecosystems held within this area (forests, grasslands, scrublands, steppes, lawns, wetlands, saline soils). The Phytocenoses found there are organized by the following tree species: Cedrus atlantica (Atlas cedar), Quercus faginea (Zeen oak), Quercus suber (cork oak), Quercus rotundifolia (green oak), Tetraclinis articulata (Barbary thuja), Juniperus thurifera (Spanish juniper), Pinus halepensis (Aleppo pine), Pinus pinaster var. maghrebiana (Maghreb maritime pine) and spiny Xerophytes.

Two major biogeographic zones are distinguished in the Middle Atlas [22]:

1. The Area occupied by vegetation adapted to the wetter climate conditions. It extends to the west and northwest of the Guigou plain and to the causes of the Middle Atlas, as well as to their northern piedmont.
2. Zones of xeric vegetation of the folded Middle Atlas and its borders (Moulouya and Sebou rivers).

In the Middle Atlas, animal husbandry is one of the main ancestral activities. It constitutes the first and, for some households, the only source of income. The pastoralists of the region, who by a way of life and adapted habitats, use in an alternating way the resources of the mountain in summer, the holm oak forests of the Dir in spring, and the pastoral space offered by the Azaghar in winter: seasonal exploitation of these domains, between which are played the transhumance and, within each of these domains, use of the various collective or private courses complementary to the forest. The practice of cereal crops and legumes is limited in the Middle Atlas to depressions rich in silt and especially to the foothills of the massif.

2.2. Methodology

In order to identify medicinal plants and gather all information on therapeutic uses by the local population of the Central Middle Atlas region, an ethnobotanical survey was conducted between 2013 and 2014 using a pre-established questionnaire sheet with specific questions about the common name of each species, the part used, methods of preparation, methods of administration and the treated disease. The towns, villages and Douars (small villages) of the surveyed area are determined through the stratified probability sampling techniques [23,24].

The study area is divided into 39 homogeneous strata: Immouzer kandar, Aïoun sename, El hajeb, Tazouta, Dar Mimoun Ou Ahmed, Tagnanaït, Bou Sedra, Adarouch, Ifrane, Skoura, Taghzout, Azrou, Boulmane, Bouchbel, Aïn Leuh, Aït Ghanem, Timahdite, M'rirt, Zaouit Ifrane, Enjil, Wiwane, Sources Oum Rabia, Bakrit, Aguelmama Sidi Ali, Taoureda, Aguelmam Azegza, Khénifra, Jnane lmes, Aït Oufella, Itzer, Kerouchen, Tighssaline, El Kebab, Ouaoumana, Sidi Yahya Ou Saad, Aghbalou N'serdane, Zaouit Cheikh, Moulay Ya'qoub and Aghbala (figure 3). By adopting a simple stratified random sampling, the survey was conducted among 1560 people, randomly selected from the population of the study area, with a sample of 40 individuals per stratum.

We carried out the taxonomic identification of the samples collected in the field at the laboratory of biodiversity and natural resources of the Ibn Tofaïl University, taking into account the recent modifications.

For the determination of the plants we used the following flora:

The Catalogue of plants of Morocco [25,26]; the Flora of North Africa [27]; the New Flora of Algeria and the Southern Desert Regions [28]; Volume 1 [29], Volume 2 and Volume 3 of the Practical Flora of Morocco.

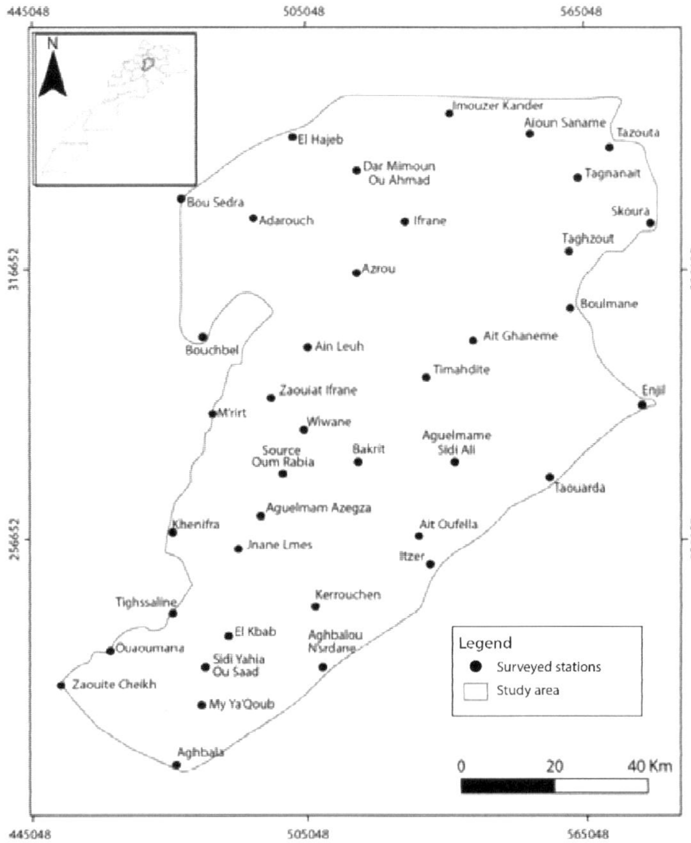

Figure 3. Map of the study area showing the surveyed stations.

3. Results and Discussion

3.1. floristic analysis of the catalog elaborated in the central Middle Atlas

The ethnobotanical study conducted in the central Middle Atlas has made it possible to establish a catalog of 267 plant species, divided into 79 botanical families and 211 genera. Most are spontaneous plants (75%), 20% are cultivated, and 5% are imported.

These species are divided into the following systematic groups:

- The Lichens are represented by a single family (Parmeliaceae), a single genera (Evernia), two species (Evernia furfuracea Mann., Evernia prunastri Ach.).
- Bryophytes are represented by two families (Orthotrichaceae, Pottiaceae) two genera (Orthotrichum, Tortula) and two species (Orthotrichum sp, Tortula muralis).
- Gymnosperms are represented by three families (Cupressaceae, Pinaceae and Taxaceae), four genera (Juniperus, Tetraclinis, Pinus and Taxus) and eight species (*Juniperus oxycedrus* L., *Juniperus phoenicea* L., *Juniperus thurifera* L., *Tetraclinis articulata* (Vahl) Masters, Cedrus atlantica (Endl.) Carr., Pinus halepensis Mill., *Pinus pinaster* Aiton and *Taxus baccata* L.).
- Chlamydosperms are represented by a single family Ephedraceae, a single genus Ephedra and a single species *Ephedra nebrodensis* Guss.
- Angiosperms comprise 72 families, 203 genera and 254 species. Among the latter, dicotyledons are dominant with 221 species, or 87 %, and Monocotyledons with 33 species, or 13%.

The medicinal flora of the central Middle Atlas is rich and very diverse because of its geographical position, its orography, its edaphic structure, its geology and its climatic conditions. Also as a result of the great diversity of habitats (great diversity of ecosystems and ecocomplexes) that this area contains (forests, pre-forests, matorrals, steppes, grasslands, wetlands, saline soils) [30].

3.1.1. Genera analysis of botanical families identified in the study area

At the genera level, as shown in the table 1, the most represented families in the central Middle Atlas are the Asteraceae with 21 genera, the Apiaceae with 19, the Lamiaceae with 13, the Poaceae and the Fabaceae with 12, the Rosaceae with 8, the Brassicaceae with 7. Some families possess 5 genera; others comprise less than 3 genera; a large number of families are monogeneric.

Table 1. Nature and genera richness of medicinal plant families used in the Central Middle Atlas area.

Families	Genera numbers	Families	Genera numbers
Asteraceae	21	Cactaceae	01
Apiaceae	19	Cannabinaceae	01
Lamiaceae	13	Capparaceae	01
Fabaceae	12	Cistaceae	01
Poaceae	12	Convolvulaceae	01
Rosaceae	08	Cyperaceae	01
Brassicaceae	07	Ephedraceae	01
Amaranthaceae	04	Ericaceae	01
Cucurbitaceae	04	Fagaceae	01
Myrtaceae	04	Gentianaceae	01
Solanaceae	04	Geraniaceae	01
Zingiberaceae	04	Globulariaceae	01
Arecaceae	03	Juglandaceae	01
Asparagaceae	03	Juncaceae	01
Caryophyllaceae	03	Linaceae	01
Euphorbiaceae	03	Musaceae	01
Lauraceae	03	Myristicaceae	01
Liliaceae	03	Nitrariaceae	01
Lythraceae	03	Orthotrichaceae	01
Malvaceae	03	Paeoniaceae	01
Ranunculaceae	03	Papaveraceae	01
Verbenaceae	03	Parmeliacees	01
Anacardiaceae	02	Pedaliaceae	01
Apocynaceae	02	Piperaceae	01
Cupressaceae	02	Plantaginaceae	01
Iridaceae	02	Polygonaceae	01
Moraceae	02	Portulacaceae	01
Oleaceae	02	Pottiaceae	01
Pinaceae	02	Resedaceae	01
Plombaginaceae	02	Salicaceae	01
Rhamnaceae	02	Santalaceae	01
Rubiaceae	02	Sapotaceae	01
Rutaceae	02	Schisandraceae	01
Scrophulariaceae	02	Tamaricaceae	01
Thymelaeaceae	02	Taxaceae	01
Amaryllidaceae	01	Theaceae	01
Aristolochiaceae	01	Urticaceae	01

IOP Conf. Series: Earth and Environmental Science 1090 (2022) 012025 doi:10.1088/1755-1315/1090/1/012025

Berberidaceae	01	Vitaceae	01
Borraginaceae	01	Zygophyllaceae	01
Buxaceae	01		

3.1.2. Species analysis of botanical families identified in the study area

At the species level, as shown in the figure 4 and table 2, the Asteraceae are always the best represented in the Central Middle Atlas with 29 taxa, the Lamiaceae with 26, the Apiaceae with 21, the Rosaceae with 14, the Fabaceae and the Poaceae with 13, the Brassicaceae with 9 and the Solanaceae with 8.

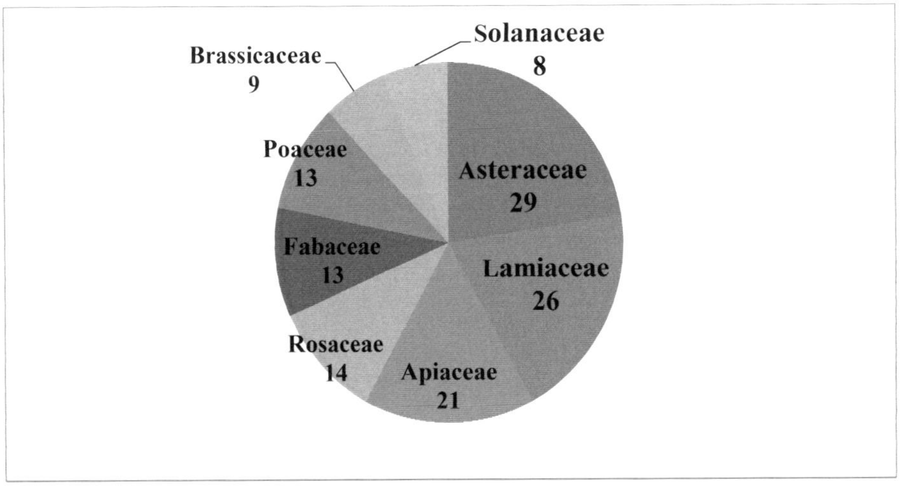

Figure 4. Number of the most represented species per family in the study area.

Table 2. Nature and species richness of medicinal plant families used in the Central Middle Atlas area.

Families	Number of species	Families	Number of species
Asteraceae	29	Aristolochiaceae	01
Lamiaceae	26	Berberidaceae	01
Apiaceae	21	Borraginaceae	01
Rosaceae	14	Buxaceae	01
Fabaceae	13	Cactaceae	01
Poaceae	13	Cannabinaceae	01
Brassicaceae	09	Capparaceae	01
Solanaceae	08	Convolvulaceae	01
Cucurbitaceae	05	Cyperaceae	01
Euphorbiaceae	05	Ephedraceae	01
Amaranthaceae	04	Ericaceae	01
Cupressaceae	04	Gentianaceae	01
Lauraceae	04	Geraniaceae	01
Myrtaceae	04	Globulariaceae	01
Zingiberaceae	04	Juglandaceae	01
Anacardiaceae	03	Juncaceae	01
Arecaceae	03	Linaceae	01

Asparagaceae	03	Musaceae	01
Caryophyllaceae	03	Myristicaceae	01
Cistaceae	03	Nitrariaceae	01
Fagaceae	03	Orthotrichaceae	01
Liliaceae	03	Papaveraceae	01
Lythraceae	03	Pedaliaceae	01
Malvaceae	03	Piperaceae	01
Oleaceae	03	Plantaginaceae	01
Pinaceae	03	Polygonaceae	01
Ranunculaceae	03	Portulacaceae	01
Rhamnaceae	03	Pottiaceae	01
Thymelaeaceae	03	Resedaceae	01
Verbenaceae	03	Salicaceae	01
Amaryllidaceae	02	Santalaceae	01
Apocynaceae	02	Sapotaceae	01
Iridaceae	02	Schisandraceae	01
Moraceae	02	Tamaricaceae	01
Paeoniaceae	02	Taxaceae	01
Parmeliaceae	02	Theaceae	01
Plombaginaceae	02	Urticaceae	01
Rubiaceae	02	Vitaceae	01
Rutaceae	02	Zygophyllaceae	01
Scrophulariaceae	02		

The most represented families on the generic level, are those which are the richest in species, and which are moreover the richest families in Morocco (figure 5).

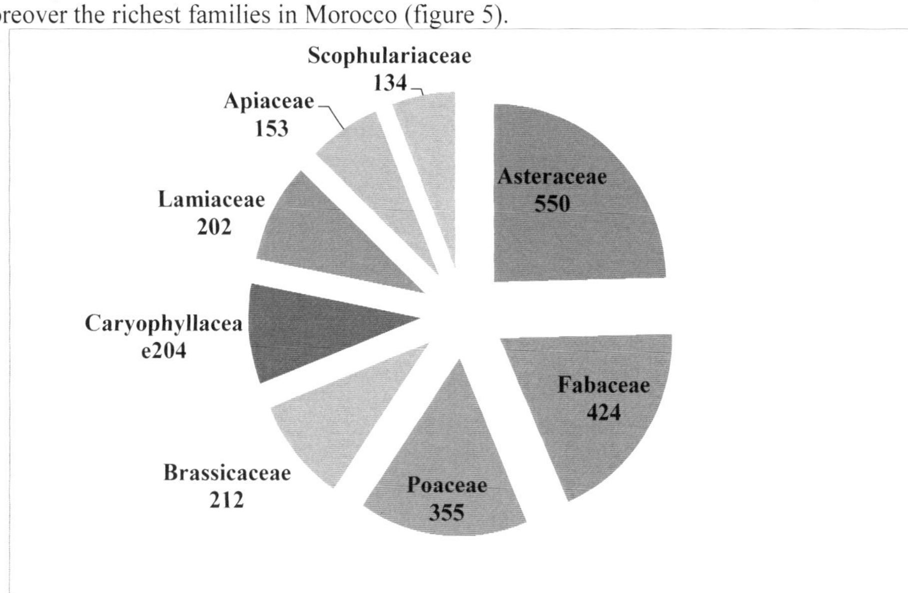

Figure 5. The richest families of Moroccan vascular flora [28].

The figure 6 illustrates the comparison of our results with the Moroccan vascular flora.

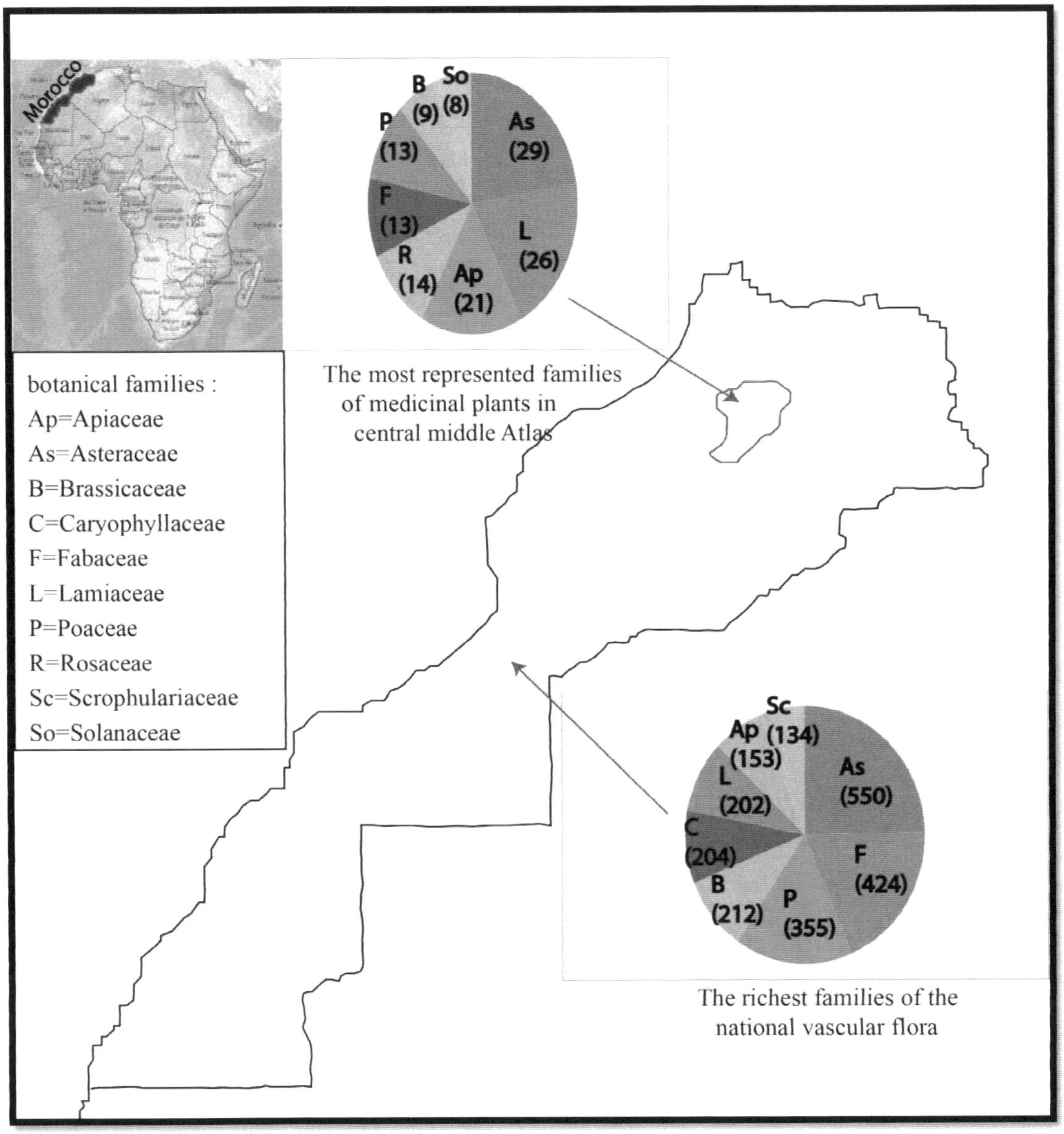

Figure 6. the most represented families of medicinal plants in the central middle atlas compared to Moroccan Vascular Flora.

3.1.3. Endemic species

In the studied area, the elaborated catalog presents endemic taxa. According to (table 3), the number of endemics is equal to 11 species, which is 1.1% of the total number of endemic species and subspecies of Morocco, these last belonging to 7 families. The Lamiaceae have 4 species, it is the family that occupies the first place with an endemic rate of 40% and the 6th rank on the species richness of the national vascular flora inventory [31].

Table 3. The endemic species in the elaborate catalog in the central Middle Atlas and their families

species	Families
Aristolochia baetica **(IA)**	Aristolochiaceae
Anacyclus pyrethrum **(IA)**	Asteraceae
Artemisia mesatlantica **(EM)**	Asteraceae
Mentha gattefossei **(EM)**	Lamiaceae
Origanum compactum **(I)**	Lamiaceae
Thymus munbyanus **(IA)**	Lamiaceae
Thymus zygis **(I)**	Lamiaceae
Armeria alliacea **(IA)**	Plombaginaceae
Argania spinosa **(EM)**	Sapotaceae
Thymelaea lythroides **(I)**	Thymelaeaceae
Zygophyllum gaetulum **(AC)**	Zygophyllaceae

(EM) endemic of Morocco (IA) Ibero-Algerian-Moroccan endemism (I) Ibero-Moroccan endemism (AC) endemic to Morocco, the Canary Islands and Algeria.

3.1.4. Rare or Endangered Species

The elaborated catalog counts 17 rare or threatened species, that is to say 6,36 % of the elaborated catalog, of which almost the third (6 species) is considered as very rare. These rare taxa are distributed in 12 families (Lamiaceae: 4; Asteraceae: 3; Caryophyllaceae: 1; Cupressaceae: 1; Fagaceae: 1; Lauraceae: 1; Ranunculaceae: 1; Rosaceae: 1; Sapotaceae: 1; Tamaricaceae: 1; Taxaceae: 1; Thymelaeaceae: 1) (table 4). For comparison, on the scale of Morocco, Astraceae, Lamiaceae, Caryophyllaceae are the richest families in rare or threatened taxa [32]. Hence the need for an urgent alarm of the danger of the intensive use of these endangered species, and to make the population aware of the ecological, biological, economic and social importance of the national plant heritage.

Table 4. Number of taxa in the elaborate catalog in the Central middle Atlas by rarity category.

Families	Species	Rarity category
Lamiaceae	*Mellissa officinalis* L.	R *
	Mentha suaveolens L	R
	Satureja calamintha (L.) Scheele.	RR
	Satureja vulgaris	RR
Asteraceae	*Artemisia absinthium* L.	RR
	Matricaria camomilla	RR
	Tragopogon porrifolius L.	RR
Caryophyllaceae	*Hernaria hirsuta* L.	R
Cupressaceae	*Juniperus thurifera* L	V
Fagaceae	*Quercus faginea* lamk.	R
Lauraceae	*Laurus nobilis* L.	R
Ranunculaceae	*Delphinium staphysagria* L.	R
Rosaceae	*Prunus avium* L.	RR
Sapotaceae	*Argania spinosa* L.	V
Tamaricaceae	*Tamarix gallica* L.	**
Taxaceae	*Taxus baccata* L.	V
Thymelaeaceae	*Thymelaea lythroides* Barr & Murb.	R

RR: very rare; R: rare; R*: Suspected rare; V: vulnerable; **: Taxon extinct or doubtful presence.

IOP Conf. Series: Earth and Environmental Science 1090 (2022) 012025 doi:10.1088/1755-1315/1090/1/012025

According to the results of (tables 3 and 4), it can be seen that the species Argania spinosa (argan tree) is both an endemic species, a vulnerable species and used by the population in herbal medicine, so the exploitation of this species must however be done in a reasonable way in order to preserve and perpetuate it. Also for the species Mentha suaveolens (the round-leaf mint) is a rare taxon and among the most used species in the study area (number of citations = 820).

3.1.5. Species of the most represented families in the study area

Table 5 present the ranking of the six richest families of the elaborate catalog in relation to their position in the national vascular flora. It indicates a clear correspondence between the positions of the richest families in the local and national catalog with slight differences in classification, with the exception of the Rosaceae family, which ranks fourth in the local flora of the study area, while it occupies the eleventh place in the national vascular flora.

This result is not surprising given the diversity of Middle Atlas ecosystems, ranging from cedar or oak forests to xerophytic pastures.

Table 5. Classification of Families by Number of Medicinal Species in the Central Middle Atlas and Other Regions

Family name	**Rank and numbers of plant species**				
	Central Middle Atlas Region	**Region of Rabat**	**Region of Zaër**	**Central Plateau Region**	**Moroccan flora**
Asteraceae	1 (29)	1 (26)	2 (24)	2 (16)	1 (550)
Lamiaceae	2 (26)	4 (21)	3 (23)	1 (20)	6 (202)
Apiaceae	3 (21)	2 (25)	1 (27)	3 (12)	7 (153)
Rosaceae	4 (14)	5(13)	5 (10)	6 (06)	11 (62)
Fabaceae	5 (13)	3 (23)	4 (15)	5 (08)	2 (424)
Poaceae	5 (13)	6(10)	5 (10)	4 (09)	3 (355)

According to (table 5 and figure 7), the six richest families of the elaborated catalog of the studied region are the same as those obtained in the other regions with a difference in the ranking, namely the study of El Hilah [33], the study of Lahsissène [34] and the study of Hseini [35]. This is due to the fact that these families are among the most important of Moroccan flora in terms of the number of genera and species [36]. These families of plants are among those that provide the most plants to the Moroccan traditional pharmacopoeia.

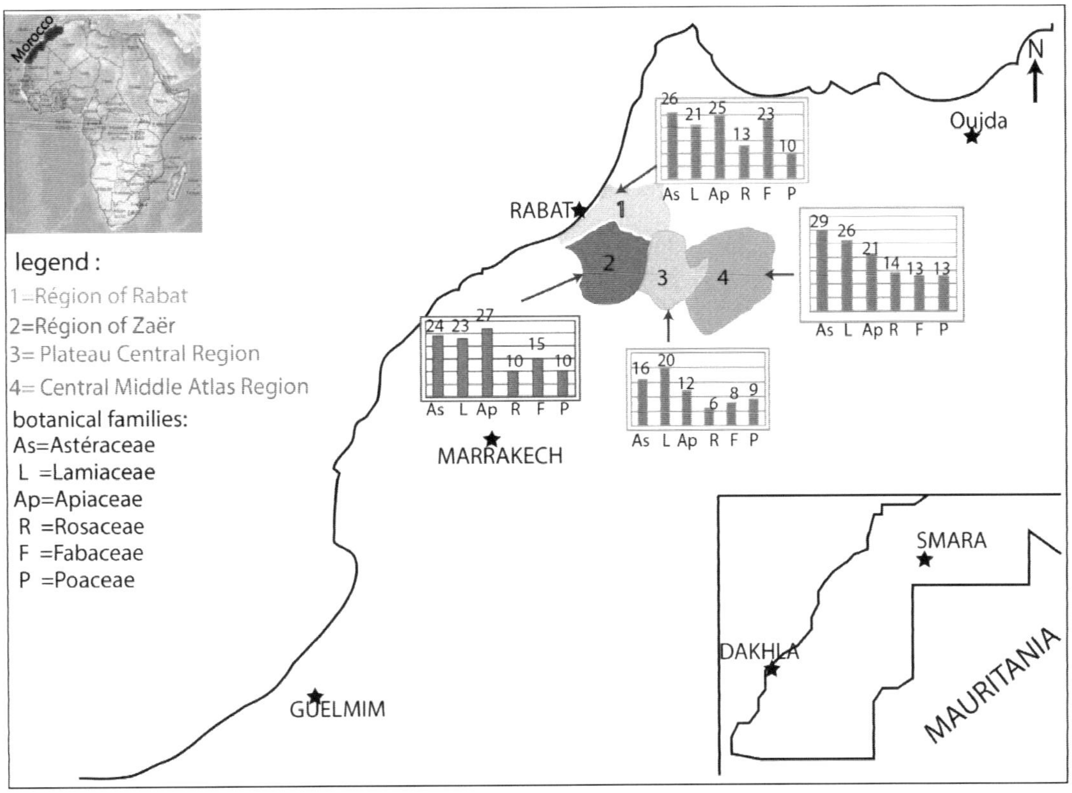

Figure 7. Comparative map of the most represented families of Medicinal plants in the Central Middle Atlas with the results of other regions [37, 38].

3.1.6. *Biological type of species encountered*

The analysis of the forms of adaptation of plants of plants adaptation allows provides a better appreciation of the ecological conditions in which they live. The biological types, by their distribution, faithfully reflect the ecological conditions of a region.

The analysis of (figure 8) reveals that 26.02% of species are therophytes (annual plants) and 18.37% of plants are phanerophytes followed by 15.31% of Geophyte and respectively by soothesteps (14,80%). , Nanophanerophytes (14.29%) and hemicryptophytes (11.22%).

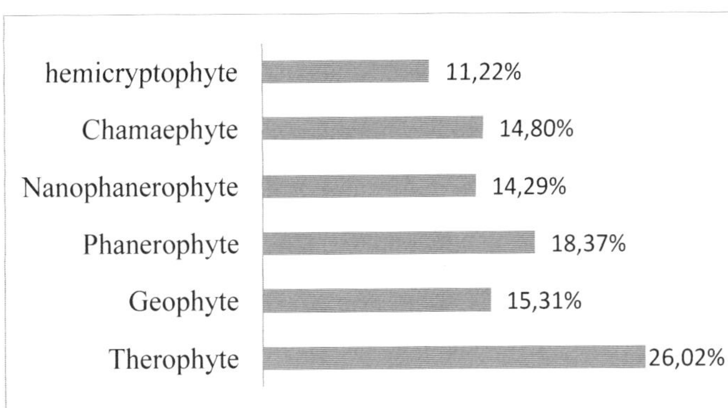

Figure 8. Distribution of the medicinal flora of the Middle Atlas region by biological type.

The abundance of therophytes can be explained by the high presence of seasonal immersion habitats, conducive to the development of germinating and fast growing annual plants), especially in wetlands of the said region.

The high frequency of therophytes attests to the disturbance of this zone (clearing, grazing, but also related to the local climate).

According to [39,40], theophyta is a form of resistance to drought as well as high temperatures in arid environments.

The significance of therophyty has been extensively debated by these authors who attribute it:

- Either to the adaptation to the cold winter constraint or to the summer drought,
- Either to environmental disturbances by grazing, crops, etc.

3.1.7. *Bioclimatic stages*

According to the results obtained in figure 9, the medicinal species listed in our ethnobotanical study belong successively to the following bioclimatic stages: semi-arid, sub-humid, arid, humid, Saharan and per-humid.

As a result, we can deduce that the surveyed population generally uses the local plants available in its surroundings, as the majority of the stations surveyed are located in the humid, subhumid and arid bioclimatic stages on The Emberger climate climagram.

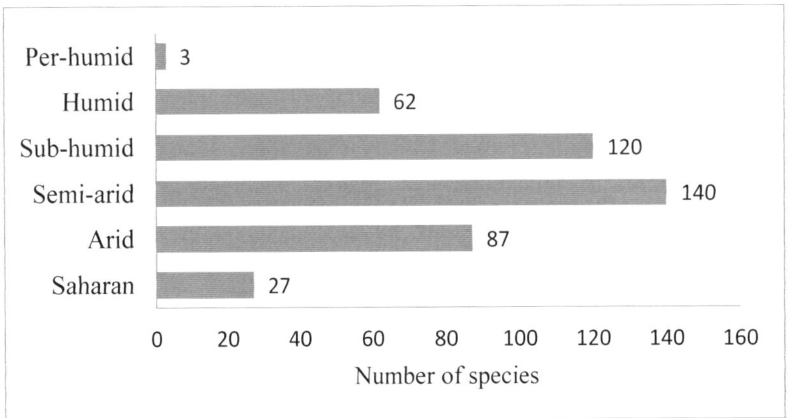

Figure 9. Number of plant species used as medicinal plants in the central middle Atlas region by bioclimatic stages.

3.1.8. *The most commonly used (species) by the population of the study area*

The Figure 10 shows that oregano (Origanum compactum) is ranked in first position (1039) among the most frequently cited plants in the survey, followed by rosemary (928) (Rosmarinus officinalis), the pouliot (922) (Mentha pulegium), round-leaved mint (820) (Mentha suaveolens), thuja (724) (Tetraclinis articulata), the artemisia (Artemisia herba-alba) (717), caralluma of europe (Caralluma europaea) (716), the chenopod (629) (Chenopodium ambrosioides), the olive tree (625) (Olea europaea), and then we find fenugreek (617) (Trigonella foenum-graecum), thyme (Thymus zygis) with a frequency of use (615) followed by the white marrube (614) (Marrubium vulgare).

This frequency of use can be explained by the therapeutic biological properties of these species, which are involved in the treatment of recurring pathologies such as fever, digestive diseases, aches, respiratory diseases, urinary diseases, inflammations, Diabetes and dermal diseases. These medicinal species are collected in the forests of the region.

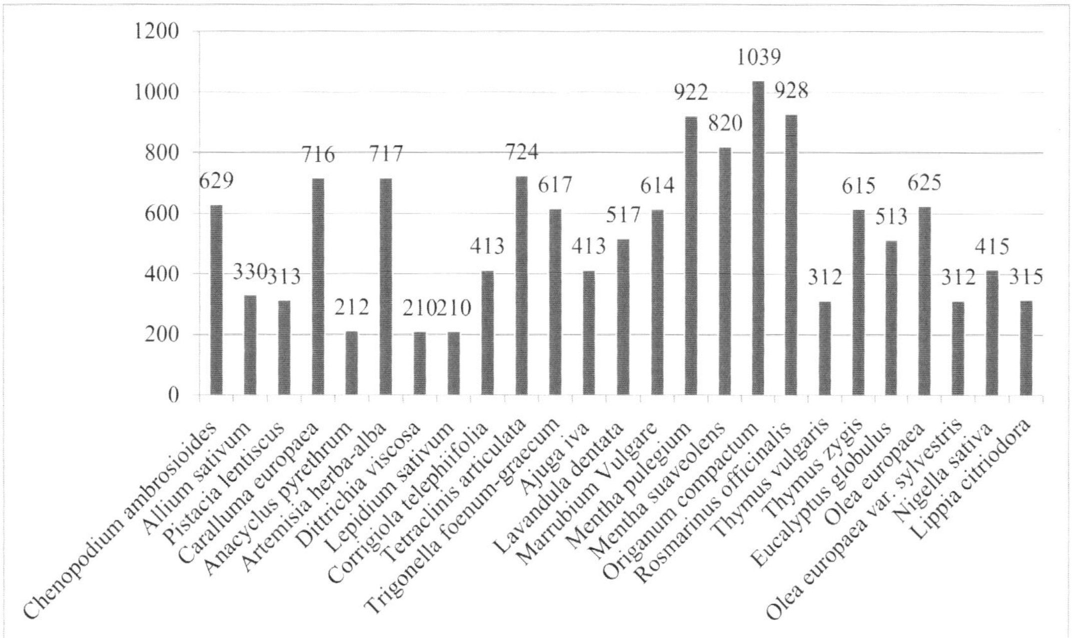

Figure 10. Classification of the most used plants by the population studied by number of citations (more than 210 citations).

3.1.9. The list of medicinal species identified in the Central Middle Atlas

The table 6 represents the species inventoried during our ethnobotanical survey. Indeed, for each plant inventoried, the monograph contains the vernaculair name, the scientific name, the family and the phylum.

Table 6. Catalog of medicinal plants identified in the central Middle Atlas region

Phylum	Family	Scientific name	Vernacular name
Lichens	Parméliacées	*Evernia furfuracea* Mann.	Lahyate Chikh / Tamert numghar
		Evernia prunastri Ach.	Lahyate Chikh / Tamert numghar
Bryophytes	Orthotrichaceae	*Orthotrichum sp*	Sanbel berri
	Pottiaceae	*Tortula muralis*	Sanbel
Gymnosperms	Cupressaceae	*Juniperus oxycedrus* L.	Taqqa
		Juniperus phoenicea L.	Araar finiqui
		Juniperus thurifera L.	Tawayt / Tawalt
		Tetraclinis articulata (Vahl) Masters.	Araar
	Pinaceae	*Cedrus atlantica* (Endl.) Carr.	L-ârz / Atgal
		Pinus halepensis Mill.	Sanawbar
		Pinus pinaster Aiton	Taïda
	Taxaceae	*Taxus baccata* L.	Îgen
Chlamydosperms	Ephedraceae	*Ephedra nebrodensis* Guss.	Timitrte

Monocotyledonous angiosperms	Amaryllidaceae	*Allium cepa* L. *Allium sativum* L.	Lbasla / Azalim Tuma / Tichert
	Arecaceae	*Chamaerops humilis* L	Doum / Thiguazdan / Ignad / Jemmar/ Aghâz (fruit)
		Cocos nucifera L.	Lkouk
		Phoenix dactylifera L	Tazdâyet / Nakhela / Tînî (fruit) / Tmar (fruit)
	Asparagaceae	*Agave americana* L.	Sabra
		Asparagus albus L.	Sekkum / Azzu
		Urginea maritima (L.) Baker.	Bsel idane
	Cyperaceae	*Cyperus rotundus* L.	Tara
	Iridacea	*Crocus sativus* L.	Zaafran hore
		Iris germanica L.	Anbar
	Juncaceae	*Juncus acutus* L.	Zriaat smâr / Agazmir
	Liliaceae	*Asphodelus microcarpus* Salzm. &Viv.	L-berwag / Înghri / Tigrad
		Colchicum automnale	Bûkbûka
		Ruscus aculeatus L	Bû-chûka
	Musaceae	*Musa paradisiaca* L.	Banane
	Poaceae	*Avena sativa* L.	Khortal
		Festuca arundinacea Schreb.	Agazmir
		Hordeum vulgare L	Echîr / Timzine
		Oryza sativa L	Ruz
		Pennisetum typhoides (Burm.) Staph. & hub.	Illan
		Phalaris canariensis L.	Zwane
		Phragmites communis Trin.	L-qesba / Äghanim
		Secale cereale L.	Adku
		Sorghum halepense (L) Pers.	Ababû
		Sorghum vulgare L.	Bachna
		Stipa tenacissima L.	Ârî / Halfa
		Triticum durum Desf.	Îrden / Gamh
		Zea mays L.	Lakbal / Dra
	Zingiberaceae	*Alpinia officinarum* Hance.	Khodenjâl
		Curcuma longa L.	Kharkûm beldi
		Elettaria cardamomum L. Maton	Qa'qolla
		Zingiber officinalis Rosc.	Skenjbir
Dicotyledonous Angiosperms	Amaranthacea	*Beta vulgaris* L.	Lbarba
		Chenopodium ambrosioides L.	Mkhinza
		Fredolia aretioides Coss. & Dur.	Sella' / Achennud
		Spinacia oleracea L.	Sabanikh
	Anacardiaceae	*Pistacia atlantica* Desf.	Lbtem / Ijj
		Pistacia lentiscus L.	Trou
		Rhus pentaphylla Desf.	**Tizrha**

	Apiaceae	*Ammi visnaga (*L.) Lam.	**Barghanisse / Bouchnikha**
		Ammodaucus leucotrichus Coss. & Dur.	Kamoun sufi
		Apium graveolens L.	Krafes
		Carum carvi L.	Karwiya
		Coriandrum sativum L	Qasbor
		Cuminum cyminum L.	Kamoun
		Daucus carota L.	Khizou
		Daucus crinitus Desf.	Arq bou zafour
		Elaeoselinum asclepium (L.) Bertol	Arq yabû
		Ferula assa-foetida L.	Hantita
		Ferula communis L.	Uffal / Awli / Klakh / Lboubal
		Ferulago lutea (Poiret) Grande	Abaghour
		Foeniculum vulgare P. Mill.	Nafaa / Lbesbas / Amssa
		Hippomarathrum libanotis L	Timjghdine
		Kundmania sicula (L.) DC	Tajguamine
		Magydaris panacifolia (Vahl.) lange	Frifra
		Petroselinum sativum Hoffm.	Ma'adnous
		Pimpinella anisum L.	Habat hlawa
		Ptychotis verticillata Dub.	Nounkha
		Ridolfia segetum (L.) Moris	Tebch
		Smyrnium olusatrum L.	Lheyyar
	Apocynaceae	*Caralluma europaea* (Guss.).	Daghmus
		Nerium oleander L.	Defla / Allili
	Aristolochiaceae	*Aristolochia baetica* L.	Berztem / Ajrarhi
	Asteraceae	*Anacyclus pyrethrum* L.	Iguntas / Tagundecht
		Anacyclus radiatus Lois.	Kra'djaja
		Artemisia absinthium L.	Chiba
		Artemisia herba-alba Asso.	Chih / Ifzi / Izri
		Artemisia mesatlantica Maire	Tizrit
		Atractylis cancellata	Tisnanine wadou
		Atractylis gummifera L.	Addad
		Calendula officinalis L.	Jemra
		Centaurea calcitrapa L	Asnane azerwal
		Chamaeleon macrophyllus Desf.	Achekmud
		Cichorium intybus L.	Âdghurru
		Cynara cardunculus L.	Lkhorchef / Tagua
		Cynara humilis L.	Timta
		Cynara scolymus L	Qôq
		Dittrichia viscosa (L.) Greuter	Terehla

IOP Conf. Series: Earth and Environmental Science 1090 (2022) 012025 doi:10.1088/1755-1315/1090/1/012025

		Echinops spinosus L.	Taskra
		Helianthus annus L.	Nouar chems
		Helianthus tuberosus L.	Btata lkasbia
		Lactuca sativa L	Khuss
		Lactuca scariola L.	Ahchlaf n'ssem
		Mantisalca salmantica	Tazemourt
		Matricaria chamomilla L.	Lbabounaj
		Pulicaria arabica (L) Cass	L-ghattasa
		Rhaponticum acaule (L.) DC	Tafgha
		Scolymus hispanicus L.	Gûrnina / Taghdiût
		Stemmacantha acaulis	Tifghwine
		Taraxacum obovatum DC.	Iwjdem / L-handba
		Tragopogon porrifolius L.	Iwjdem awtem / aliam ouyiis
		Xanthium strumarium	Zakhyadd / Mzzer izm
	Berberidaceae	*Berberis hispanica* Boiss. & Reut	Arghis
	Borraginaceae	*Borrago officinalis* L.	Iles founas
	Brassicaceae	*Alyssum spinosum*	Aguerbaz
		Anastatica hierochuntica L.	Chajarat maryem
		Brasica napus L.	Leftt
		Brasica oleracea L.	Krumb
		Brassica rapa L.	Left beldi
		Lepidium sativum L.	Hab rchad
		Nasturtium officinalis R. Br	Gernûnech
		Raphanus sativus L.	Lfjel
		Sinapis alba L.	Khardal
	Buxaceae	*Buxus balearica* Lam.	Azazer / Lbakous
	Cactaceae	*Opuntia ficus-indica* (L.) Mill	Handiya
	Cannabinaceae	*Cannabis sativa* L.	Kif
	Capparaceae	*Capparis spinosa* L.	Kabbar / Taylulut
	Caryophyllaceae	*Corrigiola telephiifolia* Pourr.	Sarghina / Tawsarghine
		Herniaria hirsuta L.	Herras lehjar
		Saponaria vaccaria L.	Tigheghsht
	Cistaceae	*Cistus albidus* L.	Boutour
		Cistus ladanifer L.	Tuzalt
		Cistus salviifolius L.	Rbibit
	Convolvulaceae	*Convolvulus arvensis* L.	Lûwaya dial na'na'
	Cucurbitaceae	*Citrullus colocynthis* (L.) Schrad.	Lhdej / Tafrzizte
		Citrullus vulgaris Schrad.	Dellah
		Cucumis sativus L.	Lkhiyâr
		Cucurbita pepo L.	Gar'a l-hamra / Takhsaït
		Lagenaria siceraria (Molina) Standl.	Slawi
	Ericaceae	*Arbutus unedo* L.	Sasnu
	Euphorbiaceae	*Euphorbia falcata* L.	Hayat en-nufûs
		Euphorbia nicaensis All.	Zeg oryiriw
		Euphorbia resinifera Berg.	Tikiwt

		Mercurialis annua L.	Harryga malsa
		Ricinus communis L.	Lkharwaa / Anguaref
	Fabaceae	*Astragalus gummifer* labill.	Ktira
		Cassia fistula	'ûd salib
		Cassia senna L.	Sana haram
		Ceratonia siliqua L.	Kharoub / Taslighwa
		Cicer arietanum L.	Lhammas
		Glycyrrhiza glabra L.	Arak sous
		Lens culinaris Med.	La'des
		Medicago sativa L.	Fessa
		Ononix natrix L.	Âfesdad
		Phaseolus vulgaris L.	Lûbya
		Soja hispida.	Soja
		Trigonella foenum-graecum L.	Lhalba
		Vicia faba L.	Lfoul / Ibawn
	Fagaceae	*Quercus faginea* lamk.	Tacht / L'afsa (gall)
		Quercus ilex L.	Keruch / Tassaft
		Quercus suber L.	Lferchi / Fernane
	Gentianaceae	*Centaurium erythraea* Rafn.	Qusset el-hayya / Ahchlaf n'tawrra
	Geraniaceae	*Pelargonium roseum* Willd.	La'tercha
	Globulariaceae	*Globularia alypum* L.	A'yen lerneb / Taselgha
	Juglandaceae	*Juglans regia* L.	Swak / Gargaa
	Lamiaceae	*Ajuga iva* (L.) Schreb.	Tûf tolba / Chendgoura
		Lavandula dentata L.	Lkhezama lbaldia
		Lavandula multifida L	Lkehila
		Lavandula stoechas L.	Lhalhal
		Marrubium vulgare L.	Mriwta / Mriwa
		Mellissa officinalis L.	Lhbak soufi / Malissa
		Mentha gattefossei Maire	Fliyou dial jbel
		Mentha pulegium L.	Fliyou
		Mentha suaveolens Ehr.	Timrsade / Timija / Merssita
		Mentha viridis L.	Na'na'
		Ocimum basilicum L.	Lhbak
		Origanum compactum Benth.	Zaater
		Origanum majorana L.	Merdedûch
		Rosmarinus officinalis L.	Yazir
		Salvia officinalis L.	Salmiya
		Salvia verbenaca L.	'Achbat kola bla / khiyyata
		Satureja alpina (L.) Scheele.	Fliyyo diâl berr
		Satureja calamintha (L.) Scheele.	Manta
		Satureja rotundifolia	Tafliyoute
		Satureja vulgaris (L.) Fritsch.	Tafliyoute
		Teucrium polium L.	Ja'da
		Thymus algeriensis Boiss. & Reut.	Adouchen / Âzukni / Zaïtra
		Thymus munbyanus Boiss. & Reut.	Adouchen / Âzukni / Zaïtra
		Thymus vulgaris L.	Adouchen / Âzukni / Zaïtra
		Thymus zygis Loefl. ex L.	Adouchen / Âzukni / Zaïtra
		Ziziphora Hispanica L.	Taflyoute / Flyou Dial L'ber
	Lauraceae	*Cinnamomum camphora* Ness.	Kafour

		Cinnamomum cassia Ness.	Qarfa
		Laurus nobilis L.	Awrak sidna moussa
		Persea gratissima Gaertn.	Avocat
	Linaceae	*Linum usitatissimum* L.	Zriaat al kettane
	Lythraceae	*Lawsonia inermis* L.	Henna
		Lythrum junceum Banks & Sol	Sabun n'wamane
		Punica granatum L.	Romane
	Malvaceae	*Abelmoschus esculentus* (L.) Moench	Lmloukhia
		Hibiscus sabdariffa L.	Karkadé
		Malva sylvestris L.	Baqûla
	Moraceae	*Ficus carica* L.	Karma / Tazarett / Chriha
		Morus alba L	Tût lbari
	Myristicaceae	*Myristica fragrans* Houtt.	L-gûza
	Myrtaceae	*Eucalyptus globulus* Labill.	Calitous
		Eugenia caryophyllata Thunb.	Qronfel
		Myrtus communis L.	Rihane
		Pimenta officinalis Lindl.	Nwiwra
	Nitrariaceae	*Peganum harmala* L.	Lharmel
	Oleaceae	*Jasminum fruticans* L.	Yasmin
		Olea europaea L.	Zitoune
		Olea europaea var. sylvestris (Mill.) Brot.	Azmour / Azbouj
	Paeoniaceae	*Paeonia corallina* L. ssp. Maroccana	Haberchich
		Paeonia coriacea Boiss.	Haberchich
	Papaveraceae	*Papaver rhoeas* L.	Belaamane
	Pedaliaceae	*Sesamum indicum* L.	Janjlane
	Piperaceae	*Piper cubeba* L.F.	L-kebbaba
	Plantaginaceae	*Plantago major* L.	L-messasa / Tibedasse / Tifrotoute
	Plombaginaceae	*Armeria alliacea* (Cav.) Hoffm	Iwdmi
		Plumbago europeae L.	Timrabdine
	Polygonaceae	*Rumex roseus*	Tasmoune / Zbib iddane
	Portulacaceae	*Portulaca oleracea* L.	Rejla / Tismammine
	Ranunculaceae	*Delphinium staphisagria* L.	Habbet ras
		Nigella sativa L.	Haba souda / Sanouj
		Ranunculus bullatus L.	Wden l'halluf
	Resedaceae	*Reseda luteola* L.	harez izwane
	Rhamnaceae	*Rhamnus alaternus* L.	Amlîles
		Rhamnus lycioides L.	L-harcha
		Zizyphus lotus (L.) Lam.	Sedra / Azar / Azougar
	Rosaceae	*Crataegus laciniata* Ucr.	Admam
		Cydonia oblonga Mill.	Sfarjel
		Eriobotrya japonica (Thunb.) Lindl.	M'zah
		Fragaria vesca L.	L-frez
		Malus communis Poir	Tefâh
		Prunus amygdalus stokes var. amara DC.	Lûz murr
		Prunus amygdalus stokes var. dulcis DC.	Lûz lahlou
		Prunus armeniaca L.	Mechmach
		Prunus avium L.	Habb el-mluk

		Prunus domestica L.	Berqûq
		Prunus persica (L.) Batsch.	Khûkh
		Rosa canina L.	Tikhfaret
		Rosa damascena Mill.	Lward
		Rubus ulmifolius Schott	Cherrmû / Tabgha / Afras
	Rubiaceae	*Coffea arabica* L.	Qahwa
		Rubia peregrina L.	Fuwwa / Tarûbya
	Rutaceae	*Citrus limon* (L.) Burm.	Lhamed
		Ruta montana L.	Lfijel / Iwrmi
	Salicaceae	*Populus alba* L.	Sefsaf
	Santalaceae	*Viscum cruciatum*. Cieber ex Boiss.	Amrires / Marzaddal
	Sapotaceae	*Argania spinosa* (L.) Skeels	Âqqa wargan
	Schisandraceae	*Illicium verum* Hook.F	Badiana
	Scrophulariaceae	*Scrofularia sambucifolia* L.	Saboune ntmgharine
		Verbascum sinuatum L.	Aberdud n-izem
	Solanaceae	*Capsicum annuum* L.	Felfla lahlowa / Teflfel
		Capsicum frutescens L.	Felfel Hârra/ soudania
		Mandragora autumnalis Bertol	Bîd al-ghûl
		Nicotiana tabacum L.	Tanfiha
		Solanum lycopersicum L.	Maticha
		Solanum melongena L	Danjal
		Solanum nigrum L.	Maticha-wuchen
		Solanum tuberosum L.	Batata
	Tamaricaceae	*Tamarix gallica* L.	Tamimayt / Takawt (gall)
	Theaceae	*Camellia thea* L.	Atay
	Thymelaeaceae	*Daphne gnidium* L.	Âlezzâz
		Thymelaea lythroides Barr & Murb.	Metnane
		Thymelaea tartonraira (L.) All.	Talzzâzt
	Urticaceae	*Urtica pilulifera* L.	Hurriga / Tissarkmaz
	Verbenaceae	*Lippia citriodora* H.B. & K.	Lwiza
		Verbena officinalis L.	Baymût
		Vitex agnus-castus L.	Ângarf / L'kharw'
	Vitaceae	*Vitis vinifera* L.	Deritt / Dalya / Adil / Zbib
	Zygophyllaceae	*Zygophyllum gaetulum* Emb. & Maire.	L'aggâya

4. Conclusion

Medicinal plants constitute an important source of health care in the world and global demand is more and more growing. This is why ethnobotany and ethnopharmacology are used to inventory identify, all over the world, plants deemed active for which it is necessary to specify the properties and validate the uses.

The aim of this study was to realize the most complete inventory of the medicinal plants used in the central Middle Atlas region and to gather information on the therapeutic uses practiced in this region.

Allowed the identification of 267 medicinal species distributed in 79 families and 211 genera with the predominance of 6 families: Asterceae (29), Lamiaceae (26), Apiaceae (21), Rosaceae (14), Fabaceae and Poaceae with 13 species.

According to the elaborate catalog, rare or threatened plants contain 17 taxa of which 6 are considered very rare. Also the elaborate catalog is characterized by the presence of endemic taxa with a number equal to 11.

This study constitutes an important database for the valorisation of medicinal plants and the determination of the health profile of Central Middle Atlas local population).

The use of medicinal plants by local populations of the said region and around the world, as well as the prevalence of diseases, prompts researchers to explore this pathway for the discovery of new bioactive molecules. The valorisation of medicinal plants used in traditional medicine also requires scientific validation to justify their use. Therefore, we will propose some recommendations.

- ✓ To elaborate a Moroccan medicinal flora.
- ✓ To adopt new management approaches for the safeguard and preservation of natural resources.
- ✓ The cultivation of endangered medicinal plants, the conservation of seeds of medicinal plants.
- ✓ Raising awareness of the population for the safeguard of the natural heritage (protection of the plant cover).

5. References

[1] Lorrain E., 2013. 100 questions sur la phytothérapie. Ed. La boétie, Italie.the Nobel Prize

[2] Lardry J.M., Haberkorn V., 2007. L'aromathérapie et les huiles essentielles. Kinesither Rev ; 61 : 14-7p.

[3] Adefa M and Abraha B., (2011).Ethnobotanical Survey of Traditional Medicinal Plants in Tehuledere District, South Wollo, Ethiopia. Journal of Medicinal Plants Research Vol. 5(26), pp. 6233- 6242.

[4] Kasparek M, al-Janabi S. 2008. Plantes médicinales - La diversité biologique au service de la santé. Fiche thématique. Deutsche Gesellschaft für Technische Zusammenarbeit (GTZ) GmbH.

[5] Dibong, S. D., Mpondo, M. E., Nigoye, A., Kwin, M. F. & Betti, J. L. 2011. Ethnobotanique et phytomédecine des plantes médicinales de Douala, Cameroun. [Ethnobotany and phytomedicine of medicinal plants sold in Douala markets] — Journal of Applied Biosciences 37: 2496 – 2507. ISSN 1997–5902. Published online at www.biosciences.elewa.org.

[6] Hmamouchi M., 1999. Les plantes médicinales et aromatiques marocaines. Utilisations traditionnelles, maché biologie, écologie, toxicologie, lexiques. Imprimerie de Fedala, 389 p.

[7] Benabid A., 2000. Flore et diversité des écosystèmes du Maroc. Évaluation et préservation de la biodiversité, Édition Ibis press, Paris, France, 236 p.

[8] HCEFLCD, SPEFI, PAGPNI, 2007. Haut- Commissariat aux Eaux et Forêts et à la Lutte contre la Désertification, Service Provincial des Eaux et Forêts d'Ifrane : Projet d'aménagement et de protection des massifs forestiers de la Province d'Ifrane.

[9] Portail de l'Agriculture Marocaine (PAM), 2002. Terre et Vie, N° 59/60, Rabat, Maroc. (htpp://marocagri.multimania.com). 4 p.

[10] Mehdioui R. et Kahouadji A: 2007. Etude ethnobotanique auprès de la population riveraine de la forêt d'Amsittène: cas de la Commune d'Imi n'Tlit (Province d'Essaouira). Bulletin de l'Institut Scientifique, Rabat, section Sciences de la Vie n°29: 11-20 p.

[11] Bellakhdar J., 1997. La pharmacopée marocaine traditionnelle. Médecine arabe ancienne et savoirs populaires. Ed. Le Fennec, Paris, 764 p.

[12] Lazos E., and Paré L., 2000. Miradas Indígenas sobre una Naturaleza entristecida: Percepciones del deterioro ambiental entre nahuas del sur de Veracruz. Mexcio, UNAM.

[13] Giovannini P., and Heinrich M., 2009. Xki yoma' (our medicine) and xki tienda (patent medicine)-Interface between traditional and modern medicine among the Mazatecs of Oaxaca, Mexico. Journal of Ethnopharmacology, 121 p.

[14] Michard A., 1976. Eléments de géologie marocaine. Notes et Mém. Serv. géol. Maroc, 252, 408p.

[15] Arboleya M.L., Teixell A., Charroud M., Julivert M., 2004. A structural transect through the High and Middle Atlas of Morocco. Journal of African Earth Sciences 39. 319–327 p.

[16] Fedan B., 1988. Evolution géodynamique d'un bassin intra-plaque sur décrochement : le Moyen Atlas durant le Méso-Cénozoïque. Travaux de l'Institut Scientifique de Rabat, N° 18, 142 p.
[17] Texier J. P., Raynal J. P. et Lefevre D., 1985. Nouvelles positions pour un cadre chronologiques raisonné du Quaternaire marocain. C. R. Acad. Sc. Paris, t, 301, Série II, n° 3, 183-188 p.
[18] Herbig H.G., 1988. Synsedimentary tectonics in the Northern Middle Atlas (Morocco) during the Late Cretaceous and Tertiary In: V. Jacobshagen, Editors, The Atlas System of Morocco, Springer-Verlag, Berlin, 321–337 p.
[19] Cavallar W., 1950. Esquisse préliminaire de la carte des sols du Maroc, professeur docteur de pédologie de l'Institut Agronomique d'Ukraine chargé de mission par le Centre National de la Recherche Scientifique au centre de Recherches Agronomiques du Maroc, Rabat.
[20] Martin J., 1981. Le Moyen Atlas central étude géomorphologique. Notes et Mémoires du service Géologique N° 258 bis Rabat Maroc 447 p.
[21] Chillasse L., Dakki M., 2004. Potentialités et statuts de conservation des zones humides du Moyen-Atlas (Maroc), avec référence aux influences de la sécheresse. Sécheresse, 15 (4) : 337-45 p.
[22] Lecompte M., 1986. Biogéographie de la montagne marocaine : le Moyen Atlas central. CNRS (éd), Paris
[23] Godron M., 1971. Essai sur une approche probabiliste de l'écologie des végétations. Thèse Doct. D'Etat, U.S.T.L., Montpellier, 247 p.
[24] Godron M., Daget Ph., 1982. Analyse de l'écologie des espèces dans les communautés Coll. Ecol. 18, Masson, 163 p.
[25] Jahandiez E., Maire R., 1931-1934. Catalogue des plantes du Maroc, 3 vols. Minera, le chevalier, Alger.
[26] Emberger L., & Maire R., 1941. Catalogue Des Plantes Du Maroc (Spermaphytes Et Ptéridophytes), Tome Iv. Supplément Aux Volumes I, Ii Et Iii. - Minerva, Alger, Lix-Lxxv + 915-1181.
[27] Maire R., 1952-1987. Flore De L'Afrique Du Nord. Tomes 1 A 16. – Paul Lechevalier, Paris, "Encyclopédie Biologique". 5559 p.
[28] Quézel P. & Santa S., 1963. Nouvelle Flore De L'Algérie Et Des Régions Désertiques Méridionales. Tome 2. - Cnrs Ed., Paris, 494 P.
[29] Fennane M., Ibn Tattou M., Mathez J., Ouyahya A. & El Oualidi J., 1999. Flore Pratique Du Maroc ; Manuel De Détermination Des Plantes Vasculaires, Vol. 1. Trav. Inst. Sci., Rabat, Sér. Botanique, 36, 558 P.
[30] Benabid A., 2002. Le Rif et le Moyen Atlas (Maroc) : Biodiversité, menaces, préservation. African Mountains High Summit Conference, Nairobi, Kenya 6-10 mai 2002.
[31] Fennane M. & Ibn Tattou M., 2012. Statistiques et commentaires sur l'inventaire actuel de la flore vasculaire du Maroc. Bull. Inst. Sci. Rabat sec. Sci. 34(1): 1-9 p.
[32] Fennane M. & Ibn Tattou M. 1998. Catalogue des plantes endémiques, rares ou menacées du Maroc. Bocconea, 8 : 1-243 p.
[33] El Hilah F., 2016. Enquête ethnobotanique des plantes médicinales auprès de la population du plateau central marocain. Thèse de Doctorat national. Faculté des Sciences de Kénitra. 156 p.
[34] Lahsissène H., 2010. Recherches ethnobotanique & floristique des plantes médicinales utilisées dans la région phytogéographique de Zaër. Thèse de Doctorat, faculté des Sciences. Rabat. 273 p.
[35] Hseini S., 2007. Etude ethnobotanique de la flore médicinale dans la région de Rabat. Thèse de Doctorat. Faculté des Sciences. Rabat. 182 p.
[36] Fennane M., 2004. Propositions de Zones Importantes pour les Plantes au Maroc (ZIP Maroc), Rabat.
[37] Sauvage Ch., 1961. Flore Des Subéraies Marocaines. (Catalogue Des Cryptogames Vasculaires Et Des Phanérogames). Travaux de L'institut Scientifique Chérifien, Série Botanique N°22, Rabat, 25 Août, 1961: X + 257 P.

[38] Ministère chargé de l'environnement, 1997. Rapport sur l'Etat de l'Environnement du Maroc. Chapitre III : Biodiversité et Milieu Naturel, 48 p.

[39] Daget PH., 1980. Sur les types biologiques botaniques en tant que stratégie adaptative, cas des thérophytes. In « Recherches d'écologie théorique ». Les stratégies adaptatives. 89- 114 p.

[40] Barbero M., Quezel P. et Loisel R., 1990. Les apports de la phytoécologie dans l'interprétation des changements et perturbations induits par l'homme sur les écosystèmes forestiers méditerranéens. Forêt Méditerranéenne. XII. 194-215 p.

Evaluation of the anti-*Candida albicans* and anti-*Cryptococcus neoformans* activity of the essential oils of three Moroccan medicinal plants

Y LAHLOU [1]*, B El AMRAOUI [1,2,3], M El Wahidi[1], S Moujabbir[3], A Aboukhalaf[3], T BAMHAOUD [1]

[1]Department of Biology, Control Quality in Bio-control Industry & Bioactive Molecules Laboratory, Faculty of Sciences, ChouaibDoukkali University, El Jadida 24000, Morocco.
[2] Department of Biology, Biotechnology, Materials and Environment Laboratory, Faculty Polydisciplinary of Taroudant B.P 271, Ibn Zohr University, Agadir B.P 8106. Morocco.
[3] Department of Biology, Laboratory of Biotechnology, Biochemistry and Nutrition. Training and Research. Unit on Nutrition and Food Sciences. Faculty of Sciences, ChouaibDoukkali University, El Jadida 24000, Morocco.

*Corresponding author: lahlouyossef@gmail.com

Abstract.

The results reported in this study provide new information about the antifungal activity of these medicinal plants and suggest their uses in the high incidence of fungal infections caused by *Candida albicans* and *Cryptococcus neoformans* is due to several factors as resistance to antifungals, a fungistatic than the fungicidal effect of these drugs and the side effects of synthetic antifungals. Then, the search for new sources of molecules effective against pathogenic fungi is a major objective. In this context, the screening of antifungal activity of essential oils of three medicinal plants from Morocco (*Lavandula angustifolia*, *Salvia officinalis*, and *Rosmarinus officinalis*) has been carried out against *C. albicans* and *C. neoformans*, using Disk-diffusion assay and micro-dilution method. The essential oils are obtained by Hydrodistillation, the characterization of the most active essential oil was carried out by Fourier transform infrared spectroscopy (FTIR). The results show that *L. angustifolia* essential oil is the only oil that showed strong fungicidal activity against *C. neoformans* and against *C. albicans*. The FTIR analysis of this oil showed the existence of several antifungal compounds, mainly linalool, linalyl acetate, terpinene-4-ol, carvacrol, and citral. *S. officinalis* essential oil has a fungicidal effect against *C. neoformans*, but a fungistatic effect against *C. albicans*. This activity is due to the presence of camphor, 1,8-cineole, and limonene. *R. officinalis* essential oil showed a fungistatic effect depending on its main chemical components, including eucalyptol, limonene, cymene, α-pinene, and camphor. The results reported in this study provide new information about the antifungal activity of these medicinal plants and suggests their uses in the extraction of active principle for the synthesis of new antifungal drugs.

Keywords: medicinal plants, antifungal activity, essential oils, FTIR analysis

1. Introduction

Candida albicans is one of the most important species of Candida pathogenic to humans. It can cause a variety of superficial and deep yeast infections. The opportunistic pathogen, it is likely to attack immunocompromised people or causes digestive mycosis as it can also attack critically ill non-neutropenic patients [1]. This commensal yeast from the genital and gastrointestinal tract is responsible for 85 to 90% of vulvovaginal candidiasis and thrush [2-4]. For immunocompetent patients, it causes thrush, redness, and itching of the skin or anus. Moreover, it causes urethritis for men and skin infections of the feet for athletes. When Candida albicans enters the bloodstream, it results in candidemia, which is characterized by a mortality rate of 40% [5].

Cryptococcus neoformans is capsulated yeast capable of causing fatal neurological infections for humans, especially for people suffering from a deficiency in cell-mediated immunity, such as cryptococcosis for patients with HIV which is represented mainly by cryptococcal meningitis, malignant hematologic tumors, cryptococcosis, meningoencephalitis, Pneumonia, and hematologic malignancies [6-9]. This yeast is also known for its resistance to several types of antifungals (multi-resistance), such as flucytosine, amphotericin B, and fluconazole [10,11]. Bird droppings are the main environmental source for *C. neoformans* in urban areas [12].

The use of synthetic antifungals in the treatment of infections caused by these two yeasts has generated drug pressure, which has resulted in the appearance of some resistant species or selected strains that are naturally resistant or not very sensitive [5,13]. In addition, the use of available antifungals such as amphotericin B, flucytosine, nystatin, and triazoles presents frequent and severe side effects, in particular, nephrotoxicity and those relating to the infusion (fever, chills) [14].

Thus, the development of new antifungals with a broad spectrum of action, more potent activity, and a few adverse effects has become a major concern. Medicinal plants that are used for a long time in traditional medicine for the treatment of several diseases, can present a reliable alternative to synthetic antifungals. A wide variety of extracts and essential oils can be used to discover new chemical classes of naturally occurring antifungal drugs. In this context, we carried out this study to evaluate the antifungal activity of essential oils of three medicinal plants known in Morocco (*Rosmarinus officinalis*, *Salvia officinalis*, *Lavandula angustifolia*) against two yeasts: *Candida albicans* and *Cryptococcus neoformans*) against two yeasts: *Candida albicans* and *Cryptococcus neoformans*.

2. Material and methods

2.1. Plant material

The studied plants were collected from El Jadida city (Morocco at Latitude of 33° 14' 0.0024" N and a Longitude of 8° 30' 0.0000" W). All plant harvests were carried out in January 2019. Botanical data, the used part, and the harvest site of each plant are shown in table1.

Each plant was freed of all foreign elements, placed in a bag, and transported to the laboratory within 24 hours of harvest. Plant samples were dried away from light and moisture at room temperature and stored carefully in a dry place for analysis. Each sample was then placed in an electric mill of the Moulinex type. The powder of each plant has been carefully stored in a dry place to maintain the color, taste, and mainly the therapeutic effects.

2.2. Preparation of essential oils

The essential oils were extracted by hydrodistillation of the Clevenger type for 4 h. The distilled essential oils were collected, filtered, and stored at 4 ° C. The extraction yield of each essential oil is calculated as the ratio between the mass of essential oil obtained after extraction (M ') and the mass of the plant material used in dry weight (M) [15]. It's expressed as a percentage and calculated according to the following formula:

$$R\,(\%) = \frac{M'}{M} x100$$

R: essential oil yield;
M ': mass of essential oil in grams of dry weight (gdw);
M: mass of plant material used in (gdw).

Table 1. Botanical data and some medicinal uses of the studied medicinal plant

Scientific name	Family	Common name	Some medicinal uses	Harvest site	Used part
Rosmarinus officinalis	Lamiaceae	Yazir, Azir	antioxidant, antibacterial, anticancer, anti-inflammatory, antidepressant, diuretic, anti-ulcer, emmenagogue, antispasmodic	El Jadida city	Leaves
Salvia officinalis	Lamiaceae	Salmiya	anticancer, cytoprotective, antidiabetic, melanogenic, antibacterial, antioxidant, virucidal, analgesic, anti-inflammatory activity	El Jadida city	Leaves
Lavandula angustifolia	Lamiaceae	Khouzama	neuroprotective, hepatoprotective, insecticide, diuretic, anti-inflammatory, spasmolytic, antiparasitic, carminative	El Jadida city	Seeds

2.3. *Antifungal activity*

2.3.1. *Fungal strains*

The antifungal activity of extracts and essential oils was evaluated against reference strains (Table 2), obtained from the Collection of Institute Pasteur (CIP), which were used as a targets strains. The method of in vitro evaluation of antifungal activity, used in this study, is the disk diffusion method in a Sabouraud medium.

Table 2. Origin and references of the fungal strains used

Fungal strains	Reference	Origin
Candida albicans	CIP 48.72	Collection of Institut Pasteur (CIP)
Cryptococcus neoformans	CIP960	

2.3.2. *Disk-diffusion method*

The yeast inoculum was prepared from the yeast colonies in the exponential growth phase, from a 24-hour culture, at 28 ° C, on Sabouraud's medium. The cell density of this fungal suspension was evaluated by counting using a Malassez cell. It was then adjusted by dilution in sterile distilled water to obtain a final concentration of 10^4 CFU / ml. Cellulose discs 6 mm in diameter, already impregnated with the essential oil in their raw state at a concentration of 20 µl / disc. The discs of the reference antifungal (Econazole and Fluconazole) at 30 µg / disc were placed on the surface of the agar and they are used as a positive control. Each test is repeated three times. The Petri dishes are maintained at 4 ° C for 2 hours and then incubated at a temperature of 28 ° C for 24 to 48 hours to allow the growth of the yeasts and their interaction with the essential oil or the dichloromethane/ ethanol extract studied. The evaluation of the antifungal activity is validated by measuring the diameters of inhibition. Each test is performed in triplicate.

2.3.3. *Determination of the minimum inhibitory concentration (MIC) and the minimum fungicidal concentration (CMF)*

The minimum inhibitory concentration (MIC) of each essential oil was carried out by the micro-dilution method used resazurin [16]. 100 µl of Sabouraud's liquid culture medium are poured into each well, then 100 µl of the essential oil diluted in Tween 80 are taken to obtain a concentration of 33% (v / v). Dilution series, ranging from 33 to 0.065% (v / v) (Table 3), are prepared in Bouillon de Sabouraud. Each well of the plate is then inoculated with 40 µl of the fungal strain (10^4 CFU / ml). The wells serving as negative control do not contain essential oil, and those serving as positive control include essential oil without yeast. The plates are then incubated at 28 ºC for 24 hours in the dark to allow the growth of the yeasts and their interaction with the essential oil or the extract.

Table 3. Distribution of dilutions of essential oils in the microplate wells

Wells	HERM (v/v)	HESV (v/v)	HELV (v/v)
1	1/3= 33.33%	1/3= 33.33%	1/3= 33.33%
2	1/6= 16.67%	1/6= 16.67%	1/6= 16.67%
3	1/12=8.33%	1/12=8.33%	1/12=8.33%
4	1/24=4.17%	1/24=4.17%	1/24=4.17%
5	1/48=2.08%	1/48=2.08%	1/48=2.08%
6	1/96=1.04%	1/96=1.04%	1/96=1.04%
7	1/192=0.52%	1/192=0.52%	1/192=0.52%
8	1/384=0.26%	1/384=0.26%	1/384=0.26%
9	1/768=0.13%	1/768=0.13%	1/768=0.13%
10	1/1536=0.07%	1/1536=0.07%	1/1536=0.07%

For the determination of the MIC 1.5 µl of resazurin 0.01% (w / v) is added to each well, the MIC corresponds to the most minor concentration of the essential oil which does not produce a change in color of the resazurin from blue to fluorescent pink, and that signifies the absence of fungal growth. MFC is the minimum concentration of essential oil capable of killing inoculum. It is determined by inoculating 10 µl aliquots taken from the wells in which the resazurin coloration is not changed, After incubation of the dishes at 28 ° C for 18 to 24 hours, the square without fungal growth corresponds to the value by CMF. In addition, the CMF / MIC ratio is calculated to assess its antifungal power.

2.4. FTIR analysis of essential oils

To study the chemical composition of the essential oils tested, each essential oil was analyzed by an IRTF infrared spectrometer of the JASCO 4000 type, equipped with a detector (TGS), the wavelength range used is between 400 - 4000 cm^{-1} and the resolution is the 2 cm^{-1}.

3. Results and discussion

3.1. Extraction yield

The evaluation of the extraction parameters showed that the extraction yields of the essential oils studied, the highest, are those of *R.officinalis* 2.5% and *S.officinalis* (SEO) 2% (Table 4).

These values are higher than those obtained by other previous studies, which found the extraction yields of 1.35% and 1.03 % for the essential oil of *R. officinalis* and *S. officinalis* respectively [17,18]. On the other hand, the extraction yield of the essential oil of *L. angustifolia* is lower, since it does not exceed 0.74% compared to other studies that have found a higher extraction yield of 3.3% [19]. These differences can be explained by several factors such as edaphic, climatic factors, harvest period, intraspecific variations, and extraction operating conditions (hydrodistillation time, drying time, water / Plant matter ratio, drying temperature, etc.) [20].

Table 4. Extraction yield (%) of the studied essential oils

The essential oils	Extraction yield (%)
R. officinalis (HERM)	2.5
S. officinalis (HESV)	2
L. angustifolia (HELV)	0.74

3.2. In vitro antifungal activity

3.2.1. Disk-diffusion method

The screening of the antifungal activity of the essential oils studied is presented in Figure 1, it indicates that all the essential oils expressed a very high antifungal activity with zones of inhibition between 16 mm and 21 mm, very close to the values recorded by Econazole 30µg. The strongest inhibition was exerted by HELV against *C. neoformans* (d = 21mm), while the weakest inhibition was exerted by HERM against *C.albicans* (d = 16mm). We also note that *C. neoformans* is the most sensitive yeast (inhibition zone ≥ 20 mm) on the other hand *C. albicans* is the most resistant yeast (inhibition zone ≤ 20 mm).

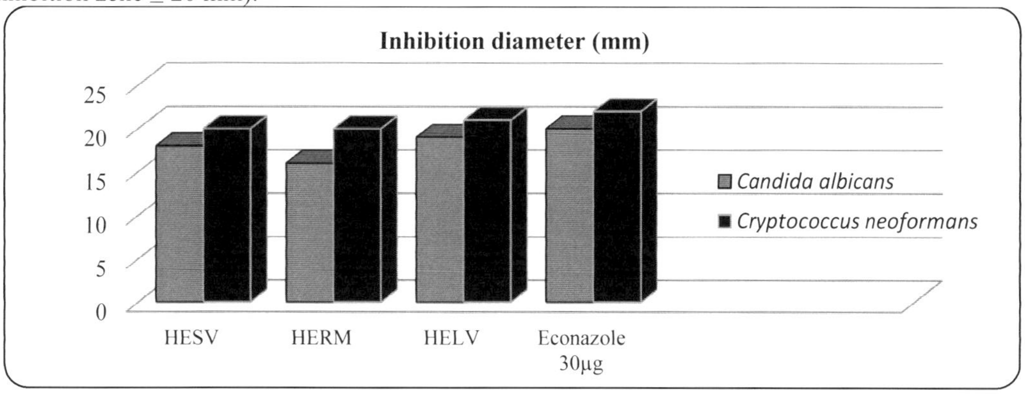

Figure 1. Inhibition diameters of the essential oils of *S. officinalis* (HESV), *R.officinalis* (HERV) and *L.angustifolia* (HELV)

3.2.2. Determination of the minimum inhibitory concentration (MIC)

The results of the tests using the method of micro-dilution in a liquid medium (Figure 2), showed that all the extracts are endowed with inhibitory activity against *C. albicans* and *C. neoformans*. The determination of MICs relating to active essential oils revealed variable levels of antifungal activity depending on the essential oil used.

Thus, the essential oil of *R. officinalis* is the most active; in fact, the MICs are of the order of 1/96 = 1.04% (v / v) against *C. neoformans* and *C. albicans*, followed by the essential oil of *L. angustifolia* whose MICs vary from 1 / 24 = 4.17% (v / v) against *C. albicans* at 1/12 = 8.33% (v / v) against *C. neoformans* and finally, the essential oil of *S. officinalis* which is more active against *C. albicans* (MIC = 1/48 = 2.08% v / v), but weakly active against *C. neoformans* (CMI = 1/3 = 33.33% v / v).

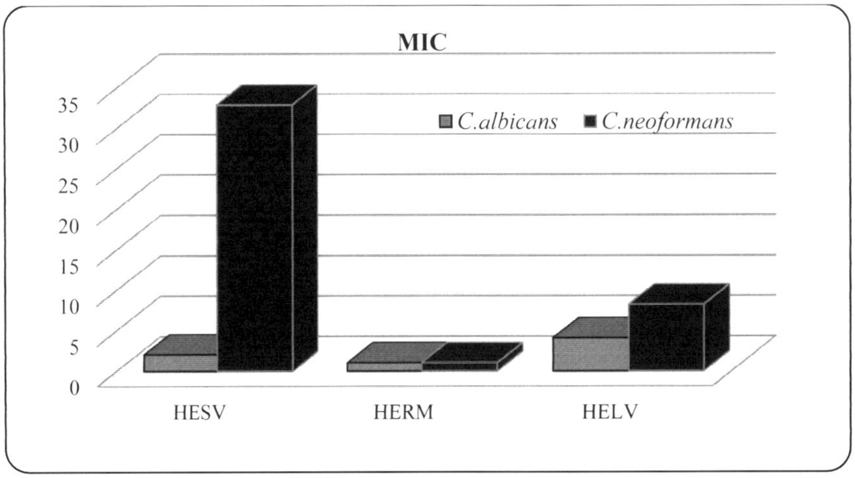

Figure 2. The MIC values of the essential oils of *S. officinalis* (HESV), *R.officinalis* (HERV) and *L.angustifolia* (HELV)

3.2.3. Determination of the minimum fungicidal concentration (MFC)

The minimum fungicidal concentration (MFC) is the parameter that determines the fungicidal effect of the essential oil studied. In this study, the essential oil of *R. officinalis* exhibits a higher antifungal activity against *C. neoformans* (CMF = 4.17% v/v) and an average against *C. albicans* (CMF = 8.33% v/v), followed by the essential oil of *L. angustifolia* which shows an average antifungal activity against *C. albicans* (CMF = 8.33% v/v) and less against *C. neoformans* (CMF = 16.67% v/v) then, the essential oil of *R. officinalis* which gives an average antifungal activity against *C. albicans* (CMF = 8.33% v/v) and a weak activity against *C. neoformans* (CMF = 33.33% v/v) (Figure 3).

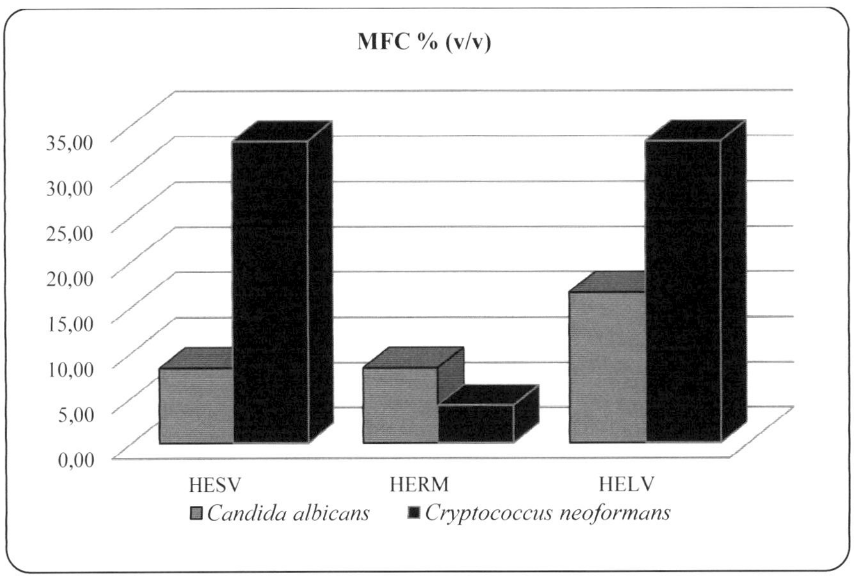

Figure 3. CMF values of the essential oils of *S. officinalis* (HESV), *R.officinalis* (HERM) and *L.angustifolia* (HELV)

3.2.4. The MFC / MIC ratio

The MFC / MIC ratio makes it possible to define the fungicidal or fungistatic nature of essential oil. When this ratio is less than 4 [21], the essential oil is considered a fungicidal. In this study, the CMF/MIC ratios of the essential oils of *S. officinalis*, *R. officinalis*, and *L. angustifolia* between 1 and 8 for all the fungal strains studied (Table 5). Only the essential oil of *L. angustifolia* seems to exert a fungicidal action against the two fungal strains tested (*C. albicans* and *C. neoformans*) with an MFC / MIC = 2, the essential oil of *S. officinalis* has a fungicidal effect against *C. neoformans* (MFC / MIC = 1) and a fungistatic effect against *C. albicans* (MFC / MIC= 4) finally, the essential oil of *R. officinalis* exerts a fungistatic power against *C. neoformans* (MFC / CMI = 4.01) and against *C. albicans* (MFC / CMI = 8.01).

So we can conclude from this analysis that the essential oil of *L. angustifolia* is the most effective, followed by that of *S. officinalis*, and finally, the essential oil of *R. officinalis* which has only a fungistatic effect.

Table 5. The MFC / MIC ratio and the fungicidal or fungistatic character of the studied essential oils

The essential oils (% v/v)	*Candida albicans*		*Cryptococcus neoformans*	
	MFC/MIC	Character	*MFC/MIC*	Character
HESV	4.00	fungistatic	1.00	fungicidal
HERM	8.01	fungistatic	4.01	fungistatic
HELV	2.00	fungicidal	2.00	fungicidal

3.3. FTIR analysis of essential oils

In this analysis, we are interested in the essential oil which has fungicidal activity against the two strains of yeasts tested, which is *L. angustifolia* essential oil (HELV).

The infrared spectrum is in the form of a series of absorption bands, thin and more or less wide, the X-axis represents the number of waves in cm^{-1}, , and the axis is oriented from the right to the left (from 400 cm^{-1} to 4000 cm^{-1}), the Y-axis is oriented upwards and represents the transmittance in%. In practice, no attempt is made to identify all the bands given their large number. The verification only concerns the presence or absence of a band characteristic of a given functional group. Based on the interpretation of IR spectra, we can determine the functional groups contained in a molecule: alcohol, aldehyde, ketone, and acid ..., as well as the bonds between the carbon atoms of a chain (saturated, unsaturated, aromatic character of a molecule) by indicating the main vibration bands based on the tables provided.

The infrared spectrum of *L. angustifolia* essential oil (Figure 4) indicates:

- An absorbance band at 1080 cm^{-1} indicates the possible existence of a C-O group of ethers in the form of eucalyptol [22].
- Two bands at 1053 cm^{-1} and 1237 cm^{-1}, and another band at 1793 cm^{-1}, indicate the presence of CO and C = O bonds of esters (Table 6), including linalyl acetate, lavandulyl acetate, neryl acetate, geranyl acetate, bornyl acetate, hexyl hexanoate, nerol acetate, hexyl crotonate, lavandulol acetate, and butanoate 2-methyl hexyl [22-24].
- An absorption band is also observed at 1324 cm^{-1}, associated with the C-N bond of aromatic amines such as linalyl anthranilate [25].
- A weak absorption band at 1644 cm^{-1}, indicates a C = C bond of alkenes, which may be related to the presence of terpenes in the essential oil of *L. angustifolia*, such as α-tuyene, pinene, camphene, α-phellandrene , δ-3-carene, limonene, cis-sabinene hydrate, β-caryophyllene, E-β-farnesene, γ-cadinene, terpinolene, and germacrene [23].
- An absorption band at 1674 cm^{-1}, may indicate the presence of C = O in ketones, such as octan-3-one, piperitone, pinocarvone, cryptone, and camphor [22,23].

IOP Conf. Series: Earth and Environmental Science 1090 (2022) 012026 doi:10.1088/1755-1315/1090/1/012026

- The essential oil of *L. angustifolia* can also contain aldehydes including α-camphonelal, cuminaldeyde, myrenal, z-citral, and geranium [23,24], with an absorption band at 2879 cm^{-1} associated with the CH of hydrocarbons aromatics also exist as cymene [22], as there is an intense band at 1451 cm^{-1} bonded to the C = C bonds of aromatic hydrocarbons.
- One band at 1168 cm^{-1}, and another band at 3447 cm^{-1}, both can be associated with the -CO and OH group of alcohols and phenols. This observation is explained by several previous works which confirm the existence of alcohols such as linalool, lavandulol, nerol, geraniol, terpinene-4-ol, α-terpineol, borneol, trans-pinocarveol, dihydrocarveol, α-bisabolol, α-cadinol, carveol [22-24], and phenols in particular methyl carvacrol and carvacrol [26] in the essential oil of *L. angustifolia* (Table 6).

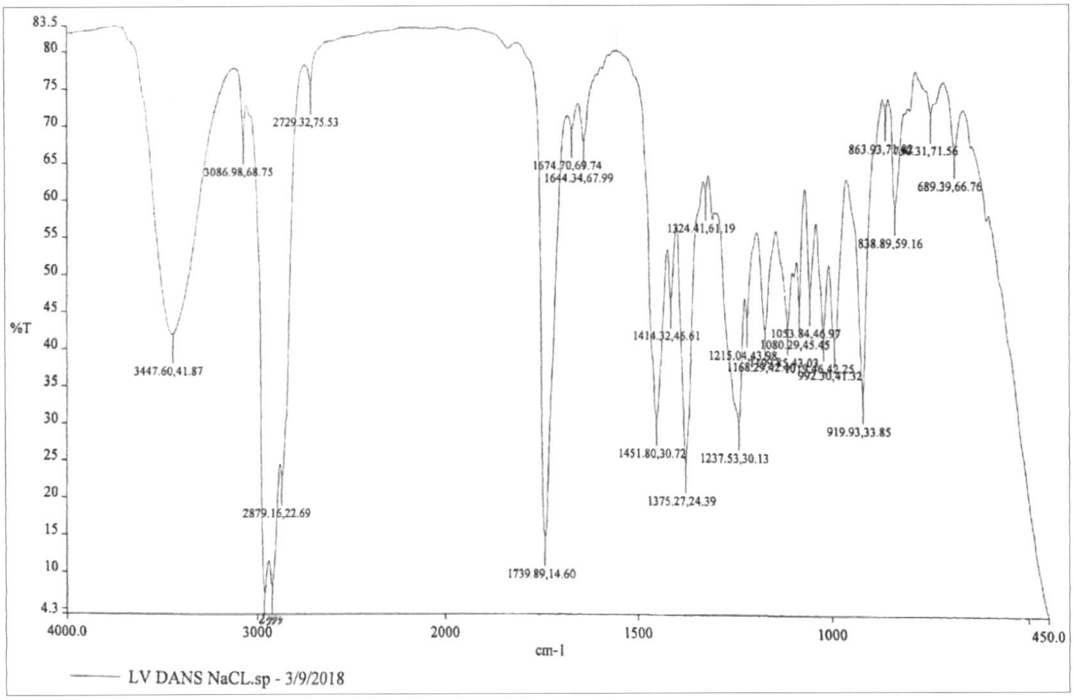

Figure 4. Infrared spectrum of *L. angustifolia* essential oil

Table 6. Infrared spectrum analysis of *L. angustifolia* essential oil

σ (cm^{-1})	Standards	Intensity	Chemical Bond	Functional Group	Interpretation
1080	1150-1020	Medium to strong	C-O	Ethers	Eucalyptol [22]
1237 et 1215	2 bandes 1300-1050	Strong	C-O	Esters	Linalyl acetate, lavandulyl acetate, neryl acetate, geranyl acetate, bornyl acetate, hexyl hexanoate, nerol acetate, hexyl crotonate, lavandulol acetate, hexyl 2-methylbutanoate[22-24]
1739	1750-1730	Strong	C=O		
1324	1360-1180	Medium to strong	C-N	Aromatic amines	Linalyl anthranilate [25]

1644	1675-1645	Medium t	C=C	Alkenes (Terpenes)	α-tuyene, α and β-pinene, camphene, α-phellandrene, δ-3-carene, sabinene, limonene, cis-sabinene hydrate, β-caryophyllene, E- β-farnesene, γ-cadinene, terpinolene, germacréne[23]
germacrène1674	1700-1670	Medium	C=O	Ketones	Octan-3-one, piperitone, pinocarvone, cryptone, carvone, camphor [22,23]
2879	2750-2850	Medium to strong	C-H aldéhyde	Aldehydes	α-camphonelal, cuminaldehyde, myrenal, Z-citral, geranial[23,24]
1451	1450-1650	Medium to strong	C=C	Aromatic hydrocarbons	p-Cymene, o-Cymene [22]
1168	1200-1050	Medium	C-O	Alcohols	Linalool, lavandulol, nerol, geraniol, Terpinene-4-ol, α-terpineol, borneol, trans-pinocarveol, dihydrocarveol, α-bisabolol, α-cadinol, cumic alcohol, Carveol[22-24]
3447	3300-3600	Medium	O-H	Alcohols and phenols	Methyl Carvacrol, Carvacrol [26]
1168	1000-1300	Medium	C-O		

3.4. Discussion

Facing the dramatic increase in the rates of infections caused by *C. albicans* and *C. neoformans*, which is due to several factors such as resistance to common antifungals. This is because antifungal drugs are cytotoxic and the sensitivity of certain species of pathogenic yeasts to these drugs is reduced. In addition, most of these drugs have a fungistatic than fungicidal effect and the patient has to consume the drug for a long time to have the desired effect on his overall health, which leads the drug to express its side effects more [27,28]. As a result, the adoption of a new therapeutic strategy against fungal infections based on green chemistry has become a preferred method. It is with this in mind that the essential oils of *L. angustifolia*, *S. officinalis*, and *R. officinalis* were tested against *C. albicans* and *C. neoformans*. Thus the study of the antifungal activity of the essential oil of *L. angustifolia* shows a strong fungicidal activity against *C. neoformans* (d = 21mm, MIC = 8.33% v / v, CMF = 16.67% v / v) and against *C. albicans* (d = 19mm, MIC = 4.17% v/v, CMF = 8.33% v / v). These results are in agreement with several later ones. Indeed, several other studies confirm these results and concluded that the essential oil of *L. angustifolia* has a strong antifungal activity also against *Hansenula anomala*, *Saprochaete capitate*, and against dermatophytes, and molds [29]. Lavender oil exhibited fungistatic and fungicidal activity with MIC of 1.04% and MFC of 1.8% against oropharyngeal strains of *C. albicans* [30].

On the other hand, the FTIR analysis of the essential oil of *L. angostifolia* has shown the existence of several compounds, among others, we find linalool, linalyl acetate, terpinene-4-ol, and carvacrol (Figure 5). In accordance with other studies that attributed the antifungal activity of this essential oil to its richness of linalool, linalyl acetate, terpinene-4-ol, and other compounds such as butyric acid and propionic acid [29,31-33]. Carvacrol has shown effective antifungal activity against *C. neoformans* [34]. Another study concluded that Citral exerted a strong antifungal effect against *Zygosaccharomyces rouxii* [35]. The recent findings demonstrated significantly high antifungal activity of these compounds, especially against *C. albicans* and *C. neoformans* [36-40].

The mechanisms of action of these compounds remain a little ambiguous. Studies of flow cytometry suggest a mechanism of action leading to the rupture of the cytoplasmic membrane (interaction of the essential oil with membrane permeabilization) and the death of the cells, or by osmotic shock, other studies have assumed the induction of oxidative stress at the same time with lipid peroxidation at the level of the yeast cell membrane, which generates reactive oxygen species that damage the fungal cell [34,41,42].

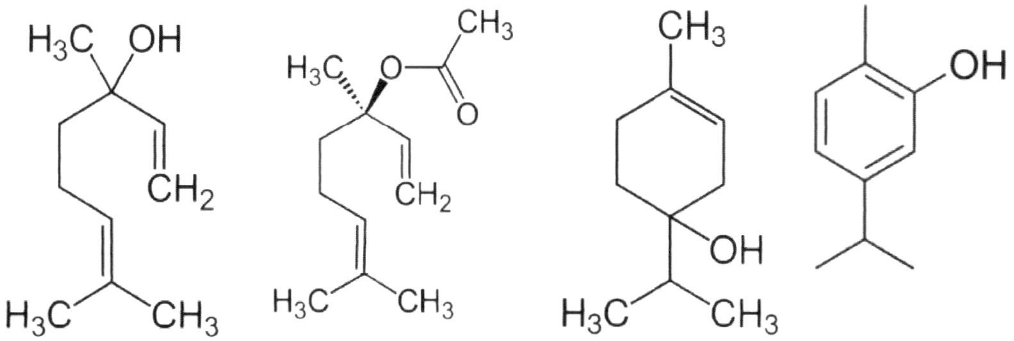

Linalool Linalyl acetate Terpinene-4-ol Carvacrol

Figure 5. The main compounds responsible for the antifungal activity of *L. angustifolia* essential oil

Regarding the antifungal effect of *S. officinalis* essential oil against *C. albicans* and *C. neoformans*, we note that this essential oils is endowed with a high antifungal activity, with a fungicidal effect against *C. neoformans* (d = 20mm, MIC = 33.33%, CMF = 33.33%) and a fungistatic effect against *C. albicans* (d = 18mm, MIC = 2.08%, CMF = 8.33%). Our results agree with those reported by other finding which demonstrated that the essential oil of sage is effective against *C. albicans* and *C. neoformans* with MIC values of 2.5 µl / ml and 1.25 µl / ml and a fungicidal concentration of the order of 5µl / ml and 2.5µl / ml respectively [43]. The essential oil of sage completely inhibited the fungal growth of *C. albicans* and *C. neoformans* with an inhibition diameter of (d = 19 mm and d = 18 mm respectively) and MIC values of 100 µg / ml for the two studied yeasts [44], this finding is very close to that obtained in our study. Other study [45] has proven that this essential oil is characterized by a good antifungal activity against *C. albicans* with MIC values of 200 µl / ml and CMF = 250 µl / ml higher than those obtained in our study. Other studies have attributed this antifungal activity to the presence of a high concentration of monoterpenes with antifungal potential, especially camphor [46,47]which gives the essential oil better fungicidal activity [48]. Other research has shown that sage essential oil with the following major compounds: cis-thujone, camphor, 1,8-cineole, camphene, α-pinene, expressed antifungal activity against *C. neoformans* (MIC = 625 µg / mL) and *C. albicans* (MIC = 625 µg / mL) [49][50]. The effectiveness of sage essential oil can also be explained by the presence of 1,8-cineole and limonene. Indeed, 1,8-cineole can modify the structure and the quantity of water of the cell membranes of yeasts, interfering with them, or by inhibiting the respiratory chain in mitochondria of pathogens cells [51], the limonene can influence the structure of the lipid bilayer of yeast [52].

Finally, regarding *R. officinalis* essential, The obtained results are close to those obtained in other work [53] when rosemary essential oil exhibited intermediate antifungal activity (MIC = 1.10 mg / mL) against *C. albicans*. Likewise, its antifungal activity has been demonstrated by previous results that found a fungistatic activity at concentrations above 8% [54,55], with lower zones of inhibition (d between 10 to 13 mm) than those obtained in our work (d = 16 mm) [54]. Another more recent study confirmed that this essential oil was active against *C. albicans* (MIC = 600µg / ml), but was inactive against *C. neoformans* (resistant strain) [67], although in our study, this essential oil showed activity against *C. neoformans* with an inhibition diameter of 20 mm, a MIC of 1.04% and a CMF of 4.17%

[56]. The antifungal activity of this essential oil confirmed in this study depends on its major chemical components, including eucalyptol, limonene, cymene, α-pinene, and camphor [56-58]. Phenols, cinnamic aldehydes, alcohols, ketones, ethers, and hydrocarbons are also responsible for antifungal activity [59]. The mechanisms of action of these compounds can be based on changes in cell morphology following the denaturation of cell structures which leads to alterations in the permeability of cell membranes, because the liposoluble nature of essential oils and their constituents facilitates their interaction with lipid cell structures, resulting in increased membrane permeability, which in turn can cause electrolyte imbalance and fungal cell death [60-64]. Other research has suggested that the decrease in the production of DNA, RNA, ergosterol, and polysaccharides is involved in the construction of the yeast cell wall following exposure of the yeasts to the active products of the essential oil of *R. officinalis* [65] .

4. Conclusion
The essential oils that were the subject of our study were extracted from three species of Moroccan medicinal plants, namely *L. angustifolia*, *S. officinalis,* and *R. officinalis*. These are the plants that are frequently used in traditional herbal medicine. The antifungal activities of these essential oils were finally tested on two yeasts, *C. albicans*, and *C. neoformans*. These yeasts are responsible for several kinds of infections in Morocco and worldwide. The three essential oils recorded very high zones of inhibition between 16 mm and 21 mm, they revealed the greater fungicidal and fungistatic activities. The essential oil of *L. angustifolia* is the only one that has a fungicidal effect against both yeasts both *C. albicans* and *C. neoformans* that is why the analysis was carried out by infrared spectroscopy, it showed the existence of certain compounds endowed with an antifungal power such as linalool, linalyl acetate, terpinene-4-ol, carvacrol, and citral. The secondary essential oil is *S. officinalis* essential oil which has shown a fungicidal effect against *C. neoformans*, but a fungistatic effect against *C. albicans*. This effect is due to the presence of camphor, 1,8-cineole, or limonene. Finally, the essential oil of *R. offocinalis* has shown a fungistatic effect against the two yeasts*, C. albicans* and *C. neoformans*, depending on its main active molecules, in particular eucalyptol, limonene, cymene, α-pinene and camphor. This study contributes to the knowledge of the antifungal potential of essential oils studied in vitro, it would also be interesting to carry out other studies to assess the antifungal potential in vivo.

5. References
[1] Eggimann P, Garbino J, Pittet D.2003.Management of candidiasis Management of *Candida* species infections in critically ill patients. *The Lancet Infectious Diseases.* 3: 772-785.
[2] Odds F C, ed. ed.*Candida* and candidosis: a review and bibliography. Bailliere Tindall; 1988.
[3] Anane S, Kaouech E, Zouari B, Belhadj S, Kallel K, Chaker E.2010.Les candidoses vulvovaginales : facteurs de risque et particularités cliniques et mycologiques. *Journal de mycologie médicale.* 20: 36-41. https://doi.org/10.1016/j.mycmed.2010.01.001.
[4] Amouri I, Abbes S, Sellami H, Makni F, Sellami A, Ayadi A.2010.La candidose vulvovaginale : revue. *Journal de mycologie médicale.* 20: 108-115. https://doi.org/10.1016/j.mycmed.2010.01.005.
[5] Kouidhi B, Ben Gaied M, Mhadhebi L, Bakhrouf A, Bouraoui A.2010.Les pompes à efflux en mycologie médicale: mécanismes moléculaires et perspectives thérapeutiques. *Journal de mycologie médicale.* 20: 304-314. https://doi.org/10.1016/j.mycmed.2010.07.007.
[6] Bamba S, Lortholary O, Sawadogo A, Millogo A, Guiguemdé R T, Bretagne S.2012.Decreasing incidence of cryptococcal meningitis in West Africa in the era of highly active antiretroviral therapy. *Aids.* 26: 1039-1041.
[7] Mazuelos E M, García A.2010.Microbiological aspects of the cryptococcosis in the post-HAART era. *Enfermedades infecciosas y microbiologia clinica.* 28: 40-45.
[8] Trpković A, Pekmezović M, Barać A, Crnčević Radović L, Arsić Arsenijević V.2012.In vitro antifungal activities of amphotericin B, 5-fluorocytosine, fluconazole and itraconazole against

Cryptococcus neoformans isolated from cerebrospinal fluid and blood from patients in Serbia. *Journal de mycologie médicale.* 22: 243-248. https://doi.org/10.1016/j.mycmed.2012.06.002.
[9] Vechi H T, Theodoro R C, de Oliveira A L, da Silva Gomes R M O, de Almeida Soares R D, Freire M G, Bay M B.2019.Invasive fungal infection by *Cryptococcus neoformans* var. grubii with bone marrow and meningeal involvement in a HIV-infected patient: a case report. *BMC infectious diseases.* 19: 220.
[10] Vanden H B, Dromer F, Improvisi I, Lozano-Chiu M, Rex J, Sanglard D.1998.Antifungal drug resistance in pathogenic fungi. *Medical mycology.* 36: 119-128.
[11] Revie N M, Iyer K R, Robbins N, Cowen L E.2018.Antifungal drug resistance: evolution, mechanisms and impact. *Current opinion in microbiology.* 45: 70-76.
[12] Nnadi N E, Enweani I B, Cogliati M, Ayanbimpe G M, Okolo M O, Kim E, Sabitu M Z, Criseo G, Romeo O, Scordino F.2016.Molecular characterization of environmental *Cryptococcus neoformans* VNII isolates in Jos, Plateau State, Nigeria. *Journal de mycologie médicale.* 26: 306-311. https://doi.org/10.1016/j.mycmed.2016.04.001.
[13] Rex J H, Rinaldi M, Pfaller M.1995.Resistance of *Candida* species to fluconazole. *Antimicrobial agents and chemotherapy.* 39: 1-8.
[14] Datry A, Bart-Delabesse E.2006.La caspofungine: du mécanisme d'action aux applications thérapeutiques. *La Revue de Médecine Interne.* 27: 32-39. https://doi.org/10.1016/j.revmed.2005.05.012.
[15] Bssaibis F, Gmira N, Meziane M.2009.Activité antibactérienne de *Dittrichia viscosa* (L.) W. Greuter. *Rev Microbiol Ind San et Environn.* 3: 44-45.
[16] Sarker S D, Nahar L, Kumarasamy Y.2007.Microtitre plate-based antibacterial assay incorporating resazurin as an indicator of cell growth, and its application in the in vitro antibacterial screening of phytochemicals. *Methods.* 42: 321-324.
[17] Elyemni M, Louaste B, Nechad I, Elkamli T, Bouia A, Taleb M, Chaouch M, Eloutassi N.2019.Extraction of Essential Oils of *Rosmarinus officinalis* L. by Two Different Methods: Hydrodistillation and Microwave Assisted Hydrodistillation. *The Scientific World Journal.* 2019: 1-6. 10.1155/2019/3659432.
[18] El Euch S K, Hassine D B, Cazaux S, Bouzouita N, Bouajila J.2019.*Salvia officinalis* essential oil: Chemical analysis and evaluation of anti-enzymatic and antioxidant bioactivities. *South African Journal of Botany.* 120: 253-260. https://doi.org/10.1016/j.sajb.2018.07.010.
[19] Gavahian M, Chu Y-H.2018.Ohmic accelerated steam distillation of essential oil from lavender in comparison with conventional steam distillation. *Innovative food science & emerging technologies.* 50: 34-41. https://doi.org/10.1016/j.ifset.2018.10.006.
[20] Fadil M, Farah A, Ihssane B, Haloui T, Rachiq S.2015.Optimisation des paramètres influençant l'hydrodistillation de *Rosmarinus officinalis* L. par la méthodologie de surface de réponse Optimization of parameters influencing the hydrodistillation of *Rosmarinus officinalis* L. by response surface methodology. *J Mater Environ Sci.* 6: 2346-2357.
[21] Eberlin T, Périlleux E, ed. ed.Les antibiotiques: classification, mode d'action, utilisation thérapeutique. Nathan; 1994.
[22] Gérard G. Abecedaire de chimie organique. In; 2019
[23] Cardia G F E, Silva-Filho S E, Silva E L, Uchida N S, Cavalcante H A O, Cassarotti L L, Salvadego V E C, Spironello R A, Bersani-Amado C A, Cuman R K N.2018.Effect of lavender (*Lavandula angustifolia*) essential oil on acute inflammatory response. *Evidence-Based Complementary and Alternative Medicine.* 2018:
[24] Kıvrak Ş.2018.Essential oil composition and antioxidant activities of eight cultivars of Lavender and Lavandin from western Anatolia. *Industrial Crops and Products.* 117: 88-96. https://doi.org/10.1016/j.indcrop.2018.02.089.
[25] Rashed M M A, Zhang C, Ghaleb A D S, Li J, Nagi A, Majeed H, Bakry A M, Haider J, Xu Z, Tong Q.2019.Techno-functional properties and sustainable application of nanoparticles-based *Lavandula angustifolia* essential oil fabricated using unsaturated lipid-carrier and

biodegradable wall material. *Industrial Crops and Products.* 136: 66-76. https://doi.org/10.1016/j.indcrop.2019.04.070.

[26] Valizadeh S, Fakheri T, Mahmoudi R, Katiraeec F, Ghajarbeygi P.2015.Phytochemical and antimicrobial properties of *Lavender angustifolia* and *Eucalyptus camaldulensis* essential oils. *Journal of Food Safety and Hygiene* 1: 46-52.

[27] Mansourian A, Boojarpour N, Ashnagar S, Momen Beitollahi J, Shamshiri A R.2014.The comparative study of antifungal activity of *Syzygium aromaticum*, *Punica granatum* and nystatin on *Candida albicans*; An in vitro study. *Journal de mycologie médicale.* 24: e163-e168. https://doi.org/10.1016/j.mycmed.2014.07.001.

[28] Hawser S, Islam K.1999.Comparisons of the effects of fungicidal and fungistatic antifungal agents on the morphogenetic transformation of *Candida albicans*. *Journal of Antimicrobial Chemotherapy.* 43: 411-413.

[29] Blažeković B, Yang W, Wang Y, Li C, Kindl M, Pepeljnjak S, Vladimir-Knežević S.2018.Chemical composition, antimicrobial and antioxidant activities of essential oils of *Lavandula* × *intermedia* 'Budrovka' and *L. angustifolia* cultivated in Croatia. *Industrial Crops and Products.* 123: 173-182. https://doi.org/10.1016/j.indcrop.2018.06.041.

[30] D'Auria F D, Tecca M, Strippoli V, Salvatore G, Battinelli L, Mazzanti G.2005.Antifungal activity of *Lavandula angustifolia* essential oil against *Candida albicans* yeast and mycelial form. *Medical mycology.* 43: 391-396. 10.1080/13693780400004810.

[31] Zambonelli A, d'Aulerio A Z, Bianchi A, Albasini A.1996.Effects of essential oils on phytopathogenic fungi in vitro. *Journal of Phytopathology.* 144: 491-494.

[32] de Rapper S, Viljoen A, van Vuuren S.2016.The In Vitro Antimicrobial Effects of *Lavandula angustifolia* Essential Oil in Combination with Conventional Antimicrobial Agents. *Evidence-Based Complementary and Alternative Medicine.* 2016: 2752739. http://dx.doi.org/10.1155/2016/2752739.

[33] Adam K, Sivropoulou A, Kokkini S, Lanaras T, Arsenakis M.1998.Antifungal activities of *Origanum vulgare* subsp. hirtum, *Mentha spicata*, *Lavandula angustifolia*, and *Salvia fruticosa* essential oils against human pathogenic fungi. *Journal of agricultural and food chemistry.* 46: 1739-1745.

[34] De Andrade Monteiro C, dos Santos J R A.2019.Phytochemicals and their antifungal potential against pathogenic yeasts. *Phytochemicals in human health.* 1-31.

[35] Cai R, Hu M, Zhang Y, Niu C, Yue T, Yuan Y, Wang Z.2019.Antifungal activity and mechanism of citral, limonene and eugenol against Zygosaccharomyces rouxii. *LWT.* 106: 50-56. https://doi.org/10.1016/j.lwt.2019.02.059.

[36] Garnier L, Penland M, Thierry A, Maillard M-B, Jardin J, Coton M, Leyva Salas M, Coton E, Valence F, Mounier J.2020.Antifungal activity of fermented dairy ingredients: Identification of antifungal compounds. *International Journal of Food Microbiology.* 322: 108574. https://doi.org/10.1016/j.ijfoodmicro.2020.108574.

[37] Ebani V V, Nardoni S, Bertelloni F, Najar B, Pistelli L, Mancianti F.2017.Antibacterial and Antifungal Activity of Essential Oils against Pathogens Responsible for Otitis Externa in Dogs and Cats. *Medicines.* 4: 21.

[38] Kim J-E, Lee J-E, Huh M-J, Lee S-C, Seo S-M, Kwon J H, Park I-K.2019.Fumigant antifungal activity via reactive oxygen species of Thymus vulgaris and Satureja hortensis essential oils and constituents against Raffaelea quercus-mongolicae and Rhizoctonia solani. *Biomolecules.* 9: 561.

[39] Francisconi R S, Huacho P M M, Tonon C C, Bordini E A F, Correia M F, Sardi J d C O, Spolidorio D M P.2020.Antibiofilm efficacy of tea tree oil and of its main component terpinen-4-ol against Candida albicans. *Brazilian oral research.* 34:

[40] Satoh M, Kusumoto N, Matsui N, Makino R, Hashida K, Arai D, Iiduka Y, Ashitani T.2021.Antitermitic and antifungal properties of enantiopure linalool and furanoid linalool

oxide confirmed in Lindera umbellata var. membranacea. *Journal of Wood Chemistry and Technology*. 1-9. 10.1080/02773813.2021.2004166.

[41] Khan A, Ahmad A, Akhtar F, Yousuf S, Xess I, Khan L A, Manzoor N.2011.Induction of oxidative stress as a possible mechanism of the antifungal action of three phenylpropanoids. *FEMS yeast research*. 11: 114-122.

[42] Zuzarte M, Gonçalves M J, Cavaleiro C, Canhoto J, Vale-Silva L, Silva M J, Pinto E, Salgueiro L.2011.Chemical composition and antifungal activity of the essential oils of *Lavandula viridis* L'Her. *Journal of medical microbiology*. 60: 612-618.

[43] Abu-Darwish M S, Cabral C, Ferreira I V, Gonçalves M J, Cavaleiro C, Cruz M T, Al-bdour T H, Salgueiro L.2013.Essential Oil of Common Sage (*Salvia officinalis* L.) from Jordan: Assessment of Safety in Mammalian Cells and Its Antifungal and Anti-Inflammatory Potential. *BioMed research international*. 2013: 538940. 10.1155/2013/538940.

[44] Ewais E, Aly M M, Ismail M, Abdel Shakour E, Hassanin M.2014.ANTIBACTERIAL, ANTIFUNGAL, ANTITUMOR AND TOXICITY OF ESSENTIAL OILS OF *SALVIA OFFICINALIS*, *THYMUS VULGARIS*, *EUGENIA CARYOPHYLLATA* AND *ARTEMISIA ABSINTHIUM*. *Scientific Journal of Flowers and Ornamental Plants*. 1: 265-274.

[45] Bozin B, Mimica-Dukic N, Samojlik I, Jovin E.2007.Antimicrobial and antioxidant properties of rosemary and sage (*Rosmarinus officinalis* L. and *Salvia officinalis* L., Lamiaceae) essential oils. *Journal of agricultural and food chemistry*. 55: 7879-7885.

[46] Pinto E, Salgueiro L R, Cavaleiro C, Palmeira A, Gonçalves M J.2007.In vitro susceptibility of some species of yeasts and filamentous fungi to essential oils of *Salvia officinalis*. *Industrial Crops and Products*. 26: 135-141.

[47] Bouaziz M, Yangui T, Sayadi S, Dhouib A.2009.Disinfectant properties of essential oils from *Salvia officinalis* L. cultivated in Tunisia. *Food and Chemical Toxicology*. 47: 2755-2760. https://doi.org/10.1016/j.fct.2009.08.005.

[48] Soković M, Glamočlija J, Ćirić A, Kataranovski D, Marin P D, Vukojević J, Brkić D.2012.Antifungal activity of the essential oils and components in vitro and in vivo on experimentally induced dermatomycoses at rats. *Digest Journal of Nanomaterials and Biostructures*. 7: 959-966.

[49] Powers C N, Osier J L, McFeeters R L, Brazell C B, Olsen E L, Moriarity D M, Satyal P, Setzer W N.2018.Antifungal and Cytotoxic Activities of Sixty Commercially-Available Essential Oils. *Molecules*. 23: 1549.

[50] Hassiotis C N.2018.The role of aromatic *Salvia officinalis* L. on the development of two mycorrhizal fungi. *Biochemical systematics and ecology*. 77: 61-67. https://doi.org/10.1016/j.bse.2018.01.004.

[51] Andriole V T.1999.Current and future antifungal therapy: new targets for antifungal agents. *Journal of Antimicrobial Chemotherapy*. 44: 151-162.

[52] Gupta A K, Katz H I, Shear N H.1999.Drug interactions with itraconazole, fluconazole, and terbinafine and their management. *Journal of the American Academy of Dermatology*. 41: 237-249.

[53] Dalleau S, Cateau E, Bergès T, Berjeaud J, Limbert C.2007.In vitro activity of essential oils and their major components against *Candida albicans* yeasts growing planktonically and as biofilms. *International Journal of Antimicrobial Agents*. 29: 147.

[54] Lima I d O, Oliveira R d A G, Lima E d O, Farias N M P, Souza E L d.2006.Antifungal activity from essential oils on *Candida* species. *Revista Brasileira de Farmacognosia*. 16: 197-201.

[55] Packer J F, Luz M.2007.Método para avaliação e pesquisa da atividade antimicrobiana de produtos de origem natural. *Revista Brasileira de Farmacognosia*. 17: 102-107.

[56] Souza E L d, Lima E d O, Freire K R d L, Sousa C P d.2005.Inhibitory action of some essential oils and phytochemicals on the growth of various moulds isolated from foods. *Brazilian Archives of Biology and Technology*. 48: 245-250.

[57] Mondello F, De Bernardis F, Girolamo A, Cassone A, Salvatore G.2006.In vivo activity of terpinen-4-ol, the main bioactive component of *Melaleuca alternifolia* Cheel (tea tree) oil against azole-susceptible and-resistant human pathogenic *Candida* species. *BMC infectious diseases.* 6: 1-8.
[58] Lorenzo-Leal A C, Palou E, López-Malo A.2019.Evaluation of the efficiency of allspice, thyme and rosemary essential oils on two foodborne pathogens in in-vitro and on alfalfa seeds, and their effect on sensory characteristics of the sprouts. *International Journal of Food Microbiology.* 295: 19-24.
[59] Saad N Y, Muller C D, Lobstein A.2013.Major bioactivities and mechanism of action of essential oils and their components. *Flavour and Fragrance Journal.* 28: 269-279.
[60] Castro R D d, Lima E d O.2010.Atividade antifúngica in vitro do óleo essencial de *Eucalyptus globulus* L. sobre *Candida* spp (In vitro antifungal activity of essential oil of Eucalyptus globulus L. on *Candida* spp). *Rev Odontol UNESP (Online).* 179-184.
[61] Hammer K, Carson C, Riley T.2004.Antifungal effects of *Melaleuca alternifolia* (tea tree) oil and its components on *Candida albicans*, *Candida glabrata* and *Saccharomyces cerevisiae*. *Journal of Antimicrobial Chemotherapy.* 53: 1081-1085.
[62] Nascimento P, Nascimento A, Rodrigues C, Antoniolli A, Santos P, Barbosa Junior A, Trindade R.2007.Antimicrobial activity of the essentials oils: a multifactor approach of the methods. *Rev Bras Farmacogn.* 17: 108-113.
[63] Ibrahim D, Osman H.1995.Antimicrobial activity of *Cassia alata* from Malaysia. *Journal of ethnopharmacology.* 45: 151-156.
[64] Cavalcanti Y W, Almeida L d F D d, Padilha W W N.2011.Anti-adherent activity of *Rosmarinus officinalis* essential oil on *Candida albicans*: an SEM analysis. *Revista Odonto Ciência.* 26: 139-144.
[65] Xia Z, Mao X, Luo Y.1999.Study on antifungal mechanism of alpha-pinene. *Hunan yi ke da xue xue bao= Hunan yike daxue xuebao= Bulletin of Hunan Medical University.* 24: 507-509.

6. Acknowledgments

We are grateful to Mr Abdelghani IBENELKIHAL for the linguistic assistance and to Dr. Karim ARIFI for the scientific collaboration.

IOP Conf. Series: Earth and Environmental Science 1090 (2022) 012027 doi:10.1088/1755-1315/1090/1/012027

Ethnobotanical study of medicinal plants in the Central Middle Atlas region (Morocco)

Maryama Hachi[1], Touria Hachi[2]*, Hamza Essabiri[3], Adnane El Yaakoubi[3], Lahcen Zidane[1] and El Hassan Abba[3]

[1] Laboratory of Botany, Biotechnology and Plant Protection, Department of Biology, Faculty of Science, BP. 133, University Ibn Tofail, Kenitra, Morocco
[2] Laboratory of Biotechnology, Environment and Quality, Department of Biology, Faculty of Science, PO Box 133, University Ibn Tofail, Kenitra, Morocco
[3] Higher School of Technology. Sultan Moulay Slimane University B.P. 170, 54000 Khénifra, Morocco.

hachitouria03@gmail.com

Abstract. This work focuses on the ethnobotanical study of medicinal plants in the Central Middle Atlas region. The collection of information using a questionnaire form was done in 2013 and 2014. The location of the different environments, ethnobotanical surveys and floristic surveys, was located by the stratified sampling method. The ethnobotanical study revealed that Modern medicine is mainly chosen by men, young people, single people and people with relatively high academic level (secondary and higher). Traditional medicine, on the other hand, is mainly chosen by women, older and younger people, married people and people who are illiterate or have a primary school education. The results of this study allowed us to draw up a catalog of 267 medicinal species, divided into 211 genera and 79 botanical families, of which the most usable by the local population are the Lamiaceae, Asteraceae, Apiaceae, Fabaceae and Rosaceae. Oregano (Origanum compactum Benth.) is the most used species. The majority of the medicinal species are used mainly against digestive diseases, with a percentage of 25%.

Key words: ethnobotany, medicinal plants, central Middle Atlas, Morocco.

1. Introduction

The use of plants for therapeutic purposes is reported in ancient Arabic, Chinese, Egyptian, Hindu, Greek and Roman literature [1]. In Africa, the therapeutic power of plants was known by our ancestors and relatives in an empirical way [2]. In Africa, more than 80% of the rural population uses them for their health care [3,4,5]. Furthermore, the World Health Organization (WHO) estimates that more than 4 billion people worldwide use traditional herbal medicines for their primary health care. Medicinal plants are also invaluable resources for the pharmaceutical industry. According to a WHO estimate, they represent about 70% of the basic materials of modern pharmaceutical products. About 25% of the drugs produced and marketed worldwide are derived from plants, and at least 35,000 plant species are used for medicinal purposes [6]. Thanks to its particular geographical situation and its very varied climate, the country knows a diversification of the flora: more than 4500 species and subspecies have been listed [7]. These species are divided into 981 genera and 155 families [8]. Moreover, Morocco is among the first five countries with the richest floristic resources in the Mediterranean Basin [9]. As for endemism, the total number of species endemic to Morocco or shared with the riparian territories: Iberian Peninsula, Algeria, Mauritania and the Canary Islands is 1551 species and subspecies, of which 920, or about two thirds, are specific to Morocco. Those shared with Algeria or with the Iberian Peninsula are relatively numerous. On the other hand, the number of species shared with Mauritania or the Canary Islands is very low [8]. Given this richness, Morocco has an ancestral knowledge for the medication by plants. Indeed, there is a real traditional therapeutic arsenal of recipes and practices within our country, stemming from traditions and beliefs, transmitted through generations and/or collected from herbalists, enriched by personal experiences or those of relatives. Morocco is a traditional producer of medicinal and aromatic plants (MAP), it is one of the main suppliers worldwide (of rosemary, verbena, rose, coriander, mint, pouliot...) and an exclusive supplier of several essential oils such as mugwort, chamomile [10]. Aromatic and medicinal plants (MAP) constitute a natural wealth that can play an important role in the socio-economic development of some economically disadvantaged regions. They could, thus, offer a viable source of income to the local population.

In this perspective, an ethnobotanical study of medicinal plants in the central Middle Atlas region was conducted in order to identify the plants used in traditional pharmacopoeia, and the exploitation of these plants in traditional medicine.

2. Material and methods

2.1. Study area

The central Middle Atlas (figure 1) constitutes the ethnobotanical and floristic study area of medicinal plants practiced by its population.

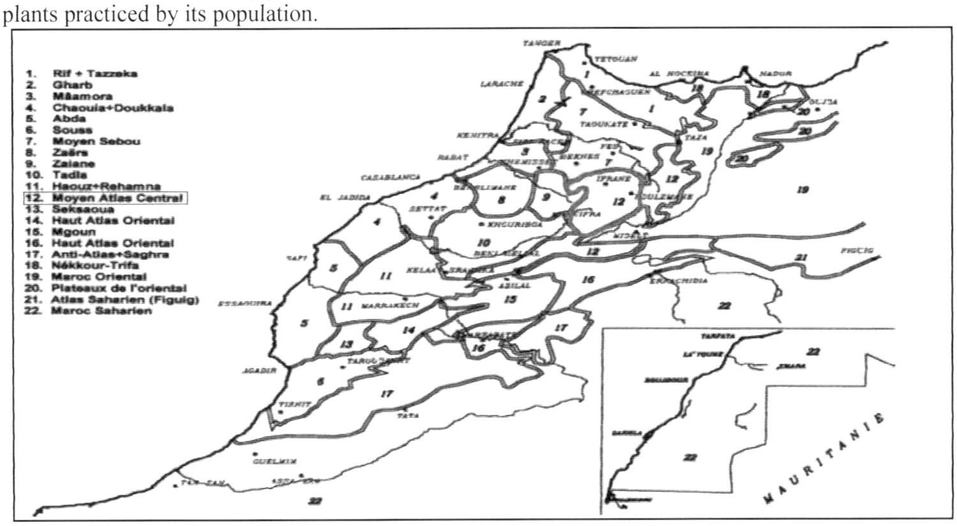

Figure 1. Map of the phytogeographical divisions of Northern Morocco [11].

The Middle Atlas, part of the domain of the Atlas ranges [12], it occupies an intermediate position between the Rif in the north, and the High Atlas in the south. Along an EW cut, the Middle Atlas is bordered by the high plateaus of the Moroccan-oran meseta in the east and the coastal meseta in the west. To the north, it is bounded by the Saïs plain (Neogene sediment) and the front of the Rifa nappe, to the northeast by the Guercif basin, to the south and southeast by the Moulouya depression [13].

Overall, the structure of the Middle Atlas corresponds to large synclinal basins with axes parallel to the chain and narrow anticlinal rifts intruded sometimes by gabbroic rocks [14]. It is formed by two structural sets separated by the North-Middle Atlas Accident [13]: the Tabular Middle Atlas or Middle Atlas Causse in the NW and the Folded Middle Atlas in the SE. The tabular Middle Atlas is formed in its major part by a Paleozoic basement covered by an important Mesozoic cover, the whole is strewn with scattered volcanic outpourings of Quaternary age [15,16]. The folded Middle Atlas corresponds to the eastern part of the chain and is distinguished from the sub-tabular causse by the presence of wrinkles that form reliefs oriented in the general direction of the chain. These anticlinal wrinkles frame synclinal depressions that continued to function during the Middle Jurassic as fairly thick marl and marl-limestone depocenters [14].

The climate that prevails in the Middle Atlas is of Mediterranean mountain type, it is characterized by a wet and cold climate. This particular climate of the Middle Atlas is mainly due to its altitudinal

IOP Conf. Series: Earth and Environmental Science 1090 (2022) 012027 doi:10.1088/1755-1315/1090/1/012027

position, its geographical location and its exposure to marine influences [17]. The Middle Atlas, water tower of Morocco par excellence, is the best watered mountain range and the richest in wetlands, including natural lakes, rivers and fresh springs that offer varied ecological habitats and promote a great biodiversity [18].

The Middle Atlas is globally very rich in plant species, due to the great diversity of ecosystems that this area contains (forests, pre-forests, matorrals, steppes, grasslands, wetlands, saline soils). The phytocenoses found there are organized by the following tree species: Cedrus atlantica (Atlas cedar), Quercus faginea (Zeus oak), Quercus suber (cork oak), Quercus rotundifolia (holm oak), Tetraclinis articulata (Berber cedar), Juniperus thurifera (thurifer juniper), Pinus halepensis (Aleppo pine), Pinus pinaster var. maghrebiana (Maghreb maritime pine) and thorny xerophytes.

Two major biogeographic zones are distinguished in the Middle Atlas [19]:

1) Zone occupied by vegetation adapted to the most humid climatic conditions. It extends to the west and NW of the plain of Guigou and on the limestone plateaus of the Middle Atlas and their northern foothills.

2) Zones of xeric plant formations of the folded Middle Atlas and its borders (Moulouya and Sebou).

2.2. *Methodology*

The ethnobotanical survey is conducted in 39 surveyed points that are determined through stratified sampling techniques [20,21], of which the surveyed territory is divided into 39 homogeneous strata derived from the superimposition of maps at the same scale (topographic, geological, climatic, and vegetation) on a transparent support (figure 2).

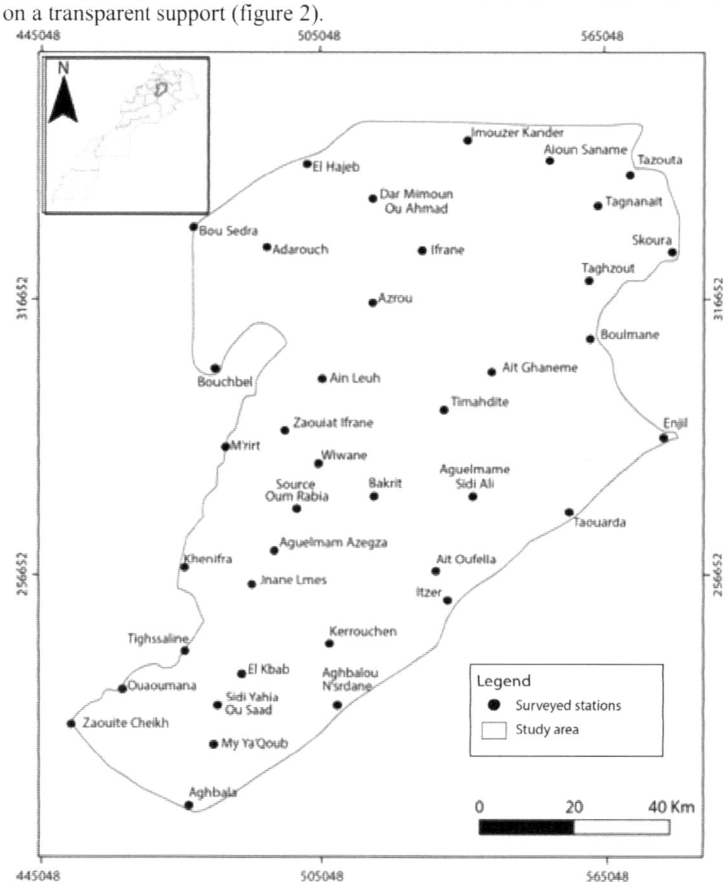

Figure 2. Map of the study area showing surveyed stations.

At the level of each stratum, a random sampling was adopted, the survey was carried out among 1560 inhabitants randomly chosen among the population of the studied region with a sample of 40 persons per stratum. In order to be able to identify the different data needed to develop a small synthesis of the pharmacopoeia of the Middle Atlas, we have developed a questionnaire form (appendix I) that we used during the surveys that are done in 2013 and 2014.

The taxonomic identification of the samples collected in the field was carried out later at the Laboratory of Biodiversity and Natural Resources of Ibn Tofaïl University, taking into account the recent modifications.

For the determination of these plants we used the following floras: the Catalogue of plants of Morocco [7,22]; the Flora of North Africa [23]; the New Flora of Algeria and the Southern Desert Regions [24]; Volume 1 [25], Volume 2 [26] and Volume 3 of the Practical Flora of Morocco.

2.3. Data processing and statistical analysis

All the data from these ethnobotanical surveys were recorded on questionnaire forms, then transcribed into a database and processed by SPSS version 21 (System Professional for Social Sciences) software.

- Three methods of statistical analysis were used:
- A descriptive analysis: for categorical variables, we used frequencies and percentages, while for quantitative variables, we used means and standard deviations.
- Bivariate analysis: During this analysis, we used the statistical test: Pearson's chi-square test for the comparison of frequencies and percentages.
- A multivariate analysis of the factors associated with the use of traditional medicine will allow us to draw conclusions and eliminate possible confounding variables.

The significance level was set at 0.05 (5%).

Difference between two means is:

Insignificant: $P < 0.05$ (*);

Significant: $P < 0.01$ (**);

Highly significant: $P < 0.001$ (***) ;

Highly significant: $P < 0.0001$ (****).

3. Results and discussion

3.1. The informant's profile and the therapeutic choice

The study conducted in the Middle Atlas region is prospective. Our study sample is composed of 1560 respondents chosen in a simple random way.

3.1.1. Informant profile

The distribution of our sample (table 1) showed that the average age of the respondents is 45.18± 0.42 years (with a minimum age of 18 years and a maximum age of 90 years). The sex ratio was balanced at 1.21 ($p<0.000$). The sample therefore meets the characteristics of the Gaussian distribution, it is a simple random distribution (skewness coefficient = 0.4; kurtosis coefficient = -0.71). However, 79% of the respondents are married versus 21% are single. According to the figure 3, 45% of the respondents are male (n=704), 36.50% of whom have never attended school and 28.98% of whom have a university degree, compared to 55% who are female (n=856), 62.15% of whom are illiterate and only 11.45% of whom have a higher education level.

Table 1. Distribution of the sample by gender.

Sex	Number N	X±Est	IC à 5%	min	max	CV%
Male	704	44,99±0,62	[42,76 ; 45,22]	18	86	37,84%
Female	856	46,16±0,58	[45,01 ; 47,31]	18	90	37,00%

X±Est: mean ± standard error;
CI at 5%: confidence interval at 5% error;
min: minimum age ;
max: maximum age;
CV: coefficient of variation

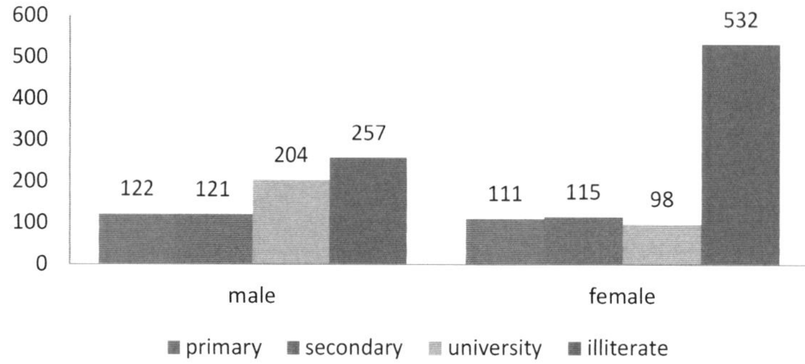

Figure 3. Distribution of sample by academic level.

The distribution of respondents according to their place of residence revealed that 69% (n=1083) were from rural areas, 21% (n=319) were from urban areas, and 10% lived a nomadic life.

In light of these results, the units constituting our sample were well aware of what we were asking of them. Filling out the survey forms was a logical response to the various questions proposed.

3.1.2. The choice between the two medicines

The table 2 and the figure 4 shows that 222 people, or 14% of the surveyed population, use traditional medicine alone; 236 people, or 15%, use modern medicine only; and 1102 people, or 71%, use both traditional and modern medicine. Thus, we can deduce that the majority of respondents are more likely to use traditional medicine than modern medicine as a therapeutic practice. These results are consistent with WHO (World Health Organization) statistics which indicate that 80% of the population in developing countries use traditional medicine as a source of health care (85%).

Table 2. Distribution of the sample according to the type of medicine chosen.

	Effectifs	Percentage
TM	222	14 %
MM	236	15 %
TM+MM	1102	71 %
Total	1560	100,0 %

TM: traditional medicine;
MM: modern medicine

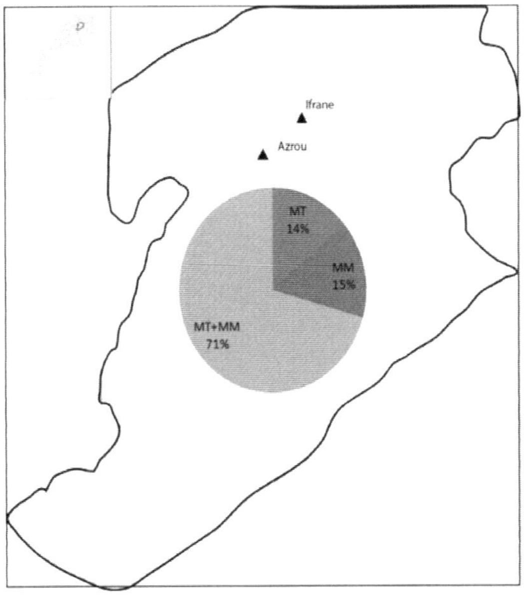

Figure 4. Pie chart distribution of the sample by type of medicine chosen.

3.2. Choice of therapeutic practice by gender, age, family status, academic level, locality.

The results obtained are shown in the following (table 3).

Table 3. Distribution of therapeutic practices in number and percentage by gender, age, family status, academic level, locality.

	Distribution	Workforce	Percentage	Choice between 2 M			Total	Khi-deux	Pvalue
				TM	MM	TM+MM			
Gender/sex	male	704	45,1 %	80	140	484	704	26	0,000*
	female	856	54,9 %	141	97	618	856		
age	[18;30[	321	20,6 %	2	108	211	321	280	0,000*
	[30;40[	318	20,4 %	26	78	214	318		
	[40;50[	330	21,2 %	41	46	243	330		
	[50;60[	285	18,3 %	70	4	211	285		
	>60	306	19,6 %	82	1	223	306		
family situation	single	325	20,8 %	0	160	165	325	394	0,000*
	married	1235	79,2 %	221	77	937	1235		
academic level	primary	233	14,9 %	6	6	221	233	699	0,000*
	secondary	236	15,1 %	3	61	172	236		
	university	302	19,4 %	0	166	136	302		
	illiterate	789	50,6 %	212	4	573	789		
locality	rural	1083	69,4 %	144	125	814	1083	239	0,000*
	urban	319	20,4 %	11	112	196	319		
	nomadic	158	10,1 %	66	0	92	158		

TM: traditional medicine;
MM: modern medicine;
*: very highly significant difference

According to the Chi-square test, there is a highly significant difference between the "Traditional medicine user" group, the "Modern medicine user" group and the "Traditional and modern medicine user" group for the "gender" variable, the "age" variable, the "family situation" variable, the "academic level" variable and the "locality" variable.

According to the bivariate analysis, the users of traditional medicine were highly significant female, married, over 30 years of age, illiterate or with a primary school education, and living a nomadic and rural lifestyle.

3.2.1. *Choice of therapeutic practice by gender*

According to the results obtained in the figure 5, it is observed that the male sex predominates the female sex in the use of modern medicine with a percentage of 19.88% (n=140) against 11.33% (n=97). In the case of both modern and traditional practices, the female sex predominates with a percentage of 72.19%. Thus for traditional medicine, the male sex is represented by a percentage of 11.36% against 16.47% for the female sex.

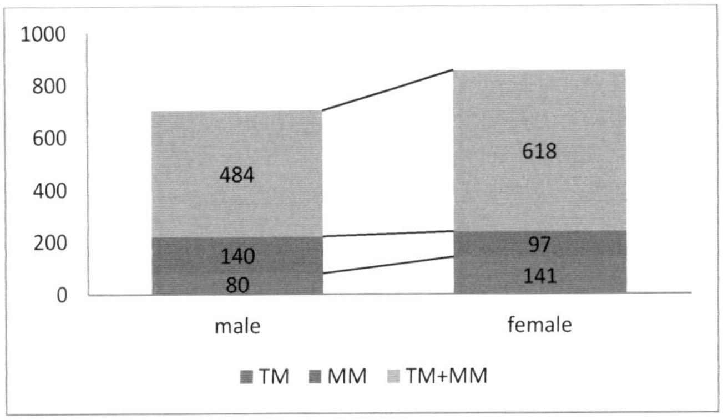

Figure 5: Treatment practice choices by gender.

3.2.2. *Choice of therapeutic practice by age*

According to the results obtained in the figure 6, the frequency of use of modern medicine alone decreases with age, with the most represented age group being between 18 and 30 years (108). At the same time, the number of users of traditional medicine increases with age. While all age groups combine the two medicines with equal importance.

This is because the elderly are mostly the heads of households and represent the family authority. The elderly are also expected to provide more reliable information, as they hold much of the ancestral knowledge that is transmitted orally and therefore will be lost with the departure of the elders.

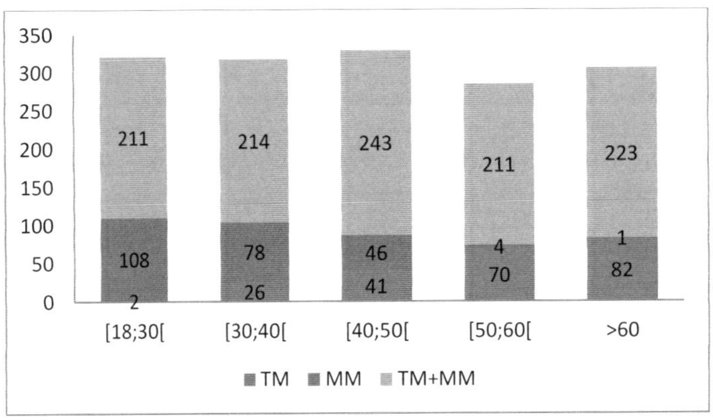

Figure 6: Treatment practice choices by age.

3.2.3. Choice of therapeutic practice by family situation
According to the figure 7, we observe that:
- the single category is the most remarkable in case of modern medicine with a percentage of 49.23% against 6.23% for the married category.
- 17.89% of the married category practice traditional medicine while the single category does not have access to this type of medicine.
- In the case of both therapeutic practices, married subjects represent 75.87%.

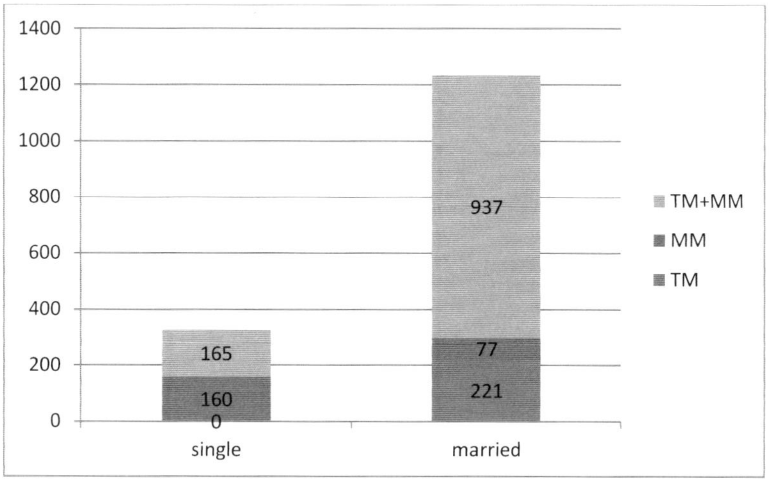

Figure 7. Treatment practice choices by family status.

3.2.4. Choice of therapeutic practice by academic level
The figure 8 shows that:
- 26.86% of illiterate subjects practice traditional medicine.
- Informants with a higher level of education prefer modern medicine with a percentage of 54.96%.
- The use of both therapeutic practices is equally remarkable among all users regardless of their academic level.

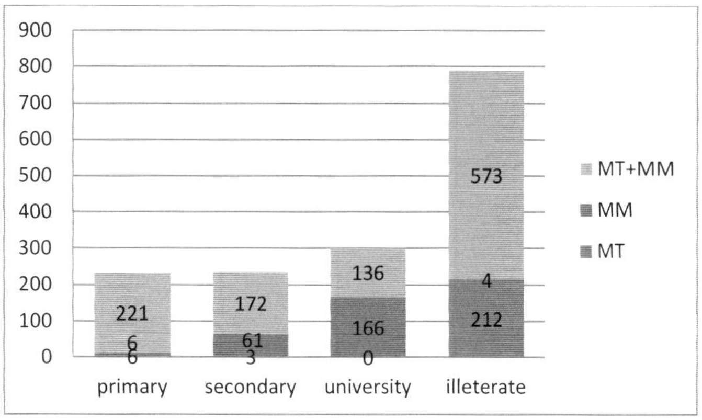

Figure 8. Therapeutic practice choices by academic level.

3.2.5. Treatment practice choices by place of residence. The analysis of the results obtained according to habitat, presented in the figure 9, shows that the number of users of traditional medicine is higher in nomadic areas (41.77%) than in rural areas (13.29%) because of geographical isolation and the absence or weakness of means of transport and infrastructure.

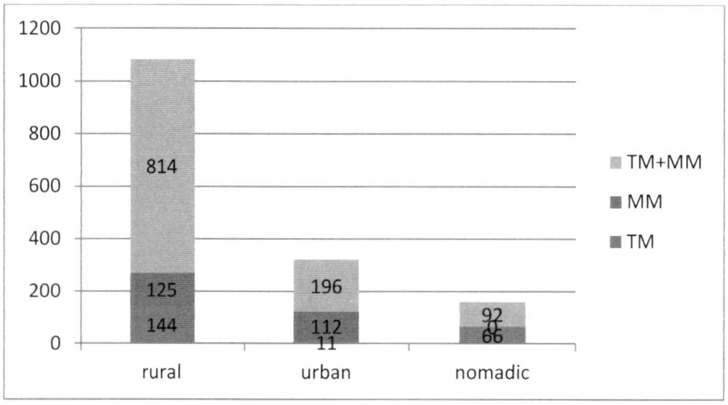

Figure 9. Choice of therapeutic practice by place of residence.

3.3. The reasons for the choice between traditional and modern medicine

The reasons of the surveys for their use of traditional or modern medicine are represented in the following table 4:

Table 4. Reasons for choosing between traditional and modern medicine

		Choice between 2 M			Total	Khi-deux	Pvalue
		TM	MM	TM+MM			
Effective TM	oui	219	0	835	1054		
	non	2	237	267	506	628	0,000*
Total		221	237	1102	1560		
MT cheaper	oui	217	0	1062	1279		
	non	4	237	40	281	1272	0,000*
Total		221	237	1102	1560		
MT acquisition	oui	218	0	1011	1229		
	non	3	237	91	331	1043	0,000*
Total		221	237	1102	1560		
ineffective medication	oui	97	0	57	154		
	non	124	237	1045	1406	341	0,000*
Total		221	237	1102	1560		
Effective MM	oui	0	92	19	111		
	non	221	145	1083	1449	426	0,000*
Total		221	237	1102	1560		
MM more precise	oui	0	69	2	71		
	non	221	168	1100	1489	388	0,000*
Total		221	237	1102	1560		
plant toxicity	oui	0	236	23	259		
	non	221	1	1079	1301	1390	0,000*
Total		221	237	1102	1560		

TM: traditional medicine;
MM: modern medicine;
*: very highly significant difference

The vast majority of respondents 82% (n=1279) who prefer traditional medicine think that it is less expensive than modern medicine ($p<0.000$). 78.78% (n=1229) of respondents who prefer traditional medicine think that the products are easy to acquire ($p<0.000$). 67.56% (n=1054) think that traditional medicine is very effective ($p<0.000$), and for 9.87% (n=154), the reason for the choice is the ineffectiveness of medicine ($p<0.000$).

The chi-square test of independence shows in a very highly significant way that there is a strong link between the choice of traditional medicine and its price, its acquisition, its effectiveness, and the non-effectiveness of the medicine.

The reasons given by the respondents for choosing modern medicine are:

16.60% of the respondents who prefer modern medicine stated that the toxicity of plants is the main cause with a very highly significant difference ($p<0.000$), for 7.11%, the treatment obtained is very safe and effective ($p<0.000$). For 4.55%, the treatment of modern medicine is very reassuring, safe and fast recovery as well as accuracy of treatment ($p<0.000$).

It appears from Table 4 that the use of traditional medicine seems to be much more related to its lower cost compared to modern medicine and the ease of its acquisition. On the other hand, the use of modern medicine is correlated with the rapid and safe healing and toxicity of medicinal plants.

3.4. *Global analysis of variables by Multiple Correspondence Analysis (MCA)*

ACM: this is a method of multi-variate analysis; it aims to reduce a large number of variables to a limited number in order to facilitate the reading of the results and to facilitate the detection of the different relationships existing between the variables.

Table 5. Validation of the questionnaire by Cronbach's test

Dimension	Cronbach's Alpha	Explained variance	
		Total (proper value)	Inertia
1	0,903	5,999	0,461
2	0,483	1,804	0,139
Total		7,802	0,600
Average moyenne	,806[a]	3,901	0,300

a. The average Cronbach's Alpha value based on the average eigenvalue.

The multiple correspondence analysis (MCA) of the different variables shows that the two dimensions alone absorb 60% of the total variation (table 5). The Cronbach's Alpha coefficient shows a high intra and inter variable reliability (0.81).

The analysis of the joint diagram of the points (figure 10) allows to classify all the modalities in two distinct groups.

The first group located on the positive side of the dimension I, defined by people coming from rural areas and also those who lead a nomadic life, who are generally at a low educational level and illiterate, most of whom are female, married, and aged 50 and over, this group uses traditional medicine either alone or in combination with modern medicine.

The second group, located on the negative side of dimension I, represented by unmarried urban males with a higher level of education who are under 30 years of age and prefer modern medicine.

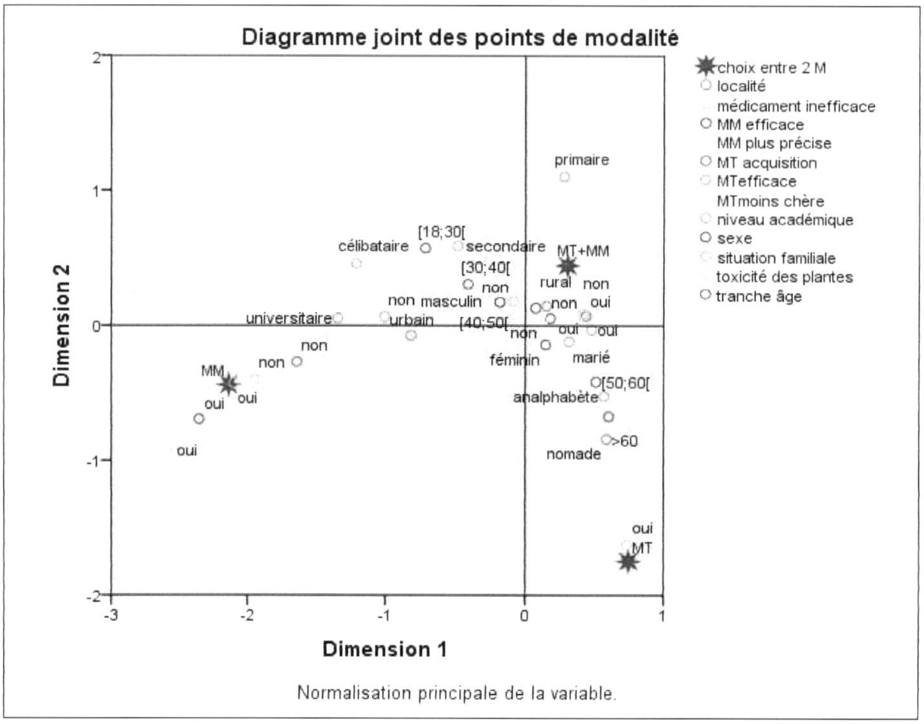

TM: traditional medicine; MM: modern medicine

Figure 10. Multiple correspondence analysis (MCA) of the set of variables that make up the data matrix

3.5. *The most used families*

According to the results of the figure 11, we can say that the Lamiaceae family is the most used (20.7 %), followed by the Asteraceae (12 %), Apiaceae (7.8 %), Fabaceae (3.5 %) and Rosaceae (3 %),

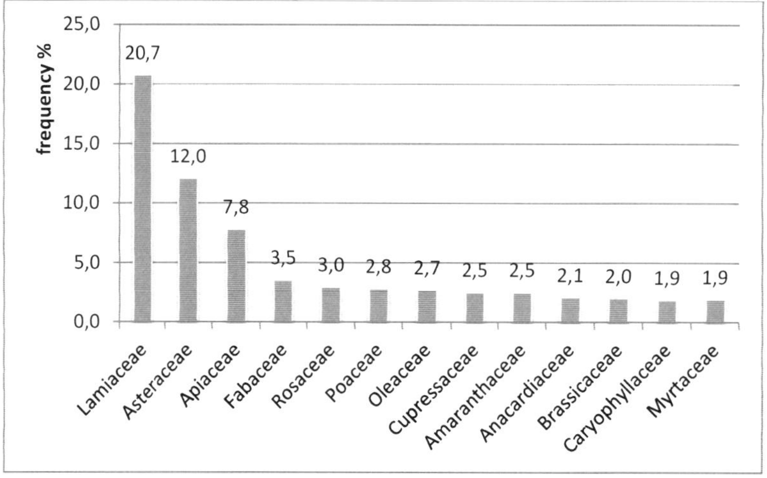

Figure 11. Distribution of families according to their percentage of species used.

This can be explained on the one hand by the dominance and abundance of these botanical families in Morocco in general, and in the Middle Atlas region in particular, and on the other hand, they are particularly well known by the population interviewed, widely used and contribute to the treatment of many diseases thanks to their richness in secondary metabolites. Also we notice that the order of classification of the botanical families the most used by the population is almost the same as the one of the specific richness of those of the catalog elaborated in the studied region.

The dominance of the most used Lamiaceae has been reported by several authors, in this case, El Hilah [27] in the Moroccan Central Plateau region, Lahsissène [28] in the Zaër region and Hseini [29] in the Rabat region.

3.6. Species most used by the population in the study area

According to the figure 12, among the most cited plants in the survey, we find that oregano (Origanum compactum) is ranked first (1039 citations), followed by rosemary (928 citations) (Rosmarinus officinalis), the pouliot (922 citations) (Mentha pulegium), round-leaf mint (820 citations) (Mentha suaveolens), cedar (724 citations) (Tetraclinis articulata) the mugwort (Artemisia herba-alba) (717 citations), Carallume d'europe (Caralluma europaea) (716 citations), the goosefoot (629 citations) (Chenopodium ambrosioides), the olive tree(625 citations) (Olea europaea), and then we find the fenugreek (617 citations) (Trigonella foenum-graecum), the thyme (Thymus zygis) with a frequency of use (615 citations), followed by the white marrube (614 citations) (Marribium vulgare).

This frequency of use can be explained by, the therapeutic biological properties of these species, intervene in the treatment of recurrent pathologies such as fever, digestive diseases, aches and pains, respiratory diseases, urinary diseases, inflammations, cold, diabetes and dermal diseases. These medicinal species are collected in the forests of the region.

It is deduced that oregano is the most used species in the Middle Atlas region, the same finding was confirmed by [30] in the Eastern High Atlas (Upper Moulouya) and by [27] in the Moroccan central plateau area.

Throughout Morocco, oregano is considered a panacea. It is widely used in popular medicine because of its multiple therapeutic effects. It is used especially in infusion or decoction in the treatment of dysentery, colitis, gastrointestinal diseases, gastric acidity and broncho-pulmonary diseases, against colds, flu, ear, nose and throat diseases and bronchitis. It is also administered in the form of fumigations. It is used in gargles against mouth diseases (canker sores and gingivitis) [31,32]. In the Middle Atlas region, the leafy stem, in infusion or decoction, is also used as a hypoglycemic [33]. Finally, it is used as a tonic, aphrodisiac and appetite stimulant [31].

However, over-harvesting of these species can lead to their extinction, resulting in the destruction of natural resources and habitats. Thus, it is necessary to protect the genetic resources of biodiversity and traditional knowledge to ensure their sustainability.

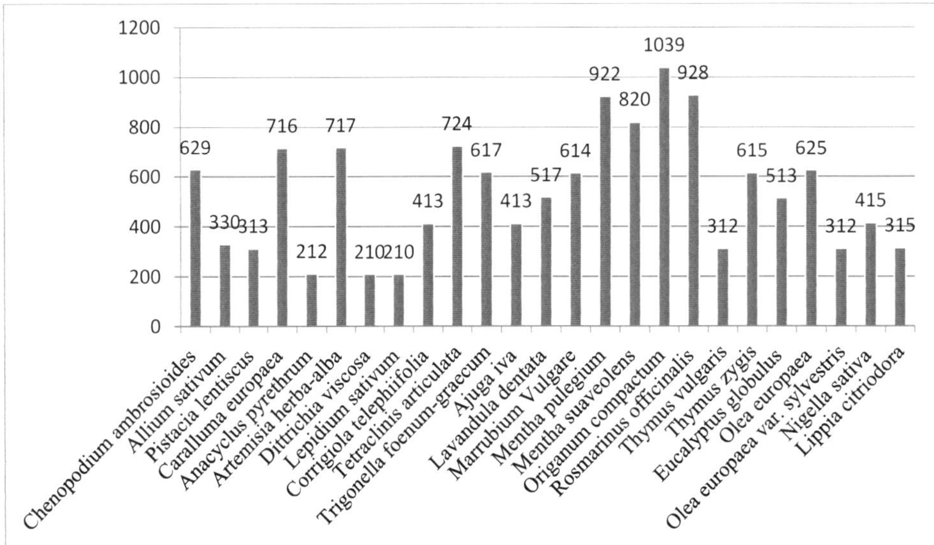

Figure 12. Ranking of the most used plants by the study population by number of citations (Over 210 citations).

3.7. Types of diseases treated by medicinal plants

Several thousand plants are used throughout the world. Their field of action is vast and their power varies. Most of them have specific effects on certain parts of the body and are known to be able to treat various cases.

The figure 13 illustrates that 49% of the plants identified are used to treat a single disease, 33% are prescribed to treat two diseases, 11% are indicated to treat both three diseases, 4% are used to treat four diseases, 2% to treat five diseases and 1% to treat both six diseases.

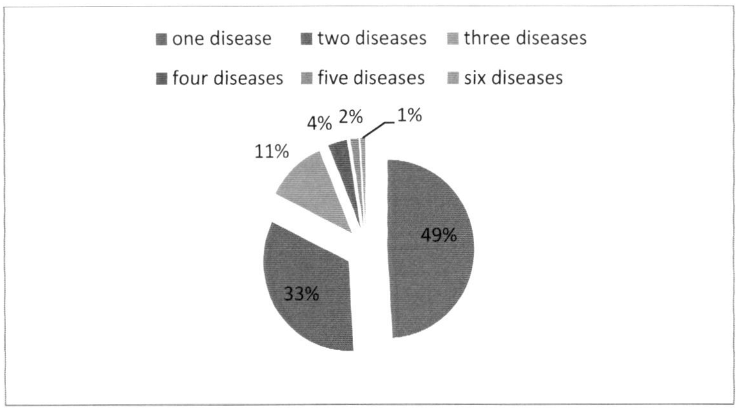

Figure 13. Presentation of the percentages of diseases treated.

Plants are then prescribed for the treatment of several kinds of diseases and symptoms, the most important of which in our region are digestive tract diseases, metabolic diseases and osteoarticular diseases ‘figure 14’. The ethnobotanical survey revealed that the majority of medicinal species are used mainly against digestive diseases, with a percentage of 25%, followed by metabolic diseases 15%, osteoarticular diseases 12%, dermal diseases 11%, genitourinary diseases 10% and respiratory diseases with the same percentage 10%, cardiovascular diseases 8%, neurological diseases with the same 5% and the category "other diseases" with a percentage of 4% (figure 14).

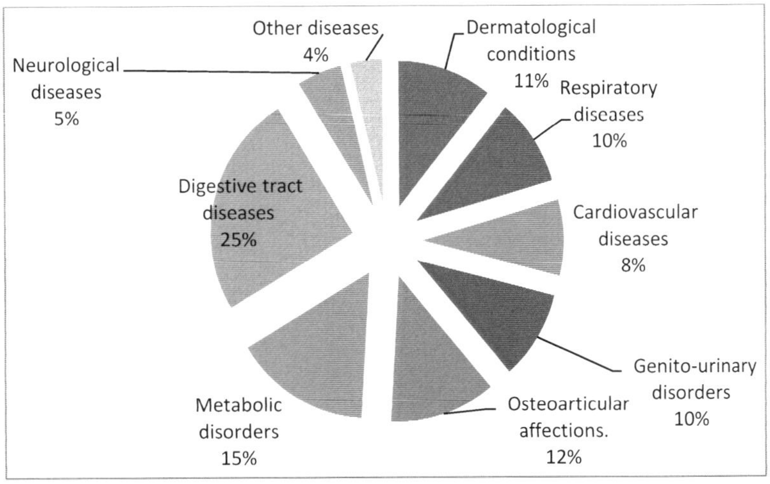

Figure 14: Percentage of plant use by condition treated.

3.8. The list of species recorded in the Central Middle Atlas

The table 6 represents the species inventoried during our ethnobotanical survey. Indeed, for each plant inventoried, the monograph contains the *scientific name, the family and the phylum.*

Table 6. Catalog of medicinal plants identified in the central Middle Atlas region

Phylum	Families	Species
Lichens	Parméliacées	*Evernia furfuracea* Mann. *Evernia prunastri* Ach.
Bryophytes	Orthotrichaceae	*Orthotrichum sp*
	Pottiaceae	*Tortula muralis*
Gymnosperms	Cupressaceae	*Juniperus oxycedrus* L. *Juniperus phoenicea* L. *Juniperus thurifera* L. *Tetraclinis articulata* (Vahl) Masters.
	Pinaceae	*Cedrus atlantica* (Endl.) Carr. *Pinus halepensis* Mill. *Pinus pinaster* Aiton
	Taxaceae	*Taxus baccata* L.
Chlamydosperms	Ephedraceae	*Ephedra nebrodensis* Guss.
Monocotyledonous angiosperms	Amaryllidaceae	*Allium cepa* L. *Allium sativum* L.
	Arecaceae	*Chamaerops humilis* L. *Cocos nucifera* L. *Phoenix dactylifera* L
	Asparagaceae	*Agave Americana* L. *Asparagus albus* L. *Urginea maritima* (L.) Baker.
	Cyperaceae	*Cyperus rotundus* L.
	Iridaceae	*Crocus sativus* L. *Iris germanica* L.
	Juncaceae	*Juncus acutus* L.

IOP Conf. Series: Earth and Environmental Science 1090 (2022) 012027 doi:10.1088/1755-1315/1090/1/012027

	Liliaceae	*Asphodelus microcarpus* Salzm. &Viv. *Colchicum automnale* *Ruscus aculeatus* L
	Musaceae	*Musa paradisiaca* L.
	Poaceae	*Avena sativa* L. *Festuca arundinacea* Schreb. *Hordeum vulgare* L *Oryza sativa* L *Pennisetum typhoides* (Burm.) Staph. & hub. *Phalaris canariensis* L. *Phragmites communis* Trin. *Secale cereale* L. *Sorghum halepense* (L) Pers. *Sorghum vulgare* L. *Stipa tenacissima* L. *Triticum durum* Desf. *Zea mays* L.
	Zingiberaceae	*Alpinia officinarum* Hance. *Curcuma longa* L. *Elettaria cardamomum* L. Maton *Zingiber officinalis* Rosc.
Angiosperms Dicotelydones	Amaranthaceae	*Beta vulgaris* L. *Chenopodium ambrosioides* L. *Fredolia aretioides* Coss.& Dur. *Spinacia oleracea* L.
	Anacardiaceae	*Pistacia atlantica* Desf. *Pistacia lentiscus* L. *Rhus pentaphylla* L.
	Apiaceae	*Ammi visnaga (*L.) Lam. *Ammodaucus leucotrichus* Coss. & Dur. *Apium graveolens* L. *Carum carvi* L. *Coriandrum sativum* L. *Cuminum cyminum* L. *Daucus carota* L. *Daucus crinitus* Desf. *Elaeoselinum asclepium* (L.)Bertol *Ferula assa-foetida* L. *Ferula communis* L. *Ferulago lutea* (Poiret) Grande *Foeniculum vulgare* P. Mill *Hippomarathrum libanotis* L *Kundmania sicula* (L.)DC *Magydaris panacifolia* (Vahl.) lange *Petroselinum sativum* Hoffm. *Pimpinella anisum* L. *Ptychotis verticillata* Dub. *Ridolfia segetum* (L.) Moris *Smyrnium olusatrum* L.
	Apocynaceae	*Caralluma europaea* (Guss.). *Nerium oleander* L.
	Aristolochiaceae	*Aristolochia baetica* L.
	Asteraceae	*Anacyclus pyrethrum* L. *Anacyclus radiatus* Lois.

IOP Conf. Series: Earth and Environmental Science 1090 (2022) 012027 doi:10.1088/1755-1315/1090/1/012027

		Artemisia absinthium L. *Artemisia herba-alba* Asso. *Artemisia mesatlantica* Maire *Atractylis cancellata* *Atractylis gummifera* L. *Calendula officinalis* L. *Centaurea calcitrapa* L *Chamaeleon macrophyllus* Desf. *Cichorium intybus* L. *Cynara cardunculus* L. *Cynara humilis* L. *Cynara scolymus* L *Dittrichia viscosa* (L.) Greuter *Echinops spinosus* L. *Helianthus annus* L. *Helianthus tuberosus* L. *Lactuca sativa* L *Lactuca scariola* L. *Mantisalca salmantica* *Matricaria chamomilla* L. *Pulicaria arabica* (L) Cass *Rhaponticum acaule* (L.) DC *Scolymus hispanicus* L. *Stemmacantha acaulis* *Taraxacum obovatum* DC. *Tragopogon porrifolius* L. *Xanthium strumarium*
	Berberidaceae	*Berberis hispanica* Boiss. & Reut
	Borraginaceae	*Borrago officinalis* L.
	Brassicaceae	*Alyssum spinosum* *Anastatica hierochuntica* L. *Brasica napus* L. *Brasica oleracea* L. *Brassica rapa* L. *Lepidium sativum* L. *Nasturtium officinalis* R. Br *Raphanus sativus* L. *Sinapis alba* L.
	Buxaceae	*Buxus balearica* Lam.
	Cactaceae	*Opuntia ficus indica* (L.) Mill
	Cannabinaceae	*Cannabis sativa* L.
	Capparaceae	*Capparis spinosa* L.
	Caryophyllaceae	*Corrigiola telephiifolia* Pourr. *Herniaria hirsuta* L. *Saponaria vaccaria* L.
	Cistaceae	*Cistus albidus* L. *Cistus ladanifer* L. *Cistus salviifolius* L.
	Convolvulaceae	*Convolvulus arvensis* L.
	Cucurbitaceae	*Citrullus colocynthis* (L.) Schrad. *Citrullus vulgaris* Schrad. *Cucumis sativus* L. *Cucurbita pepo* L.

		Lagenaria siceraria (Molina) Standl.
	Ericaceae	*Arbutus unedo* L.
	Euphorbiaceae	*Euphorbia falcata* L. *Euphorbia nicaensis* All. *Euphorbia resinifera* Berg. *Mercurialis annua* L. *Ricinus communis* L.
	Fabaceae	*Astragalus gummifer* labill. *Cassia fistula* *Cassia senna* L. *Ceratonia siliqua* L. *Cicer arietanum* L. *Glycyrrhiza glabra* L. *Lens culinaris* Med. *Medicago sativa* L. *Ononix natrix* L. *Phaseolus vulgaris* L. *Soja hispida.* *Trigonella foenum-graecum* L. *Vicia faba* L.
	Fagaceae	*Quercus faginea* lamk. *Quercus ilex* L. *Quercus suber* L.
	Gentianaceae	*Centaurium erythraea* Rafn.
	Geraniaceae	*Pelargonium roseum* Willd.
	Globulariaceae	*Globularia alypum* L.
	Juglandaceae	*Juglans regia* L.
	Lamiaceae	*Ajuga iva* (L.) Schreb. *Lavandula dentata* L. *Lavandula multifida* L *Lavandula stoechas* L. *Marrubium vulgare* L. *Mellissa officinalis* L. *Mentha gattefossei* Maire *Mentha pulegium* L. *Mentha suaveolens* Ehr. *Mentha viridis* L. *Ocimum basilicum* L. *Origanum compactum* Benth. *Origanum majorana* L. *Rosmarinus officinalis* L. *Salvia officinalis* L. *Salvia verbenaca* L. *Satureja alpina* (L.) Scheele. *Satureja calamintha* (L.) Scheele. *Satureja rotundifolia* *Satureja vulgaris* (L.) Fritsch. *Teucrium polium* L. *Thymus algeriensis* Boiss. & Reut. *Thymus munbyanus* Boiss. & Reut. *Thymus vulgaris* L. *Thymus zygis* Loefl. ex L. *Ziziphora Hispanica L.*

	Lauraceae	*Cinnamomum camphora* Ness. *Cinnamomum cassia* Ness. *Laurus nobilis* L. *Persea gratissima* Gaertn.
	Linaceae	*Linum usitatissimum* L.
	Lythraceae	*Lawsonia inermis* L. *Lythrum junceum* Banks & Sol *Punica granatum* L.
	Malvaceae	*Abelmoschus esculentus* (L.) Moench *Hibiscus sabdariffa* L. *Malva sylvestris* L.
	Moraceae	*Ficus carica* L. *Morus alba* L
	Myristicaceae	*Myristica fragrans* Houtt.
	Myrtaceae	*Eucalyptus globulus* Labill. *Eugenia caryophyllata* Thunb. *Myrtus communis* L. *Pimenta officinalis* Lindl.
	Nitrariaceae	*Peganum harmala* L.
	Oleaceae	*Jasminum fruticans* L. *Olea europaea* L. *Olea europaea var. sylvestris* (Mill.) Brot.
	Paeoniaceae	*Paeonia corallina* L. ssp. Maroccana *Paeonia coriacea* Boiss.
	Papaveraceae	*Papaver rhoeas* L.
	Pedaliaceae	*Sesamum indicum* L.
	Piperaceae	*Piper cubeba* L.F.
	Plantaginaceae	*Plantago major* L.
	Plombaginaceae	*Armeria alliacea* (Cav.) Hoffm *Plumbago europeae* L.
	Polygonaceae	*Rumex roseus*
	Portulacaceae	*Portulaca oleracea* L.
	Ranunculaceae	*Delphinium staphisagria* L. *Nigella sativa* L. *Ranunculus bullatus* L.
	Resedaceae	*Reseda luteola* L.
	Rhamnaceae	*Rhamnus alaternus* L. *Rhamnus lycioides* L. *Zizyphus lotus* (L.) Lam.
	Rosaceae .	*Crataegus laciniata* Ucr. *Cydonia oblonga* Mill. *Eriobotrya japonica* (Thunb.) Lindl. *Fragaria vesca* L. *Malus communis* Poir *Prunus amygdalus stokes var. amara* DC. *Prunus amygdalus stokes var. dulcis* DC. *Prunus armeniaca* L. *Prunus avium* L. *Prunus domestica* L. *Prunus persica* (L.) Batsch. *Rosa canina* L. *Rosa damascena* Mill. *Rubus ulmifolius* Schott
	Rubiaceae	*Coffea arabica* L.

		Rubia peregrina L.
	Rutaceae	*Citrus limon* (L.) Burm. *Ruta montana* L.
	Salicaceae	*Populus alba* L.
	Santalaceae	*Viscum cruciatum*. Cieber ex Boiss.
	Sapotaceae	*Argania spinosa* (L.) Skeels
	Schisandraceae	*Illicium verum* Hook.F
	Scrophulariaceae	*Scrofularia sambucifolia* L. *Verbascum sinuatum* L.
	Solanaceae	*Capsicum annuum* L. *Capsicum frutescens* L. *Mandragora autumnalis* Bertol *Nicotiana tabacum* L. *Solanum lycopersicum* L. *Solanum melongena* L *Solanum nigrum* L. *Solanum tuberosum* L.
	Tamaricaceae	*Tamarix gallica* L.
	Theaceae	*Camellia thea* L.
	Thymelaeaceae	*Daphne gnidium* L. *Thymelaea lythroides* Barr & Murb. *Thymelaea tartonraira* (L.) All.
	Urticaceae	*Urtica pilulifera* L.
	Verbenaceae	*Lippia citriodora* H.B. & K. *Verbena officinalis* L. *Vitex agnus-castus* L.
	Vitaceae	*Vitis vinifera* L.
	Zygophyllaceae	*Zygophyllum gaetulum* Emb. & Maire.

4. Conclusion

From the results of this analysis, it can be concluded that there are different factors that can influence people's choice, including gender, age, education level, family status and place of residence.

According to the people interviewed, the choice of traditional medicine is justified on the one hand by the lower cost of the products used, by its effectiveness and its acquisition, and on the other hand, by its low toxicity compared to the side effects of synthetic drugs. For the others, they chose modern medicine which seems to them more effective and avoid medicinal plants because of their toxicity. The rest use both medicines, they think that it is necessary to diagnose the disease first and then choose the treatment either based on medicinal plants or medication.

The ethnobotanical analysis revealed that the medicinal plants most used by the local population are: Origanum compactum, Rosmarinus officinalis, Mentha pulegium, Mentha suaveolens, Tetraclinis articulata, Artemisia herba-alba, Caralluma europaea, Chenopodium ambrosioides, Olea europaea, Trigonella foenum-graecum, also the majority of the listed medicinal species are used mainly against digestive diseases, with a percentage of 25%.

5. References

[1] Anonyme, 1974. Encyclopédie-Le Grand Médical. L'histoire de la médecine et de la chirurgie, l'avenir de la médecine, les prix Nobel. Edition Service S.A., Genève (Suisse), 397 pp.

[2] Nacoulma-Ouédraogo O, 1996. Plantes médicinales et pratiques médicales traditionnelles au Burkina Faso: cas du Plateau central, Thèse de Doctorat ès Sciences Naturelles, Université de Ouagadougou, (Burkina-Faso), 605 p.

[3] Fokunang CN, Jiofack RB, Ngameni B, Asongalem E, Guedje NM, Kechia F, Tembe-Fokunang E, Ngoupayou J, Torimiro NJ, Gonsu KH, Sielinou V, Ndikum V, Tabi O, Ngadjui BT, Ngogang J, Asonganyi T, Abena OM, Nkongmeneck, Lohoue J, Colizzi V and Angwafor III F, 2010. Traditional Medicine: Past, Present and Future Development Prospects and Integration in the

National Health System of Cameroon. Accepted for publication by African Journal of Traditional, Alternative and Complementary Medecine (AJTCAM).

[4] Jiofack T, Ayissi I, Fokunang C, Guedje N and Kemeuze V, 2009. Ethnobotany and phytomedicine of the upper Nyong valley forest in Cameroon. African Journal of Pharmacy and Pharmacology, **3(4):** 144- 150 p. http://www.academicjournals.org/ ajpp

[5] Jiofack T, Fokunang C, Guedje N, Kemeuze V, Fongnzossie E, Nkongmeneck BA, Mapongmetsem PM and Tsabang N, 2010. Ethnobotanical uses of medicinal plants of two ethnoecological regions Cameroon. International Journal of Medicine and Medical Sciences, **2(3):** 60- 79 p. http://www.academicjournals.org/ ijmms

[6] Kasparek M, al-Janabi S, 2008. Plantes médicinales - La diversité biologique au service de la santé. Fiche thématique. Deutsche Gesellschaft für Technische Zusammenarbeit (GTZ) GmbH.

[7] Jahandiez E and Maire R, 1931-1934. Catalogue des plantes du Maroc, **3** vols. Minera, le chevalier, Alger.

[8] Fennane M & Ibn Tattou M, 2012. Statistiques et commentaires sur l'inventaire actuel de la flore vasculaire du Maroc. Bull. Inst. Sci. Rabat sec. Sci. **34(1)**: 1-9 p.

[9] Quezel P and Médail F, 1994. La région circum-méditerraniéenne, centre mondial majeur de biodiversité végétale. Actes 6éme Rencontre de l'A.R.P.E. : 152-161, Gap, France.

[10] El Khanchoufi A, 2013. Les plantes aromatiques et médicinales au Maroc, une richesse peu exploitée et un impact socio-économique très limité. Journal Lemag. Publié le Jeudi 31 Janvier 2013. http://www.lemag.ma/Les plantes-aromatiques- et médicinales- au Maroc_a66633.html.

[11] Ministère chargé de l'environnement, 1997**.** Rapport sur l'Etat de l'Environnement du Maroc. Chapitre III : Biodiversité et Milieu Naturel, 48 p.

[12] Michard A., 1976. Eléments de géologie marocaine. Notes et Mém. Serv. géol. Maroc, 252, 408 p.

[13] Arboleya M.L, Teixell A, Charroud M and Julivert M, 2004. A structural transect through the High and Middle Atlas of Morocco. Journal of African Earth Sciences **39**. 319–327 p.

[14] Fedan B., 1988. Evolution géodynamique d'un bassin intra-plaque sur décrochement : le Moyen Atlas durant le Méso-Cénozoïque. Travaux de l'Institut Scientifique de Rabat, N° **18**, 142 p.

[15] Texier J. P, Raynal J. P et Lefevre D, 1985. Nouvelles positions pour un cadre chronologiques raisonné du Quaternaire marocain. C. R. Acad. Sc. Paris, t, 301, Série II, n° **3**, 183-188 p.

[16] Herbig H.G, 1988. Synsedimentary tectonics in the Northern Middle Atlas (Morocco) during the Late Cretaceous and Tertiary In: V. Jacobshagen, Editors, The Atlas System of Morocco, Springer-Verlag, Berlin, 321–337 p.

[17] Martin J, 1981. Le Moyen Atlas central étude géomorphologique. Notes et Mémoires du service Géologique N° 258 bis Rabat Maroc 447 p.

[18] Chillasse L, Dakki M, 2004. Potentialités et statuts de conservation des zones humides du Moyen-Atlas (Maroc), avec référence aux influences de la sécheresse. Sécheresse, 15 (4) : 337-45 p.

[19] Lecompte M, 1986. Biogéographie de la montagne marocaine : le Moyen Atlas central. CNRS (éd), Paris.

[20] Godron M, 1971. Essai sur une approche probabiliste de l'écologie des végétations.Thèse doct. d'Etat, U.S.T.L., Montpellier, 247 p.

[21] Godron M et Daget Ph, 1982. Analyse de l'écologie des espèces dans les communautés Coll. Ecol. 18, Masson, 163 p.

[22] Emberger L & Maire R., 1941. Catalogue Des Plantes Du Maroc (Spermaphytes Et Ptéridophytes), Tome Iv. Supplément Aux Volumes I, Ii Et Iii. - Minerva, Alger, Lix-Lxxv + 915-1181.

[23] Maire R, 1952-1987. Flore De L'Afrique Du Nord. Tomes 1 A 16. – Paul Lechevalier, Paris, "Encyclopédie Biologique". 5559 P.

[24] Quézel P. & Santa S., 1963. Nouvelle Flore De L'Algérie Et Des Régions Désertiques Méridionales. Tome 2. - Cnrs Ed., Paris, 494 P.

[25] Fennane M, Ibn Tattou M, Mathez J, Ouyahya A & El Oualidi J, 1999. Flore Pratique Du Maroc; Manuel De Détermination Des Plantes Vasculaires, Vol. **1**. Trav. Inst. Sci., Rabat, Sér. Botanique, 36, 558 P.

[26] Fennane M, 2004. Propositions de Zones Importantes pour les Plantes au Maroc (ZIP Maroc) Rabat.

[27] El Hilah F, 2016. Enquête ethnobotanique des plantes médicinales auprès de la population du plateau central marocain. Thèse de Doctorat national. Faculté des Sciences de Kénitra. 156 p.
[28] Lahsissène H, 2010. Recherches ethnobotanique & floristique des plantes médicinales utilisées dans la région phytogéographique de Zaër. Thèse de Doctorat, faculté des Sciences. Rabat. 273 p.
[29] Hseini S, 2007. Etude ethnobotanique de la flore médicinale dans la région de Rabat. Thèse de Doctorat. Faculté des Sciences. Rabat. 182 p.
[30] Benlamdini N, Elhafian M, Rochdi A et Zidane L, 2014. Étude floristique et ethnobotanique de la flore médicinale du Haut Atlas oriental (Haute Moulouya). Journal of Applied Biosciences 78: 6771 – 6787.
[31] Bellakhdar J, 1997. La pharmacopée marocaine traditionnelle. Médecine arabe ancienne et savoirs populaires. Ed. Le Fennec, Paris, 764 p.
[32] Lahsissene H, Kahouadji A, Tijane M et Hseini S, 2009. Catalogue des plantes médicinales utilisées dans la région de zaër (Maroc occidental). Lejeunia, BE ISSN 0457-4184 p.
[33] Hachi M, Benkhnigue O, Hachi T, El Bouhaddioui M, Bouabadi I, Rochdi A & Zidane L, 2016. Contribution to the ethnobotanical study of antidiabetic medicinal plants of the Central Middle Atlas region (Morocco). *Lazaroa* **37** 2016: 1-11.

Appendix I

Questionnaire

Medicinal plants and herbal medicine

Date :……………………………………………………………………………………

Region :…………………………………………………………………………………

Commune :…………………………….. …………………………………………

Place said :……………………………….. ……………..

Card number :…………………………………………………………….............

Informant :

- Age :……………… ………
- Family situation : Single □ married □
- Gender : Male □ Female □
- Academic level : Primary □ secondary □ Univerity □ Illiterate □
- Locality : Douar □ Village □ city □ Nomad □

****

Therapeutic practices.

➢ When you feel sick, you turn to:

- To traditional medicine □, Why : Effective □ Cheaper □ Acquisition □ Ineffective medication □
- To modern medicine □, Why : Effective □ More precise □ . Plant toxicity □

****

Plant material.

- Vernacular name : ……………………….
- Scientific name : ……………………….
- Type of plant :

- Wild □ Cultivated □ Imported □

▪ Use of the plant :
  - Therapeutic □ Cosmetic □ Other □

▪ Plant alone □ Possible association (of plants) □ ……………
……………………………………………………

▪ State of the plant: Fresh □ Desiccated □
  - If dried out, drying method:
    Exposed to the sun □ In the shade □

▪ Parts used: Stem □ Flowers □ Fruits □ Seed □
  Bark □ Rhizome □ Bulb □ Leaves □
  Entire plant □ Other combinations □ :………………………

▪ Form of use : Herbal tea □ Powder □ Essential oils □ Extract □

▪ Method of preparation : Infusion □ Decoction □ Cataplasm □ Cru □
  Cooked □ Others □ :………………………………..

▪ Dose used :
  - Pinch □ handle □ Spoonful □
  - Precise dose :
    - Quantity in g / glass:…………………..
    - Quantity in g/ liter :…………………..
    - Others :…………………..

▪ Method of administration : Oral □ Massage □ Rinse □ Brushing □
  Others □ :……………….

****

Use:

▪ Disease type :
  - Dermatological conditions □
  - Respiratory diseases □
  - Cardiovascular diseases □
  - Genito-urinary disorders □
  - Osteoarticular affections. □
  - Metabolic disorders □
  - Digestive tract diseases □
  - Neurological diseases □
  - Other diseases □

▪ Diagnosis By :
  Himself □ The doctor □ The herbalist □ Others □ : …………………

▪ Results : Healing □ Improvement □ Ineffective □

▪ Toxicity :………………………………………………………………

IOP Conf. Series: Earth and Environmental Science 1090 (2022) 012028 doi:10.1088/1755-1315/1090/1/012028

Water Uses in Agriculture and Preservation Measures: What types of Relationships Between Them? Case of El Hajeb Province

A El Allaoui [1] M Aafir [2] and M Elyaagoubi [1]

[1] Space, Dynamics, History and Sustainable Development Laboratory, Faculty of Letters and Human Sciences, Sidi Mohamed Ben Abdullah University, Fez, Morocco.
[2] Research professor at the Faculty of Languages, Arts and Human Sciences, Ait Melloul University Campus. Ibn Zohr University, Agadir, Morocco

E-mail: majidou1978@yahoo.fr

Abstract. Large scale forces are reshaping the world of water, like climate change, human settlements, and human economic activities. These forces make water management a challenging issue that has become more and more complicated with multiplicity and increasing need to use water. That is why sustained water management requires coordination between preservation measures and the different ways of water uses. In this article, we investigate and approach these two components through the expansion of agriculture as a main human activity in the Province of El Hajeb, to underline and analyze the types of relationships existing between water uses and conservation measures.

Keywords: water uses- conservation-relationships

1. Introduction

With 650 m3 per ha per year, Morocco is in a situation described as water scarcity. If the current rate continues, water resources would be less than 500 m3 per ha per year in 2030. This worrying situation requires urgent reforms, particularly in the agricultural sector where water loss in irrigated plots is estimated at 40 % (EESC, 2020). In general, agriculture is the most consuming water in Morocco: « Irrigation accounts for 85% of the water use compared to 10% for domestic use, and 5% for industry » [1]. In the Province of El Hajeb, our study area, agriculture covers about 88 percent of the province's area, in 1977 it covered only 46%. This expansion of agricultural area was accompanied by a shrinking of the forested area and increasing exploitation of water beds. Although there is a strong relationship between the abundance of water and the level of precipitation in the province [3], agriculture consumes more water than that obtained through rain, which will raise the water supply problem in the future, this observation indicates that a major reduction in water supply could harm the level and value of agricultural production in the Province, that why «we must set in motion new strategies governing the way we live and interact with our environment to ensure that there will be enough water to support, rather than constrain, future generations» [2]

2. Problematic question and methodology

Both agricultural and water systems are complex and the relationships between them make a mesh of many dimensions. Not least there are many sorts of agricultural system and a plethora of natural waters and communities (Moss, 2015). Controlling the relationship between agricultural activity and the exploitation of natural resources, especially water is one of the most urgent requirements, whether at the national, regional, or local level. But before this, we must define the nature of this relationship and its development over time, and then, its future implications according to the perspective of sustainable agriculture. To reach this goal, we combined several techniques: At first, we collected a set of data available from some institutions interested in the subject, completed by direct interrogation of many actors in the sector, and then employing analysis of aerial images taken by the Landsat satellite, in the Province of El Hajeb, during separate periods of time in order to monitor the spatial changes that interest us in this research.

3. Geographical location of the study area

The territory of the Province of El Hajeb covers an area estimated at 214000 hectares. As depicted in figure 1, this area combines a part of the fertile Sais Plain on an area of approximately 98156 hectares, which represent 45.87 % of the province's area, and the front of the Middle Atlas Mountains on an area of 63045 hectares which represent 29.46 %, and the Central Plateau on an area of 52799 hectares, or 24.67%. The plains in fertile soil, the abundant water represented by the groundwater, the constantly and intermittent springs, and the significant amounts of annual precipitation, as well as the forests located on the south of the province, are the most important natural resources in which they are located.

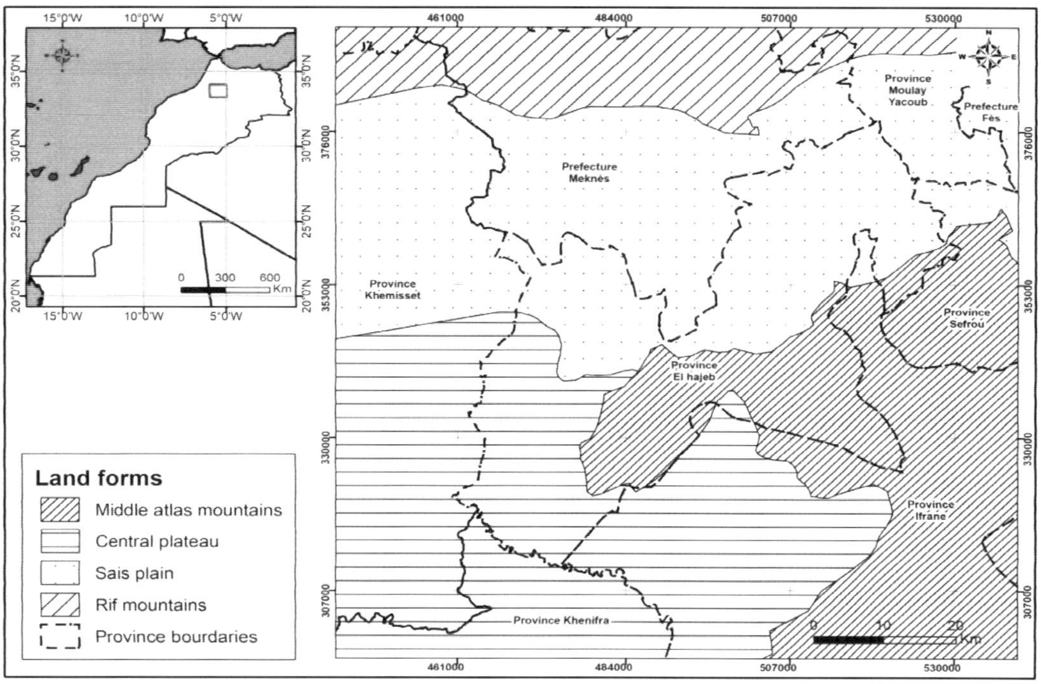

Figure 1. Geographical location of El Hajeb Province.

El Hajeb is located entirely within the left side of the Sebou River Basin, whose area is estimated at about 4 000 000 hectares and therefore the province occupies 5.35 % of its area, It is interspersed with several valleys: El kell, Elouaar, and Boujir valleys, are subsidiaries of Sebou River (Figure 2).

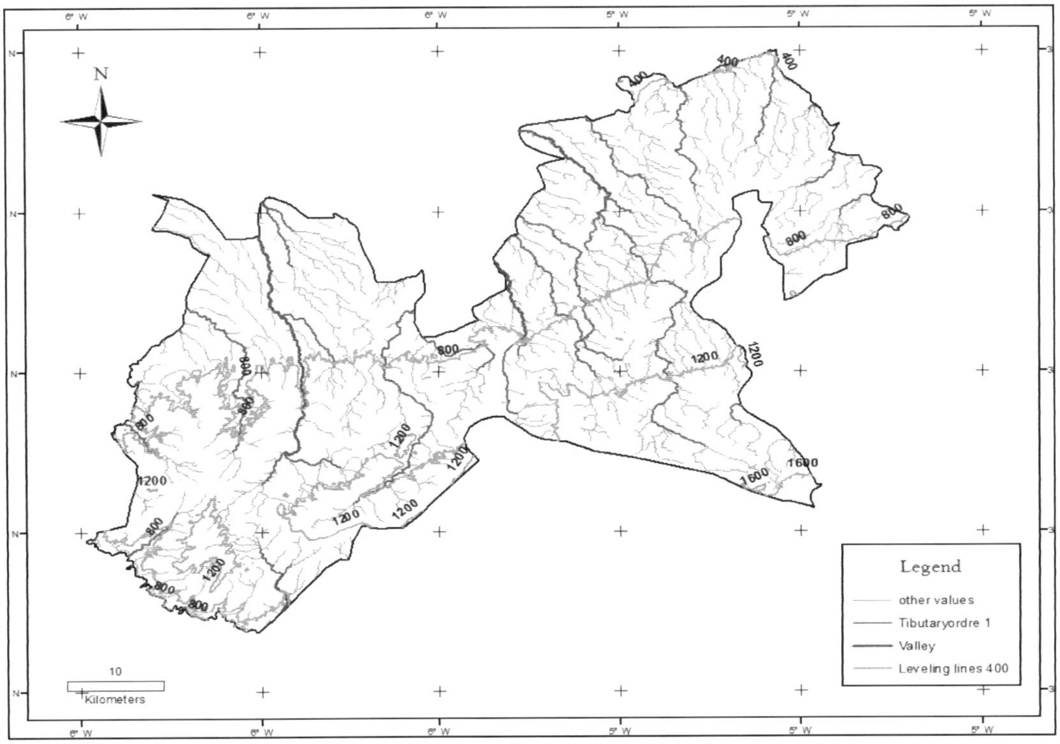

Figure 2. Waterways in the El Hajeb Province.

The province is characterized by the predominance of agricultural activity which occupies more than 60% of the local population and it is oriented to the market, this activity has a great impact on water resources.

4. Large context of water management and main results

Launched in 2008, the Green Morocco Plan has set very promising goals: economic efficiency, social equity, and preservation of natural resources within the framework of sustainable agriculture which was established as the main lever of the economy. The water dimension was present in the genesis of the aforementioned Plan through the consideration of climate change, incentive pricing of water, progressive delegation management of irrigation water, generalization of modern techniques.

The achievements were very convincing in economics terms; today the agricultural sector contributes by 13 % in the Good Domestic Product (GDP), and represents 13% of the national exports value, it accounts for 72% of rural employment with 10 million people who are related, but on the other hand, the environmental consequences were very worrying, especially water resources, both at the national level and at the local level.

4.1 Agricultural area extends and water problems.

At the Province level, water resources have been depleted, and forests have receded in narrow areas. In 1977, the agricultural area represents only 46 % of the province areas, most of the agricultural lands were located on the Sais Plain; the part of Central Plateau and Middle Atlas on which the Province is located were covered by forests and pasture (figure 3).

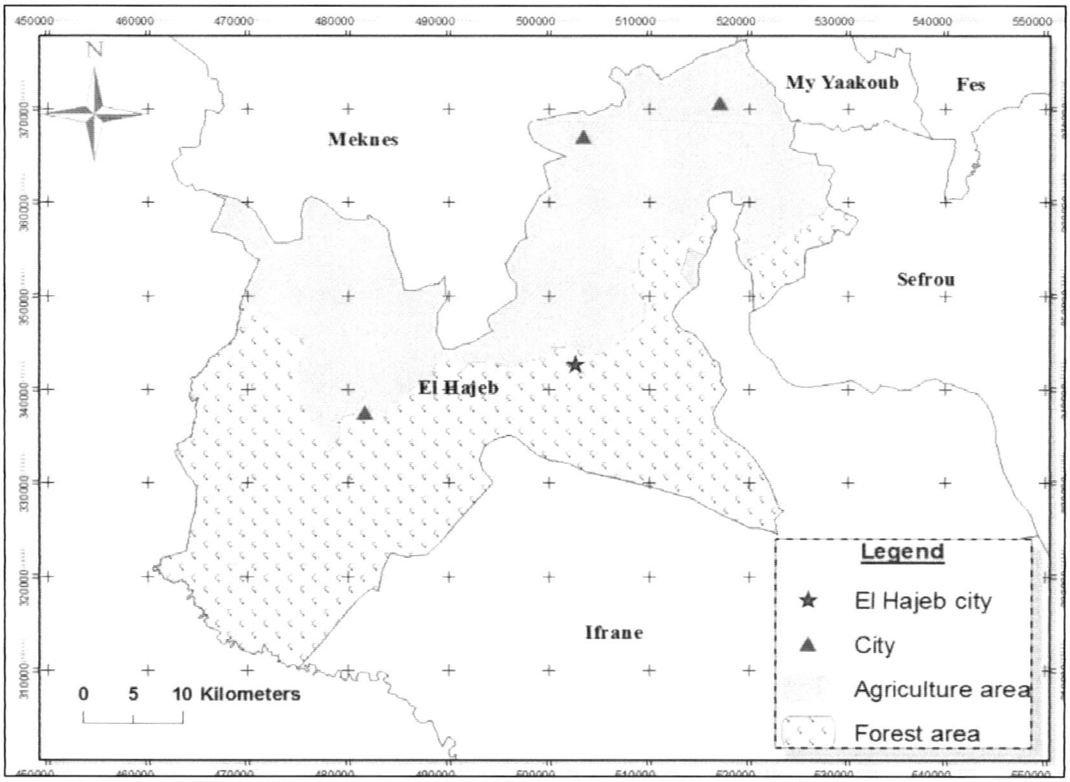

Figure 3. The agricultural area of El Hajeb Province in 1977.

During that period there were several characteristics that should not be overlooked:

- The space in which wells could be drilled was limited.
- Surface water was important and only a few meters needed to be dug to find water.
- Agriculture was primarily oriented to meet family or local needs.
- Also, the techniques and machines were limited.

Because of many factors as for example the population growth, multiplication of needs, Moroccan agricultural plans, the agricultural area were extended more and more throughout the years. And so, after about half a century, the agricultural area has expanded significantly, covering an area of about 88% (figure 4), and extended to cover the whole part of the Sais Plain on which the province is located, and vast areas on the Central Plateau and Middle Atlas.

IOP Conf. Series: Earth and Environmental Science 1090 (2022) 012028 doi:10.1088/1755-1315/1090/1/012028

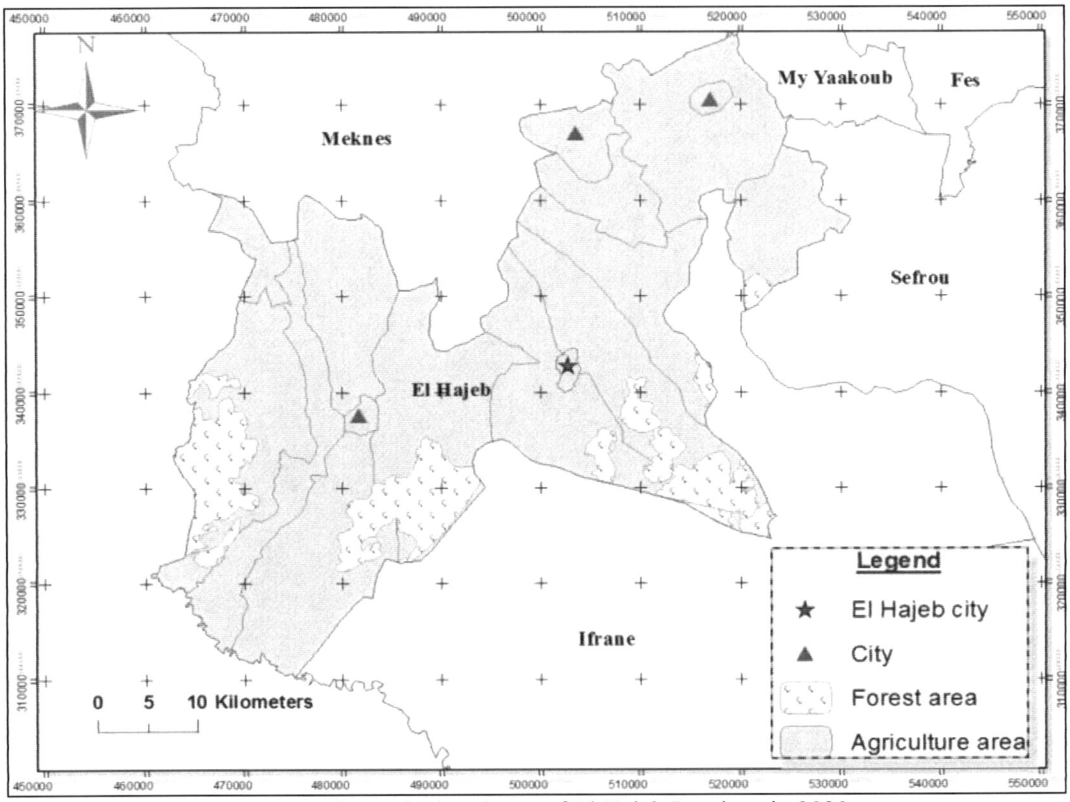

Figure 4. The agricultural area of El Hajeb Province in 2020.

The field investigation highlights the scattered nature of the distribution of trees in the southern part of this province located in front of the Middle Atlas Mountains, and the areas that suffer from deterioration were evident, and they are those that border or lie on the plain area of the region, and there are some species on the way to disappear, while the High Commissioner For water and forests is trying reforestation of some areas. However, this process is plagued by many problems, perhaps the most important of which is overgrazing, which eliminates young saplings as soon as they are planted, and other matters related to the modalities of executing deals related to this process. Because of the withdrawal of the forest space in the province the mode of exercise of pasture has changed too, instead of exercising in the prairie and within the forest, the shepherds find themselves condemned either to exercise it on their own farms or to have recourse to the purchase of fodder. This increasing expansion of agricultural area had, as one of the most negatives effects, the increasing demand of surface and underground water. And perhaps the most important indicator of the high demand was the increase in the number of wells in the El Hajeb province.

4.2 The number of wells and water holes has increased and bad distributed.

The number of wells was around 3715 in 2008, the start date of Morocco Green Plan, they are used to irrigate more than 15900 hectares, besides 75 water springs used to irrigate 5729 hectares, the agricultural area at that time was around of 146000 hectares, that means that about 11% of this area was dependent on water extracted from wells, water holes were not common at that time. In 2020 the agricultural area was extended to 188320 hectares, in this time more than 12000 wells and holes are in the province, due to the secret drilling of wells, this number is only an approximation, this rapid increase of wells and holes as well as their random distribution has certainly led to the depletion of water. And so the subterranean water became deeper in 2016.

As it is shown in figure 5, in some areas water has become deeper with about 9 meters in only one year. This is because of many factors as the bad distribution of wells and holes through the territory of the province. Perhaps what is happening in the province is only a part of what is taking place in broader level which is the Basin of Sebou

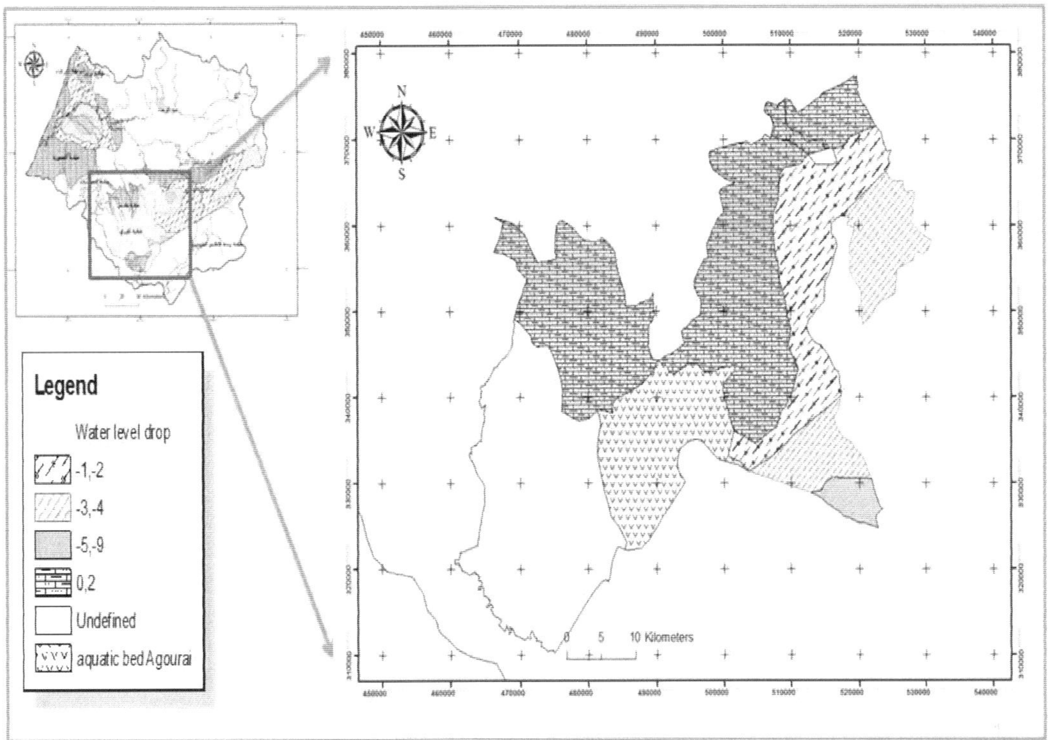

Figure 5. Increasing depth of waterbed by regions in 2016 in El Hajeb Province (by meters).

4.3 ***The subterranean water became deeper.***

The depth of the waterbed did not exceed several meters during the seventies and eighties. Today it is necessary to dig more than 100 meters to obtain water in many places of the river Basin.

As it is well shown in Figure 6, the piezometric level of water decrease following a function with negative coefficient, thus after 36 years the water level decrease by 2.88 meters per years. Same situation occurs in the Meknes plateau.

Let us consider the regression line presented in figure 7, after 2008 the year of the lunch of the Green Morocco Plan, the curve becomes below the line, which is very significant, the water level becomes quickly deeper, that means that the situation becomes more dangerous through time. This situation is exacerbated by the increasing quantities of pollutants reaching the surface and underground waters through agriculture activity and other sources: 9166 t of total nitrogen ant 1570 t of phosphorus coming from agriculture in the river Basin.

The unique approach designed for a very large territory with a thousand faces that is Morocco means that certain places were not all predisposed to receive the major agricultural projects in the same way, the neglect of local particularities harms these small territories, both on the natural level as well as on the social level, this is how environmental components have been badly affected, as we have seen above. By its multidimensional nature, the focus on profitability starts from a conception considering agriculture as an economic activity similar to that of industry or services, yet the latter is based on ancestral social ties combined by peasant practices localized.

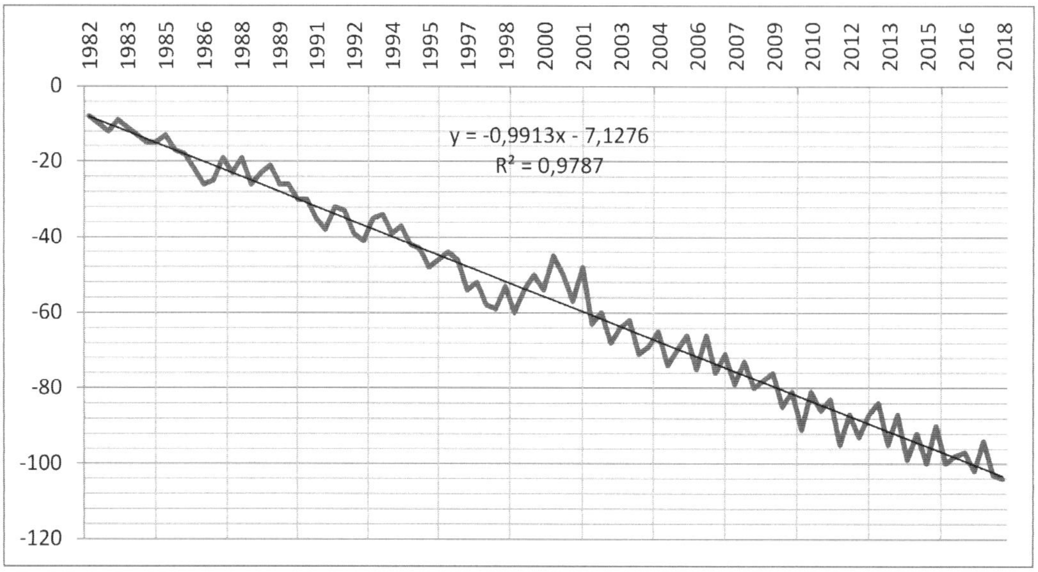

Figure 6. Evolution of the piezometric level of water between 1968 and 2016 in the Sebou River Basin.

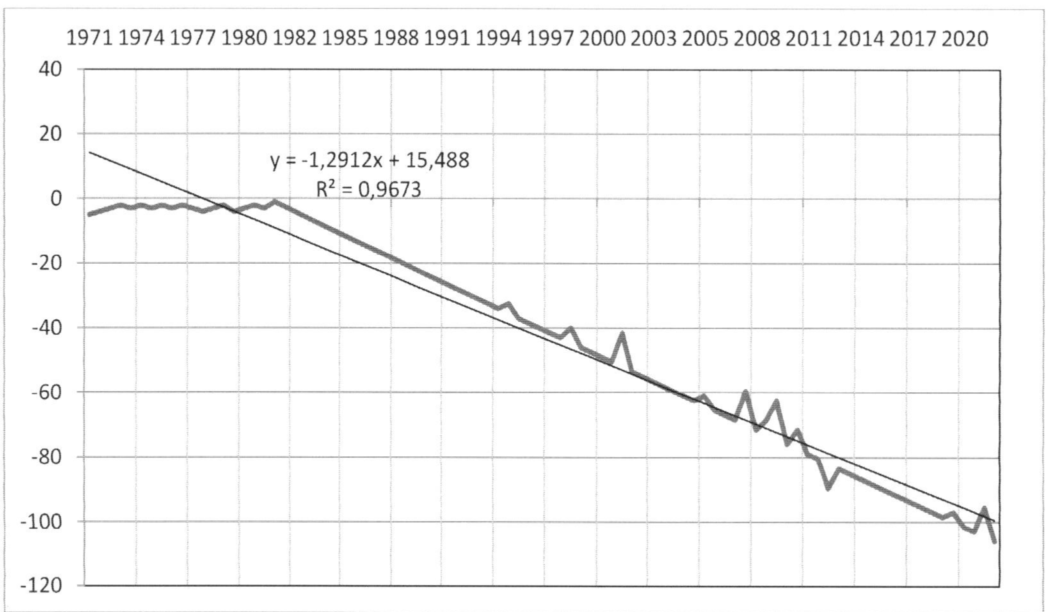

Figure 7. Evolution of the piezometric level of water between 1982 and 2020 in the aquatic bed of Meknes plateau.

Some measures come up in official discussions as solutions to certain problems created by the GMP, such as the installation of water meters, if this measure seems effective in terms of monitoring, dosing and quantification of the water used, it is based on a pragmatic approach rather than a problem-solving approach, because even with the absence of a bill for water used in agriculture, there is a certain level of inaccessibility to water in some categories of farmers, for example in Breeched, the coastal town located south of Casablanca "farmers who have neither the capital nor a property to rent out nor the know-how that makes them essential for tenants are condemned to exclusion, their exclusion stems from an inability to participate in cooperative arrangements" [6]. Until today this type of problem does not arise in our province thanks to the good rainfall it receives, however there are farmers who cannot dig boreholes of more than 50 m, therefore it seems necessary to distinguish between farmers who exercise this activity as investors and those who exercise it for surviving.

The Hydraulic Basins of Sebou (BH), in which the province is located have rich and updated data of the geographical territories located in their field of competence, the most important are the data concerning the zones which record an alarming piezometric retreat, and those whose water pollution and soils reach worrying levels, for this the notion of "hydrological or environmental rest" seems very useful, through the establishment of Territorial Plans or Schemes for the development or rehabilitation of the affected areas, accompanied by a set of substitution projects for the local population. On the other hand, climate change, especially for a country like Morocco which experiences severe years of cyclical drought, doubles the challenge of sustaining agricultural activity.

5. Conclusion

The situation presented above highlights the great discordance between protection measures and preservation goals fixed by the different programs and plans, it also shows how effective are the different laws and water strategies. The most relevant question we must ask ourselves is: What must we do?

1. If the Moroccan Green Plan had great results in economic terms the environmental consequences were very worrying, especially the water resources, both at the national level and at the local level, that is why we must review the main orientations of this Plan.
2. As we have seen, the province is interspersed by several valleys and receives a great amount of rain per year, with an annual rate of more than 512 mm^3 that is why we need to strengthen the surface water storage policy in the province.
3. For the most affected areas we must think to implement a hydrological break policy.
4. As many countries, pricing water can be a useful way to implement sustained water management, but it must be part of an approach of equity and equal access to avoid the problems that we have highlighted above.

6. References

[1] Ait Kadi M and Ziyad A 2018 Integrated Water Resources Management in Morocco. World Water Council (ed), *Global water security , water resources, development and management.* 143-163.Singapore. https://doi.org/10.1007/978-981-10-7913-9_6

[2] Ait Kadi M et al (2014b) Book. Water and the future of humanity- Revising Water security. Co-author, Calouste Gulbenkian Foundation, springer, Berlin

[3] Amraoui.F, Razack.M ,Bouchaou.L (2003), Turbidity dynamics in karstic systems. Example of Ribaa and Bittit springs in the Middle Atlas, *Hydrological Sciences–Journal–des Sciences Hydrologiques,* 48(6), Morocco

[4] Food And Agriculture Organization of The United Nation 2017 A global water quality crisis and the role of agriculture. Executive summary :*Water pollution from agriculture: a global review.*

[5] Moss B 2015 Water pollution by agriculture: a Journal .Phil.Trans.R.Soc.B.*Sustainanble agriculture* I. 659-666.

[6] Ouassissou.R, Kuper.M,Dugué.P, El Amrani.M, Hammani.A, Ameur.F (2019). Rivalités et arrangement cooperatifs pour l'accés à l'eau souterraine dans la plaine de Berrechid au Maroc. Cah.agric.2019.vol28.n°4. https://doi.org/10.1051/cagri/2019006

IOP Conf. Series: Earth and Environmental Science 1090 (2022) 012029 doi:10.1088/1755-1315/1090/1/012029

Seasonal diet of *Salmo trutta macrostigma* in a mountain stream (Ifrane Morocco)

E H Abba[1]*, T Hachi[1], A El Moujtahid[2], H Essabiri[1], C Laiboud[1], O Boumalkha[1], S Zerkani[3] and N Laadel[4]

[1]Higher School of Technology of Khenifra (EST-Khenifra), University of Sultan Moulay Slimane, BP 170, Khenifra 54000 Morocco.
[2]Faculty of Sciences and Technologies. University of Sultan Moulay Slimane.
[3]Faculty of sciences. University Moulay Ismail. Meknès
[4]National Center of Hydrobiological and Fish Farming. B.P.11, 53100 – Azrou Morocco

corresponding author: e.abba@usms.ma

Abstract. The diet of *salmo trutta macrostigma* in Sidi Rachid wadi was conducted on 94 individual ranging in weight from 3 to 575 g with a mean of 110.38 g and a standard deviation of 112.94. Crustaceans mainly characterize the annual food spectrum and insects constitute more than 98%. The study of the variations of the diet spectrum shows that the nature and the importance in number of the ingested items of prey change according to the seasons, with an important relative predominance of crustaceans during all the seasons except winter when insects are the more important food source. In addition, the different taxa can be classified into two categories, a first group of preys always present in the stomach contents during all seasons and whose relative abundance is generally high, and a second group whose presence in the stomach contents was limited to one or two seasons. The most commonly used taxa are amphipoda, Ephemeroptera, and Diptera, with these three taxa forming at least 77% in summer and 90% in winter.

Key words: *Salmo trutta macrostigma,* seasonal food spectrum, Sidi Rachid wadi, Morocco

1. Introduction

The fish biodiversity of the North African ecoregion is characterized by a poor richness in freshwater fish, but reveals an interesting diversity of cyprinid species [1,2]. Salmonids species are distributed over a wide area covering important rivers in northern Norway, northeastern Russia, the southern Atlas Mountains of North Africa and from east to west from Afghanistan to the Aral Sea. River trout, *Salmo trutta,* is one of the important fish species with a high economic value [3]. In Morocco, in addition to the native cyprinids that form the most important family of rivers, the fresh waters of high altitude contain the common trout that is an endemic species *Salmo trutta macrostigma*, Dumeril, 1858, [4,5]. In flowing waters, habitat structure and hydraulic conditions form the parameters of community structuring [6,7]. The management of fishery resources requires a good knowledge of the biological bases such as growth, diet and ecological, namely the habitat in all its forms [8]. The study of the diet of animal species is one of the first steps in their biology, especially for farmed fish [8, 9]. The analysis of the feeding regime of fish species is one of the first phases in their ecology. [10]. Indeed, analysis of the diet can partly explain predator-prey relationships in the studied environment [11] as well as inter- and intra-specific competition processes [12] In the same way, the diet of a species can explain variations in growth as well as certain biological and economic phenomena and physiological aspects of all fish species, such as reproduction [13-14] . Studying the feeding habits of the common trout (Salmo trutta

L.) is one of the fundamental ways to understand its biology. The analysis of the fish's diet not only reveals its trophic requirements, but also provides indirect information on its feeding mode and its interaction with other species [15]. The objective of this study was to determine the seasonal diet spectrum of the endemic fish (Salmo trutta macrostigma) in a mountain stream Sidi Rachid wadi of Middle Atlas between May 2007 and April 2008, to note that managers of fisheries resources especially in livestock whose diet is natural can use the results of this study.

2 Material and methods

2.1 The study area

Sidi Rachid oued is a small mountain stream; it is located at an altitude of 1620m. Its geographical coordinates are 5°9'N and 33°28'W. It is one of the few permanent wadi in the region. Water of the Wadi come in a permanent way from the springs called Ayoun Sidi Rachid from which it takes its name, and from the springs of Ras El Ma, which feed the basins of the salmon farming station of Ras El Ma, which is part of National Centre of Hydrobiology and Pisciculture. The stream flows over a distance of several kilometres before dividing into several small canals intended for the irrigation of agricultural land in the Ougmes region belonging to the rural commune of Ait Ali Ouyakoub (figure 1).

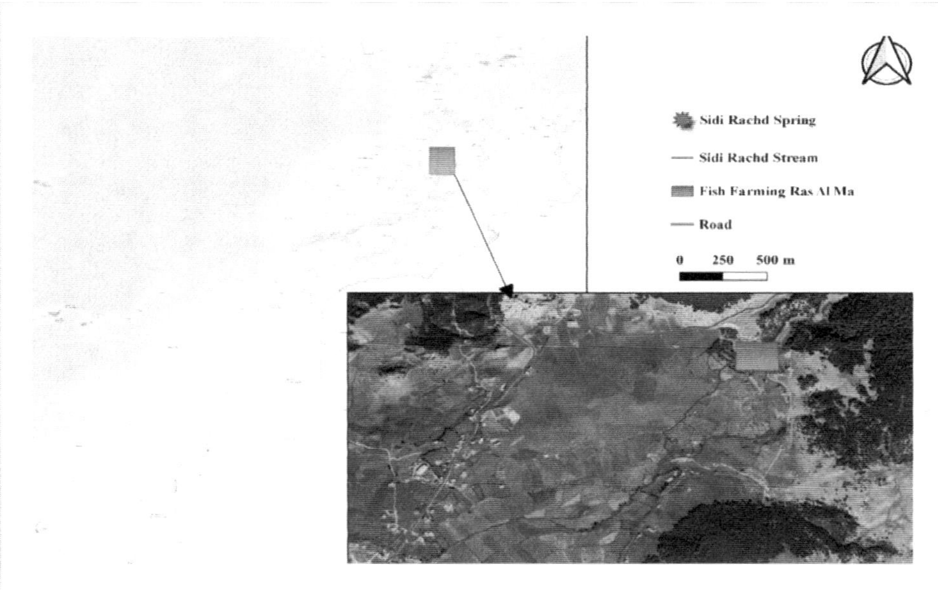

Figure 1: location map of Sidi Rachid River and Fish farming station of Ras Al Ma.

2.2 Water physico-chemical parameters of the study site.

During the study period, the physico-chemical parameters of water quality were monitored in situ, namely temperature (T°C), dissolved oxygen (O_2 mg/L), pH, and Electrical conductivity (EC µS/Cm) using portable instruments. These four environmental parameters are the main elements that affect the biology and ecology of the fish and are related to other physiological parameters of aquatic plants such as photosynthesis and respiration of the different organisms in the environment and the degree of mineralization for the electrical conductivity [14].

2.3 Biological material used in the study

Determining the feeding regime of a given species requires qualitative and quantitative analysis of its stomach contents. This approach is still the most widely used because it gives data on feeding behavior and consumption rate [16,17]. In fact, the analysis of stomach contents also provides information on the relative contribution of prey ingested by a fish in the hours preceding its capture. However, it is generally recognized that this approach provides a biased representation of the fish feeding diet because the time of prey digestion is variable, resulting in higher proportions of prey that take longer to digest compared to small, easily digested prey. An additional potential source of error associated with this method is the inclusion of prey ingested but not assimilated by the predator [18]. Electric fishing, a sampling method that is particularly well suited to certain particular environments (small rivers, streams, ponds, edges, rapids) that are inaccessible to other sampling techniques, captured the specimens used in this study. It also allows the capture of a large number of species, in particular small species that are generally difficult or impossible to capture by other methods [4] . In this study, 94 fish authorized by the water and forestry administration was used to establish the diet of this species during different seasons. The size and the weight of the specimens varied between 6.3cm (minimum weight 3g) as a minimum value and 37cm (as a maximum weight; 575g) as a maximum value during a one-year cycle. After collection, the specimens were fixed immediately by injection of 10% formalin in order to stop the digestive activity and consequently a good preservation of the stomach contents, which are then collected and fixed in the laboratory [19]. Then, each stomach is opened by a longitudinal incision and emptied of its contents by washing with a pizzette, the stomach contents are collected in a Petri dish and are determined qualitatively and quantitatively under binocular magnifying glasses.

3. Results and discussion

3.1 Water quality of Sidi Rachid stream

The maximum temperature recorded during the study period is 15.7°C for the month of May, compared to only a minimum temperature of 11.7°C during February. The average annual temperature is 13.93°. For dissolved oxygen, the minimum concentration recorded is 6.9 mg/L against 12.5 mg/L as the maximum value. As for the hydrogen potential (pH), the values oscillate between 8.5 to 6,3 with an average pH of 7.4. For electrical conductivity, the results obtained vary between 116 and 551.5 µs/cm with an annual average of 295.5 µs/cm. With regard to salmonids waters, the values of the physicochemical parameters recorded meet the requirements of salmonids waters [20,21,22,4].

3.2 food spectrum of salmo trutta macrostigma

The annual food spectrum of common trout was determined by the relative abundance of prey in the stomach contents. Analysis of the data shows that gamarids represent 45.97%, followed by Ephemeroptera 25.85% and Diptera with 12.68%. These results show that brown trout are an opportunist species, because, the abundance of prey in the stomach correlates with the relative abundance of prey in its ecological habitat. Therefore, common trout tend to select the most abundant prey and the most vulnerable taxa in the habitat [23,24]. In addition, these and other preys are the most preferred taxa for brown trout because they have a number of characteristics such as exposure on the substrate and/or in the water column [25,26,27] high mobility [28,29,30,31]; without morphological defences [32-33], or chemical [34] and intermediate size between 5 and 20 mm [34,35] (figure 2 and 3).

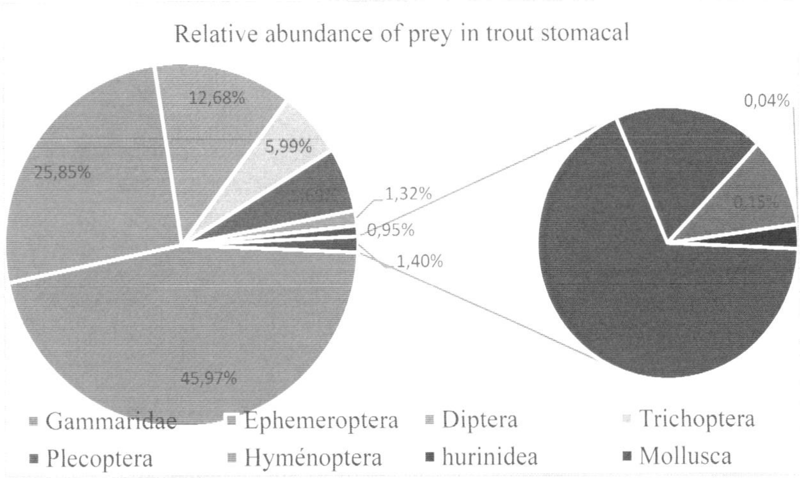

Figure 2. Relative abundance of prey in trout stomachal.

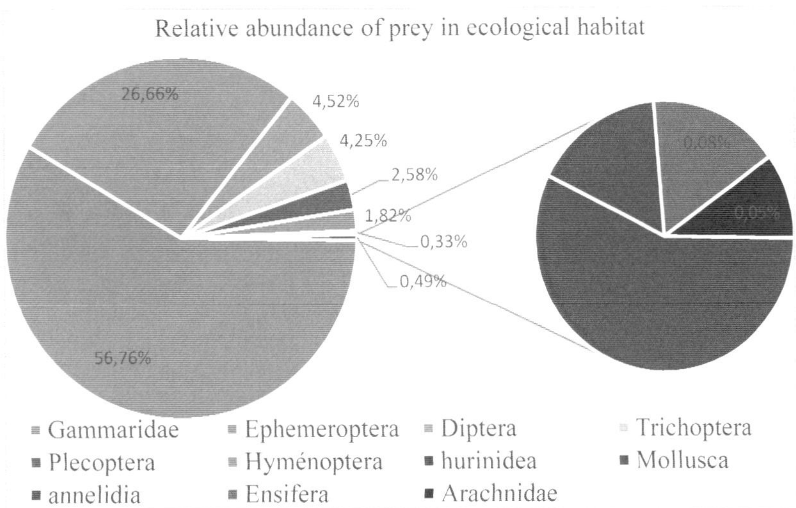

Figure 3. Relative abundance of prey in ecological habitat

The analysis of the variations of the food spectrum of the fario trout during the investigated period (figure 4) shows that the type and the number of ingested preys changes according to the season.The results obtained corroborate those determined by [36], with a significant relative predominance of amphipoda (gammaridae) present in large quantities in the watercourse [24]. Amphipoda are the most consumed during all seasons except winter, when these prey items have given way to mayflies Similarly, the various taxa can be classified into two categories. A first group of prey that are constantly present in the stomach contents in all four seasons and whose relative abundance is generally high, and a second group whose presence in the stomach contents is limited to a few seasons, generally one or two, in which case the relative abundance of prey items is less than 2%. However, it should be noted that some taxa are completely absent during some seasons, such as ensifera and annelidiae during the summer and spring seasons, unlike the other two seasons. This can be interpreted by the fact that the trout resorts to burrowing prey to satisfy its needs following the reduction of the food web during the autumn and winter or during periods of flooding when the water carries away macro invertebrates downstream. It should also be noted that some terrestrial items such as arachnids are found in the stomach contents of trout during a single season with a percentage lower than 0.2%.

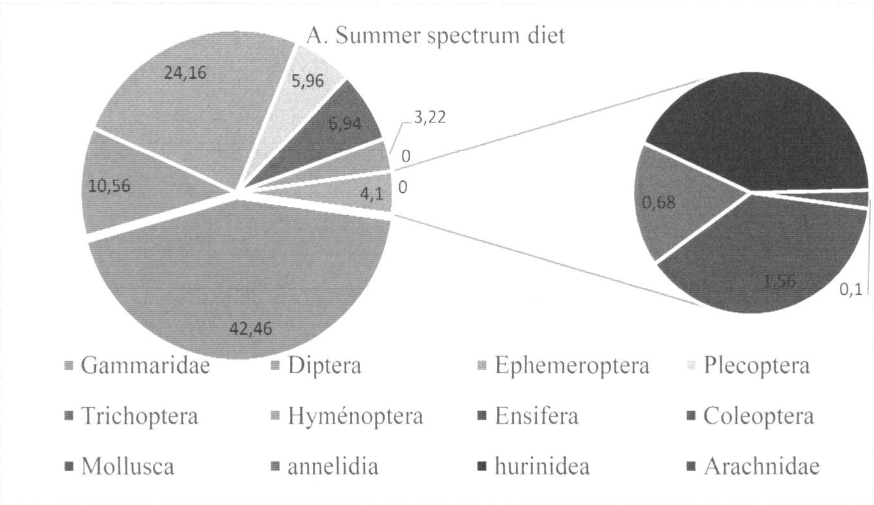
A. Summer spectrum diet
24,16
5,96
6,94
3,22
0
0
4,1
10,56
42,46
0,68
1,56
0,1
Gammaridae
Diptera
Ephemeroptera
Plecoptera
Trichoptera
Hyménoptera
Ensifera
Coleoptera
Mollusca
annelidia
hurinidea
Arachnidae

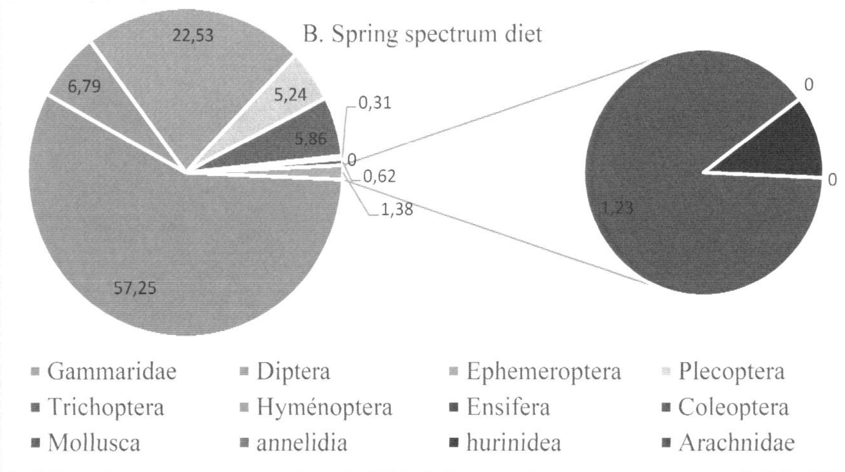
B. Spring spectrum diet
22,53
6,79
5,24
0,31
5,86
0
0,62
1,38
57,25
0
0
1,23
Gammaridae
Diptera
Ephemeroptera
Plecoptera
Trichoptera
Hyménoptera
Ensifera
Coleoptera
Mollusca
annelidia
hurinidea
Arachnidae

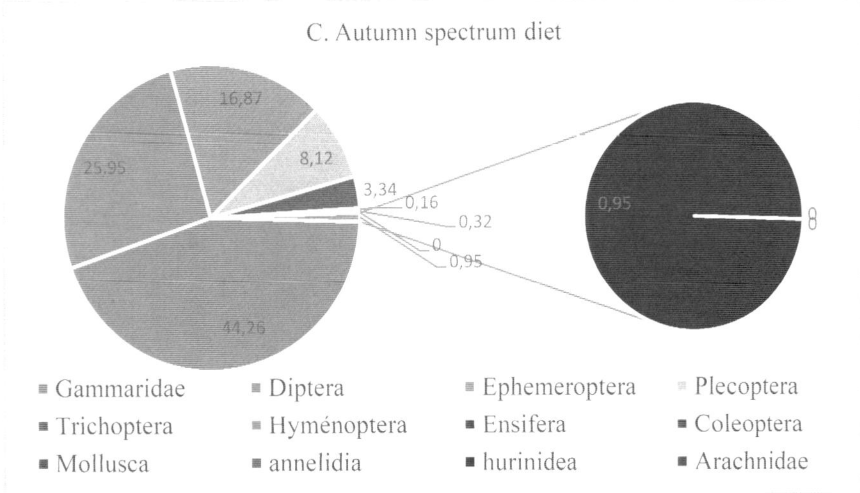

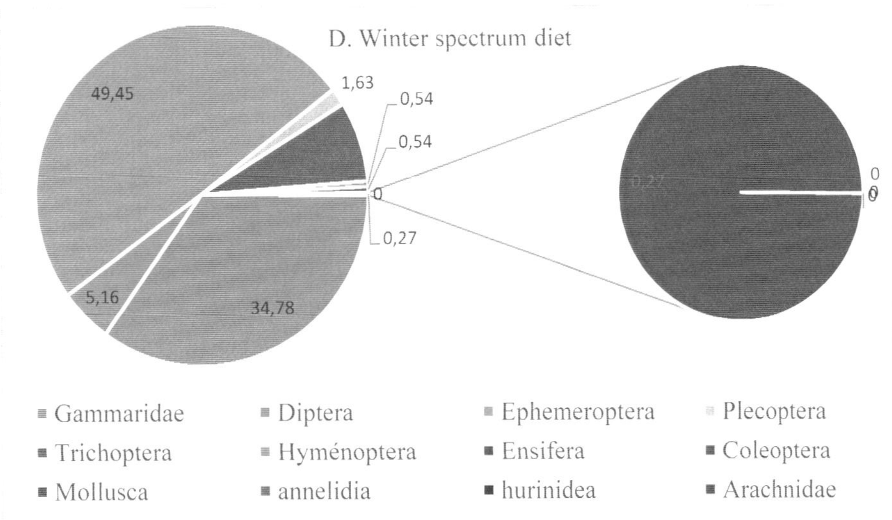

Figure 4: Seasonal spectrum diet of *Salmo trutta macrostigma* (A, B, C and D).

4. Conclusion

The common trout (*Salmo trutta macrostigma,* Dumeril, 1958) is an endemic fish species of Morocco, the study of its diet shows that very varied according to the period of the year. The common trout tends to select the most abundant and vulnerable prey in the environment. Depending on the characteristics of the environment, the season, and the characteristics of the trout itself, the most abundant and vulnerable prey are selected. The most used taxa are amphipods, ephemeroptera and diptera preys, these three taxa constituting at least more than 77% of the trout's diet in summer and 90% in winter.

5. References

[1] Azeroual A, Crivelli AJ, Yahyaoui A and Dakki M. 2000. L'ichtyofaune des eaux continentales du Maroc. Cybium, 24(Suppl. 3) : 17–22.

[2] Mimeche F, Biche M, Ruiz-Navarro A and Oliva-Paterna FJ. 2013. The population structure, age and growth of Luciobarbus callensis (Cyprinidae) in a man-made lake in the Maghreb (NE Algeria). Limnetica, **32 (2)**: 391-404 2013.

[3] Arzu Z, Ozvarol B, Ozvarol Y, Gozacan MN, Bektas S and Yildirim A. 2011. The diet of the brown trout (Salmo trutta L.) in different streams of River Coruh in Turkey. Journal of Food, Agriculture & Environment, Vol.**9 (1)**, January 2011
[4] Abba. E.H. 2011. Etude ecologique et biologique de la Truite Commune (Salmo Trutta Macrostigma, Dumeril, 1858) de L'Oued Sidi Rachid (Ifrane-Maroc). Thèse de Doctorat. Es Sc. Université Ibn Tofail. Fac.Sc. Kénitra. Maroc
[5] Fekhaoui M, Yahiaoui A, Perea S and Doadrio I. 2016. Fragilité hydrologique et biodiversité piscicole et aquacole des plans d'eau des Parcs Nationaux de la cédraie de l'Atlas. Travaux de l'Institut Scientifique, série Zoologie N° 50. 2016.
[6] Hart DD and Finelli CM. 1999. Physical-biological coupling in streams: The pervasive effects of flow on benthic organisms. Annu Rev Ecol Syst 30:363–395. 11
[7] Thomson JR, Clark BD, Fingerut JT and Hart DD. 2004. Local modification of benthic flow environments by suspension-feeding stream insects. Oecologia (2004) 140: 533–542
[8] Essamlali I, Mesfioui M et Bazairi H. 2005. Etude du régime alimentaire du Merlu (Merluccius Merluccius,Linnaeus 1758) et du Pageot (Pagellus Acarne, Risso 1826) de la côte Nord Atlantique Marocaine. 3ème Journées Internationale des Géosciences de l'Environnement. Maroc. El Jadida les 8, 9 et 10 juin.
[9] Boujarda T,. Gélineaua A, Covèsb D, Corrazea G, Duttob G, Gassetb E and Kaushika S. regulation of feed intake, growth, nutrient and energy utilisation in European sea bass fed high fat diets. Aquaculture, 2004, Vol. **231**, No. 1/4, pp. 529- 545
[10] Kaiser JM and Hughes NH.1993. Factors affecting the behavioural mechanisms of diet selectionin fishes. In Bihavioural Ecology of Fishes (F.A. Huntingfordt & P.Torrecilli,eds). Erice, Sicily (Italy): Marine Behaviour and Physiology.
[11] Sih A and Wooster DE.1994. Prey Bihaviour, prey dispersal, and predator impacts on stream prey. Ecology 75, 1199-1207.
[12] De Pirro M, Marchetti GM and Foraging G.1999. Interactions among three benthic fish in a Posidonia oceanica reef lagoon along the Tyrrhenian coast. Journal of Fish Biology **54**, 1300-1309.
[13] Rosecchi E et Nouaze Y.1987. Comparaison de cinq indices Alimentaires Utilisés dans l'analyse des Contenus Stomacaux. Rev. Trau. Inst. Pêches marit.49 (3 et 4) : 111-123.
[14] Kestemont R, Jourdan S, Houbart H, Mélard C, Paspatis M, Fontaine P, Cuvier A, Kentouri M and Baras less E. 2003. Aquaculture Size heterogeneity, cannibalism and competition in cultured predatory fish larvae: biotic and abiotic influences. Biology. Aquaculture.
[15] Ruginis T 2008. Diet and Prey Selectivity by Age-0 Brown Trout (Salmo Trutta L.) in Different Lowland Streams of Lithuania. Acta Zoologica Lituanica
[16] Bagliniere JJ et Maisse G.1990. La croissance de la truite commune (Salmo trutta L) sur le bassin du Scorff. Bulltin Français de pêche et de pisciculture, 318,89-101.
[17] Bagliniere JJ et Maisse G 1990. La truite, biologie et écologie, éditions ENRA.
[18] Peron C.2006.. Contribution des épibiontes au régime alimentaire des poissons associées aux piliers des jetées dans les canaux de la Gold Coast, Queensland, Australie. Mémoire de Master 2Université de La Rochelle.
[19] Dietoa YM, Gourene G et Ouattara A. 2007. Habitudes alimentaires de Brycinus longipinnis dans le complexe fluvio-lacustre de la Bia, Côte d'Ivoire. Belg. J. Zool., **137 (1):** 3-9.
[20] Bagliniere JJ. 1999. Introduction: The brown trout (Salmo trutta): its origin, distribution and economic and scientific significance: In Biology and ecology of the brown and sea trout, Baglinière J.L. et Maisse . (Eds), 1-12. Springer-Praxis, Chichester,UK Schlumberger.O.2002. Memento de pisciculture d'étang. CEMAGREF, 4e édition, 237p.
[21] Schlumberger.O.2002. Memento de pisciculture d'étang. CEMAGREF, 4e édition, 237p.
[22] Vandeputte M et Launney S.2004. Quelle gestion génétique de la domestication chez les poissons? INRA Prod.Anim., 17. 237-242.
[23] Macneil C, Elwood RW and Adick JT. 1999. Predator- prey interactions between brown trout Salmo trutta and native and introduced amphopods, their implications for fish diets. Ecology 22, 686-696.
[24] Macneil C, Elwood RW and Adick JT.2000. Factors influencing the importante of Gammarus spp. (Crustacea: Amphipoda) in riverine salmonid diets. Archiv fur Hydrobiologie 149, 87- 107.

[25] Mclaughlin RL, Noakes DLG and J.W.A. 2000. Living with failure: the prey capture success of young brook charr in streams. Ecology of Freshwater Fish 9, 81-89.

[26] Lehane BM, Walsh B, Giller PS and Halloran JO. 2001. The influence of small-scale variation in habitat on winter trout distribution and diet in an afforested catchment. Aquatic Ecol., 2001; 61: 61-71.

[27] Kara C and Alp A. 2005. Feeding Habits and Diet Composition of Brown Trout (Salmo trutta) in the Upper Streams of River Ceyhan and River Euphrates in Turkey. Turk J Vet Anim Sci 29 (2005) 417-428 13.

[28] Ware D.M 1973. Risk of epibenthic prey to predation by rainbow trout (Salmo gairdneri). Journal of the Fisheries Research Board of Canada 30: 787–797.

[29] Ivrine JR and Northcote TG. 1983. Selection by youg raibow trout (Salmo gairdneri) in simulated stream environments for live and dead prey of different sizes.Canadian Journal of Fisheries and Aquatic Sciences 40, 1745-149. Otto.C.1984 Adaptative head coloration in case-making caddis larvae. freshwater Biology14,317-321.

[30] OTTO.C. 1984.Adaptative head coloration in case-making caddis larvae. Freshwater Biology14,317-321.

[31] Feltmate BW, Williams DD and Montgomerie A. 1992. Relationship between diurnal activity patterns, cryptic coloration, and subsequent avoidance of predaceous fish by perlid stoneflies. Canadian Journal of Fisheries and Aquatic Sciences 49, 2630-2634.

[32] Otto C and Siostrom P. 1983. Cerci as antipredatory attributes in stonfly nymphs.Oikos 41, 200-2004.

[33] Wootton JT, Parker MS and Power ME.1996. Effects of disturbance on river food webs. Science 273,1558-1561.

[34] Scrimshaw S and Kerfoot WC. 1987. Chemical defenses of freshwater organisms: beetles and bugs .in predation: direct and indirect impacts on aquatic communities. (W.C.Kerfoot & A.Sih,eds),pp240-262. Haniver: University of Pr. New England Allan, 1981.

[35] Rader RB. 1997. A functional classification of the drift: traits that influence invertebrate availability to salmonids. Canadian Journal of Fisheries and Aquatic Sciences 54, 1211-1234

[36] Neveu, A. and M. Thibault. 1977. Comportement alimentaire d'une population sauvage de truites fario (Salmo trutta L.) dans un ruisseau des Pyren´ ees ´ Atlantiques, le Lissuraga. Ann. Hydobiol., 8: 111-128.

Mapping the natural habitats of a continental wetland: Dayet Er Roumi lake (Khémisset, Morocco)

M Sadki[1*], A Chahlaoui[1], A Maliki [1], M Khaffou [2]and S Lahrouz [3]

1. Natural Resources Management and Development Team, Biology Department, Moulay Ismail University, Faculty of Sciences of Meknes, BP 11 201 Zitoune, Meknes, Morocco.
2. Sultan Moulay Slimane University, Superior School of Technology, Khénifra, Morocco.
3. Training Center for Education Inspectors, Street of Ibn Aicha, Akkari, Rabat.

*sadkimohcine@gmail.com

Abstract. The Dayet Er Roumi lake is located at 15 km from the city of Khémisset, is the only permanent natural continental lake of low altitude in Morocco, and is a zone of biological and ecological interest (SIBE) of priority n°1 and which must be raised to the rank of effectively protected areas (Natural Reserve). The mapping of natural habitats of the Dayet Er Roumi wetland was based on MedWet criteria. This classification proposes to identify the wetland in four hierarchical levels (system, sub-system, class and sub-class), based on three parameters of water characterization (hydrological regime, salinity and artificialization). This study showed that 68.80% of the mapped area of the wetland is occupied by surface water whose depth exceeds two meters; this habitat constitutes a favorable aquatic ecosystem for the colonization of aquatic birds. However, the lake is subject to several anthropogenic activities that hinder the development of natural habitats and their interaction with aquatic birds.

Keywords: Dayet Er Roumi, Wetland, Mapping, Natural habitats, Khémisset, Morocco

1. Introduction

Wetlands are any environment that is completely or partially immersed in water, eitherpermanently or temporarily. They are among the richest environments in terms of biodiversity. Freshwater wetlands are home to more than 40% of our planet's species and 12% of all animal species [1]. These wetlands play a major role in the water cycle and are vitally important to human survival because of the countless benefits and ecological services they have provided for thousands of years.

In wetlands, water is the most important factor for the functioning of these natural areas as well as for animal and plant life. The physical (bathymetry, substrate...), chemical (salinity, temperature, pH...), and biological characteristics of water undergo daily, seasonal or annual fluctuations. These variations depend on climatic conditions, the location of the area within the hydrographic basin and the geomorphological context (geography, topography).

In this context, the mapping of natural habitats plays a key role in the management and preservation of these wetlands [2]. However, in Morocco this mapping started only towards the end of the 20th [ème]century and is based on two types of internationally recognized classifications, that is, the classification of the Ramsar Convention and the classification of theMediterranean initiative MedWet. Thus, several wetlands at the national scale havebeen mapped, namely: the Merja of Fouawarate [3], Smir marsh [4] and Oued Mellah marsh [5].

The overall objective of this work is to map the different types of natural habitats of Dayet Er Roumi, in order to understand its functioning through the distribution of its habitats, as well as their interactions with the existing aquatic avifauna, in order to contribute to the support of the natural resources managers to elaborate an adequate development and management plan for this wetland.

2. Materials and methods

2.1 Study area

2.1.1. *Geographical location.* The Dayet Er Roumi lake is located at 15 km from the city of Khemisset from the national road n°1. Administratively, it comes under the province of Khemisset and is attached to three rural communes namely Houderrane, Ait Ourible and Ait Ouahi. It extends over 2 km in length and from 400 to 700 m in width, the surface area of the water body is about 90 ha with a maximum depth of about 14 m [6], representing the only permanent natural lake of low altitude in Morocco (Figure 1). Dayet Er Roumi is an area of biological and ecological interest (SIBE) of priority n°1and should be elevated to the rank of effectively protected areas (Nature Reserve).

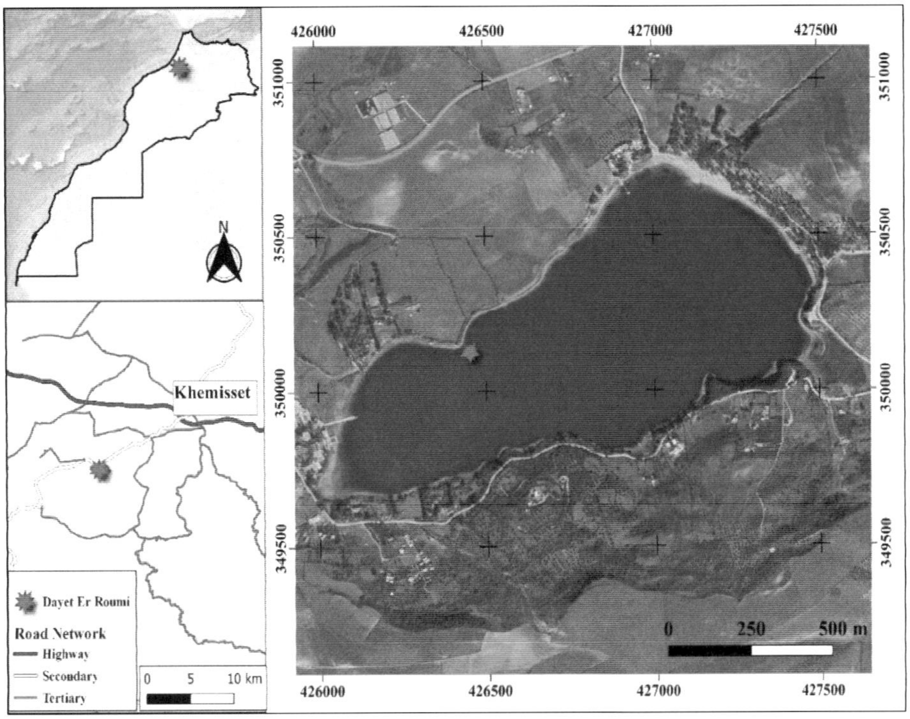

Figure 1: Geographic location of the study area

2.1.2. Soil, geological and hydrological framework. At Dayet Er Roumi Lake, the soils show great spatial variation depending on localenvironmental factors, but can generally be considered as fersiallitic interspersedwith dark red to pink coloured calcimagnesic soils [7].

The study site is part of the so-called central meseta biogeographical zone, occupying the centre of a hilly landscape of more or less bare Mio-Pliocene marl marked by a multitude of funnel-shaped sinkholes. The Triassic evaporiticrocks are partly responsible for these sinkholes, and as a result the waters of thelake are slightly salty [6]. The study site is located on Mesozoic limestone covered by thick deposits of Quaternary alluvium [8].

The Dayet Er Roumi lake is characterized by the existence of an exchange of flows with the groundwater through the fluvio-lacustrine formations and the faults which cross the lake. The groundwater flow converges towards the Daya,which contributes to its supply. As a result, the Dayet Er

Roumi lake has a considerable natural water potential ranking it as the most important lake in theprovince of Khémisset [9].

2.1.3. *Climatic framework.* The study area is subject to a semi-arid bioclimate with temperate winters and a Mediterranean-type rainfall regime. It is characterized by uncertainties and irregularities of rainfall during the same year and by seasonal variations of temperatures. The lake itself has a sub-humid microclimate, contrasting with the vast semi-arid zone that surrounds it [10].

2.2 Methodology

In order to better understand the functioning of our aquatic system, the study of the natural habitats that characterize the Dayet Er Roumi lake is essential. In our work, the cartography of the habitats was based on the MedWet classification, since the study site is located in the Mediterranean region [11].

This classification proposes to identify the wetland in four hierarchical levels (system, sub-system, class and sub-class), based on three water characterization parameters (hydrological regime, salinity and artificialisation). This typology also allows criteria relating to hydrological parameters (rate of submersion, bathymetry and origin of water), sedimentology (granulometry) and vegetation(biological type) to be taken into consideration.

In the present study, the habitat map was made using satellite images, topographic map and the specific management plan of Dayet Er Roumi. Severalfield visits were carried out to identify the habitats and their boundaries, and the presentation of the data was done using GIS (QGis 3.18).

The mapping of habitats was schematized following the MedWet classificationcriteria, where each habitat type is characterized by a nomenclature and an appropriate code (Figure 2).

3. Results and discussion

The mapping of habitats according to the MedWet classification has allowed Dayet Er Roumi to be classified as a lacustrine system (L), with a depthexceeding two metres, over an area of about 76ha (surface water and aquatic bed) (Figure 2 and Table 1). This system is divided into two subsystems: limnetic (M) and littoral (L).

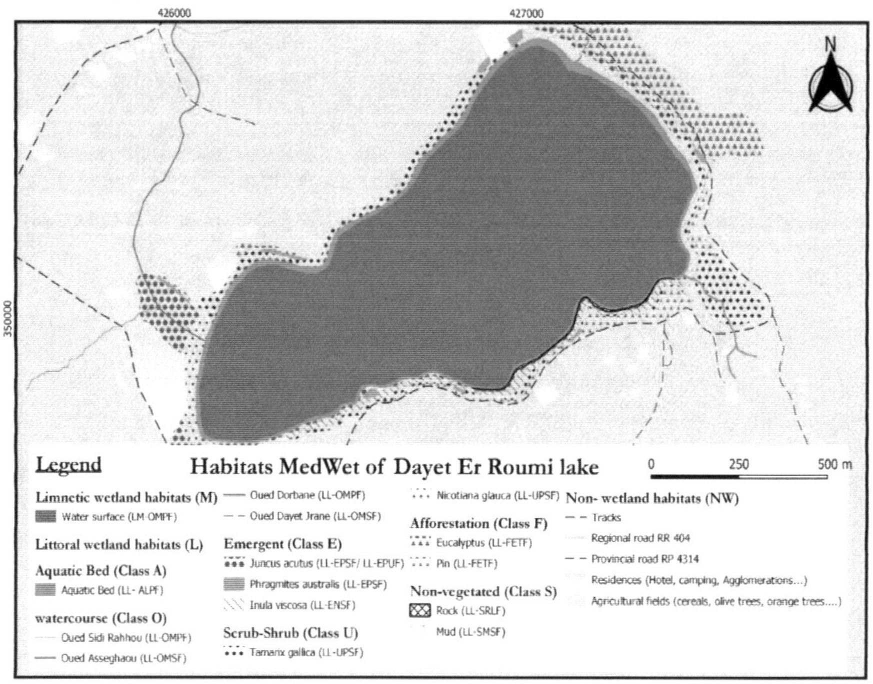

Figure 2: Map of natural habitats of Dayet Er Roumi Lake

Table 1. Different types of habitats characterizing the Dayet Er Roumi lake (MedWet typology)

Habitats				Area(ha)	% Wetland
Wetland habitats (lacustrine L)	Limnetic (M) :	Water surface (LM-OMPF) :		71,88	68,80
	Littoral (L) :				
		Aquatic Bed (LL- ALPF)		4,28	4,10
		Emergent (E) :	*Juncus acutus* (LL-EPSF/ LL-EPUF)	4,68	4,48
			Phragmites australis (LL-EPSF)	0,49	0,47
			Inula viscosa (LL-ENSF)	1,35	1,29
		Scrub-Shrub (U) :	*Tamarix gallica* (LL-UPSF)	6,17	5,91
			Nicotiana glauca (LL-UPSF)	1,34	1,28
		Forested (F) :	Eucalyptus (LL-FETF)	9,06	8,67
			Pin (LL-FETF)	2,5	2,39
		Non-vegetated (S) :	Rock (LL-SRLF)	0,3	0,29
			Mud (LL-SMSF)	2,43	2,33
		Watercourse (O)			
Non- wetlandhabitats (NW)	Agricultural fields	Cereals, olive trees, orange trees			
	Residences :	Hotel, camping, Agglomerations...			

3.1 Inventory of limnetic wetland habitats (M)
In this limnetic subsystem, only one class is distinguished.

3.1.1. Water surface (class O). Covering about 71.88 ha, it is the most dominant habitat, resembling a depression,with water depths exceeding two meters, under muddy substrate (M) and permanent flooding (P) from precipitation and continental waters (Figure 3). The lake is classified as a freshwater wetland (F).

68.80% of the mapped area of the wetland is occupied by surface water, this habitat constitutes a favourable aquatic ecosystem for the colonisation of Rallidae (such as the Coot), Anatidae (e.g. Mallard), Phalacrocoracidae (e.g. Great Cormorant) and Podicipedidae (e.g. Great Crested Grebe).

Figure 3. Water surface (LM-OMPF)

3.2 Inventory of littoral wetland habitats (L)
This subsystem includes all wetland habitats of the lake system extending from the shoreline to a depth of two meters below low water [8]. It isrepresented by five (05) classes.

3.2.1. Aquatic bed (Class A). Corresponds to a thin surface of about 4.28 ha around the water surface, its surfaceis more or less variable depending on the amount of precipitation, it is characterized by the presence of floating type plants (L) (e.g. *Potamogeton pectinatus* and *Utricularia vulgaris*). This type of habitat is regularly flooded (P).

The aquatic bed representing 4.10% of the total surface is considered as a potential place for the colonization of waders (such as the White Stilt), Ardeidae (such as the Little Egret) and Ciconidae (such as the White Stork).

3.2.2. Emergent vegetation (Class E). Occupies an area corresponding to 6.24% of the surface of the lake system and iscomposed of three formations of persistent (P) and non-persistent (N) type.

***a.** Formation of (Juncus acutus L).* A rush of 4.68 ha, corresponding to 4.48% of the total surface of the wetland, surrounds almost the entire water body, it is of the persistent and seasonally submerged type (LL-EPSF). The major part of this reedbed dries up in summer,but towards the NW and SE parts and in contact with two marshy areas under theeffect of the watercourses that feed the lake (Oued Asseghaou, Oued Sidi Rahhouand Oued Dorbane), this reedbed seems to be saturated (LL-EPUF) (Figure 4).

Figure 4. Formation of *Juncus acutus* (LL-EPSF/ LL-EPUF)

***b.** Formation of Phragmites australis Cav.* This vegetation is in the form of small persistent mosaics (P) scattered around thesite as a result of illegal cutting by the neighbouring population for domestic use (Figure 5). This formation occupies an area of 0.49 ha and is seasonally submerged(LL-EPSF).

Figure 5. Formation of *Phragmites australis (LL-EPSF)*

c. *Formation of Inula viscosa* L. Occupies the SE part of the site, bordering the rush to the edge of the Tamarix stand (Figure 6). This non-persistent invasive vegetation of 1.35 ha is characterized by seasonal flooding (LL-ENSF).

Figure 6. Formation of *Inula viscosa* (LL-ENSF)

The vegetation formations based on evergreen and non-evergreen emergents (Jonchaie, Phragmitaie and slimy Inula) represent only 6.24% of the surface ofthe wetland, and the limit between these habitats is fluctuating and in some places form a mixed habitat. These emergent formations form areas used by some bird species for feeding and nesting.

The presence of nitrophilous plants, particularly the genus Rumex (*Rumex conglomeratus* and *Rumex pulcher*), shows that Lake Dayet Er Roumi is subjectto increased nitrate pollution [12].

3.2.3. *Scrub-Shrub (Class U).* This evergreen habitat type (P) consists of two main formations.

a. Formation of (Tamarix gallica L). Forming a high tree expanse around the lake of 6.17 ha and not exceeding 10 m in width, but at the NE end of the lake, where there is a swampy depression, this vegetation occurs as a dense stand of 2.5 ha in area (Figure 7). The tamarack standis flooded during the rainy season (LL-UPSF).

Figure 7. Formation of *Tamarix gallica* (LL-UPSF)

b. Formation of (Nicotiana glauca Graham). This type of habitat forms a more or less dense isolated mass towards the SE endof the lake, opposite the Daya hōtel, its surface area is estimated at 1.34 ha and isseasonally submerged (LL-UPSF) (Figure 8). Its presence in the vicinity of the hotel can bebeneficial through its use as an insecticide. However, it is very dangerous for human health due to its toxic nature [13].

Figure 8. Formation of *Nicotiana glauca* (LL-UPSF)

-

The natural formations of Scrub-Shrub (Tamarack and Tobacco) cover 7.19% of the wetland area, this vegetation serves as a dormitory and nesting place for some species of birds (Ardeidae and Ciconiiformes).

3.2.4. *Forested (Class F)* This habitat class is presented in the form of artificial reforestations located on small landforms with a high risk of erosion. It consists of two afforestations.

a. Afforestation of (*Eucalyptus Gomphocephala* DC). It is an artificial reforestation of Eucalyptus gomphocephala of an area equal to 9.06 ha, temporarily flooded. It is located in the NE part of the lake, passing the RP4314 along a stream coming from a swampy area located 1-2 km NE of the lake (LL-FETF). Towards the SW part of the lake, there are two naturally regenerating Eucalyptus gomphocephala clumps bordering a tourist residence (Figure 9).

Figure 9. Afforestation of *Eucalyptus Gomphocephala* (LL-FETF)

b. Afforestation of Pinus halepensis Mill. It takes the form of two artificial reforestation plots of Aleppo pine with a total area of 2.5 ha (Figure 10). They are located in areas of temporary flooding (LL- FETF).

Figure 10. Afforestation of *Pinus halepensis* (LL-FETF)

IOP Conf. Series: Earth and Environmental Science 1090 (2022) 012030 doi:10.1088/1755-1315/1090/1/012030

More than 80% of the land around Dayet Er Roumi is cultivated with cereals and legumes and natural forest is absent [8]. Artificial vegetation is composed essentially of Eucalyptus and Pine reforestation and islocated on small reliefs at high risk of water erosion.

3.2.5. Non-vegetation (class S)
a. Mud. Located in the northern part of the wetland on a muddy substratum and submergedduring the rainy season (LL-SMSF), this type of habitat with a surface area of 2.43 ha is the result of degradation of the rush as a result of anthropogenic factors.
b. Rock. It corresponds to the parts that are located towards the East of the lake on rocky substratum and semi regularly flooded (LL-SRLF). This area is very frequented by fishing enthusiasts.

3.2.6. *Watercourse (class O).* Lake Dayet Er Roumi occupies the bottom of an elongated doline in WSW-ENEdirection, fed by groundwater and surface runoff.

From the SE, a permanent stream (Oued Dorbane) feeds the lake (LL-OMPF).It is regularly flooded in winter and shows a flourishing and varied vegetation (rush, Tamarix, Acacia, Plantain horn of stag).

At the NW end of the lake, where a stream (Oued Asseghaou) feeding the lakeseasonally emerges (LL-OMSF), a platform corresponding to the outlet of thelake joins it to a small permanent stream (LL-OMPF) (Oued Sidi Rahhou), thewhole forms a meadow based on a mixture of rushes, goosefoot and dock, characterized by the presence of dead trees of Eucalyptus.

To the NE of the lake is a drainage channel from a swampy depression (Frog Lake) 1-2 km NE of the lake, forming a seasonal stream (LL-OMSF).

3.3 Inventory of non-wetland habitats (NW)
These are the plant formations scattered around the wetland and composed mainlyof scattered trees of Atlas Pistachio (*Pistacia atlantica* Desf), Acacia (*Acacia cyanophylla* Lindl. and *Acacia horrida* L.), White Poplar (Populus alba), Elm- leaved Bramble (*Rubus ulmifolius* Schott) and False Pepper (*Schinus molle* L.).

In addition, artificial habitats (crop fields, olive trees, roads, pites, residences) were mapped, given their importance in the analysis, and understanding of the functioning and degradation of the wetland.

4. Conclusion

The mapping of wetland habitats allowed the classification of the Dayet Er Roumi lake in the continental freshwater lacustrine system. This system includes relatively diverse natural habitats and with a remarkable similarity to other freshwater wetlands in northwest Morocco such as the Merja de Fouawarate [3]. However, the Dayet Er Roumi wetland seems to be in progressive regression due to the long periods of drought that the area has experienced, the installation of arboriculture around the lake (olive trees and orange trees) and also to the intensity of the anthropic factor.

In the wetland, the distribution of habitats is affected by three main parameters,namely exposure, water factor and anthropogenic factor. The densest and mostdiversified formations are found at the northern exposure of the lake and in areas of permanent or seasonal submersion, while the anthropic factor such as the location of visitors with their vehicles are devoid of vegetation.

The presence of vascular plants, which are powerful integrators of biotopic andanthropic factors. They are subsequently the best bioindicators of the state of health of the environment, its degree of naturalness or disturbance [14].

In addition, at the level of the wetland, the main pressures and threats to the environment are anthropogenic and natural. The natural constraints are expressed by the seasonal and interannual variation of the climate, which is a determining factor in the evolution of the lake's ecosystems (prolonged periodsof drought). Whereas the anthropic constraints that are posed to the sustainability of the functions and role of the natural ecosystems of the site are

- Too many crops on the edges and hills overlooking the lake.
- Pasture: overgrazing of wet meadows.

- Disturbance of birds by water sports and fishing gear.
- Pollution from visitors (uncontrolled tourism).
- Hunting and poaching.
- Erosion due to the cultivation techniques adopted.
- Total destruction of shoreline vegetation through overgrazing andexploitation.
- Increase in the human population along the shoreline.
- Accelerated siltation of the lake.
- Accelerated degradation of water quality by the hotel, café(s), campsiteson the lake shore.

Various methods and techniques exist to identify the different natural habitats of wetlands. The present work has focused on the MedWet classification, whichis an efficient inventory tool adapted to the wetlands of the Mediterranean area [12]. This method has an advantage over other classical methods because the expected data are in the form of nomenclatures and codesappropriate to each habitat type and can be classified according to predefined criteria. However, the application of the MedWet classification at the level of the Dayet Er Roumi lake made it possible to criticize the criteria of identification of the habitats of the lacustrine system since the persistentemergents, as well as the trees and shrubs are excluded from this system. To this end, we propose to widen the criteria of definition of the lacustrine system to that only the biotic and abiotic parameters can be the principal keys of identification.

5. References

[1]. Royal C. G., Max F C 2018 Global wetland outlook: the state of the world's wetlands and the services they provide to humanity.

[2]. Sessi G, Gérard N, Dèhouégnon T, Akoeugnigan I, Alain S, Marcel R and Brice S 2016 Effectiveness of protected areas in the conservation of priority favorable habitats of valuable woody species in Benin, Bois et forêts des tropiques, 2016, n°328 (2).

[3]. Lahrouz S, Dakki M and Gmira N 2013 Mapping of natural habitats of a peri- urban wetland: the Merja of Fouwarate (Kenitra, Morocco). Bulletin de l'Institut Scientifique, Rabat, section science de la vie, 2013, n°35, 27-34.

[4]. Dakki M, Hamman F and Hammada S 2005 Mapping the natural habitats of a Mediterranean coastal wetland: the Smir marshes (Tetouan region, Morocco). In: Bayed A. & Scapini F. (Eds). Sensitive coastal ecosystems of the Mediterranean: the case of the Smir coastline. Trav. Inst. Sci, Rabat, General Series, 4, 9-15.

[5]. Tahiri S. 2004. Contribution to the phytoecological study of the biodiversity of the wetland of the oued Mellah and the anthropic activities on its natural environment and the environment (Region of Mohammedia of Mohammedia, Morocco). Mém. 3ème cycle, ENFI, 198 p.

[6]. Administration des Eaux et Forêts et Conservation des sols (AEFCS) 1996 Plan directeur des aires protégées du Maroc. Unpublished report. Rabat: AEFCS;BCEOM-SECA; Inst.Sci; EPHE, Sd; vol.1-5.

[7]. Duchaufour P 1970 Précis de Pédologie, 3rd èmeedition, Masson, Paris.

[8]. Flower R J, Stevenson A C, Dearing J A, Foster I D L, Airey A, Rippey B, Wilson J P F and Appleby P G 1989 Catchment disturbance inferred from paleolimnological studies of three contrasting sub-humid environments in Morocco; Journal of Paleolimnology 1:293-322, 1989.

[9]. EL ABIDI K, AMRAOUI F, AMAR M and BENSALMIA A 2015. Contribution of bathymetry to the hydrogeological study of Dayet Roumi. 5thème edition of the international congress Eaux, Déchets et Environnement Un défit millénaire (EDE5). Faculty of Science and Technology Settat - University Hassan I Morocco November 24-26, 2015.

[10]. Emberger L, 1955 A biogeographical classification of climates. Rev. Trav. Lab. Bot. Sci. Montpellier, 7. p. 3-43.

[11]. Farinha J C, Costa L T, Zalidis G C, Mantzavelas A L, Fitoka E N , Hecker N and Tomas V P

1996 Mediterranean Wetland Inventory: Habitat Description System. MedWet Publication IV, MedWet/EKBY/ICN/WI, 84 p.
[12]. Benabid A 2000 Flore et écosystème du Maroc : évaluation et préservation de la biodiversité. Ibis Press, Paris, 357 p.
[13]. Victoria F, Moshe H, Noa S, Gabriel S and Shoshana Z Nicotiana glauca (Tree Tobacco) Intoxication-Two Cases in One Family. Journal of medical toxicology: official journal of the American College of Medical Toxicology, 2011, no. 7:47- 51.
[14]. Chkhichekh A 2015 Etude de la diversité des groupements végétaux du Plateau Central Marocain : cas du Site d'Intérêt Biologique et Ecologique (SIBE) d'El Harcha, Maroc. PhD thesis; Faculty of Sciences Rabat - Morocco.

IOP Conf. Series: Earth and Environmental Science 1090 (2022) 012031 doi:10.1088/1755-1315/1090/1/012031

Evaluation of the metallic and bacteriological quality of well water in Khenifra province (Morocco).

Y El hammioui[1]*; H Anarghou[1]; ML Belghiti[2] ; T Hachi[3]; D Elhamdouni[4] ;C Laiboud[1];H Essabiri[1]; O Boumalkha[1]; M Khaffou[5] and E H Abba[1, 5]

[1]Biological engineering laboratory, Faculty of Sciences and Technics, Sultan Moulay Slimane University.Campus Mghila, B.P 523, 23000 Béni-Mellal, Morocco.
[2]Natural Resources Management and Recovery Team, Department of Biology, Faculty of Sciences, Moulay Ismail University, Meknes Morocco.
[3]Faculty of Science, University Ibn Tofail , Kenitra Morocco.
[4]Water ressources management and valorization and remote sensing team, department of earth sciences, Faculty of Sciences and Technics, Sultan Mouly Slimane University. Campus Mghila, B.P 523, 23000 Beni-Mellal. Morocco
[5]Heigher School of Technology,Sultan Moulay Slimane University.B.P.170, 54000 Khénifra.Morocco.

*Corresponding author : youssef.elhammioui@usms.ma

Abstract. Water contaminated by microorganisms and chemical molecules is a cause of disease for humans and the environment. For this reason, we have assessed the microbiological quality and metallic trace elements of groundwater used by the rural population of the Khenifra province, Morocco. Two groundwater sampling campaigns were carried out during the period February/August 2021, studying physico-chemical parameters such as temperature, electrical conductivity, dissolved oxygen, and ph. Then, the metallic trace elements were analyzed: lead (Pb), copper (Cu), nickel (Ni), manganese (Mn), iron (Fe), and zinc (Zn). In addition, the samples taken for the bacteriological study were filtered and introduced in a growth medium for the research and enumeration of total aerobic mesophilic germs vary from 1 to 30.10^3CFUs/ml, total coliforms (1 to 3x10^2 CFUs/100ml), Escherichia Coli(1 to 3x10^2 CFUs/100ml)and intestinal enterococci(1 to 290 CFUs/100ml). The analysis was carried out according to the recommendations of Moroccan standards. It is interesting to note that the groundwater in the Khenifra region is contaminated with pollutants and can cause a health risk for the inhabitants.

Key words: Groundwater, bacteriological quality, health risk, khenifra.

1. Introduction

In Morocco, groundwater is an important part of the country's hydraulic heritage [1]. Groundwater, often geologically protected, is exposed to agricultural, industrial, or urban pollution. The water table in the study area represents the main water supply resource of the region. Groundwater has a primary place among the resources used for human nutrition. Furthermore, the protection and conservation of these resources is a primary concern at the national and international levels. These resources are subject to daily perturbations, which lead to a deterioration of their hygienic quality [2,3]. It is of interest to note that the future of the region's economy is strongly conditioned by the quality of water resources and how they are exploited, this region has experienced in recent decades significant agricultural development and significant population growth. The water demand has consequently strongly increased, mainly at the level of the large agglomerations. This vital need for water, which affects the region's economy and its population, could easily have been met by the importance of groundwater resources in the region. The gap observed between supply and demand, the competition in the drinking water supply sector (agriculture, water pollution, water-borne diseases, etc.) are all problems posed not only to developing countries but also to industrialized countries. However, the issue of water remains a crucial subject: supplying cities with drinking water has always been an absolute priority and covers several aspects: economic, environmental, technical, political, and socio-cultural [4]. The potabilization of water intended for human consumption must benefit from particular attention. This water must contain neither dangerous chemical substances nor harmful pathogenic germs susceptible to the health of individuals in the more or less long term.[5,6]. Morocco is experiencing a water shortage due to several factors. First, the country is characterized by an arid to semi-arid climate [7] with episodic droughts to which is added the irrational management and use of water [8]. Groundwater in the study area is generally shallow and its continued use and exploitation are not permitted. Demographically, the population of this area was 100.000 in 2004, and the census data has been adjusted to 2009 administrative divisions[9]. Overall, the present study aims is to evaluate the physico-chemical quality, metallic trace elements, and microbiological quality of groundwater from different wells used by the rural population in the Khenifra province Morocco.

2. Material and methods

2.1. Study area

Khenifra province is located in the northern part of the Beni Mellal-khenifra region. It is one of twelve regions created by territorial division as part of advanced regionalization (figure 1). Administratively, the province is divided into 22 municipalities directly under the Central Government, of which 2 are in cities and 20 are in rural areas. According to the September 2014 census , its population is estimated to be 371,145 inhabitants [10]. The province's climate is a continental Mediterranean mountain range, which is characterized by cold and rainy winters, snow in the mountains, hot and dry summers, and stormy periods[11].

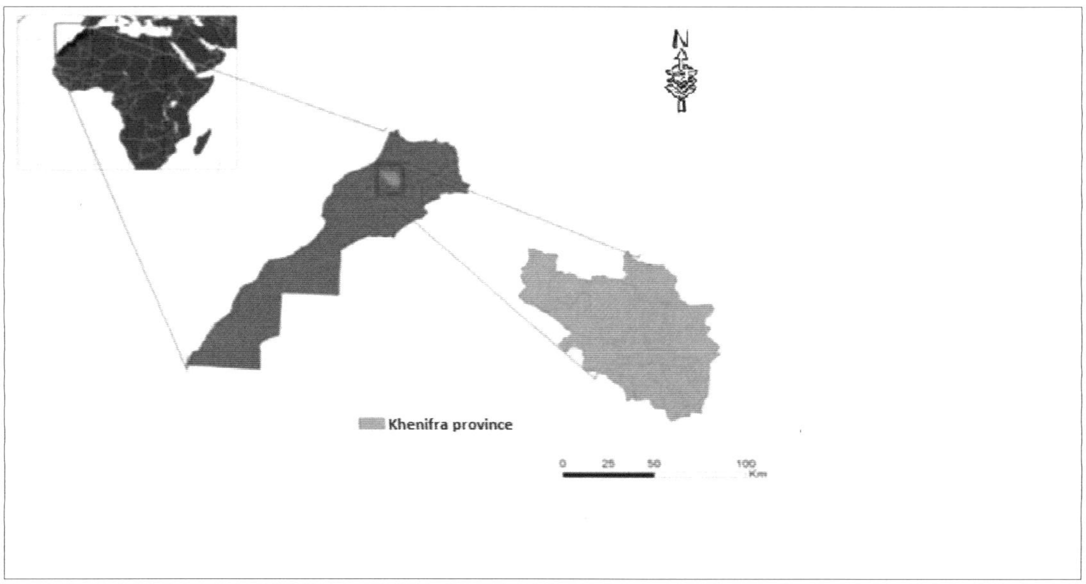

Figure 1. Map of study area

IOP Conf. Series: Earth and Environmental Science 1090 (2022) 012031 doi:10.1088/1755-1315/1090/1/012031

2.2. Methodology

For this study, the sampling of well water was done to cover the whole Khenifra province to have a correct vision of the physicochemical and bacteriological quality. To know the bacteriological quality of water, two campaigns of water sampling were carried out during the year 2021. This allowed the collection of 28 water samples from 14 wells (figure 2). Before use, the vials were thoroughly washed and then rinsed with distilled water. Then, the vials were sterilized. These samples were collected in autoclaved 500 ml vials, labeled, and stored in a cooler at 4°C during transport to the laboratory, accompanied by a collection sheet with all necessary information such as origin and date of collection.

For water sampling for bacteriological analysis, we used 500 ml bottles with a string. At the time of sampling, the bottle is opened and inserted into the well at a depth of 30 cm, taking care not to contaminate the sample. Then, the water-filled vial is removed. The string is detached and the bottle is closed under the required aseptic conditions until the moment of analysis.

To carry out the sampling of the geographical coordinates(table1), we had used GPS (of mark GARMIN GPSMAP 64S), which allowed us to take the geographical coordinates.

Table 1. Geographic coordinates of the studied stations

Well	Name Commune	Altitude(m)	Geographic cordinate	
			N	W
P1:bouchbel	El Hammam	954	33°18’11.9’’	5°35’22.3’’
P2:dar telmid bouchbel	El Hammam	943	33°18’17.6’’	5°35’20.1’’
P3:aouam tighza	El Hammam	1040	33°10’38.6’’	5°38’32.5’’
P4:el borj	El Borj	894	33°01’06.8’’	5°37’40.8’’
P5:aguelmam	Aguelmam	1500	32°58’36.8’’	5°27’14.6’’
P6:ait nouh	Moha ou Hammou Zayani	1032	32°54’25.3’’	5°35’01.3’’
P7:lenda	El Kbab	980	32°45’56.0’’	5°33’47.6’’
P8:laabar	Kahf Nssour	789	32°56’42.0’’	5°55’59.9’’
P9:laguaida	Sidi Hcine	1085	33°03’01.4’’	5°59’54.7’’
P10:timdghas	Tighssaline	1710	32°40’06.0’’	5°30’16.5’’
P11:sidi yahya ou saad	Sidi Yahya	1487	32°40’54.4’’	5°34’34.4’’
P12:souk d'aguelmous	Aguelmous	1166	33°09’40.6’’	5°49’55.3’’
P13: tnin ait boukhyou	Sebt Ait Rhou	1019	33°20’33.6’’	6°16’39.7’’
P14:foughal	Mly Bouazza	1084	33°11’13.6’’	6°08’18.5’’

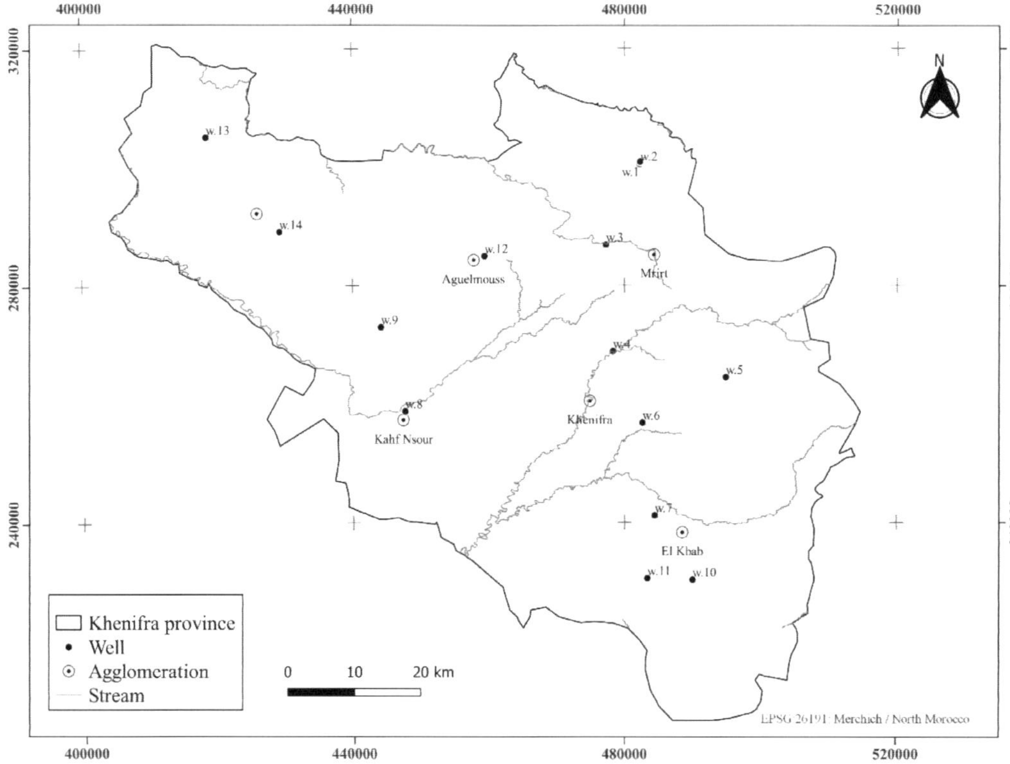

Figure 2 . Map of khenifra province, Morocco, area of sampling and the sampling points.

Some Physico-chemical parameters are measured in situ (temperature, dissolved oxygen, hydrogen potential (pH), and electrical conductivity) by a portable measuring device; model HQ40d -multiple with two channels.

2.3. Analysis of heavy metals

We proceeded to the analysis of the following elements: lead (Pb), copper (Cu), nickel (Ni), manganese (Mn), iron (Fe), and zinc (Zn). The apparatus used is the Atomic Absorption Spectrophotometer with flame model Agilent AA 200 at the level of the Mining Company Touissit in Tighza Morocco.

2.4. Bacteriological analysis

The bacteriological quality of the water was determined by the research and enumeration of bacteria subject to Moroccan standards: total aerobic mesophilic flora at 22°C and 37°C, total coliforms, intestinal enterococci, and Escherichia. Coli (table2).

Table 2: Methods of bacteriological analysis in the studied well waters

Germs counted or researched [12]	**Reference of the analysis method**	**Maximum values Admissible (NM 03.07.001, 2006)**
Total mesophilic germs 22°C and 37°C	ISO 6222, NM 03.7.005, 2007	20/1 ml at 37°C 100/1 ml at 22°C
Total coliforms	ISO 9308-1, NM 03.7.003, 2007	0/100 ml
Escherichia. Coli	ISO 9308-1, NM 03.7.003, 2007	0/100 ml
Intestinal Enterococcus	ISO 7899-2, NM 03.7.006, 2007	0/100 ml

IOP Conf. Series: Earth and Environmental Science 1090 (2022) 012031 doi:10.1088/1755-1315/1090/1/012031

3. Results and discussion

3.1. Temperature

The in-situ measurements of the temperature in the water of the controlled wells vary according to the seasons, in the wet season the temperature is lower than 20°C with a minimum of 11°C (well P11), while in the dry season it is higher than 20°C with a maximum of 30°C (wells P2 and P6) a maximum of 30°C (wells P2 and P6) (figure 3). These values remain acceptable for the Moroccan standards of potability.

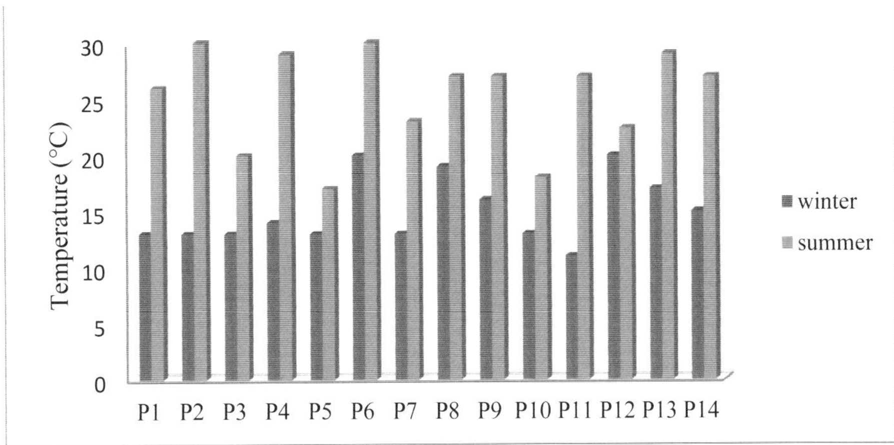

Figure 3. Seasonal variation of temperature in the studied well-water P1-P14

3.2. Hydrogen potential (PH)

The pH conditions a large number of Physico-chemical balances between water, dissolved carbon dioxide, carbonates, and bicarbonates which constitute buffered solutions that give aquatic life a favorable development [13]. This parameter depends on the origin of the water, the geological nature of the substrate, and the watershed cross [14].

In the study area, the recorded pH values are in line with human food standards (6.5 < pH < 8.5). They vary between 7.01 (well P13) and 8.09 (well P03) during the dry season and from 7.1 (well P3) to 7.9 (well P7) during the wet season (figure 4). This indicates a slight alkalinity during the period of heavy rainfall.

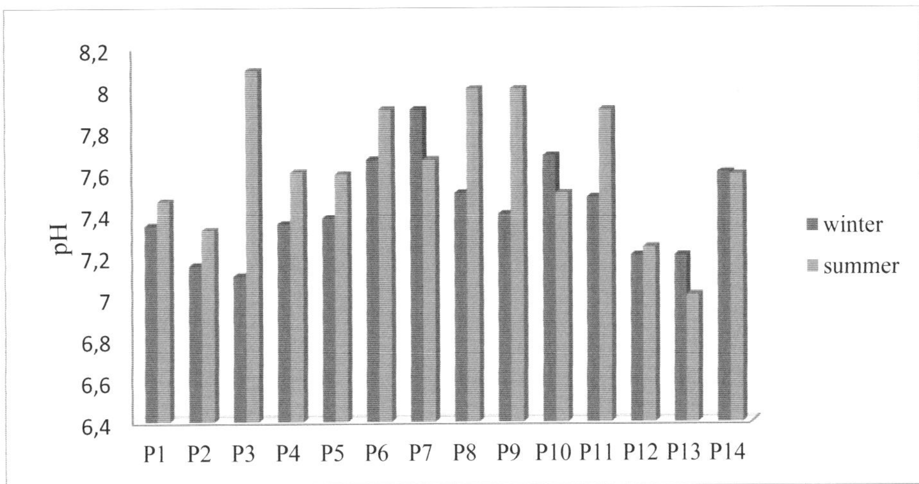

Figure 4. Seasonal variation of pH in the studied well-water P1-P14

3.3. Conductivity

Conductivity is proportional to the quantity of dissolved ionizable salts and its measurement is a good indication of the degree of mineralization of the water [15]. The results showed that the conductivity values of the water in the wells studied ranged from a minimum of 200 µs/cm to a maximum of 1800 µs/cm (figure 5). Our results confirm other studies that have shown (461 and 1587 µs/cm) and (500 and 1400 µs/cm)[16,17]. The recorded values are within human food standards (2700 µS/cm)[12].

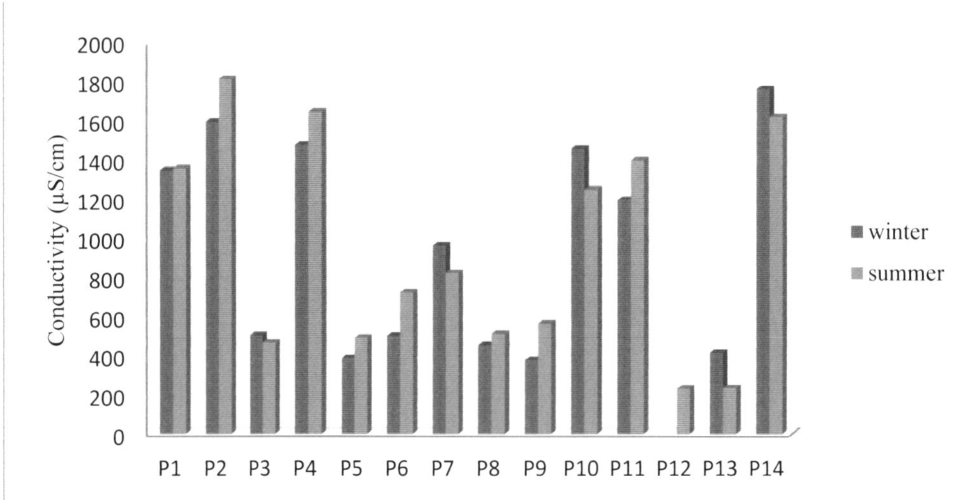

Figure 5. Seasonal variations of electrical conductivity in the studied *well water P1-P14*

3.4. dissolved oxygen

The concentration of dissolved oxygen in water depends on several factors, in particular temperature, atmospheric pressure, and salinity. Dissolved oxygen comes either from the atmosphere by diffusion or from the photosynthesis of autotrophic plants. It is consumed during the respiration of animals and plants, the decomposition of organic matter by aerobic microorganisms, and the oxidation of certain chemical substances. During the two study periods, the dissolved oxygen values in all the water points sampled (figure 6) ranged from 2.9 mg/l and 9.66 mg/l. The maximum value is observed at well P11. The other wells are less oxygenated and this could be related to the polluted water that infiltrates towards the water table or could also be explained by the absence of water-atmosphere contact and the perpetual renewal of the water [18].Winter seems to improve the oxygen content of well water somewhat by providing well-oxygenated water to the groundwater and by the cold weather which inhibits bacterial growth and therefore well-oxygenated water to the water table and by the cold which inhibits bacterial proliferation and thus minimizes their oxygen consumption. The results obtained show that the rate of non-compliance is 35%, whereas 65% of the wells comply with human food standards ($5 \leq O2 \leq 8$ mg/l) [12].

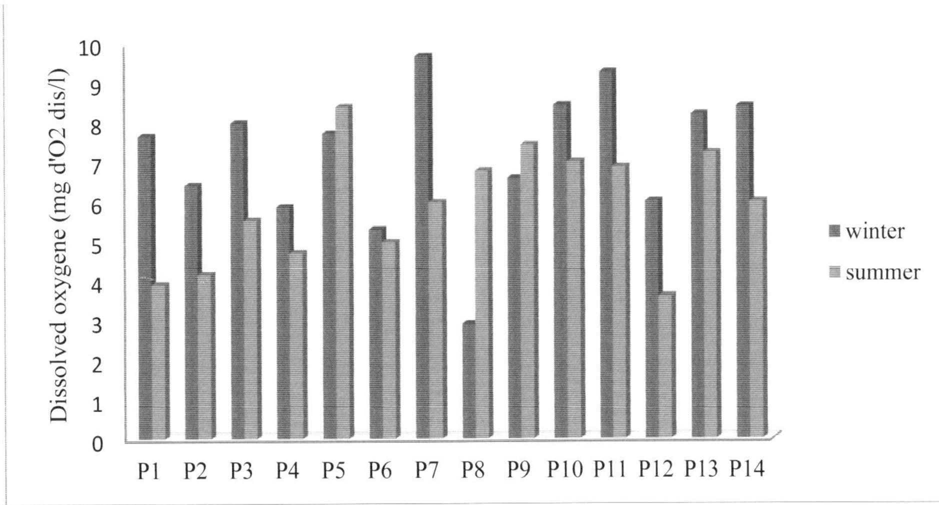

Figure 6. Seasonal variation of dissolved oxygen in the studied well-water P1-P14

3.5. Metallic trace elements

3.5.1. Lead. Lead (Pb) is a widely distributed element in the earth's crust and is one of the most widely used metals in industry. It is the most widely used metal in industry. Until the year 2000, lead, in the form of tetraethyl, was widely used in fuels as an anti-knock agent. In the last decade, lead decade, lead emissions to the environment have decreased significantly due to the widespread use of unleaded petrol [19].

The results showed that at the level of the wells studied the lead (figure 7) values vary between 0 and 0.32 µg/l. The lead levels recorded in the water of the wells studied are below the below the maximum acceptable limit for water intended for human consumption (10µg/l) Moroccan standards.

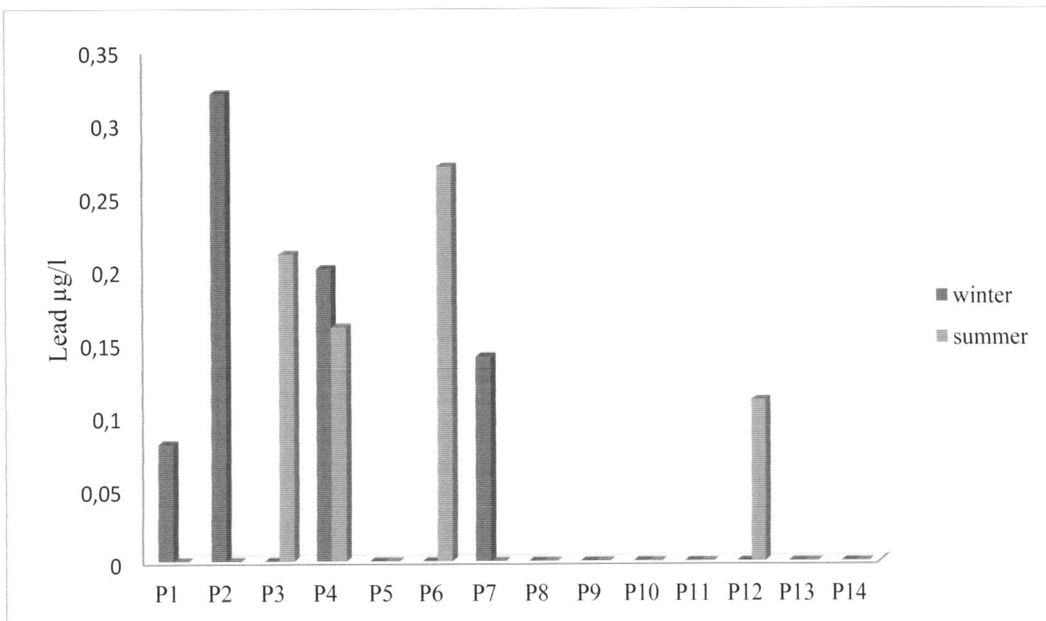

Figure 7. Seasonal variations of lead concentrations in the studied well-water P1-P14

3.5.2. Copper

The copper analysis showed that the values vary between 0 and 420 µg/l (figure 8). This does not exceed the maximum accepted limit for water intended for human consumption (2000 µg/l) Moroccan standards.

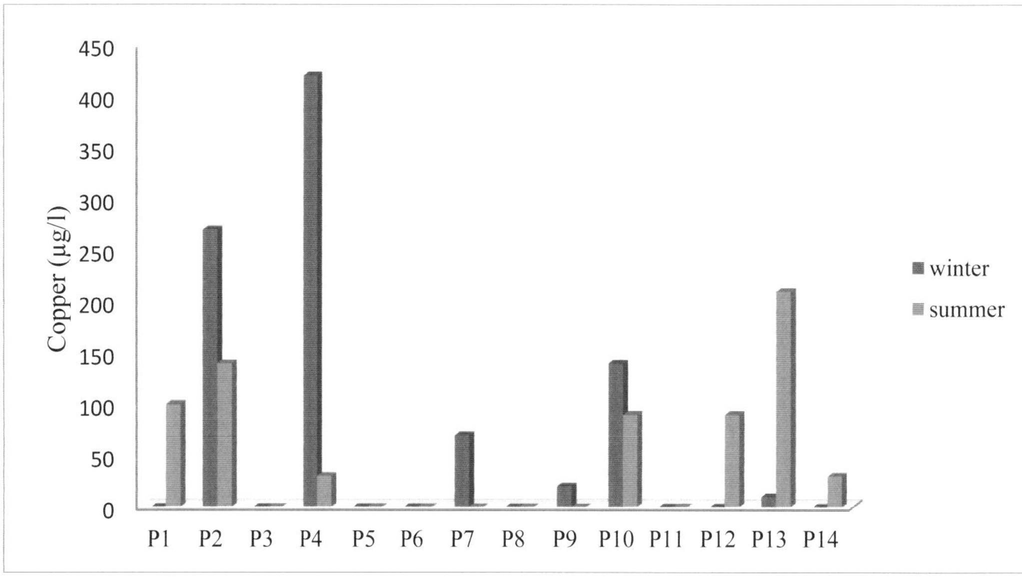

Figure 8. Seasonal variations of copper concentrations in the studied well-water P1-P14

3.5.3. Zinc

The analysis of zinc revealed that the values vary between 0 and 180 µg/l at the wells (figure 9). Level of the wells. This is below the maximum permissible value (3000µg/l) according to Moroccan standards.

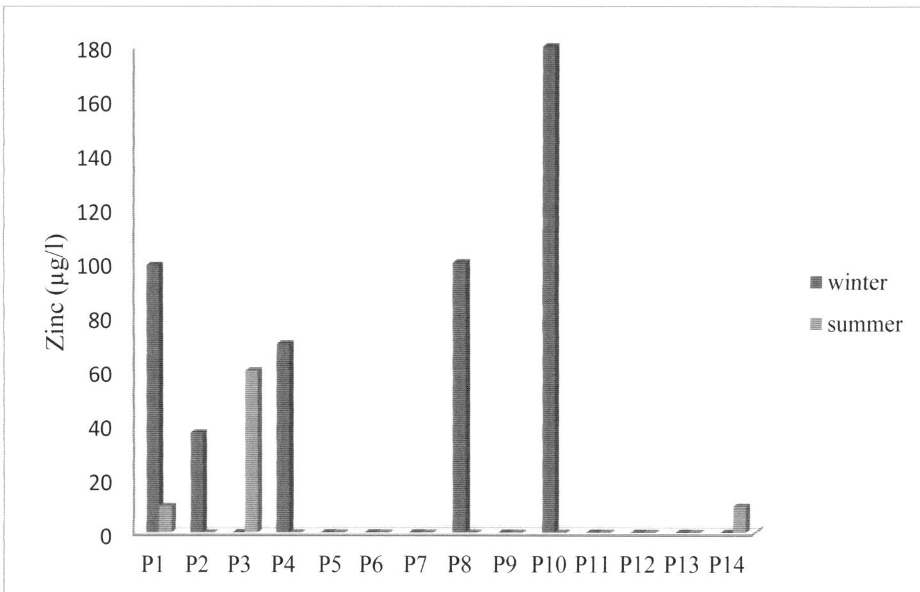

Figure 9. Seasonal variations of zinc concentrations in the studied well-water P1-P14

3.5.4. Manganese

Most of the manganese contents in the water from the different wells during both seasons (figure 10) are below the maximum permissible value for drinking water, which is 500µg/l. Except for well P3 where the content (520µg/l) exceeds the maximum permissible value. The presence of manganese in the water of well P3 would be linked to the nature of the soil, wastewater, and domestic waste [20].

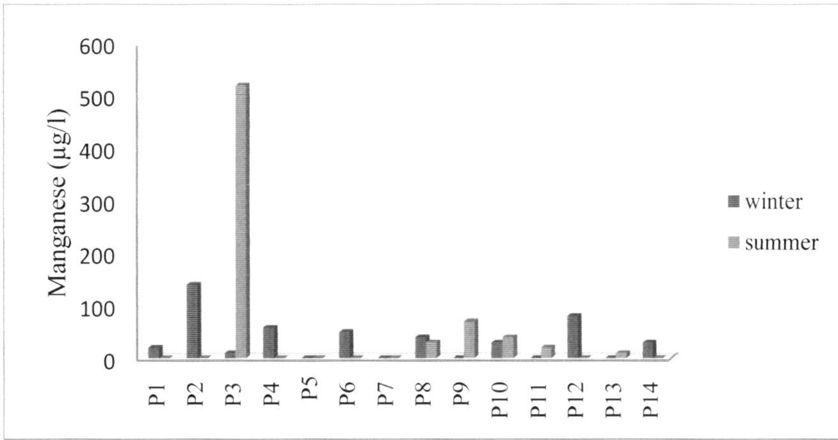

Figure 10. Seasonal variations of manganese concentrations in the studied well-water P1-P14

3. 5.5. Iron

The iron levels in the wells studied in the study area (figure 11) vary from 0 to 4920µg/l, with 42% of the wells analyzed having iron concentrations below the recommended standard, and 58% of the wells having iron concentrations above the maximum permissible value for drinking water, which is 300µg/l [12] (P1,P2,P3,P4,P6 ,P8,P10,P14).The contamination of these well waters with iron is natural or anthropogenic through agricultural activities. This contamination is also influenced by the depth of the wells and the way each well is protected. The presence of iron in groundwater has multiple origins, iron is commonly associated with sedimentary rocks deposited in reducing environments (marl, clay) and metamorphic rocks[21].

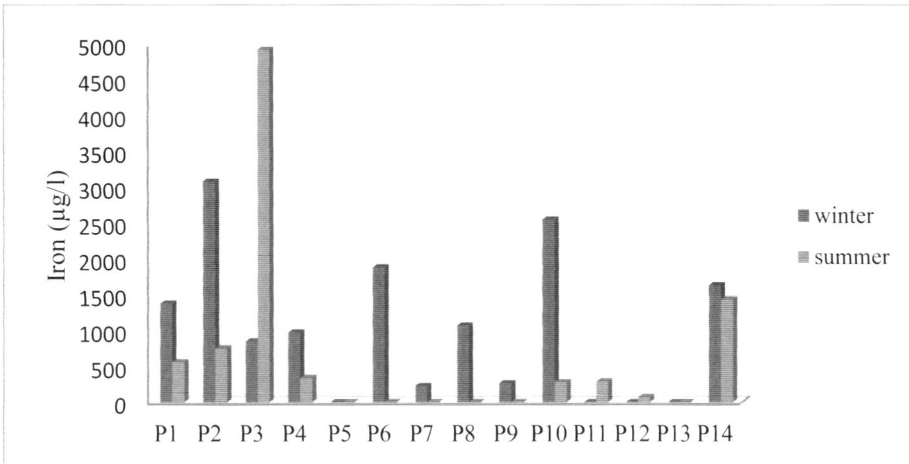

Figure 11. Seasonal variations of iron concentrations in the studied well water P1-P14

3.5.6. Nickel
The analysis of nickel showed that the values vary between 0 and 0.14 mg/l in the water of the wells studied (figure 12). By comparison with the maximum admissible value (20 µg/l) of drinking water according to the Moroccan standards, our study indicates that the nickel contents of the water from all the wells analyzed during the two seasons are lower than this maximum admissible value .

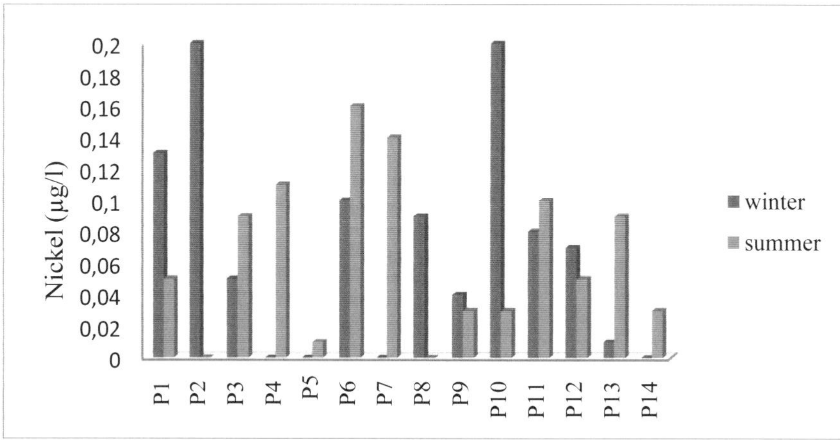

Figure 12. Variations of iron concentrations in the studied well-water P1-P14

3.6. Bacteriology
3.6.1. Total aerobic mesophilic flora. Total aerobic mesophilic flora or total germs are the set of microorganisms (bacteria, yeasts, molds, etc.) growing under normal cultural conditions and representing the average bacterial content of a natural resource [22]. This flora indicates the degree of overall bacterial contamination [23].

The concentration of total aerobic mesophilic flora at 37°C for all wells during both study periods ranged from 40 CFU /ml to 6000CFU/ml (figure 13). The concentration of total aerobic mesophilic flora at 22°C for all wells during both study periods ranged from 1CFU/ml to 30.10^3CFU/ml (figure 14). The majority of the waters analyzed far exceeded Moroccan standards. In agreement with several recent studies [24,25].

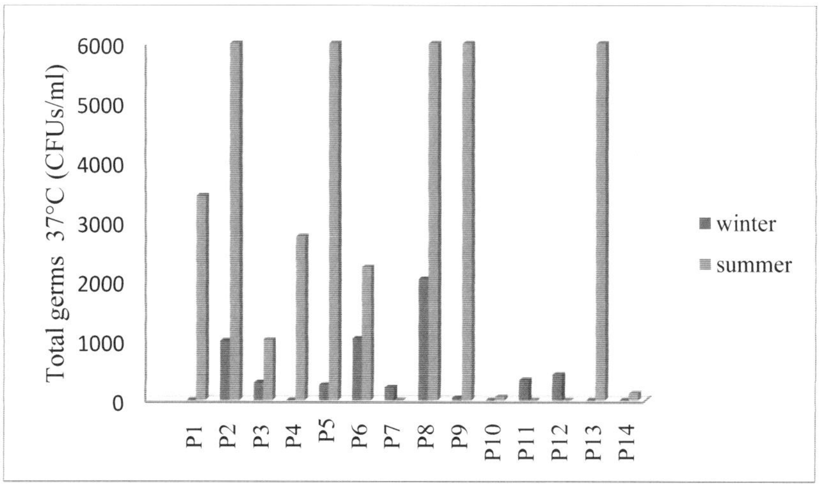

Figure 13. Seasonal variations of total mesophilic germs at 37°C of the studied well-water P1-P14

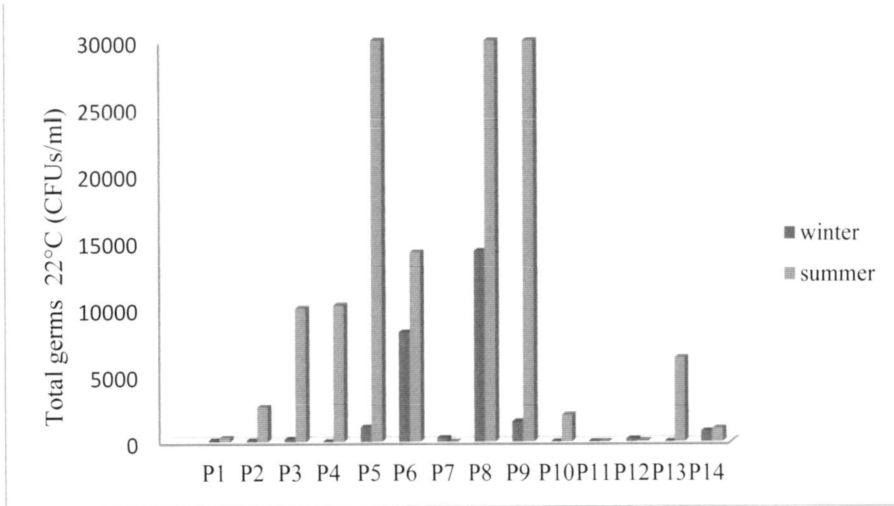

Figure 14. Seasonal variations of total mesophilic germs at 22°C of the studied well-water P1-P14

3.6.2. Total coliforms

Total coliforms are enterobacteria that include bacterial species that live in the intestines of homeothermic animals, but also in the general environment (soil, vegetation and water). This bacterial group is used as an indicator of microbial water quality. quality of water because it contains bacteria of fecal origin, such as Escherichia coli, and fecal bacteria, such as Escherichia coli (E. coli) [26].

In the figure 15, most of the wells have total coliforms in their water, with values ranging from 1 CFU /100ml to 3.10^2 CFU/100ml. These values are high in comparison with Moroccan standards (<0 CFU.100ml^{-1}). These results corroborate those of [27] in wells and springs in the fez region.

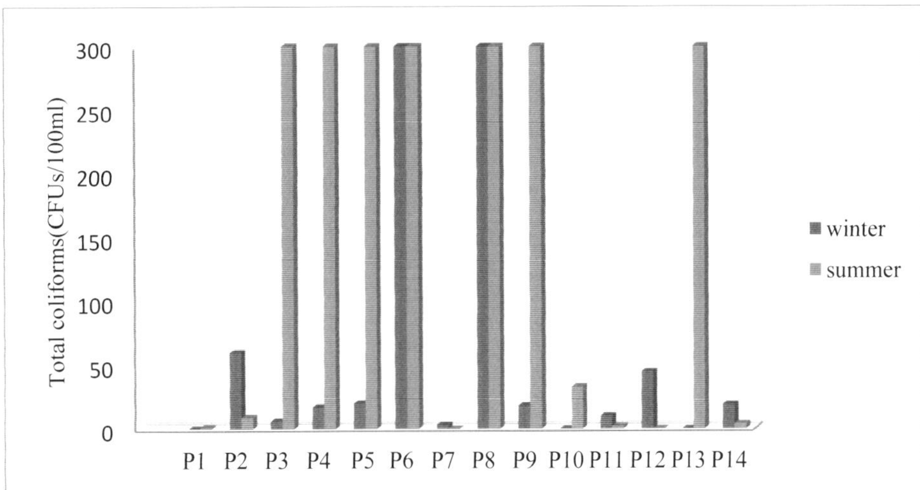

Figure 15. Seasonal Variations of total mesophilic coliforms in the studied well-water P1-P14

3.6.3. Fecal coliforms (Escherichia Coli)

In the study area, the faecal coliform count shows that during the dry season, 43% of the wells monitored are free of these faecal contamination germs, whereas they are present in the water of the other wells with numbers ranging from 1 CFU/100 ml recorded in well P6 to 300 CFU/100 ml in well P5. This high concentration of faecal coliforms (E. coli) during this period is close to that found in the Abengourou water table in Ivory Coast and Fes in Morocco [27,28].

During the wet season, the water from most of the wells studied complies with the potability standard and is free of these germs (0 CFU/100ml).standard of potability, they are free of these germs (0 CFU/100ml), only the five wells (P2.P4.P8.P11 and P14) that are polluted (figure 16).According to our surveys of the well sites, this contamination is due to manure, septic tanks, latrines ,and waste of all kinds in the areas surrounding the wells. The land surrounding the wells. We also noted a decrease in the number of fecal coliforms during the wet season in most of the wells in our study, due to a strong dilution of coliforms caused by the recharge of the water table [29].

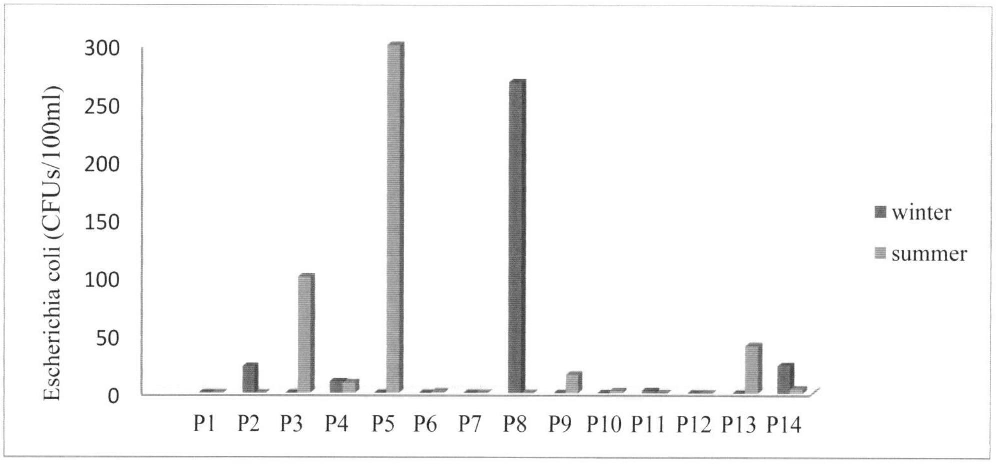

Figure 16. Seasonal variations of Escherichia coli in the studied well-water P1-P14

3.6.4. Intestinal Enterococcus

The detection of enterococci in groundwater should suggest contamination of fecal origin and the contamination of fecal origin and the presence of enteropathogenic micro-organisms Indeed, the presence of intestinal enterococci must be accompanied by the presence of fecal coliforms. The presence of intestinal enterococci must be accompanied by the presence of fecal coliforms to be certain of fecal contamination of a water supply.

Bacteriological analyses of well water samples show high levels of intestinal enterococci in 68% of the wells monitored and the remaining 32% are of good quality. As can be seen in figure 17, the number of intestinal enterococci varied from 1 CFU/100ml to 290 CFU/100ml during the two observation periods. observation periods. The minimum concentration is obtained in well P7. While the maximum concentration is obtained at well P3 in the dry period and P4 in the wet period. This number of intestinal enterococci exceeds the Moroccan standards (<0 CFU.100ml^{-1}). According to the studies that have shown the presence of large numbers of intestinal enterococci in the well water, this attests to the contamination of the water by the fecal matter stored in the latrines [25,30].

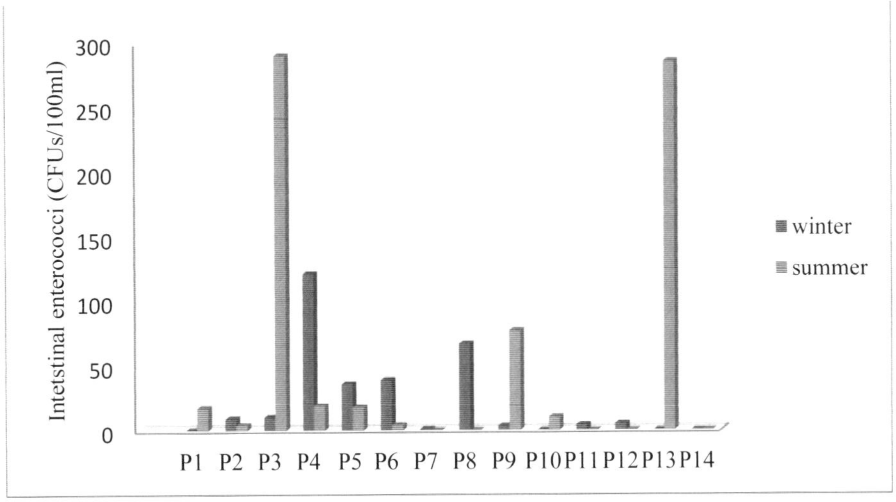

Figure 17. Seasonal variations of intestinal enterococci in the studied well-water P1-P14

4. Conclusion

The results of the Physico-chemical analysis and the analysis of metallic trace elements of the groundwater in rural areas in the study area obtained during February 2021 / August 2021 can be considered acceptable and do not present any danger for human health. From a bacteriological point of view, the wells studied present very high concentrations of fecal contamination in all wells without exception and vary from one well to another, which undoubtedly constitutes a threat to the inhabitants who draw the water necessary for the major part of their needs from the water of these sites [24]. The main cause of this pollution is anthropogenic activities that continue to present a major risk to humans and their environment, especially to water resources. Therefore, regular monitoring and chlorination of groundwater could significantly reduce the degree of bacteriological contamination of well water in the studied area [31] .Overall, this study has also shown the importance and usefulness of multi-varied analysis techniques to obtain the necessary information on the hygienic quality of water and thus prevent all kinds of pollution originating from human domestic, agricultural, and/or industrial activities. Indeed, the results of this study will be of great interest to the health services which will be led to set up preventive and curative actions to avoid possible serious sanitary risks. Moreover, they will allow the decision-makers to establish policies and programs of Information, Education, Communication, and Awareness intended for the rationalization, preservation, and development of the water resources at the level of this region.

5. References

[1] Ministry of Spatial Planning, Water and the Environment (2001). Report on the state of the environment of Morocco, Department of the Environment National Observation of the Environment of Morocco,pp 296.

[2] M. Boulal 2002 Phreatobiological research in the Souss and neighboring regions of Western Morocco: Water quality Biodiversity Ecology and historical Biogeography of stygobian species, PhD thesis Faculty of Science Semlalia of Marrakech.

[3] M. H. Boubekraoui 2005 Environnement et société des espaces oasiens marocains: cas du Tafilalet et de Marrakech, Mém. d'habilitation, Faculté des Sciences Semlalia de Marrakech .

[4] P. B. Mwanza, J. P. Katond and P.Hanocq 2018 Evaluation of the physico-chemical and bacteriological quality of well waters in the spontaneous district of Luwowoshi (Democratic Republic of the Congo) *Tropicultura*, vol. 37, n° 2, p. 1-15.

[5] WHO 2003 Situation du réseau EPA au 31 Décembre 2002 .

[6] John P.H. and Donald A.K. 2010 Microbiologie, 3ème Édition,p 1216 .
[7] S. Chaoua, S. Boussaa, A. El Gharmali, and A. Boumezzough 2019 Impact of irrigation with wastewater on accumulation of heavy metals in soil and crops in the region of Marrakech in Morocco , *J. Saudi Soc. Agric. Sci.*, vol. 18, n° 4, p. 429‑436.
[8] K Bouhoum 2009 Impact sanitaire de la réutilisation des eaux usées en agriculture à marrakech sur les parasitoses intestinales des enfants - Collection Environnement/Aspect sanitaire des eaux usée/ Traitement et réutilisation des eaux usées : impact sur la santé et l'environnement . IAV Edition, pp.55-72.
[9] S. Baroud, D. Belghyti, F. Aziz, and M. Sadi 2015 Contribution of the Principal Component Analysis (PCA) to the evaluation of the physico-chemical pollution of raw wastewater from the city of Khenifra Morocco, vol. 6, n° 9, p. 2583‑2595.
[10] RGPH 2014 Note sur les premiers résultats du Recensement Général de la Population et de l'Habitat . Rapport, province Khénifra – Maroc .
[11] D. Elhamdouni, A. Arioua, I. Karaoui, A. Baaddi,and K. A. Ouhamchich 2019 Household solid waste sustainable management in the Khenifra region, Morocco *Arab. J. Geosci.*, vol. 12, n° 23.
[12] Norme Marocaine 03.7.001 2006 Qualité des eaux d'alimentation humaine élaborée par le comité technique de normalisation des eaux d'alimentation humaine éditée et diffusée par le service de normalisation industrielle Marocaine (SNIMA), p14.
[13] L. Belghiti, A. Chahlaoui, E. L. M. R, and D. Bengoumi 2013 Contribution a L' Etude De La Qualite Bacteriologique De L ' Eau Des Sources Dans La Ville D ' Elhajeb (Region De Meknes -Maroc), *Larhyss J.*, vol. 14, p. 37‑47.
[14] E. Abba, I Idrissi, Y Bennani, A El Yaacoubi, S Cherroud, T Ainane1, and M Khaffou 2021. Assessment of water quality of oum er rabia river by microbiological quality index and water quality index *Pollution*, vol. 7, n° 3, p. 561‑571.
[15] J. Rodier B Legube and N Merlet 2009 L'Analyse de L'Eau Dunod Paris. p. 1579.
[16] M. L. Belghiti, A. Chahlaoui,and D. Bengoumi 2013 Etude De La Qualite Physico ‑ Chimique Et Nappe Plio-Quaternaire Dans La Région De Meknès *Larhyss J.*, vol. 14, n° March 2009, p. 21‑36.,
[17] S Omrania,N El khodrani,A Zouahri,A Douaik,H Iaaich,M Lahmar,and S El hajjaji 2019 « Evaluation of the Groundwater Quality of the M'nasra, Kenitra (Morocco) », *Mater. Today Proc.*, vol. 13, p. 1092 ‑1101.
[18] Y Nuto and W Guenda 2015 Application Des Méthodes Statistiques À L ' Étude Hydrochimique Des Eaux D ' Un Hydrosystème Tropical : Cas Du Bassin Versant De La Rivière Zio (Togo) », vol. 11, n° 14, p. 204‑225.
[19] D K L Maldonado 2009 Caractérisation et origine des métaux traces, hydrocarbures aromatiques polycycliques et pesticides transportés par les retombées atmosphériques et les eaux de ruissellement dans les bassins versants séparatifs péri-urbains these doctorat Ecole Centrale de Nantes France.
[20] C Merbouh, K Belhsaien, A Zouahri, and N Iounes 2020 Evaluation De La Qualité Physico-Chimique Des Eaux Souterraines Au Voisinage De La Décharge Contrôlée Mohammedia-Benslimane : (Étude Préliminaire) European Scientific Journal February 2020 edition Vol.16, No.6 .
[21] H Taybi, K Bargach, M Boulfia, and F Lamchouri 2019 Etude géologique et qualités physico-chimique et bactériologique des eaux thermales de la source " Ain Hamra " de la Province de Taza (Rif Oriental - Maroc) , n° 04, p. 443‑451.
[22] A BenMoussa , A Chahlaoui and M Chahboune 2018 Typologie Des Eaux Superficielles de l'Oued Khoumane. Moulay Idriss Zerhoun, Maroc *Faunistic Entomology* vol 71
[23] H Robert 1999 Qualité microbiologique des eaux brutes distribuées par BRL: Exigences et conception d'un suivi adapté , Mémoire de l'Ecole Nationale de la Santé Publique, p. 81.
[24] A Malek, M Kahoul, and H Bouguerra 2019 « Groundwater's physicochemical and bacteriological assessment: Case study of well water in the region of Sedrata, North-East of Algeria », *J. Water L. Dev.*, vol. 41, n° 1, p. 91‑100.
[25] H.El haissoufi S. Berrada, M. Merzouki, M. Aabouch, L.Bennani, M. Benlemlih, M. Idir, A.

Zanibou, Y. Bennis, A. Elouali Lalami 2011 Pollution Des Eaux De Puits De Certains quartiers de la ville de FES, MAROC *Rev. Microbiol. Ind. San Environn.*, vol. 5, n° 1, p. 37-68.

[26] Health Canada 2017, *Total coliforms. In Drinking Water and Health Fact Sheets Hum. Inst. Nat. Public Health of Quebec pdf.*

[27] A El Ouali Lalami, F El-Akhal, S Berrada, L Bennani, N Raiss,and S Maniar 2014 Evaluation de la qualité hygiénique des eaux de puits et de sources par l'utilisation d'une analyse en composantes principales (ACP): Une étude de cas de la région de Fès (MAROC) », *J. Mater. Environ. Sci.*, vol. 5, p. 2333-2344.

[28] N Aka, S B Bamba, G Soro,and N. Soro 2013 Étude hydrochimique et microbiologique des nappes d'altérites sous climat tropical humide : Cas du Département d'Abengourou (Sud-Est de la Côte d'Ivoire) , *Larhyss J.*, vol. 16, p. 31-52.

[29] W Ayad and M Kahoul 2016 Assessment of physico-chemical and bacteriological quality of Well water in the region of El-Harrouch (N.E-Algeria)] *J. Mater. Environ. Sci.*, vol. 7, n° 4, p. 1288-1297.

[30] J G T Youmbi, R Feumba, V T Njitat, G De Marsily and G E Ekodeck 2013 Pollution de l'eau souterraine et risques sanitaires à Yaoundé au Cameroun *Comptes Rendus - Biol.*, vol. 336, n° 5-6, p. 310-316.

[31] M H Benajiba,Y Saoud,A Lamribah,M Ahrikat, N Amajoud,and O Ouled-Zian 2013 « Assessment of microbial water quality of groundwater in Martil (Morocco), *Rev. des Sci. l'Eau*, vol. 26, n° 3, p. 223 -233.

IOP Conf. Series: Earth and Environmental Science 1090 (2022) 012032 doi:10.1088/1755-1315/1090/1/012032

Assessment of the impact of density on the growth of fry (Oncorhynchus mikiss) at the salmon farming station in the Middle Atlas Morocco

CH Laiboud* [1, 2], **T Hachi**[1] , **H Essabiri**[1,2], **O Boumalkha**[1,2], **Y Elhammioui**[2], **A El Moujtahid,**[2] **and E H Abba**[1].

[1] Khénifra Higher School of Technology. Sultan Moulay Slimane University. Morocco

[2] Sciences and technologies Faculty. Beni Mellal. University of Sultan Moulay Slimane.

chaimaalaiboud@gmail.com

Abstract. The objective of this study is to determine the impact of the load of the nursery ponds (density) on the growth of the rainbow trout fry (Oncorhynchus mikiss), this study was conducted at the fish farming station of Ras Al Ma Ifrane, Morocco. The results obtained for the densities D1 (7.48 Kg / m3), D2 (9.31 Kg / m3) and D3 (11.21 Kg / m3) during this nursery phase which lasted 30 days, the results indicate that the zootechnical parameters evaluated vary according to the density of the fry in different nursery ponds. In average, the final weight of the fry for D1 is 12.73g, for D2, the final average weight is 11.39g against only 9.56g for ponds with density D3. As such, the growth in size of the fry evolves in the same way as the growth in weight.

Key words: *Oncorhynchus mikiss*, nursery, density, zootechnical parameters.

1. Introduction

In the recent years, the continental aquaculture sector has witnessed a remarkable development due to the strategy adopted by the High Commission for Water, Forests and Combating Desertification during 2015 and 2024. As a matter of fact, the consumption of fish products has increased due to its high nutrient contents and diversified flavors. Many scientific studies support that Fish contains low cholesterol level but high in terms of mono or polyunsaturated fatty acids that play a protective role against cardiovascular diseases [1]. Rainbow trout is among the fish species used in aquaculture (Oncorhynchus mikiss) which was introduced to Morocco in 1925 [2]; [3]; for commercial purposes and to enhance Moroccan ethnological biodiversity. In fact, the growth rate of farmed rainbow trout (Oncorhynchus mikiss) is much higher than the wild populations, as a consequence of on the diet, habitat and water temperature. At the constant temperature of 15 ° C, freshwater rainbow trout are also weighing between 800g and 1kg one year after spawning [4]. Hence, the stocking density of fish is an important indicator that determines the technical and economic viability of the aquaculture business. It also greatly influences the growth and maturity of fish regardless of the cultivation system used. In order to obtain optimal yields and obtain maximum profit, it will then be necessary to breed the fish on stocking densities.

2. Material and method

2.1. Presentation

The Ras El Ma fish farm is a unit for breeding cold-water fish, under the National Center of Hydrobiology and Fish Farming of Azrou (CNHP), It was created in 1957 and located at an altitude of 1,550 m, and 9 km from the city of Azrou. The selection of the Ras El Ma site was determined by the geological location of the land also the physical and chemical quality of the water that feeds the station. Among the activities carried out in the Ras El Ma fish farm station are the production of fry of Brown Trout and Rainbow Trout for the enrichment of the national ichthyological heritage where sport fishing is practiced [5], The station has also implemented actions for the benefit of students and researchers in order to promote awareness of the main ecological problems of fish.

2.2. Geographic location

The Ras El Ma salmon farm is located in the geographical sector of the Middle Atlas, the latter is between the northern latitudes 32 ° 55 'and 34 ° as well as the longitude 5 ° 05'05 "[6]. It is located at an altitude of 1650m and 9km from the city of Azrou on the national road (RN8) connecting Azrou to Ifrane.(Figure 1).

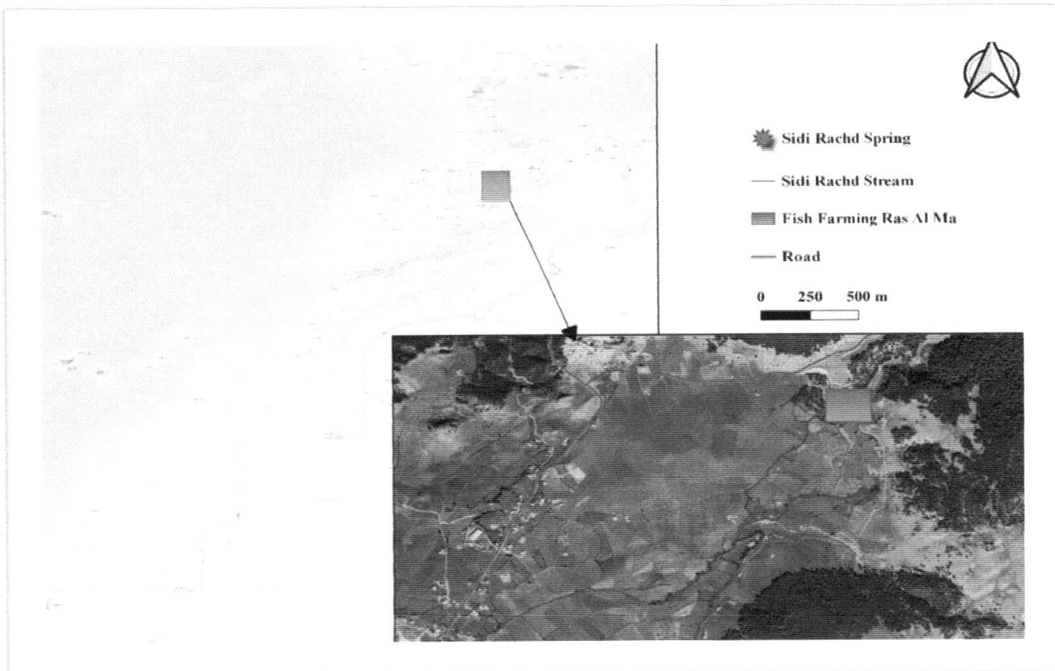

Figure 1. Geographical location of the Ras El Ma salmon farming station

This study is carried out at the Ras El Ma Ifrane Maroc fish-farming station. The rainbow trout fry used come from the Ras El Ma Ifrane fish farm.

2.3. Experimental material

In this study we used three ponds with a volume of 2.74 m3 with deferent densities; D1 (7.48 Kg / m3), D2 (9.31 Kg / m3) and D3 (11.21 Kg / m3) during 30 days. On the day of the sampling, a precision scale (Ohaus® Scout Pro) of 0.01g was used to weigh the rainbow trout fry. For the size, a ruler ensures the length measurements after anesthesia with clove nail oil to avoid stress. For the work methodology, every 10 days samples of rainbow trout fry were measured (height and weight). The food used has been formulated to meet the nutritional needs of trout it comes from a French cooperative (Le Gouessant) of the granulated type it consists of 40% crude protein, 23% crude fat, 1.62% crude fiber, 8% crude ash, 1.90% calcium, 1.60% total phosphorus, and 0.70% sodium plus additives (Vitamins, trace elements, etc.).

The daily food ration is calculated by the following method:

The daily food ration = (number of individuals * average weight * feeding rate coefficient) / 100.

The food was distributed manually 3 to 4 times a day (9h00, 12h00, 16h00) with a duration of 15 to 20 min to have a good distribution of the food in the ponds and to avoid waste.

In fact to evaluate the growth ,we used the growth assessment method:

Average weight gain = final weight - initial weight.

Average height gain = average final height - initial size.

3. Results and Discussion

3.1. Results

The results of the effect of the stocking density of rainbow trout fry on their zootechnical performance are presented in tables 1 and 2, also in figures 2 and 3. An increase in weight is observed. average of fry with density D1 (7.48 kg / m3) went from 2.04 g to 12.73 g, for density D2 (9.31 kg / m3) weighing 1.99g to 11,39g and for density D3 (11.21kg / m3) the average weight has gone from 1.77g to 9.56 g. Regarding to the average height of the density D1 (7.48 kg / m3) has been increased from 7.5 cm to 12cm, for the density D2 (9.31 kg / m3) from 7.7 cm to 11.64 cm and for density D 3 (11.21 kg / m3) the size expanded from 6.4 cm to 10.11 cm. According to the statistical analyzes carried out we can notice that the density D1 (7.48kg / m3) has a better growth compared to the other densities D2, D3.

Table 1. weight growth

	D1	**D2**	**D3**
1	2,04	1,99	1,77
2	5 ,83	5,78	5
3	7,6	7,67	6,31
4	12,73	11,39	9,56

Table 2: Linear growth

	D1	**D2**	**D3**
1	7 ,5	7,7	6,4
2	10	10,1	9,25
3	11	10,85	9,8
4	12	11,64	10 ,11

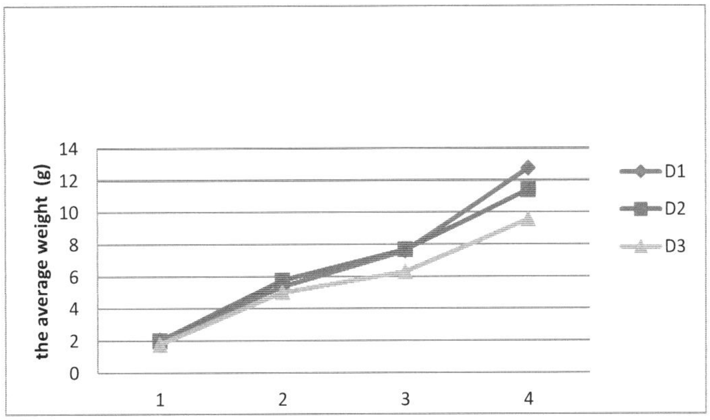

Figure 2. weight evolution of rainbow trout fry as a function of density

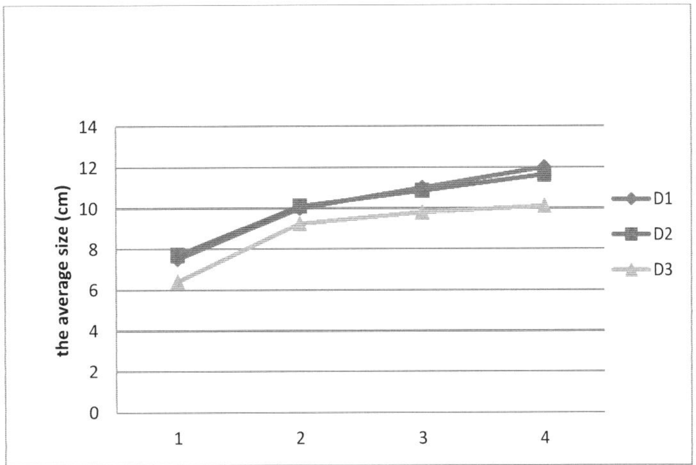

Figure 3. Linear evolution of rainbow trout fry as a function of density

3.2. Discussion

The stocking density of fish is a determining factor in aquaculture. It is an important indicator that determines the technical and economic viability of an aquaculture business [7];[8]. Indeed, it significantly affects the growth and maturation of fish regardless of the used farming system [9]; [10]; [11]; [12]. Because detecting food can be difficult when large numbers of fish are blocking sight and passage. According to several studies [13];[14]; [15], this behavior causes a reduction in the weight of individuals [16]; [17]; [18]; also, an increase in stress and a consequently decrease in growth [19]. Therefore, better growth is observed when it is a question of low density of the body growth which represents an economic advantage. On the other hand, high densities cause growth inhibition which results in loss of economic means. In order to have optimal productions and to derive maximum profit, it will then be necessary to breed the fish on optimal stocking densities.

4. Conclusion

According to the results found in this study, carried out in the Ras El Ma fish farming station on the Evaluation impact of density on the growth of fry (Oncorhynchus mikiss) It is indicated
that the best density with the highest zootechnical performance is 7.48kg / m3, because the breeding environment is optimal in comparison with the other two environments.

However, the optimal density of fish depends on the species, and on the strain for metabolic reasons. The age and size of the fish are also factors to be taken into account [20]. There are only a few studies that have been conducted on the spatial requirements of farmed fish. It is very useful to underline that the development of the aquaculture sector in Morocco requires even more research in aquaculture.

5. References

[1] Cahu C, Salen P, de Lorgeril M. 2004. Farmed and wild fish in the prevention of cardiovascular diseases: assessing possible differences in lipid nutritional values. Nutrition Metabolism and Cardiovascular Diseases. **14** : 34-41.

[2] Mouslih M. 1996 : Contribution à l'amélioration des techniques de salmoniculture de repeuplement en vue de restaurer la truite fario (Salmo trutta macrostigma, Dumeril, 1858) dans les rivières des montagnes marocaines. IAV. Rabat. Maroc.

[3] Aba M. 2006 : Contribution à l'étude ichtyoparasitologique de la truite arc en iel (Oncorhyncus mykiss) dans le plan d'eau Amghas II. Province d'Ifrane. Maroc. Memoire de DESA. Uni. Ibn Tofail. Kenitra. Maroc

[4] FAO, walbaum 1792 **:** Truite arc-en-ciel - Oncorhynchus mykiss (Walbaum, 1792) [Salmonida]

[5] Abba H. 2008. Contribution à l'étude de la qualité physicochimique des eaux d'un écosystème aquatique du Moyen Atlas : Dayet Aoua. Mémoire de DESA. Fac. Sci. Univ. Ibn Tofail. Kenitra. Maroc.76p.

[6] Anonyme 2007 : Rapport Haut-Commissariat aux Eaux et Forêts et à la Lutte Contre la Désertification. Maroc.

[7] Pangni K. B. C. Atsé and N. J. Kouassi. 2008: Effect of stocking density on growth and survival of the African catfish Chrysichthys nigrodigitatus, Claroteidae (Lacépède 1803) larvae in circular tanks. Livestock Research for Rural Development, **20 (7), 56 - 61.**

[8] Gao Y, He Z, Vector H, Zhao B, Li Z, He J, Lee JY, Chu Z. 2017: Effect of Stocking Density on Growth, Oxidative Stress and HSP 70 of Pacific White Shrimp Litopenaeus vannamei. Turkish Journal of Fisheries and Aquatic Sciences 17: 877-884. DOI: 10.4194/1303-2712-v17_5_04

[9] Rahman MM , Shahidul Islam Md , Halder GC , Tanaka M .2005: cage culture of sutchi catfish, Pangasius sutchi Fowler 1937 : effects of stocking density on growth, survival, yied and farm profitability". Aquaculture Research, volume **37**, pp.33.

[10] Celada JD, Aguilera A, Carral JM, Saez – Royuela M, Melendre PM, Perez JR. 2007: Effects of stocking density on survival and growth of juvenile tench (Tinca Tinca L). Aquaculture International. **15**: 461 – 465.

[11] Narejo NT. 2010: Effect of stocking density on growth and survival rate of labeo rohita (Hamilton) fed with formulated feed. Sindh Univ. Res. Jour. (Sci. Ser.) Vol. 42 (1) 35-38

[12] Hannibal MC, Amparo LF, Aurelio AC. 2011: Effect of Stocking Density on Growth Performance, Survival and Production of Silver Pompano, Trachinotus blochii, (Lacépède, 1801) in Marine Floating Cages, Asian Fisheries Science. **24**:321-330.

[13] Jenkins ThM, Diehl, S, Kratz, KM. and Cooper, SD. 1999: Effects of population density on individual growth of brown trout in streams. Ecology **80,** 941– 956

[14] Lobón-Cerviá J. 2005a: Spatial and temporal variation in the operation of density dependence on growth of stream-living trout (Salmo trutta L.). Canadian Journal of Fisheries and Aquatic Sciences 62, 1231– 1242. doi: 10.1139/F05–034.

[15] Jobling M .1995: simple indices for the assessement of the influences of social environment on growth performances, exemplified by studies on Arctic char: Aquaculture International. volume **3,** pp,60-65.

[16] Holm JC.1990: the effect of fish density and feeding regimes on individual growth rate and mortality in rainbow trout (oncorhychus mykiss) aquaculture ,**89 :**225-232

IOP Conf. Series: Earth and Environmental Science 1090 (2022) 012032 doi:10.1088/1755-1315/1090/1/012032

[17] Kindschi GA .1991: Performance of Two Strains of Rainbow Trout Reared at Four Densities with Supplemental Oxygen
[18] Zoccarato I.1994: Differences in performance, flesh composition and water output quality in relation to density and feeding levels in rainbow trout, Oncorhynchus mykiss (Walbaum), farming.
[19] Kebus MJ.1992: Effects of Rearing Density on the Stress Response and Growth of Rainbow Trout
[20] Piper RG, McElwain, IB, Orme, LE., McCraren, JP, Fowler, LG. and Leonard, JR. 1982: Fish Hatchery Management, US Department of Interior, Fish and Wildlife Service, Washington D.C., 517 pp.

Characterization and spatialization of the Physico-chemical quality of the waters of a natural lake of the Middle Atlas: the case of the Ramsar site lake Aguelmam Sidi Ali

N Laadel[1*], R Elouhabi[2], Y Baladia[1], H El Idrissi[1], M Sadiq[1], EH Abba[3] and H Achiban[4, 5]

[1] National Center of Hydrobiology and Fish Farming, Azrou- BP 11. Morocco
[2] Agronomic and Veterinary Institute Hassan II –Rabat- Morocco
[3] Sultan Moulay Slimane University, Béni Mellal Morocco
[4] Sidi Mohamed Ben Abdallah University, Fez-Morocco.
[5] Ifrane National Park

* Aderj.nezha@gmail.com

Abstract. The physicochemical quality of the water represents a major stake in front of the development of the piscicultural life in the natural lakes. The present work has for objective the characterization of the waters of the two parts of the Ramsar site of the lake Aguelmam Sidi Ali as well as the definition of their future vocations, as far as they present currently a great biological, ecological, and tourist interest. A follow-up of the evolution of the physicochemical quality of the two parts of the lake (small lake and large lake), via monthly samples for analyses in the laboratory and in-situ measurements of the principal physicochemical parameters, was carried out. The samplings were spread out over the period from April to December 2019. In light of the results obtained, the waters of the two parts of the lake aguelmam Sidi Ali did not present the same degrees of mineralization and hardness, and therefore, they show a clear difference in water quality. Despite this difference, the quality of the waters of Aguelmam Sidi Ali is generally good and thus favorable to the development of the percids introduced since 1939.

1. Introduction

Morocco is one of the Mediterranean countries with a great geological, climatic and orographic diversity. The central Middle Atlas has the most favorable climatic conditions on a national scale in terms of rainfall. Moreover, It is home to a very large number of wetlands that play crucial hydrological, socio-economic and ecological roles on the national level [2,3,4], offering diversity in terms of continental hydro systems. Aguelmam Sidi Ali is one of the most important lakes of the Middle Atlas, given its size, it is permanent hydrological regime, and it is undeniable ecological and socio-economic role. This lake has been classified as a site of biological and ecological interest since 1996 and as a RAMSAR site in 2005. It is a wintering and resting place for several migratory waterbirds [4,5] which offers a favorable habitat for endemic invertebrates and freshwater fish.

Currently and as a result of climate change and strong anthropic pressure, the sustainability of ecosystem services of this hydro system is one of the major concerns of the National Center for Hydrobiology and Fish Culture.

The assessment of the quality of its waters as well as their capacity to self-purify is based on the measurement of physicochemical parameters and bacteriological analysis. In this work, we are interested in the study of the quality of the waters of the lake Aguelmam Sidi Ali, through the determination of the physicochemical characteristics at the level of the small and the big lake.

2. Presentation of the study area

Aguelmam Sidi Ali, is the largest natural lake in the Middle Atlas, both by its area and by its depth. It is a permanent lake located in the territory of the province of Midelte and 110 km south of the city of Fez. It is a mountain lake (2085 m above sea level) that extends over a total area of 300 ha with a maximum depth of 36 m. Its drainage basin, which is part of the large unit of the folded Middle Atlas, covers an area of 16.1km2 and shows an elongated shape and an endoreic drainage system (figure 1, figure 2).

Figure 1. Panoramic photo of the big lake of Aguelmam Sidi Ali

From a geomorphological point of view, the wetland is structured in two lakes (large and small lakes), and a large marshy area crisscrossed by a sinuous hydrographic network fed by resurgent springs (Figure 1). The large lake to the east represents 92% of the total wetted area while the small lake represents only 8% of this area. Hydrologically, the lake is fed by meteoric waters, springs located upstream of the lake (Aghbalou or Mançour and Aghbalou Tadadat), and resurgences located at the bottom of the lake itself [1]. Currently, this lake is confronted with many dysfunctions, mainly manifested by a decrease in water level and ecological disruptions generated by human activities and accentuated by drought [6]. Its origin is mainly structural since it borders the overlapping fold of Jbel Sidi Ali, and it is limited by two faults oriented in the NE-SW direction, inherited from the Alpine orogeny [7] and [8]. It is a natural dam formed following the clogging of a karstic valley by basaltic flows of the quaternary [9]. From the lithological point of view (figure 2), this zone is formed mainly by red saliferous clays intercalated by doleritic basalts of the Triassic, limestones and dolomites of the Lias, marls and limestones of the Bajocian and finally basalts of the Quaternary which cover the totality of the bottom of the small lake and part of that of the large lake [7] and [10]. On the shores of the lake, a layer of whitish sand has been deposited, testifying to the strong alteration of the marly and dolomitic carbonate rocks.

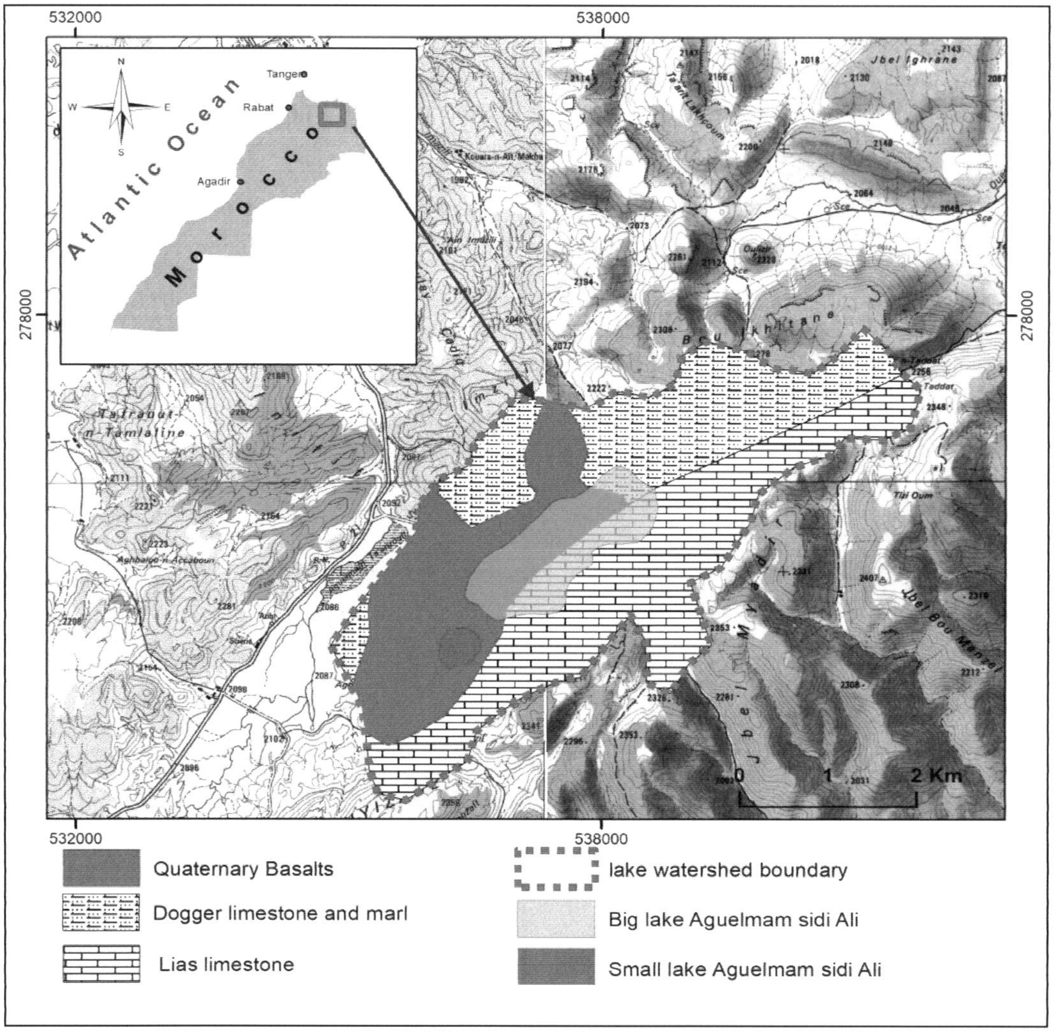

Figure 2. geographical location and lithological context of the Aguelmam Sidi Ali lake wathershed.

3. Material and methods

3.1. Sampling technique

The study lake was subject to physico-chemical monitoring during a period extending from April 2019 to December 2019, with a monthly frequency. Six samples were taken at the level of the lakes studied (figure 3). The first two samples were taken in the center of the small lake, the first (station 1), was taken at a depth of 0.5m while the second (station 2), was taken at a depth of 14m using the Van Dorn bottle. The other four samples were taken at the level of the large lake; the 1st (station 3) at a depth of 9m, the 2nd (station 4) at a depth of 20m and a little further from the previous one, the 3rd (station 5) at a depth of 10m and the last one (station 6) is located on the right bank of the lake, at a depth of 0.3m.

Figure 3. Situation of sampling points sampling and analysis methods

3.2. In-situ measurement

Four physicochemical parameters were measured in situ, using portable devices specific to each parameter, and calibrated at each campaign (Picture 1). Water temperature, which is an important factor that governs almost all physical, chemical, and biological reactions in water [12], was measured with a thermometer; Hydrogen potential (pH) which depends mainly on the geological nature of the reservoir [13] and [14], was measured with a pH-meter NEOTEK-PONSEL; Electrical conductivity which teaches about the degree of mineralization of the water [15], was measured using a NEOTEK-PONSEL conductivity meter; Dissolved oxygen was measured using an oxymeter.

Picture 1. Taking of measurements in-situ.

3.3. Laboratory analyses

The physico-chemical analyses were conducted following the standard procedures described by [16]. Water samples collected at the site were transported to the laboratory in a cooler to prevent spawning and were kept refrigerated. An analysis of the major elements was carried out, starting with the parameters likely to be modified quickly. These are nitrites, nitrates, Orthophosphates and Sulfates were determined by spectrophotometry. Hydrotimetric titer, often due to Ca^{2+} and Mg^{2+} ions, Chlorides.

4. Results and discussion

4.1. The temperature of the water

From the point of view of the thermal regime, the waters of the two lakes Aguelmam Sidi Ali are characterized by a stratification that spreads out in early summer-autumn (Figure 4). The temperature of the waters studied is subject to net variations directly depends on the seasonal variability of the climate [17]. For the small lake, overall, the temperature values oscillate between 5.44°C and 24°C on the surface, and between 4.73°C and 14°C at the bottom. In the large lake, they vary between 10.3°C and 23°C on the surface, and between 13°C and 17°C at a depth of 9m. These differences are very minimal in the cold season it should be noted that [1] found the same fact. Note that the temperature in both lakes meets the requirements of the perch, which can tolerate temperatures between 4°C and 31°C with an optimum between 6°C and 14°C.

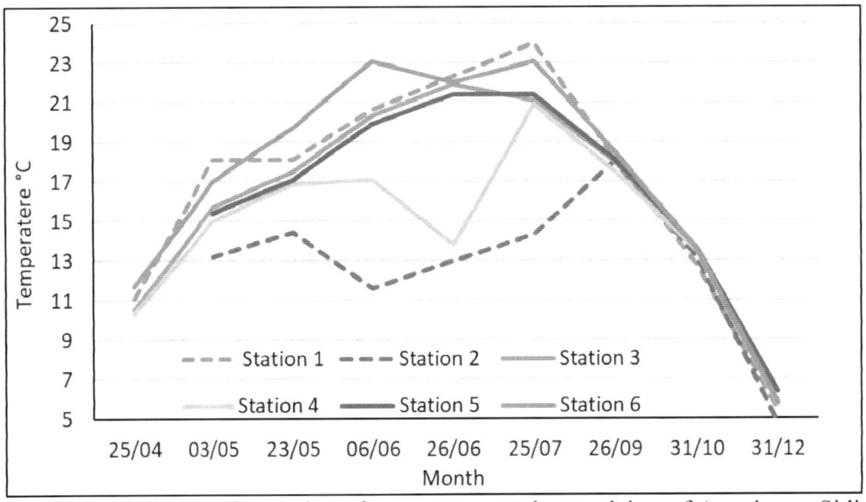

Figure 4. Spatio-temporal variation of temperature at the two lakes of Aguelmam Sidi Ali

4.2. The hydrogen potential

The hydrogen potential values obtained (figure 5), show that the waters of the two lakes are very basic. Indeed, the pH values vary globally between 8.28 and 9.85, which is probably related to the permanent contact of the waters with the basaltic formations considered as basic substrates. The temporal variation of pH was marked by increasing in summer [17], while the lowest values, and because of dilution by meteoric waters, are recorded in the winter period. In addition, and taking into account the depth, the variations are important during the stratification period.

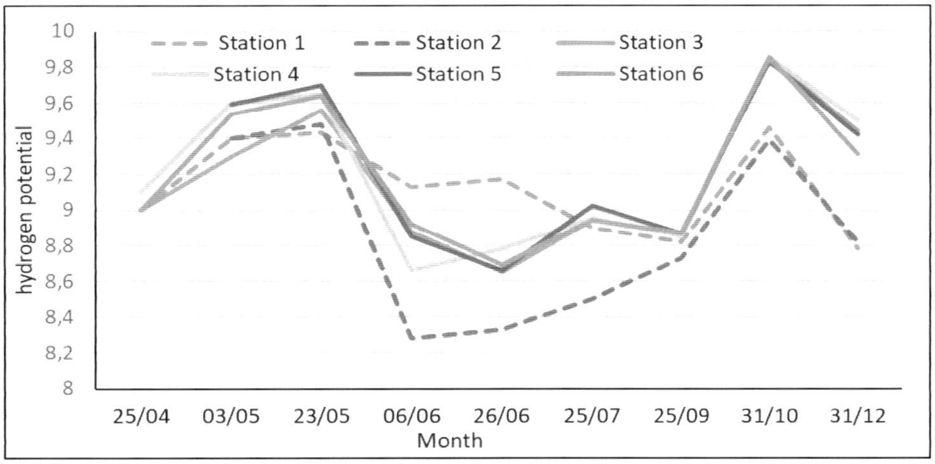

Figure 5. Spatio-temporal variation of hydrogen potential at the two lakes of Aguelmam Sidi Ali

4.3. The dissolved oxygen

The results obtained (figure 6) show that the dissolved oxygen contents of the surface waters of the small and large lake are generally good and superior to those obtained at depth. This is mainly related to the importance of gas exchanges with the atmosphere, and in particular during the valley and slope breezes, especially in the large lake where the surface area is larger. The results also show that the oxygen content decreases from the surface to the depth, [17] also reported similar observation, and from the center of the lake to the banks. This decrease has been linked not only to increased temperature that limits oxygen solubility, but also to microbial activity that breaks down organic matter at the water column and sediment level [18].

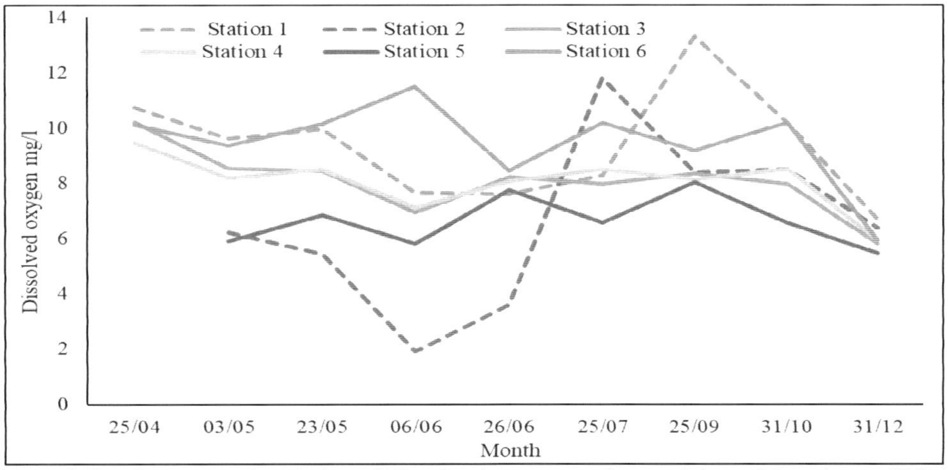

Figure 6. Spatio-temporal variation of dissolved oxygen at the two lakes of Aguelmam Sidi Ali

4.4. The electrical conductivity

The electrical conductivity tell about the degree of mineralization of the water. The results obtained (figure 7), highlighted a clear difference between the small and the large lake. Indeed, the values obtained in the latter are much higher, which is mainly related to the abundance of calcium and magnesium cations resulting from the alteration of limestones, dolomites, and marls that dominate the substratum of the large lake, contrary to the small lake, dominated by basalts. It should be noted that the minimum values recorded for both lakes are observed in May (high water period), while the maximum values are recorded in summer (low water period). In addition, and according to the results obtained, the values are homogeneous on the whole water column, similar results were reported by [17], at the level of each lake and over the whole study period.

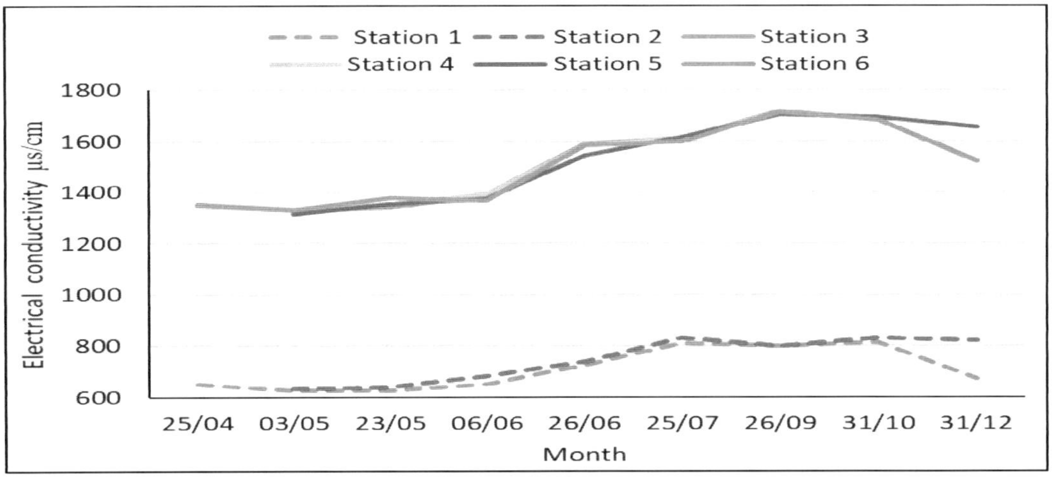

Figure 7. Spatio-temporal variation of electrical conductivity at the two lakes of Aguelmam Sidi Ali

4.5. The total hardness

The results obtained (Figure 8), show that the total hardness of water recorded at the level of the large lake is very high; it is the strongly hard water whose average is higher than 60°F. This result reflects the impact of carbonate formations of Jbel Sidi Ali on the abundance of cations responsible for calcimagnesian hardness.

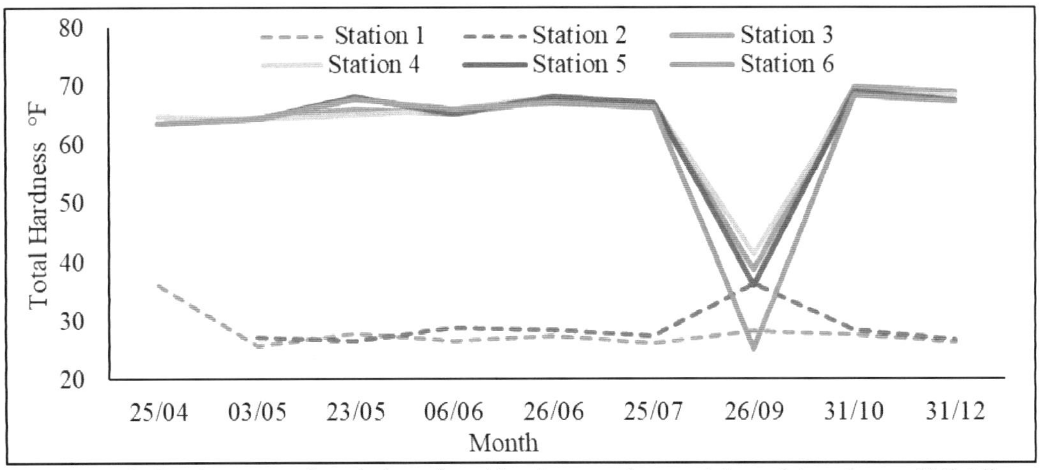

Figure 8. Spatio-temporal variation of total hardness at the two lakes of Aguelmam Sidi Ali

4.6. The Chlorides and sulfates

Chloride and sulfate levels (Figures 9 and 10), are high in the large lake with successive averages of 239.09mg/l and 30.59mg/l against only 133.7mg/l and 16mg/l recorded in the small lake. This large difference is mainly related to the direct contact of the large lake with layers of Triassic clays saturated with salt and gypsum. Chlorides exist in almost all waters at variable contents [19]. They are not adsorbed by geological formations, do not combine easily with chemical elements and remain very mobile [20]. Sulfate values in both lakes are within the water quality standards [21].

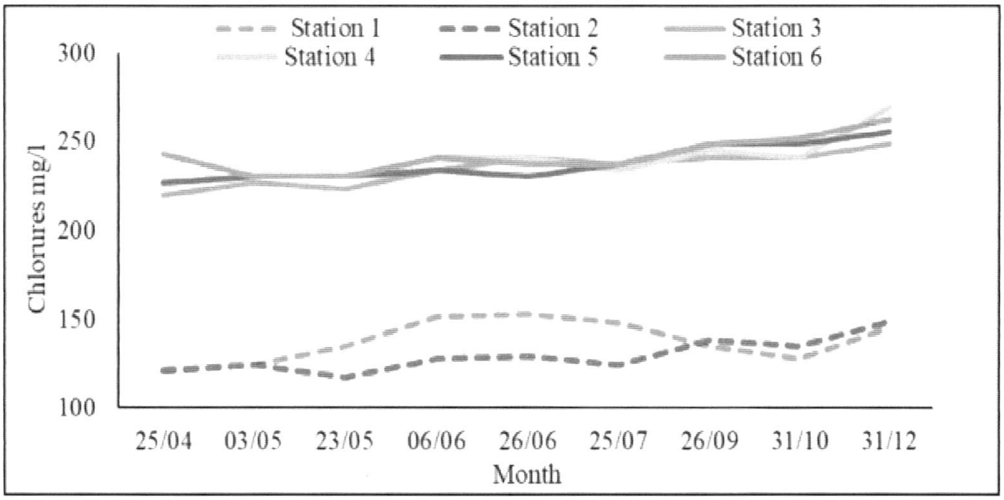

Figure 9. Spatio-temporal variation of chlorides at the two lakes of Aguelmam Sidi Ali

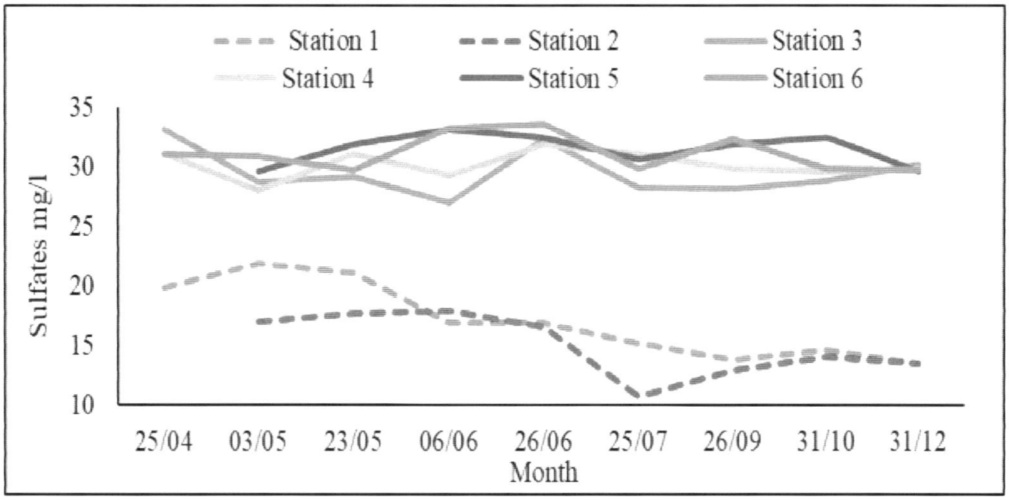

Figure 10. Spatio-temporal variation of sulfates at the two lakes of Aguelmam Sidi Ali

4.7. The orthophosphate

The orthophosphate values (figure 11), are recorded at the level of the two lakes between June and October which coincides with the period when strong pastoral pressure is exerted around the lake. Indeed, this activity generates a large quantity of organic waste.

IOP Conf. Series: Earth and Environmental Science 1090 (2022) 012033 doi:10.1088/1755-1315/1090/1/012033

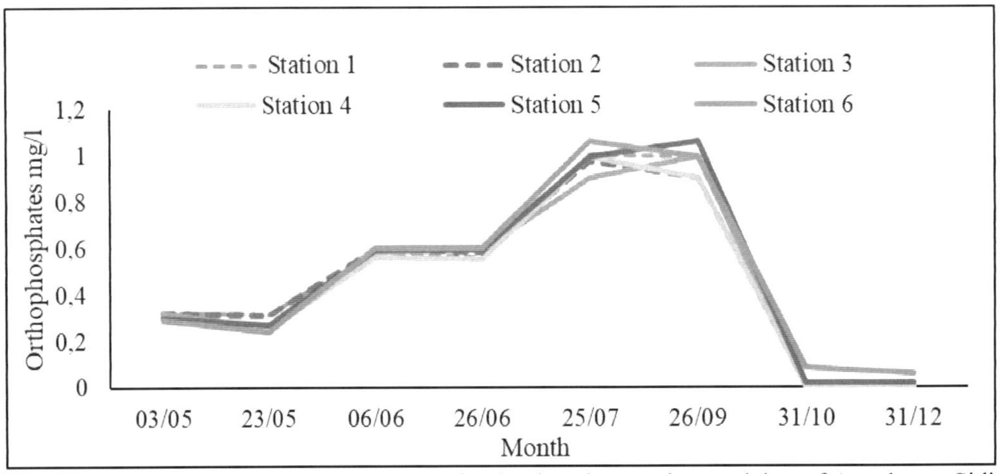

Figure 11. Spatio-temporal variation of orthophosphate at the two lakes of Aguelmam Sidi Ali

4.8. Nitrates and Nitrites

The concentrations of nitrates and nitrites (figures 12 and 13) have highlighted the low values of nitrogenous compounds, at the level of the two lakes. This result is justified by the absence of any agricultural activity in the lake watershed. It should be noted that the values of nitrites in the two lakes conform with [21] fixing the standards of quality of water which stipulate those nitrites should not exceed 0,5mg/l.

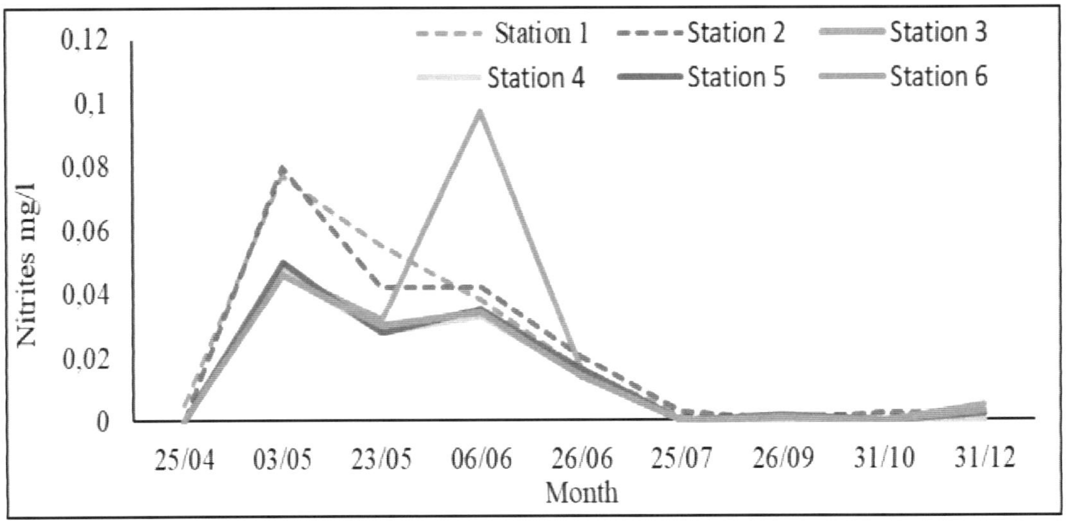

Figure 12. Spatio-temporal variation of nitrates at the two lakes of Aguelmam Sidi Ali

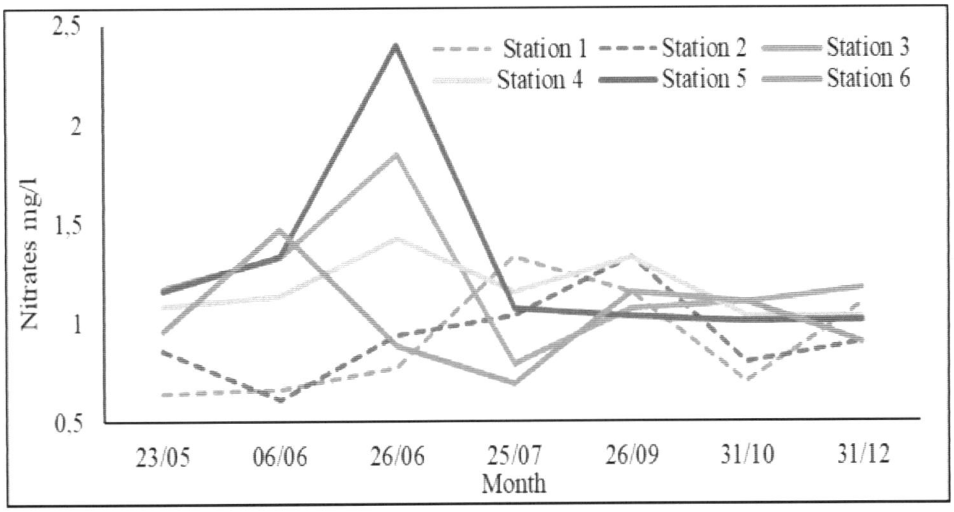

Figure 13. Spatio-temporal variation of nitrites at the two lakes of Aguelmam Sidi Ali

5. Conclusion

Lake Aguelmam Sidi Ali is one of the largest natural lakes in Morocco which has a strong fish farming potential, hence the importance of monitoring the quality of its waters. The present study was devoted to the Spatio-temporal monitoring of the physicochemical parameters of the large and small lake Aguelmam Sidi Ali. Based on the values obtained, it can be concluded that the quality of the water of the lake Aguelmam Sidi Ali is generally good to be exploited in fish farming. According to the Moroccan regulation on the quality of fish farming waters, the basic waters of the stations sampled at the great lake of Aguelmam Sidi Ali are strongly mineralized and very rich in magnesium. The thermal study revealed a stratification of the waters of the lake which lasts from the beginning of summer to the end of autumn. In parallel, a stratification in oxygen is necessary. The other Physico-chemical parameters such as nitrates, nitrites, sulfates, and chlorides were within the desirable limits.

The results of this study are incomplete, as they cover a limited period. However, the data acquired provide a characterization of the physicochemical quality of the waters of the two lakes during the annual cycle (2019), constituting a reference for comparison with previous and future studies. In perspective and in order to understand the dynamics of the waters of the two lakes of Aguelmam Sidi Ali, more detailed studies on the seasonal and interannual variation of water quality as well as on the biogeochemical cycle of sediments are necessary.

6. References

[1] Amhaouch A 1995 Etudes physico-chimique et ichthyologique des lacs du Moyen Atlas : cas d'Ifrah et Sidi Ali. Mémoire 3e cycle, École nationale forestière des ingénieurs, Salé p130.
[2] Azeroual A, Crivelli A J, Yahyaoui A, and Dakki M 2000 The ichthyofauna of the continental waters of Morocco. Cybium 24(3):17-22.
[3] Chillasse L, Dakki M, and Abbassi M 2001 Ecological values and functions of the wetlands of the Middle Atlas (Morocco). Humedales Mediterráneos 1:139-146.
[4] Chillasse L, Dakki M., 2004 Potentialities and conservation status of wetlands in the Middle Atlas (Morocco), with reference to drought influences. Sécheresse 15(4):337-345.
[5] Khaffou M, Chahlaoui A, and Samih M, 2013 Habitats used by the Ruddy Shelduck (Tadorna ferruginea) in the wetland of Aguelmam Sidi Ali - Ramsar site - Middle Atlas - Morocco.

International Journal of Biological and Chemical Sciences 7(2):598-606.
[6] Menjour F, Amraoui F, and Remmal T, 2016 Assessment of the spatio-temporal evolution of Aguelmam Sidi Ali lake using multitemporal landsat imagery (Middle Atlas - Morocco). The 6 th International Conference on Cartography and GIS, pp. 588-597.
[7] Baali, A. 1998 Genèse et évolution au Plio-Quaternaire de deux bassins intramontagneux en domaine carbonaté méditerranéen. Les bassins versants des dayets Afourgagh et Agoulmam (Moyen Atlas, Maroc). Doctoral thesis, Univ. Sidi Mohammed Ben Abdallah, Fès, p326. [In French].
[8] Hinaje S 2004 Tectonique cassante et paléochamps de contraintes dans le Moyen Atlas et le Haut Atlas central (Midelt-Errachidia) depuis le Trias jusqu'à l'actuel, Doctoal Thesis (Mohammed V University Rabat Morocco) p 363. [In French].
[9] Collet LW 1925 les lacs leur mode de formation. Leurs eaux. Leur destin. Eléments d'hydro-géologie Revue de Géographie Alpine 13 777-78[In French].
[10] Martin J 1981 Le moyen Atlas central étude géomorphologique (Rabat : Editions du Service géologique du Maroc) p 445. [In French].
[11] Sayad A, Chakiri S, Martin C, Bejjaji Z and Echarfaoui H 2011 Effet des conditions climatiques sur le niveau du lac Sidi Ali (Moyen Atlas, Maroc) PhysioGéo 5 251-68. [In French].
[12] Chapman D and Kimstach V 1996 Selection of water quality variables. Water quality assessments: a guide to the use of biota, sediments, and water in environment monitoring (Cambridge: UNESCO/WHO/UNEP) p 609.
[13] Dussart G B J 1966 Limnologie l'étude des eaux courantes (Paris : Ed. Herman) p 250.
[14] Brémond R and Vuichard R 1973 Paramètres de la qualité des eaux. Ministère de la protection de la nature et de l'environnement. Documentation francaise. Paris 11 22-35.
[15] Derwich E, Benaabidate L, Zian A, Sadki O and Belghity D 2010 Caractérisation physico-chimique des eaux de la nappe alluviale du haut Sebou en aval de sa confluence avec oued Fès LARHYSS Journal 8 101-112.
[16] Rodier J, Legube B, Merlet N, Brunet R and Coll E 2009 L'analyse de l'eau. 9e édition. Dunod, Paris, 1579 p.
[17] El Morabet I, El Fadili L. R, Taybi A. F, Mabrouki Y, Bouhaddioui A, Bahhou J, 2022 Assessment of physicochemical and biological parameters of Aguelmam Sidi Ali Lake, a threatened Ramsar Site, Morocco. AACL Bioflux 15(1):573-583.
[18] Ahangar I A, Saksena D, Mir M, Ahangar M, 2012 Seasonal variations in physicochemical characteristics of Anchar lake Kashmir, India. Journal of Advanced Biological Research 3(2):352-357.
[19] Nehme N 2014 Évaluation de la qualité de l'eau du bassin inférieur de la rivière du Litani, Liban: approche environnementale », Université de Lorraine.
[20] Layachi G 2008 Influence de la lithologie et des conditions climatiques sur la variation des paramètres physico–chimiques des eaux d'une nappe en zone semi aride, cas de la nappe de meskiana nord-est algerien. », PhD Thesis, Université Badji Mokhtar de Annaba.
[21] MCATEE 2003 Arrêté conjoint du Ministre chargé l'Aménagement du Territoire, de l'Eau et de l'Environnement n°2028-03 du 10 Ramadan 1424 (5 Novembre 2003) fixant les normes de qualité des eaux piscicoles," Bull. Off. n° 5196 du 18/03/2004, 5196, pp. 5–7, 2003.

Impact of climate change on extremes events in Morocco

W HAMMOUDY[1*], R ILMEN[2], and M SINAN[2]

[1] The Directorate General of Meteorology (DGM), BP 8106-Casa Oasis,Bd Mohamed TaïbNaciri Hay Hassani, Casablanca, Morocco
[2]Department of Hydraulics, Environment and Climate (HEC), Hassania School of Public Works (EHTP), Km 7, Road d'ElJadida, BP. 8108, Casablanca, Morocco

[*] E-mail: wahibhammoudy@gmail.com

Abstract. Today, the increase in the evolution of certain types of extreme events in Morocco has the potential to produce impacts and damages on socio-economic sectors. An increase in heat and cold extremes manifested by strong heat waves and cold waves are clearly increased in terms of their intensity, frequency and duration. This required an in-depth study to understand the link between the two. So, this study under the following theme: "Impact of climate change on extremes events in Morocco" is devoted to studying the relationship between climate change and extreme events and the impact of one to the other, their occurrence and intensity; then it's structured around one main phase, that aims to evaluate observed changes in climate extremes from 1984 to 2018 based on robust methods to estimate trends with Sen's Slop test and their significance with Mann-Kendall test. The basic data obtained from 29 meteorological stations covering all climatic zones have passed quality control tests to ensure reliable detection of any changes. The results show that Morocco has experienced a significant increase in hot extremes and also in cold extremes almost generalized over our Country Morocco. As for rainfall extremes, the trends generally indicate no significant change.

Keywords: Climate change, extreme events, climate extremes, hot extremes, cold extremes, rainfall extremes, global warming.

1. Introduction

Since the beginning of the 20th century until today, our planet is experiencing a global warming of atmosphere accompanied by a disruption of its climate at all levels. Its spatio-temporal variability is becoming greater and greater, favouring the accentuation of extreme climatic events in several regions. Current studies confirm that climate change will modify the intensity, frequency and duration of extreme events. Indeed, the IPCC found that these changes have been affected the number of these events, which has been translated into an increase in heat extremes and in cold extremes, also in heavy rainfall events in several regions [1].

Africa is considered to be the least continent responsible for greenhouse gas emissions (GGE). Nevertheless, it's one of the candidates to bear the brunt of the effects of climate extremes [2]. The 2016 edition of "Climate Risk Index" published during COP22, showed that Africa is the continent that has been most affected by this type of phenomena in 2015: "Floods have particularly affected the host continent of the climate summit".

Being part of North Africa, Morocco, which is a low emitter of greenhouse gas emissions (CGE), is also experiencing the effects of climate change in terms of its diversity, intensity, and occurrence. The meteorological extremes that Morocco is experiencing today represent inescapable effects. In particular, in the form of higher average and extreme temperatures, more extensive and frequent heat waves, lower annual precipitation in several regions, and changes in the seasonal distribution of rainfall.

Towards the end of the 20th century, Morocco was confronted with several events. Most notably, the drought, which became increasingly severe. Other regions have experienced violent and unusual floods, for example: Ourika in 1995 and 1999; Casablanca-Mohammedia in 1996; Tetouan in 2000; Mohammedia in 2002; Er-Rachidia in 2006; the northern regions, Al Gharb, Al Haouz and Sous in 2009 and 2010; Casablanca in 2010, the South in 2014, Ksarsghir in 2021; and finally, Casablanca in 2021 [3]. It's in this sense that our study, entitled "Impact of climate change on extreme events in Morocco", is registered.

2. Data and Methodology

2.1. Database

2.1.1. Data Observation

The first database used in this study consists of daily observations of maximum and minimum temperatures, and rainfall accumulation from the network of the General Directorate of Meteorology (DGM). The latter has a well-distributed observation network throughout the Moroccan territory, whose meteorological information is provided on a regular and continuous basis. In order to cover all areas of the country, it was decided to work with 29 weather stations including the maximum possible regions to include (Figure 1).

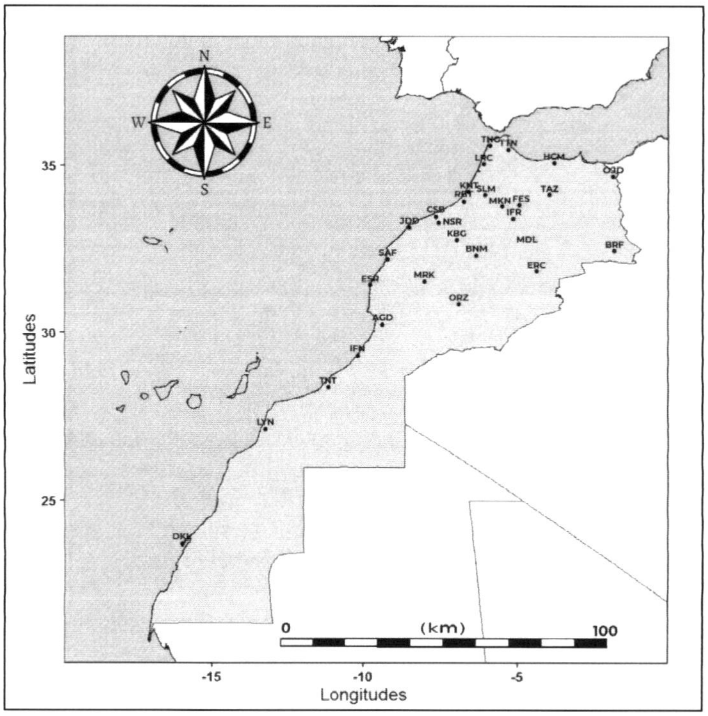

Figure 1. Weather stations study in Morocco.

IOP Conf. Series: Earth and Environmental Science 1090 (2022) 012034 doi:10.1088/1755-1315/1090/1/012034

The table 1 summarizes the information on the period of the data series of each weather station and the climate region:

Table 1. Information on the period covered (common) by 29 stations studied, climatic regions are based on the classification of Knippertz [4].

Station Name	Abbreviation	Studied Period	Climate Region
Laâyoune	LYN	1984-2018	Region IV
Sidi-Ifni	IFN	1984-2018	Region IV
Dakhla	DKL	1984-2018	Region IV
Tanger	TNG	1984-2018	Region I
Larache	LRC	1984-2018	Region I
EL-Hoceima	HCM	1984-2018	Region I
Oujda	OJD	1984-2018	Region II
Kenitra	KNT	1984-2018	Region I
Taza	TAZ	1984-2018	Region I
Rabat	RBT	1984-2018	Region I
Sidi-Slimane	SLM	1984-2018	Region I
Fez	FES	1984-2018	Region I
Meknes	MKN	1984-2018	Region I
Casablanca	CSB	1984-2018	Region I
Nouasser	NSR	1984-2018	Region I
Ifrane	IFR	1984-2018	Region I
El Jadida	JDD	1984-2018	Region I
Khouribga	KBG	1984-2018	Region I
Safi	SAF	1984-2018	Region I
BeniMellal	BNM	1984-2018	Region I
Midelt	MDL	1984-2018	Region I
Bouarfa	BRF	1984-2018	Region III
Er-Rachidia	ERC	1984-2018	Region III
Marrakech	MRK	1984-2018	Region I
Agadir	AGD	1984-2018	Region I
Ouarzazate	ORZ	1984-2018	Region III
Tan-Tan	TNT	1984-2018	Region IV
Tetouan	TTN	1984-2018	Region I
Essaouira	ESR	1984-2018	Region I

2.2. Methodology

2.2.1. Methodology

This work has been carried out initially on a large part that consists of calculating more than four climate indices (defined in the following part) on all 29 observed synoptic stations and based on the following three meteorological parameters: daily maximum temperature, daily minimum temperature, and precipitation [5]. After a calculation of the climate extremes over the period 1984 to 2018, we measure their trends using the new Sen's Slope method, as well as their significance using the Mann-Kendall test. Before starting to detail our study, it should be noted that the data have passed a process quality control based on the most popular methods including Boxes plots, thermal amplitude…All this using the R language, which is equipped with several statistical tools to facilitate this treatment. Finally, the trends were pointed by a triangle indicated in the legend on the maps using this programming language with the shapes files extracted from the ARCGIS software.

2.2.2. The Climate indices

In order to study extreme weather events, the WMO CCl Commission and the ETCCDI Group have defined more than twenty precipitation and temperature climate indices, developed for the understanding of phenomena [6] [7]. The calculation of the indices (table 2) is based on thresholds which can be fixed or calculated over a chosen base period. This calculation was done using the ClimPACT2 libraries used R (Open source and available on the website (https://github.com/ARCCSS-extremes/climpact2) [8], these indices are defined as follows:

Table 2. Lists of indices used together with their definitions and units.

indices	Definition	unit
TXm	The monthly mean value of the daily maximum temperature.	°C
TNm	The monthly mean value of the daily minimum temperature.	°C
Rx1day	Maximum cumulative rainfall in a single day.	mm
CDD	The maximum duration of the dry period or the maximum number of consecutive days with precipitation below strictly 1mm.	day

2.3. Statistical Tools

2.3.1. Sen's Slope Trend

After calculating the indices at each station, we calculated the trend using non-parametric statistics, including the Sen's Slope trend, which is a very robust adjustment method, makes it possible to calculate trends in univariate time series. It is defined as follows:
(i,x_i) a set of pairs where$x_i$is a time series, the slope of Sen's Slope is the median of all slopes calculated between each pair of points in the series **[9]:**

$$\text{Sen's Slope}=\text{Median}\left(\frac{x_j-x_i}{j-i};i<j\right)(1)$$

2.3.2. Mann-Kendall Test

After calculating the trend for each climate indices and at each observation station, we used the Mann-Kendall significance test (Mann, 1945; Kendall, 1975) to assess the trend in our study. Indeed, it's a non-parametric test to identify the monotonous trends identifiable in a time series, strongly recommended by the World Meteorological Organization (WMO) and defined as follows:

We assume the hypothesis H0: «that there is no trend». The initial value of quantity S is initiated at 0, then it's incremented or decremented by 1, the final value of S makes it possible to conclude on the existence or absence of the trend [10].

$$S_i=\sum_{i<k}^{n} sig(x_k-x_i) \text{ and } S=\sum_{k=1}^{n-1} S_k=\sum_{k=1}^{n-1}\sum_{j=k+1}^{n} sig(x_j-x_k)(2)$$

$$\text{With:} sig(x_j - x_k) = \begin{cases} 1 \text{ si } x_j - x_k > 0 \\ 0 \text{ si } x_j - x_k = 0 \\ -1 \text{ si } x_j - x_k < 0 \end{cases} \quad (3)$$

However, a probability must be calculated and compared with the "α" % (P value) threshold. The hypothesis H0 will be rejected if the probability is below the "α" % threshold. This is done using the statistical test Z which is defined as follows:

$$Z = \begin{cases} \frac{S-1}{(var(S_k))^2} & \text{if } S_k > 0 \\ 0 \text{ if } & S_k = 0 \\ \frac{S+1}{(var(S_k))^2} & \text{if} S_k < 0 \end{cases} \quad (4)$$

With Z a normal random variable of mean 0 and standard deviation 1 and the p value ("α") is taken equal to 0.05.

3. Results

In this section, we are interested in the observed changes of climatic extremes in Morocco relative to the parameters of precipitation and the maximum and minimum temperature. In addition, a data quality check was carried out to ensure reliable detection of any changes. Then we mapped the trend observed at the 29 weather observation stations

3.1. Analysis of Trend

3.1.1. HeatIndex

According to the figure 2, the average maximum (TXmean) and minimum (TNmean) temperatures show a marked an increase almost generalized on all stations during the period 1984-2018. The maximum temperature varied from 0.02°C/year (Nouasser) to 0.08°C/year (Taza), which is statistically significant. A non-significant decrease is identified at Dakhla station. As for minimum temperatures, the trend varied between 0.02°C/year at Meknes and 0.06°C/year at Nouasser over the study period. At the Rabat and El-Hoceima stations, there were decreases, although not significant (tables 3 and 4).

Table 3. Observed trends (Sen's Slope) statistically significant TXmean index between 1984 and 2018 for a few stations.

Station	**AGD**	**ERC**	**KNT**	**LRC**	**LYN**	**ORZ**	**TNG**	**TNT**	**NSR**	**MRK**
Trend	0.05	0.06	0.04	0.03	0.04	0.05	0.04	0.03	0.02	0.05
Station	**MDL**	**FES**	**CSB**	**BNM**	**IFR**	**MKN**	**TAZ**	**OJD**	**SLM**	**BRF**
Trend	0.03	0.05	0.03	0.07	0.03	0.05	0.078	0.05	0.07	0.05

Table 4. The observed trends (Sen's Slope) statistically significant TNmean index between 1984 and 2018 of a few stations.

Station	**AGD**	**ERC**	**BNM**	**JDD**	**LYN**	**KBG**	**TNG**	**NSR**	**SAF**	**MRK**
Trend	0.06	0.05	0.02	0.04	0.03	0.05	0.03	0.06	0.04	0.03
Station	**TAZ**	**FES**	**CSB**	**BNM**	**IFN**	**MKN**	**OJD**	**BRF**		
Trend	0.04	0.04	0.06	0.02	0.03	0.02	0.05	0.04		

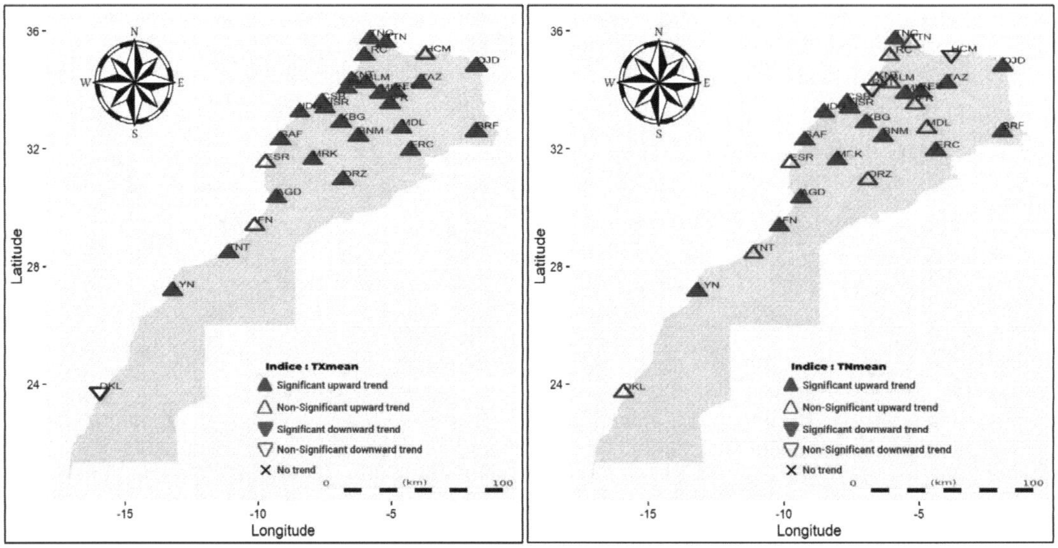

Figure 2. Observed trends (Sen's Slope) in mean maximum (TXmean: right) and minimum (TNmean: left) temperatures of the 29 weather stations over the period 1984-2018.

3.1.2. Rainfall index

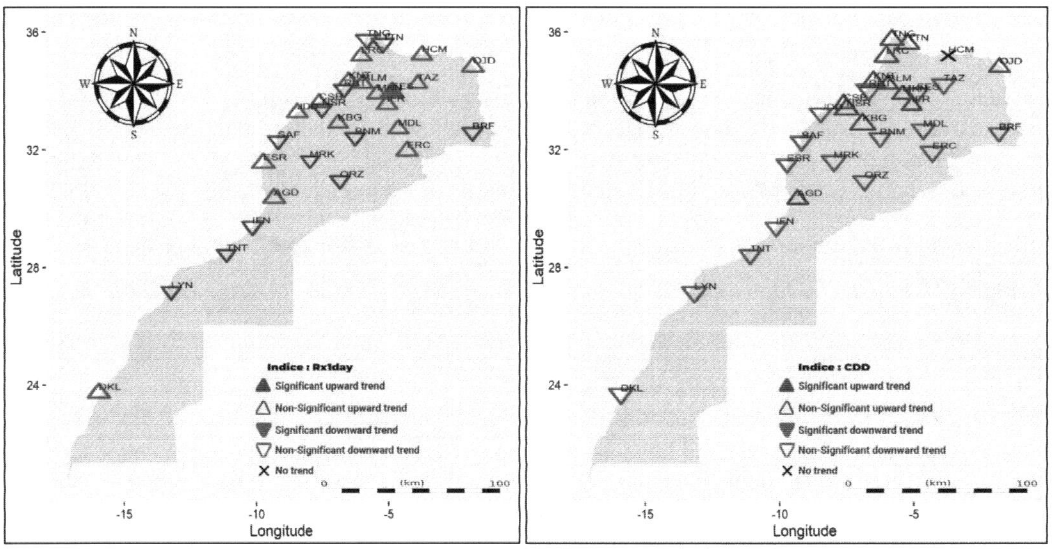

Figure 3. Observed trends (Sen's Slope) in total precipitation accumulation in a single day (Rx1day: right) and maximum duration of dry days (left: CDD) of the 29 weather stations over the period 1984-2018.

The analysis of the evolution of the Rx1day indice (figure 3), defined as the maximum of precipitation accumulation in a single day, allows us to notice an increase in precipitation quantities associated to 59% of the studied stations, which the only significant one according to Mann-Kendall test is Fez, it experienced an increase in precipitation equivalent to 0.48mm/year. In addition, 41% of the stations located in our country have experienced non-significant negative trends. As for the evolution of the CDD indice, reflecting the maximum dry period, between 1984 and 2018 is characterized by spatially variable and non-significant trends, between decrease generally observed towards the South, the center and the South-East regions (Sidi-Ifni, Tan-Tan, Laâyoune and Dakhla) and generally noted increases towards the Northwest regions (Tanger, Tetouan, Kenitra, Casablanca, Nouasser, Meknes and Ifrane).

4. Discussion

Many studies have shown changes in extreme events either thermal or rainfall, in proportion to the current warming, including increases in maximum and minimum temperature [11][12]. This increase in the study area from 20 N to 40 N is similar to the observed global warming reported in the 6th IPCC report. Nevertheless, the trends of changes observed are not significant everywhere only in Fez. Most studies homogenize the observed data before using it. In this study, we used raw data to take all the existing climate factors that can affect long-term trends, for this reason, we used the Sen's Slop trend which is a very robust adjustment method used to calculate trends in uni-varied time series. The lonely result in Fez station is showing an increasing trend of the maximum of precipitation accumulated in a single day. This shows that heavy rainfall events have been on the rise in recent years.

However, the different results emphasize changes over time in the trend of climatic extremes, it comes back to the use of the new trend Sen's Slope, a test was done using linear trend tests, the results are half non-significant, it comes back to the non-homogenizations of the observed data.

It should be noted that these conclusions confirm the results predicted by [16] on the increase in trend of hot extremes. Nevertheless, the latest results of this article have just confronted the conclusions of previous research on rainfall indices in Morocco, in fact, the increase in the greenhouse effect caused by human which acts on the redistribution of air masses coupled with the human impact on forest resources probably contributes in the growth of extreme events. The NAO regimes affect [11] directly the rainfall extremes, the episodes of heavy precipitation have experienced an increase in recent decades, contributing to a huge flood throughout some regions in Morocco [17].

5. Conclusion

The objective of this study is to assess the impact of climate change on extremes events in Morocco. An assessment of the observed trends in thermal and rainfall extremes was carried out. It's based on the observed data from 29 weather stations over the period 1984 to 2018 covering all the climatic zones of Morocco and calculating the trend using the Sen's Slop test as well as their significance using the Mann-Kendall test. Analysis of the observed trends of climate extremes over the period 1984-2018 at the weather station level shows a general increase in maximum and minimum temperatures over most of the country regions. This increase is significant for most stations. In terms of extreme rainfall, they generally show no significant trend over almost the whole of our Country. A decrease was noted on 41% of the stations in contrast to the other stations reflecting an increase that is significant only in Fez with an increase of 0.48 mm/year or 16.8mm over the study period from 1984 to 2018.Overall, additional study in future provides an analysis of changes observed at the National level of Morocco based on the trend of some climate indices reflecting drought (SPI, SPEI…etc) as defined by the WMO Commission on Climatology Expert Team [13][14][15]. They will be done to integrate other climate indices and shift to the regional scale by analysing the observed changes for each climate region as well as future changes, based on climate models that will be evaluated by the observed data.

References

[1] D Qin, G-k Plattner, M Tignor, SK Allen, J Boschung, A Nauels, Y Xia, V Bex, PM Midgley, LV Alexander, NL Bindoff, F-m Bréon, Ja Church, U Cubasch, S Emori, P Forster, P Friedlingstein, N Gillett, JM Gregory, DL Hartmann, E Jansen, B Kirtman, Knutti, K Krishna Kumar, P Lemke, J Marotzke, V Masson-Delmotte, Ga Meehl, Ii Mokhov, S Piao, V Ramaswamy, D Randall, M Rhein, M Rojas, C Sabine, D Shindell, LD Talley, DG Vaughan, S-p Xie, D Wuebbles, D Chen, MC Facchini, D Frame, N Mahowald, J-g Winther, Amg Klein Tank, M Rusticucci, S Brönnimann, Y Charabi, FJ Dentener, Ej Dlugokencky, DR Easterling, A Kaplan, BJ Soden, PW Thorne, M Wild, PM Zhai, SR Rintoul, S Aoki, E Campos, D Chambers, Ra Feely, S Gulev, GC Johnson, Sa Josey, A Kostianoy, C Mauritzen, D Roemmich, and F Wang, 2013. The Physical Science Basis. Contribution of Working Group I.

[2] M. Neil Ward, Peter J. Lamb, Diane H. Portis, Mostafa El Hamly, and Rachid Sebbari, 1999. Climate Variability in Northern Africa: Understanding Droughts in the Sahel and the Maghreb. In Beyond El Niño. Springer Berlin Heidelberg, Berlin, Heidelberg. Pages119–140.

[3] Stour, L. et Agoumi2008. Sécheresse climatique au Maroc durant les dernières décennies. Hydroécol, 2008. Tome 16, pp. 215-232.

[4] P. Knippertz, M. Christoph, and P. Speth, 2003. Long-term precipitation variability in Morocco and the link to the large-scale circulation in recent and future climates. Meteorology and Atmospheric Physics. 83(1-2):67–88. Mar (DOI: 10.1007/s00703-002-0561-y).

[5] Lisa V. Alexander, X. Zhang, T. C. Peterson, J. Caesar, B. Gleason, A. M.G. Klein Tank, M. Haylock, D. Collins, B. Trewin, F. Rahimzadeh, A. Tagipour, K. Rupa Kumar, J. Revadekar, G. Griffiths, L. Vincent, D. B. Stephenson, J. Burn, E. Aguilar, M. Brunet M. Taylor, M. New, P. Zhai, M. Rusticucci, and J. L. Vazquez-Aguirre, 2006. Global observed changes in daily climate extremes of temperature and precipitation. Journal of Geophysical Research Atmospheres. 111(5):1–22 (DOI: 10.1029/2005JD006290).

[6] Lisa Alexander, Nicholas Herold, 2016. ClimPACT2. Indices and software. A document prepared on behalf of The Commission for Climatology (CCl) Expert Team on Sector-Specific Climate Indices (ET-SCI). (February):46.

[7] Thomas R. Karl, Neville Nicholls, and Anver Ghazi, 1999. CLIVAR/GCOS/WMO Workshop on Indices and Indicators for Climate Extremes Workshop Summary. In Weather and Climate Extremes. Pages 3–7. Springer Netherlands. Dordrecht (DOI: 10.1023/A: 1005491526870).

[8] Uwe Schulzweida. 2017. *CDO User Guide*. (October):1–206.

[9] Sen, P.K. (1968) Estimates of the regression coefficient based on Kendall's Tau. Journal of the American Statistical Association, 63, 1379–1389 (DOI: 10.1080/01621459.1968.10480934).

[10] Fatima Driouech, GIL Mahé, MICHEL Déqué, Claudine Dieulin, 2010. Evaluation d'impacts potentiels de changements climatiques sur l'hydrologie du bassin versant de la Moulouya au Maroc. Global Change: Facing Risks and Threats to Water Resources. Proc. of the Sixth World FRIEND Conference, Fez, Morocco, October 2010, 340.

[11] Driouech F., El Rhaz K., 2017. ALADIN-Projections climatiques pour la région arabe. Rapport d'évaluation du changement climatique arabe, Commission économique et sociale des Nations Unies pour l'Asie occidentale « ESCWA ». Rapport principal, E/ESCWA/SDPD/2017/RICCAR/Rapport.

[12] Fatima Driouech, Hafid Stafi, Abdou Khouakhi ,Sara Moutia, Wafae Badi, Khalid ElRhaz, Abdelghani Chehbouni, 2020. Tendances climatiques observées récemment à l'échelle Nationale au Maroc.Le journal international de la climatologie (DOI : 10.1002/joc.6734).

[13] Virginia Murray and Kristie L Ebi, 2013. IPCC Special Report on Managing the Risks of Extreme Events and Disasters to Advance Climate Change Adaptation. SREX. Sep.

[14] Thomas B Mckee, Nolan J Doesken, and John Kleist, 1993. The relationship of drought frequency and duration to time scales. Technical report.

[15] Sergio M. Vicente-Serrano, Santiago Beguería, Juan I. López-Moreno, 2010. A Multi scalar Drought Index Sensitive to Global Warming: The Standardized Precipitation Evapotranspiration Index. Journal of Climate, 23(7): 1696–1718. Apr (DOI: 10.1175/2009JCLI2909.1).

[16] Born K., Christoph M., Fink A.H., 2008. Moroccan climate in present and future: combined view from observational data and regional climate scenarios. Springer Press, 29-45 (DOI: 10.1007/978-3-540-85047-2_4)

[17] Ilmen Rachid, Hassan Benjelloun, Lahcen Ouahmane, Mounir Hakkou, and Mourad Benghazi, 2016. Étude préliminaire des effets de l'Oscillation Nord Atlantique d'hiver sur les cernes du cèdre de l'Atlas, Cedrusatlantica, au Maroc. Bulletin de l'Institut Scientifique, Rabat, Section Sciences de la Vie, n° 38, 43-50.

IOP Conf. Series: Earth and Environmental Science 1090 (2022) 012035 doi:10.1088/1755-1315/1090/1/012035

The problem of access to drinking water in the city of Guercif (Northeast of Morocco): challenges and prospects

M ELYAAGOUBI[1] M AAFIR[2] and A ELALLAOUI[1]

[1] Ph. student at Space, dynamic history and sustainable development laboratory, faculty of letters and human sciences, Sidi Mohamed Ben Abdellah Fez Sais University, Morocco.

[2] Research professor at the Faculty of Languages, Arts and Human Sciences, Ait Melloul University Campus. Ibn Zohr University, Agadir, Morocco.

Email: Medyagoubi12@gmail.com

Abstract. In a climatic context, such as that one in our study area, water is a major environmental issue. The combination of several significant causes already explains the scarcity of water in the town of Guercif: lack of rainfall, climate change, and overexploitation of groundwater. In addition to this structural data, it should be remembered that the sustained dynamics of the area from the point of view of urban planning, demographics, as well as industrial and agricultural development accentuate the demand for a matter as vital as water. The need for water in Guercif will follow the trend of population growth and water-consuming activities. In this context, the deficit recorded during 2019 for the water supply, threatened the urban development of a city newly promoted to the status of the chief town of the province. This study was conducted in two phases: an exploratory phase marked by semi-directive interviews with institutional actors, the service concerned with the production and distribution of drinking water in Guercif, complemented by the second phase of a survey with a sample of heads of households. A mass of data from the field surveys during the summer of 2019 guided the analysis of the impacts of the water shortage in Guercif, as well as the behavior of the urban population towards water consumption and saving. This study describes the satisfaction scales of a sample of urban households based on a questionnaire. The results obtained from the surveys carried out during 2019 confirm the vulnerability of the water sector in Guercif, where the local water potential remains largely dependent on external contributions.

1. Introduction

No life can be envisaged without water; it is an essential material for living beings and therefore the most precious resource on earth. However, it is important to remember that water is a scarce resource, hardly renewable, and unequally distributed in the world: in effect, 9 countries monopolize 60% of the renewable natural resources of freshwater [1]. Prospective studies on the water in the world are alarming and our country is particularly concerned. The fifth report of the Intergovernmental Panel on Climate Change in 2013 places Morocco in the most rainfall-deficient zone for this 21st century. Indeed, urban water security in Morocco is threatened by significant challenges such as population growth, urbanization, economics, and climate change [2]. Thus, the problem of water scarcity is not only a common general aspect where the deficit would reach -33% by the year 2050 [3]; but shared by the city of Guercif whose emerging urban realities have induced an inordinate pressure on water resources in perpetual regression. The struggle to manage and contain the problem persists [4].

IOP Conf. Series: Earth and Environmental Science 1090 (2022) 012035 doi:10.1088/1755-1315/1090/1/012035

The objective of this article is to stimulate reflections on the temporal scarcity of water in Guercif to establish policies to guarantee the best possible distribution of water in the urban perimeter, as well as its preservation to avoid the undesired consequences of a future of water insecurity

2. Material and methods

To carry out this work, we first reported the observation of the areas to be surveyed, then we conducted interviews with the heads of households before developing the questionnaire which allowed us to extract quantitative data relating to the issue under study. In addition, to complete this work, we proceeded to the exploitation of the documents made available to us by the technical services of the national office of electricity and drinking water-water branch-.

2.1. Study area

The city of Guercif, the capital of the province, is located on a plain of the same name at 378 m above sea level. It is located in the North-East of Morocco at a crossroads between the corridor of Taourirt in the East and Taza in the West (Figure. 1). The continentality and the presence of a mountainous contour have resulted in pronounced aridity of the climate to which are added the problems of drinking water supply and sanitation [5]. The province of Guercif was created by the dismemberment of the province of Taza following the administrative division of 2009, and since 2015, it is part of the oriental region.

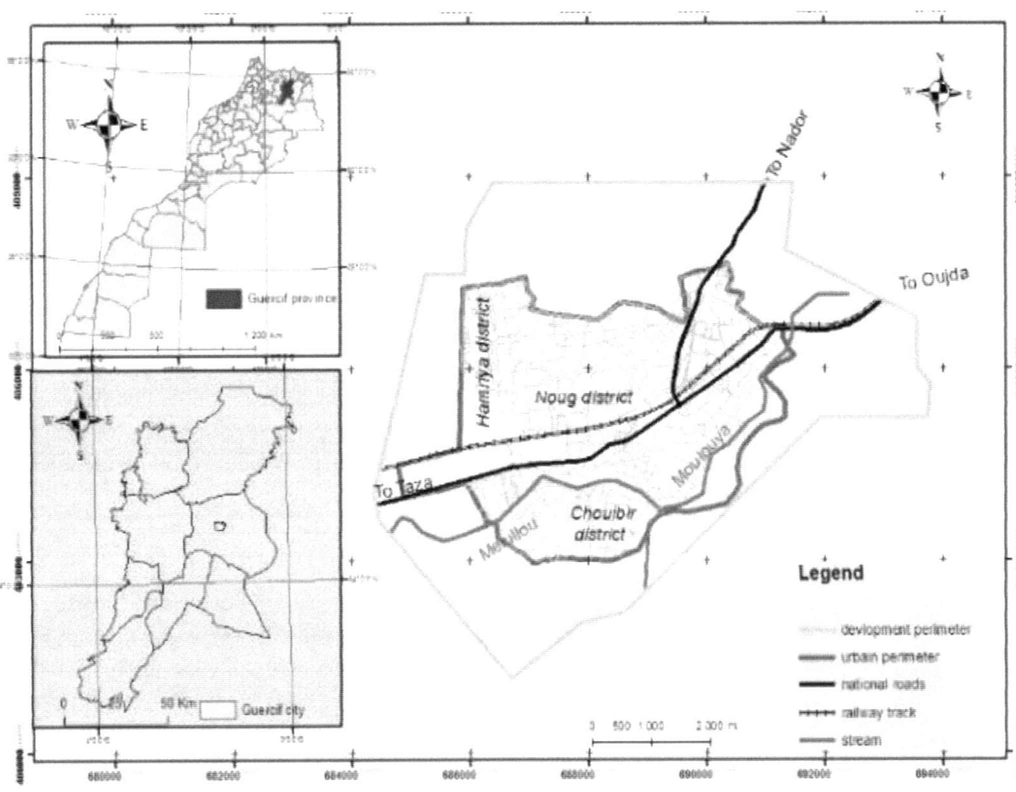

Figure 1. The geographical location of Guercif city in the national and the provincial context

Furthermore, the climatic data relating to rainfall records spread over three decades (1989-2019) at the Guercif station, provided by the agricultural advisory center, show that rainfall has been low and irregular. The annual average reached 161 ml and the wettest season was recorded in 2009-2010 with 211.5 ml; while the driest was in the 2015-2016 season with only 63.5 ml. The Guercif basin is characterized by the presence of groundwater which corresponds to the main plains [6]. Groundwater from these aquifers is used not only in extensive agricultural activities but also for the drinking water supply of the city.

On the morphological level, the city is included in a transition zone of the Moulouya basin, and on the hydrographic level, it is crossed by two rivers: Moulouya and Meloullou, threatened by evaporation which amplifies the disadvantages of the climate and accentuates the regression of the water potential [7]. The hydrological service at Guercif station estimates the amount of water that evaporates each year at around 2000 to 3000 ml. The hydrogeology of the area shows the presence of several water tables (Tafrata, Jal, Sangal, and Maarouf). Indeed, the geology of the area (Figure. 2) is characterized by sedimentary rocks with high aquifer potential.

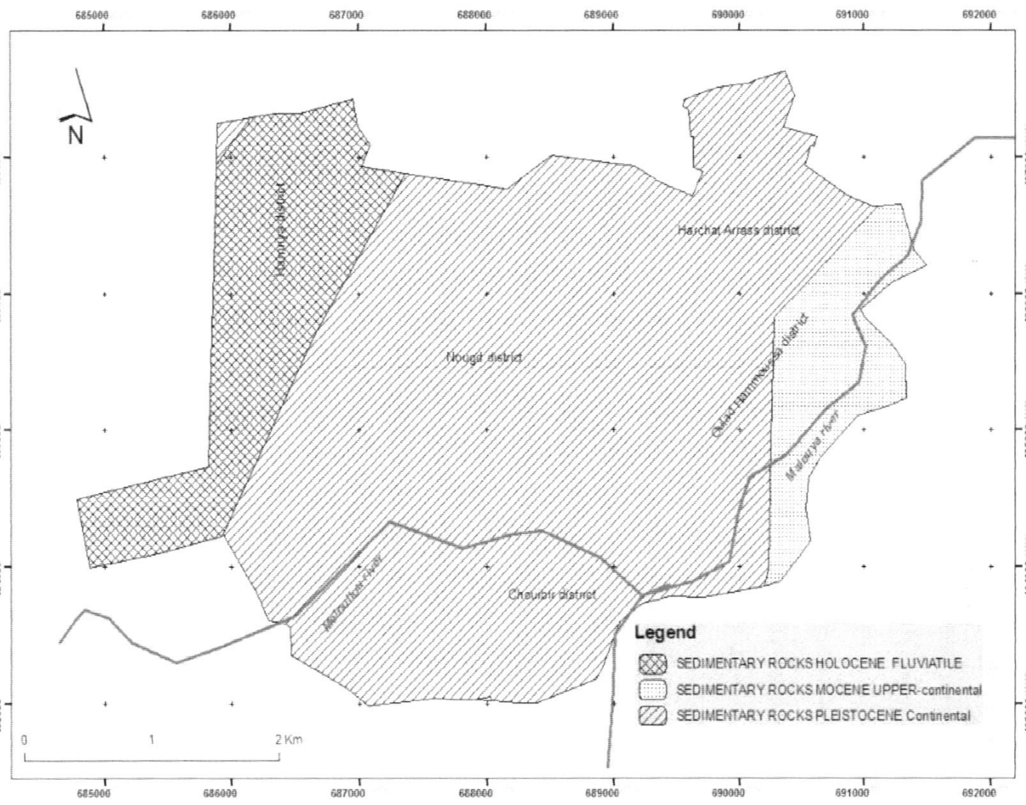

Figure 2. Geological map of Guercif

2.2. Some characteristics of the surveyed sample.

The results of the survey will allow a better understanding of the local water context and measure the state of knowledge related to the problem of water shortage in Guercif while raising awareness of the population on this subject. The survey was conducted among a sample of 240 people (heads of households) distributed according to the hydraulic sectors (Figure. 3) established by the National Office of Electricity and Drinking Water in Guercif at a rate of 40 households per sector.

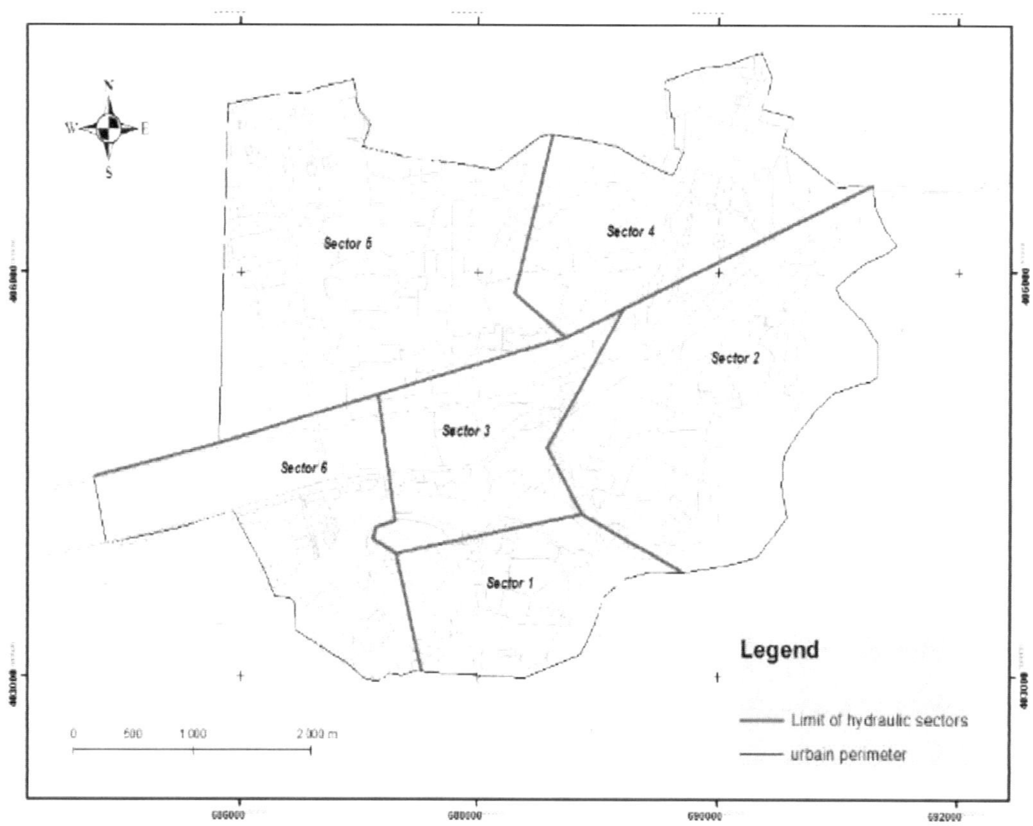

Figure 3. Distribution of households surveyed by the hydraulic sector

The perimeter of the study is the city of Guercif, which in 2019 had a total population of 109.942 inhabitants [8]. The heads of households interviewed are male and the most encountered are between 50 and 60 years old (Figure 4). The level of study of the surveyed sample varies from primary (46%), college (37%), qualifying (9%), and university level (8%). Features deemed relevant for this study relate to habitat type, water supply resources, water use and quality, and water management and saving.

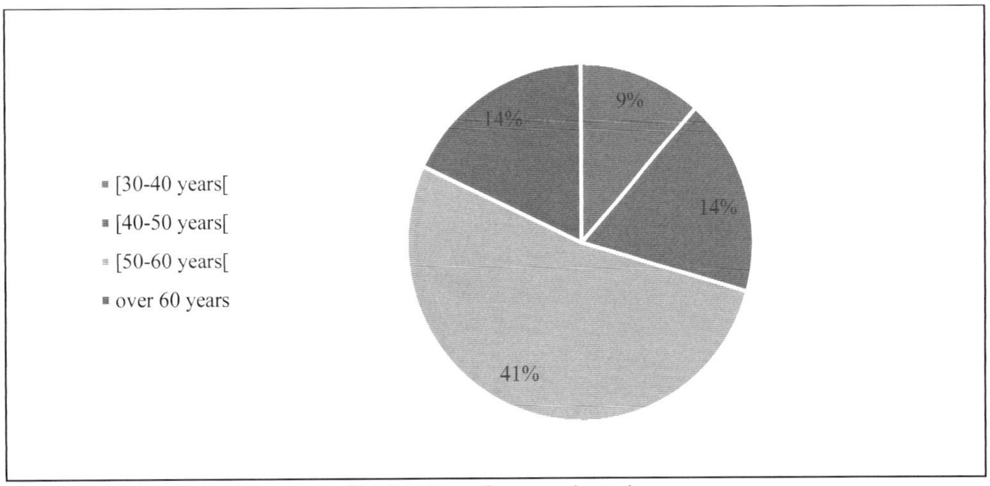

Figure 4. distribution of respondents by age group

The water shortage in Guercif is not felt in the same way. Indeed, the type of housing provides information on the severity of this problem, which is more important for those who occupy an apartment in a building than those who live in a house on the ground floor. The low pressure in the distribution network limits access to water. In addition, some of the respondents use a well to get water. However, the origin of the water supplied by the public network is relatively unknown to the interviewees. Water from the public network is consumed by all the respondents, even those who have wells. However, the quality of the water in some wells does not allow for consumption or use.

3. Results and discussion

The results relating to the variables selected to illustrate the problem of water scarcity in Guercif will be presented in the light of the analysis of the observations drawn from the survey.

3.1 Prioritization of household water needs

In the context of water shortage, the supply of water to the population is essential. All the households surveyed are connected to the public drinking water distribution network. Those who live in a house represent 83% of the overall sample and 25% of them have a well. The quantitative analysis of the coverage of water needs does not provide elements for measuring the quality of the service. To obtain it, it is necessary to fall back on the elements which one has on the degradation of the infrastructures and the frequency of the interruptions in the water supply [9]. The frequency with which households are supplied with drinking water (Figure. 5) depends on the type of housing, the density of the area, and the pressure.

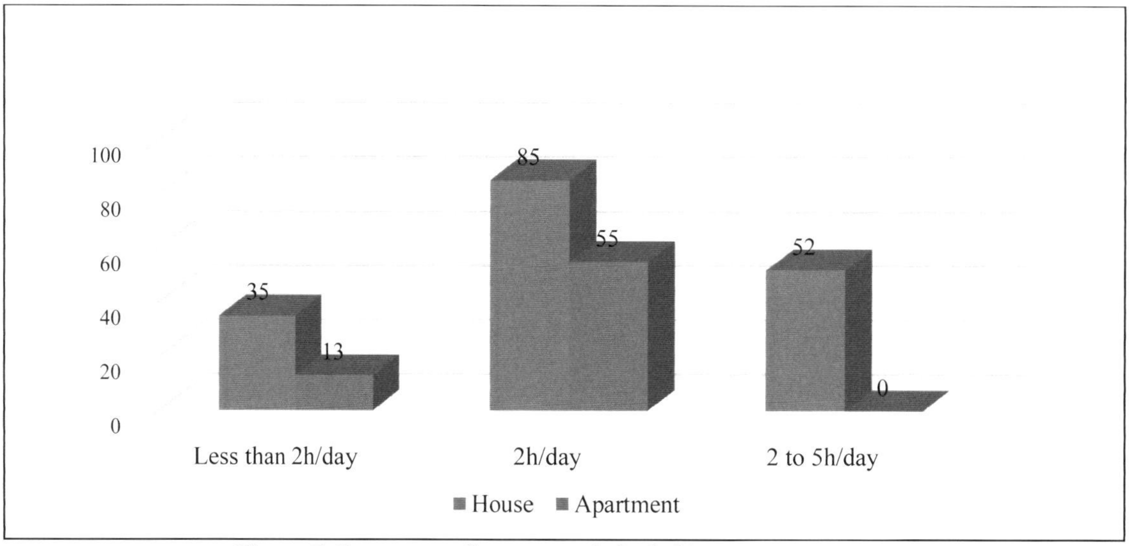

Figure 5. Frequency of water supply by habitat type

Households living in houses are better served during this period of water crisis as 55.83% of respondents in 2019 reported being supplied with water for 2 hours a day, while only 11.25% of apartment dwellers enjoy the same time envelope to have water in their homes. The apartments on the upper floors are almost devoid of water, and in the rare moments of availability of this vital material, the flow is very low not allowing any other use requiring a determined pressure. Moreover, the acuteness of the shortage is strongly felt for 60% (Figure.6) of our sample represented particularly by households residing in apartments or the houses not having another water point (well or drilling). The problem is less acute for households that have a well to facilitate their access to water.

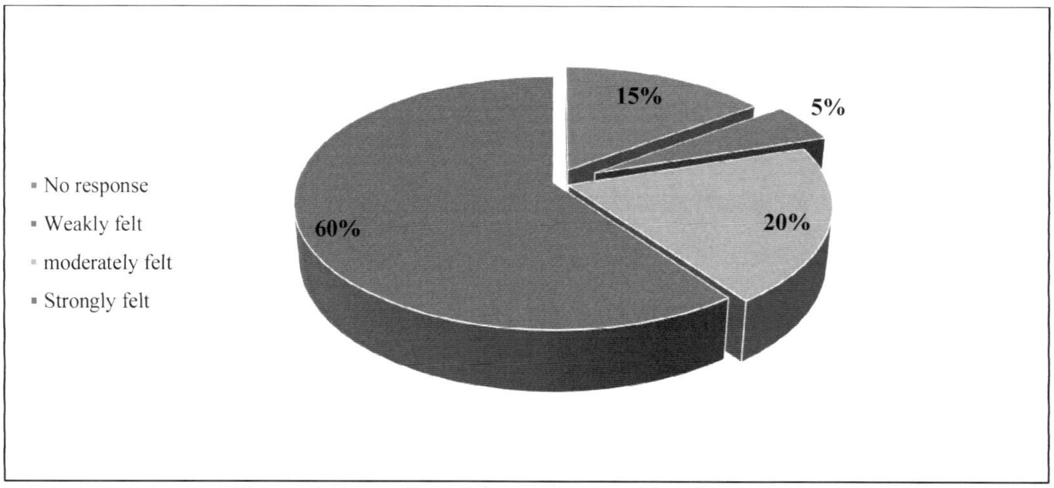

Figure 6. Perception of the acuteness of the shortage

The water shortage is strongly felt by 60% of the surveyed public, it is mainly the residents of the apartments who cannot benefit from water at home, even during the supply time reserved for their sector. The low pressure in the public network limits their access to water. On the other hand, this shortage is weakly felt by 5% of our survey sample represented especially by households equipped with a well. Regarding the water quality assured by the public network, the satisfaction rate for the criteria evaluated shows that 85% (Figure. 7) of the respondents in 2019 stated that they were quite satisfied with the pressure provided by the drinking water distribution service. The existing network needs to be technically revised to ensure the required pressure for any new developments that may be suggested by the next city development plan.

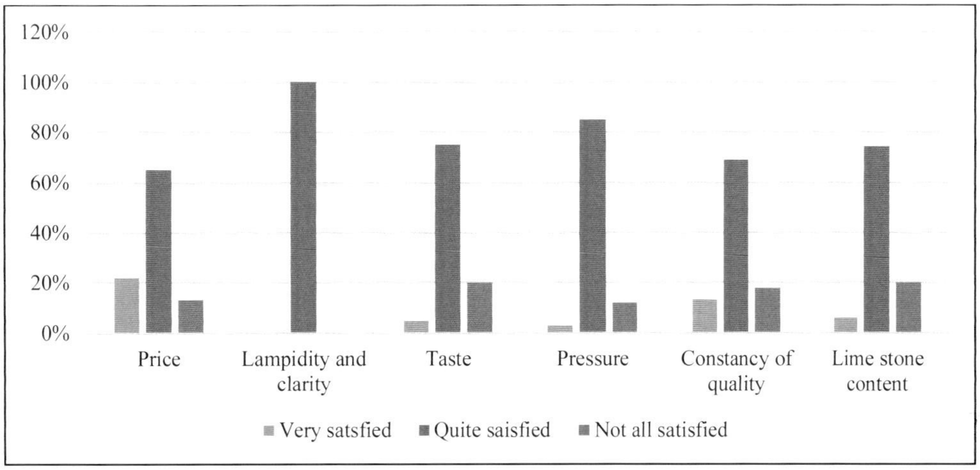

Figure 7. Water quality satisfaction ratings

Overall, the public questioned said they were quite satisfied with 5 of the 6 quality criteria used in this study. The clarity and limpidity of the water are criteria for which all the public surveyed said they were quite satisfied. The price charged by the National Electricity and Drinking Water Office, the exclusive distributor, is considered expensive by 56% of the interviewees met particularly in sectors 1 and 5, where professional activity is dominated by small businesses.

3.2Water management and conservation

Efficient management of drinking water must take into consideration the expectations expressed by the respondents (Figure. 8), of which the urbanization and extension of the city are pointed out as a deterrent to good water management in the city.

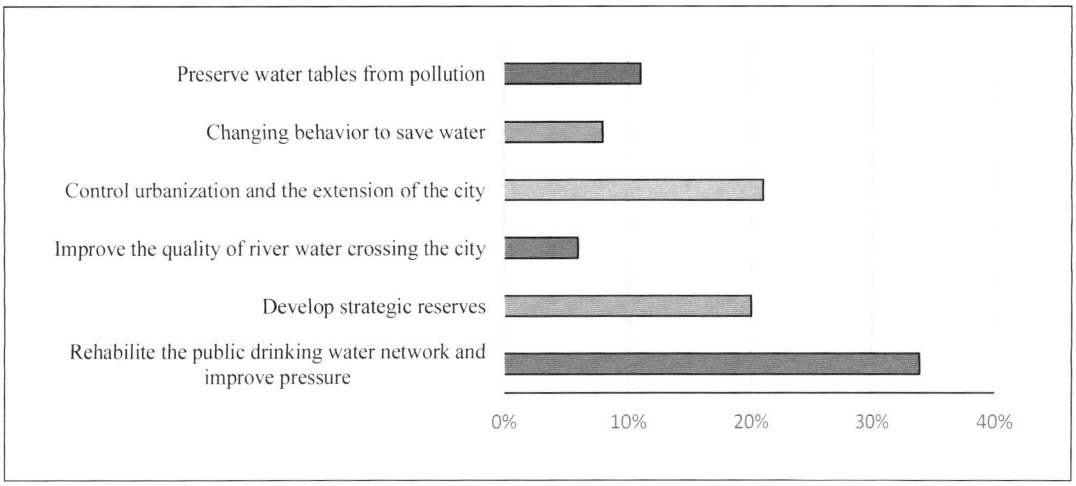

Figure 8. Respondents' expectations for efficient water management in Guercif

Efficient water management calls for the adoption of actions and behaviors that promote water conservation. More than 60% of our survey sample claim to adopt actions against water waste, specifically the repair of water leaks in their homes.

4. Conclusion

The results obtained from this study confirm the water deficit in Guercif. Water scarcity is linked to the natural potential of fairly limited resources, but it is often aggravated by natural and anthropogenic factors; hence the imbalance between near-constant supply and growing demand. A prospective study initiated by the regional directorate of the National Office for Electricity and Drinking Water in 2009 predicted a deficit of 11% in 2025 and which could reach 25% by 2030 [10]. The sustainability of the drinking water supply depends on the ongoing developments in the hydraulic sector, in particular the investment of 37 million dirhams [11], as an urgent measure to secure the drinking water supply in the municipality of Guercif. The realization of the "Targua Madi" dam project (90 km south of the city), the empowerment of the various stakeholders called upon to produce a strategic vision whose objective is the adequacy between the preservation of water resources and the guarantee of the various needs urgent for urban functioning, are alternatives capable of avoiding the vagaries of water insecurity in our study area.

5. References

[1] Shiklomanov. I 1996, Assessment of water resources and water availability in the word (state hydrological Institute of Russia/ UNESCO, February, 127 P).
[2] World bank document 2017 Management of urban water scarcity in Morocco, Report.
[3] Imerzeel W and al. 2011 Water outlook for the Middle East and North Africa. World Bank, Washington DC.
[4] Richard. D and al 2019 Quality unknown: The invisible water crisis. World Bank, Washington DC.
[5] Ministry of Land, Water and Environment 2001 National Charter for Land Use Planning. Rabat, Okad, pp. 15-16.
[6] CARLIER. P and SIMONAT. M, 1971 Water resources of Morocco, the Guercif Basin, Geological service of Morocco Editions, Rabat, P 264.
[7] El-Fellah B 1993 Guercif depression: manifestations of deterioration of the natural environment, Soil Dynamics Seminar February 18-19 Department of Geography Publications of the Faculty of Letters and Human Sciences Beni Mellal Cadi Ayyad University, Marrakech, p.79.
[8] High Commission for Planning 2017 Population projections of regions and provinces 2014-2030 Demographic Study and Research Center, May 2017.
[9] Ferragina E and Quagliarotti D, Poverty and water crisis in the Mediterranean. The market's response Revue Tiers Monde, 2010/3 (n ° 203), p. 181-200. DOI: 10.3917 / rtm.203.0181. URL: https://www.cairn.info/revue-tiers-monde-2010-3-page-181.htm.
[10] ONEP. DR5 2009 General study of the distribution of the center of Guercif. Sub-mission I.2 Rehabilitation actions to upgrade the drinking water situation. P 54.
[11] NOEDW. 2022 Commission for Basic Infrastructure, Energy, Mines and the Environment in parliament. The water resource crisis. March 1, 2022.

The impact of the precipitation on the vegetation and ecological quality in the River of Oued Guigou, Morocco

S. Zerkani*[1], E.H. Abba[2], H. Zerkan[3], T. Zair[3] and N-E. Zine[1]

[1] Laboratory of environment, ecology and health, Faculty of Science, Moulay Ismail University, BP 4010 Zitoune, Meknes Morocco
[2] Higher School of Technology, Sultan Moulay Slimane University. B.P. 170, 54000 Khenifra Morocco
[3] Chemistry of Bioactive Molecules and the Environment, Faculty of Science, Moulay Ismail University, B.P. 11201, Zitoune Meknes, Morocco

*S.zerkani@edu.umi.ac.ma

Abstract. Understanding the impact of climate change on vegetation is essential to manage a new adaptation strategy and preserve the ecology impacted by the flow of suspended matter towards the watercourse. The main purposeof the present article is to study the relationship between the normalized difference vegetation index (NDVI) and the precipitation (P), in the Guigou basin, in the Middle Atlas of Morocco. The analysis of the relationship between NDVI and P was carried out in four phases using GIS and remote sensing tools. The NDVI was derived from Landsat TM 5 (1990), LANDSAT ETM + 7 (2000), LANDSAT TM 5 (2010), and LANDSAT OLI 8 (2020). The data on precipitation were homogenized on 2 monitoring stations of AitKabbach and Aitaissa, from 1990 to 2020, using the regional vector method and the variability. The response of vegetation to climatic parameters was represented using zonal statistics. The results show a decrease in annual precipitation during the period (1990-2020) with an increase in the year 2020. In the annual scale, linear regression reported a positive relationship between NDVI and precipitation. Thus, during the study period, the vegetation growth decreased by approximately 90%, between the years 2000 and 2010, in response to the observed climatic variation. The growth of vegetation is affected by climatic variability, in particular the precipitation. Consequently, the pedo-climatic context, favorable to erosive action and runoff, generates the transfer of eroded particles responsible for the clogging of spawning grounds, where salmonid eggs are deposited, in the Guigou wadi.

Keywords : Basin ; NDVI ; Remote sensing ; SIG ; Climate change ; precipitations ; salmonids.

1. Introduction

Climate variability has a significant impact on plant growth [1-3]. In reality, the most important determining elements for vegetation development are precipitation and temperature [4]. The Normalized Difference Vegetation Index has enhanced the identification of vegetation dynamics using satellite data in recent decades (NDVI). In urban ecology [5], the NDVI index is a commonly used metric of photosynthetic activity that is used to track crop growth, farmland management, land use, and agricultural production. It's been frequently used to track and predict crop development, as well as research the effects on plant cover [6-8]. The NDVI is also interested in the relationship between landscape components and runoff and the transfer of suspended materials caused by soil erosion. It also entails evaluating the influence of agricultural and rural development activities on breeding quality deterioration [9-12]. NDVI has also been used to calculate other variables such as the leaf area index (LAI) [13], agricultural productivity [14], and evapotranspiration linked with climate change [15]. Considering that changes in agricultural land management and intensification of cultivation practices, when combined with specific climatic and pedological conditions, exacerbate the phenomenon of water erosion, resulting in changes in the profile of watercourses and changes in the quality of surface water [16]. The effects of blockage of the substrate on the development of salmonid fish eggs and fry can also be exacerbated by water erosion [17].

The NDVI indicator may also be used to classify vegetation growth at regional and global scales [18,19], and it is often regarded as the best indicator available for greenery monitoring and assessing the influence of long-term climatic variability [20]. Index of NDVI is calculated by dividing the total of near infrared (NIR) and red (R) radiation emitted by plants. This index, developed by Rouse in 1974 [21], is based on the spectral characteristics of chlorophylls A and B, and it is still the most often used measure of photosynthetic activity in ecological studies [22].

More specifically, NDVI data are frequently used to monitor the sensitivity of vegetation growth to rainfall variability in ecologically oriented regions, such as the Guigou basin, while the correlation between NDVI and rainfall is confirmed by previous studies [23,24].

The results of the correlation suggest that NDVI and precipitation have a substantial connection [25]. Furthermore, the association is much larger in African sub-regions with semi-arid and dry climates.

Where the same relationship is proved in the semi-arid zones of the center-south of Tunisia [26,27] and also shows that the correlation between NDVI and precipitation is important during the growing season in arid and semi-arid lands of Kenya [28]. In Jordan, a strong linear relationship was also found between annual precipitation and NDVI between a 10-day period and monthly [29]. In Asia, precipitation has been the limiting factor in vegetation growth in arid lands, plains [30,31] and in grassy and mountainous areas of Beijing, China [32].

Other studies indicate that precipitation is not a strong predictor of growing vegetation. This has been demonstrated in refs. [33,34] in the arid region of northwest China and transition zones in humid, arid and semi-arid environments. It shows that the relationship between precipitation and NDVI differs from region to region. This relationship can be affected by local conditions such as land use, soil and erosion [35,36].

The Guigou basin is threatened by the degradation induced by several factors, including climate variability. In fact, in recent decades, the environment of the large Sebou basin has started to change due to climate variability, which is accelerated by multifaceted interactions such as pollution, industrialization, scarcity of natural water resources and significant losses of agricultural activities.

Despite the many threats to the Guigou Basin, no scientific research has been conducted to study the sensitivity of vegetation to climate variability under local environmental conditions, in order to address better environmental management.

The main objective of this work is to study the variability of precipitation to better understand how this parameter affects vegetation growth based on the extracted NDVI data from Landsat images.

2. Materials and methods

2.1. Study area

The study area is located in the Guigou watershed, which constitutes the upper part of Sebou River basin. Its total area is 1904.58 Km^2 (190,400 ha) (figure 1).

The Guigou basin is a part of an original morphostructural context between the Middle Atlas Causse to the north and the folded Middle Atlas to the south. Oued guigou has appeared following the median accident of the Middle Atlas before draining a vast sedimentation basin, which is located on the right bank of Lake Aguelmam Sidi Ali. Before passing the syncline of BouAngueur, the Oued Guigou becomes lost in the basalts at the valley's bottom. The Aberchanne spring, just upstream of the foumKheneg, supplies it on the right bank. The Guigou plain is characterized by several shallow depressions in the basalts downstream. Before closing off at AitKhebbach, it is flanked on the right bank by the alluvial cone of the Maleh wadi.

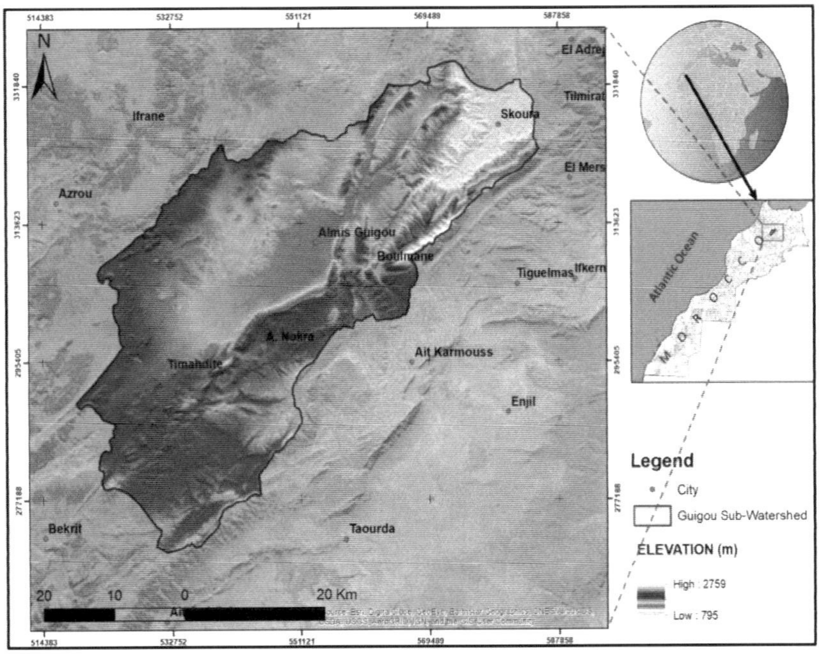

Figure 1. Localization of the Guigou basin in Moroccan Middle Atlas

2.2. *Study area geology*

The Meso-Cenozoic lithostratigraphic series of the study area ranges from the Upper Triassic to the Quaternary [37,38] (figure 2).

The topography is marked by dolomitic reefs and polje which cut across the surface. In the North-West, it is capped with volcanic domes which are at the origin of the large basaltic outflows of the Guigou plain. This was closed to AitKhebbach by the North Middle Atlas accident, is in fact a high intra-mountain plateau (1500 m altitude) belonging to the northern synclinal zone. The units of Timahdite and Guigou, present plio-Quaternary detrital deposits in the Timahdite unit and carbonate deposits from the Lias in the Guigou unit.

The northern part of the BouAnguersynform structure is characterized by the Cretaceous and Eocene structures, which are developed on the southern and eastern flanks of this structure. The southwest termination of the Ain Nokra shape structure was separated from that of BouAnguer by the Felledi accident. Most of the outcrops of this structure are of Paleogene age where Cretaceous deposits emerge [39,40].

2.3. *Climate and hydrology of the study area*

In terms of climate, the Guigou basin has a moderate and humid Mediterranean climate with a harsh winter and a cool summer. With a humid and semi-arid bioclimatic stage, the variation in height generates substantial differences between the mountain and the plain. It is marked by relatively high and uniform precipitation over time, ranging from 350 mm in the south to 1100 mm in the north, which might exacerbate the problem of increasing soil saturation and therefore runoff [42].

The hydrogeological context of the Guigou basin shows that the groundwater resources available in the region are located in the fluvio-lacustrine and volcanic formations of plio-quaternary age, and in liasic limestones. The rainfall data from the two main meteorological stations, of AitAissa and AitKabbach, in the study area over the period 1990-2020 are collected from the Sebou Hydraulic Basin Agency (ABHS).

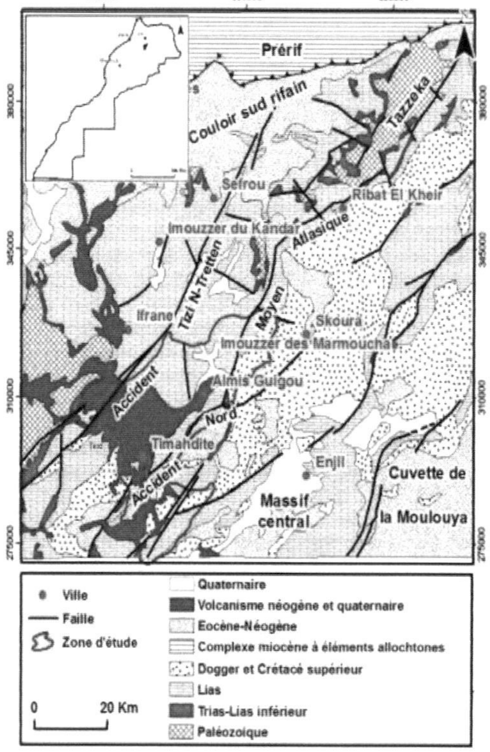

Figure 2. Geological framework of the Guigou basin in the Middle Atlas [41].

2.4. *Land use*

The precipitation flow rate that conditions a watershed's water resource, in this case the Guigou basin, is redistributed to the soil surface according to the soil and plant cover parameters [43]. In the Guigou basin, the nature of the soil and its occupation is decisive from the point of view of the water balance. In fact, it determines, in particular, the values of evapotranspiration which are very different between irrigated agriculture and agriculture labor or natural vegetation but also the spaces from which water is taken from the watercourse or the water table. Then the land use is studied as a "distributing power" and as a qualitative means of evaluating human pressure on the watershed. In a prospective framework, the occupation of land in the Guigou plain is a fundamental variable for regional planning as well as for the study and understanding of the environment [44].

2.5. *NDVI Data*

The normalized difference vegetation index (NDVI) is calculated from the visible red channel (R) and the near infrared channel (NIR) and highlights the difference between these two bands according to the following formula:

$$NDVI = \frac{(PIR - R)}{(PIR + R)}$$

The principle of NDVI is focused on "green" leaves absorbing radiation at red wavelengths (640-670 nm) due to the presence of chlorophyll pigments, while radiation at near-infrared wavelengths (700-1100 nm) is diffused according to the internal structure of the leaf [45]. In contrast, the bare soil surface has a higher reflectance than red wavelengths and a reflectance less than near-infrared wavelengths [46].

For the present study, the NDVI values were grouped into 4 classes (table 1).

Table 1. NDVI classes and values used in map processing.

Classes	*NDVI values*	Gridecode
Negative	negative values or less than 0	Red
Low	values between 0 and 0.3	Yellow
Moderate	values between 0.3 and 0.6	Green
High	values> = 0.5	Dark green

The NDVI classification values (table 1) are between -1 and +1, the negative values correspond to surfaces other than the vegetation cover, such as snow, and water for which the reflectance in the red band is greater than that of the near infrared. For bare soils, the reflectance being roughly of the same order of magnitude in the red and near infrared, the NDVI shows values close to 0. The vegetation cover has positive NDVI values, generally between 0.13 and 1.0, with higher values corresponding to the densest cover.

2.6. *Satellite Images*

This study is based on the integration of reflectance data derived from optical sensors into a Geographic Information System (GIS) in order to analyze various geographic parameters, as well as the processing and interpretation of satellite imagery using appropriate spectral and spatial analysis methods. The satellite images used for this study are of the type Landsat TM 5 (1990-07-11), LANDSAT ETM + 7 (2000-07-14), LANDSAT TM 5 (2010-07-02), and LANDSAT OLI 8 (2020-07-13). The Universal Transverse Mercator (UTM) projection technique is used to geometrically rectify the satellite pictures. At the same time, a radiometric correction is employed to maximize the visual quality of the colored compositions by removing the influence of particles from the environment.

2.7. *Rainfall and NDVI*

The statistical study of the NDVI vegetation index for the period 1990-2020 was used to examine vegetation dynamics. The vegetation indices are used to analyze the green plant mass covering the research area's soil and to describe the plant cover's phenological condition. The rainfall stations utilized in the study are from various bioclimatic zones. The (S1) AitAissa and (S2) AitKebbach stations, with annual averages of 381.9 mm and 369 mm, respectively, are examples of this.
As a result, the NDVI reports the photosynthetic activity of vegetation and creates a direct link with meteorological factors such as rainfall which is decisive in the growth cycle of plants, therefore the variation in the intensity of chlorophyll activity measured by NDVI can be related to the distribution of precipitation over the year.

3. Results and discussion

3.1. *Temporal dynamics of rainfall*

The use of average statistics to disguise possibly more frequent severe occurrences in the future is usually resumed in the spatiotemporal development of rainfall. Furthermore, precipitation is influenced by two separate mechanisms in the context of climate change: The frequency and severity of precipitation would be affected by rising temperatures, which would allow the air to hold more moisture. Warming of the atmosphere is also more relevant at higher altitudes.

There is also a little geographical variety and an emphasis on yearly rainfall disparities (figure 3). As a result, during the 1990s to 2010, precipitation increased, with particularly intense wet spells. Between 1999 and 2008, Morocco had a significant drought, characterized by the country's longest dry stretch in recent history, highlighted by a drop in precipitation and a strong trend of rising temperatures. This drought pattern is accompanied by substantial warmth as a result of global warming [47].

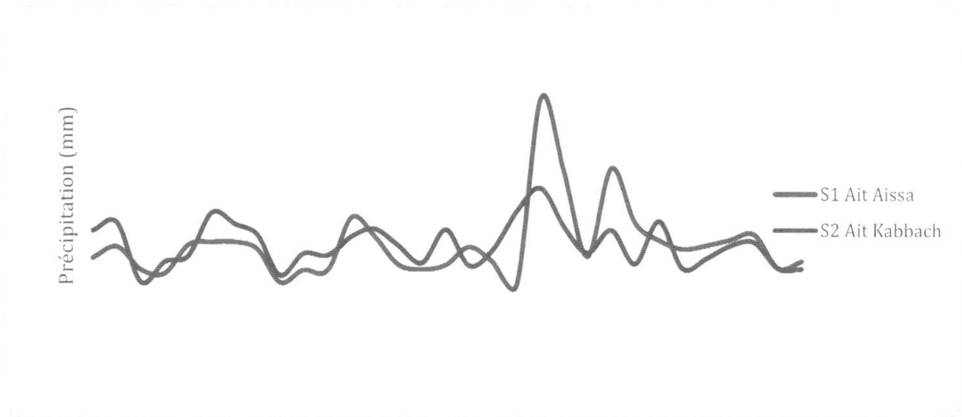

Figure 3. Variation of annual rainfall in the Guigou basin between the period 1990-2020

3.2. Variations in NDVI between the period 1990-2020

The spatio-temporal changes which take place between the four compared images (1990 and 2020) are clearly observed in the form of NDVI maps (figures 4, 5, 6 and 7) which recapitulate the data of a decimal period between (1990 and 2020).

During the first ten years (figure 4), the "low" class of the NDVI index between 0 and 0.3 occupies the largest area of the basin (83.4%), followed by the "middle" class of (0.3 to 0.6) which occupied 15.62% of the total area of the Guigou basin. While the "high" class of the NDVI index corresponds to the densest vegetation occupying the smallest area of the watershed, with 0.86% of the total area of the basin.

The map (figure 5) shows a low vegetative occupation at the level of the Guigou basin, hence the "low" class of (0 to 0.3) settles the largest area around 96.85% (table 2 and 3), While the "high" class, which presents both plant density, presents only 0.01% of the total area.

Focusing on the "high" class greater than 0.5 in 2010 (figure 6), we note almost a disappearance of the plant cover along the Guigou basin, this class is almost zero compared to the middle class between (0.3 and 0.6), which itself occupies a small area of around 3.38%.

Finally the map (figure 7), the rainfall was more or less important which influenced the land use throughout the Guigou basin, through this rainfall increase we note higher values of the "Average" class and the "high" class which occupied an area of 26.37% and 1.09%, respectively.

Table 2. NDVI density and area classes (in ha) calculated during the period 1990-2020

NDVI density classes	1990		2000		2010		2020	
	Ha	%	Ha	%	Ha	%	Ha	%
negative (Red) values or less than 0	68,313	0,04	2414,63	1,27	251,058	0,13	66,026	0,03
low values (yellow) between 0 and 0.3	158897,0 0	83,45	184411,0 0	96,8 5	183768,0 0	96,52	138091,0 0	72,53
average values (Green) between 0.3 and 0.6	29731,60	15,62	3616,9	1,90	6442,76	3,38	50209,1	26,37
high values (dark green)> = 0.5	1644,04	0,86	18,4042	0,01	0,1252	0,0000 7	2082,8	1,09

Table 3. Difference in the density classes of the NDVI index between 1990 and 2020.

NDVI density classes	Average rates of change between (1990 and 2020)	
	Ha	%
negative (red) values or less than 0	-2,29	-100
low values (yellow) between 0 and 0.3	-20806,00	-10,93
average values (green) between 0.3 and 0.6	20477,50	10,75
high values (dark green)> = 0.5	438,76	0,23

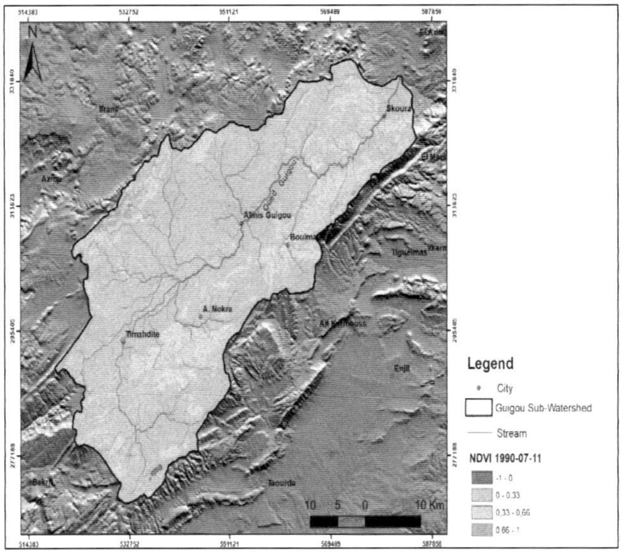

Figure 4. Land use map in Guigou basin in 1990

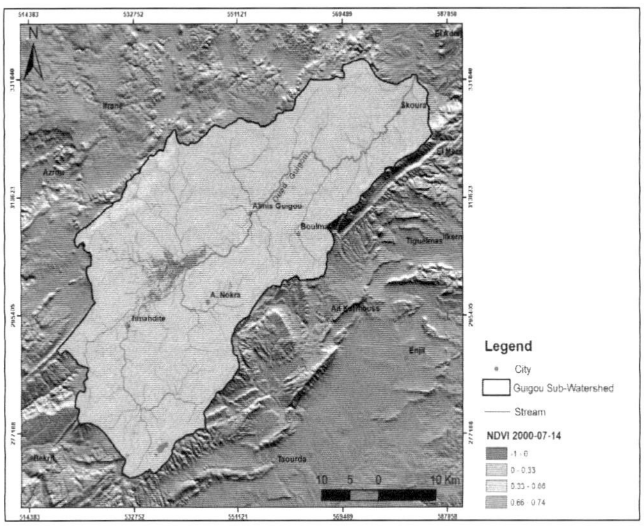

Figure 5. Land use map in Guigou basin in 2000

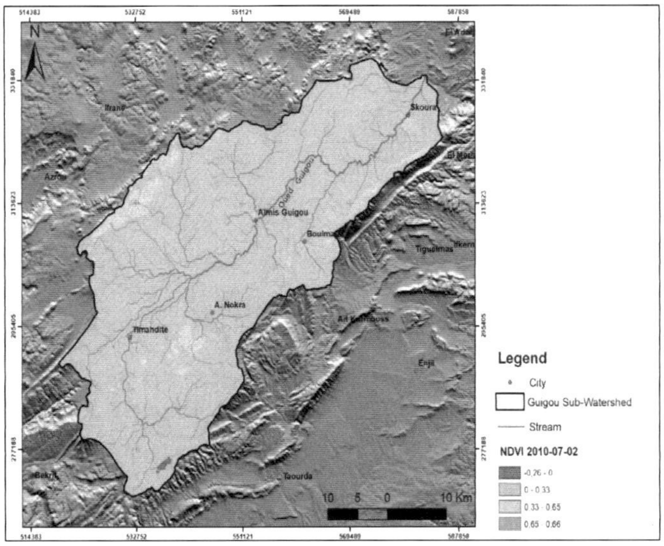

Figure 6. Land use map in Guigou basin in 2010

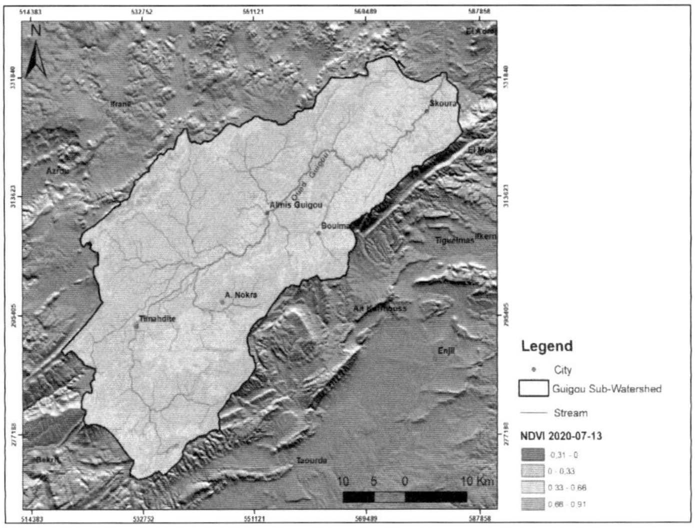

Figure 7. Land use map in Guigou basin in 2020

4. Discussion

Environmental disturbances, particularly human-induced changes like exploitation, pollution, fragmentation, and climate change, have an impact on the stability of ecological systems and, as a result, on resource sustainability. This is especially critical for aquatic ecosystems and biological diversity.

Salmonid species are the focus of a slew of management initiatives aiming at examining the effects of climate change on these species in the context of competing conservation and exploitation goals. The Oued Guigou is a favored habitat for the reproduction of the common trout in this research (Salmo-trutta). As a result, the embryo-larval phases are the most sedentary and pollutant-sensitive [48,49].

Figure 3 summarizes the key statistical features of the rainfall series of the two stations analyzed (AitAissa and AitKebbach). The yearly rainfall, notably at AitAissa, is characterized by a considerable fluctuation, according to the research of interannual climate variability. Annual precipitation has been trending to decrease at both locations (figure 3), especially in the previous ten years. Previous studies have shown that from 1896 to 1996 Morocco experienced eleven periods of generalized droughts whose intensity was moderate to strong and others less generalized quite strongly, which correlates very well with these study outcomes (figure 3) that shows the fluctuations of drought during the period studied, due to the high spatial variability of precipitation.

The environmental variables (rainfall and land use) meeting the ecological requirements of the species are relatively delicate, both for reasons of data availability and for statistical reasons. Figures 4, 5, 6 and 7 show the correlation between the annual average NDVI and climatic variables. Although the period analyzed is short, it clearly shows the geographic distribution of the NDVI-precipitation relationship.

A positive correlation between NDVI in 1990 and rainfall variability is revealed through the comparison of figures 3 and 4, which is significantly affected by drought in the years 1999 and 2009. Areas with low to medium negative NDVI classes coincide well with years of low rainfall, occurring mainly in semi-arid areas far from Oued Guigou or Lake Sidi Ali.

In semi-arid zones, precipitation is thought to be the most important factor influencing plant development. Between 2008 and 2012, there was a notable association between high class NDVI positive regions and wet years. Figure 7 clearly demonstrates a considerable occupancy of the plant cover, despite the decrease in rainfall in recent years between 2010 and 2020, which does not explain the association between NDVI and rainfall in the Guigou basin.

The average change in NDVI shows that between 1990 and 2020, the "high" class relating to the densest vegetation is the class whose area has increased slightly with an annual average of 0.23% compared to the total area of the basin of Guigou (table 3).

In terms of average vegetation, the "middle" class has seen a large growth in area, with an annual average of 10.75 percent (table 3). This shows that in 2010, the climatic conditions or the abundance of precipitation in the Guigou basin were favorable for the over-development of the plant cover.

As a result, agricultural activity is widely practiced there, and it has a "positive impact" on the development of the photosynthetic activity of plants. Conversely for soils and wetlands, with low retention capacity thus promoting very rapid water runoff, this results in more or less low NDVI values. The plant density can also be explained by the use of drip irrigation instead of gravity irrigation in the majority of agricultural land [50].

In addition, certain physicochemical variables can generally be taken into account in the environmental requirements which narrowly characterize the niche of the salmonid species in Oued Guigou. On the other hand, environmental combinations that are favorable to the species will not be taken into account, biasing the estimated relationships between its occurrence (salmonid species) and environmental variables (precipitation, temperature, dissolved oxygen, etc.) [51].

The robustness of the NDVI approach relies on the representation of a generic mechanism and not on correlations, which do not necessarily imply a cause-and-effect relationship. In addition, they make it possible to integrate individual-environment interactions (interesting, among others, to directly take into account the effect of climate change on population dynamics) as well as an explicit representation of the environment [52,53].This relationship/interaction can be explained by the effect of erosion, therefore the presence of suspended solids (SS) in Oued Guigou.

Suspended matter (SS) contributes to the degradation of aquatic ecosystems by increasing the turbidity of the water, by clogging the spawning grounds of Salmonids, and by conveying many pollutants. The modification of agricultural practices and land use planning (regrouping, uprooting of hedges) is at the origin of increased production of SS.In livestock regions, transfers to surface water of suspended matter (SS) linked to soil erosion may partly explain these problems [54]. These regions are also affected by soil erosion which is just as harmful to the balance of ecosystems. This erosion, caused by the reduction in plant cover and drought, results in the transfer of soil particles torn up during runoff and carried towards the thalwegs [55]. The streams, which flow over these watersheds, show an increasingly deteriorating fish state. The analysis of the evolution of fish populations and aquatic macrophytes in rivers shows signs of degradation of this environment with clogging of the river bottom and in particular of salmonid spawning grounds [56].

Some studies confirm that deoxygenation and the clogging of substrates in watercourses are likely to decrease the survival and growth of trout during the first months of the juvenile post-emergence period, in altering the swimming activities of the fry at the exit of the spawning ground, making them potentially more vulnerable to predation.

5. Conclusion

In our present study, the NDVI was used to monitor the state of the vegetation in the Guigou basin, in order to highlight the impacts of climate change on the ecology and also to validate the hypothesis of the pollution of Oued Guigou by suspended matter, which poses a risk to the fish state and spawning grounds of Salmonids.

As well, the study on the driving factors of vegetation change is emphasized at the regional scale, which is favorable to production and ecological restoration and is important for the study of climate change. The other objective of this study is therefore to select factors of interest for further study (such as the determinants of anthropogenic pollution and also the factors that disrupt the growth of common trout).

This study also proposes an analysis of the cartography results, which can be used as an aid for the local actors, with a view to possible measures to restore the good quality of the rivers.

6. References

[1] ZHOU C, ZELINKA M, DESSLER A and KLEIN S 2015 The relationship between interannual and long-term cloud feedbacks *Geophys. Res. Lett.* 42 10 463–469.

[2] ESHAM M and GARFORTH C 2013 Climate change and agricultural adaptation in Sri Lanka: a review Climate and Development 5 66-76.

[3] XUE-JIE G, MEI-LI W and GIORGI F 2013 Climate change over China in the 21st century as simulated by BCC_CSM1. 1-RegCM4. 0 *Atmospheric and Oceanic Science Letters* 6 5 381-386.

[4] MISRA M 2012 Applied Urban Ecology A Global Framework.

[5] RICHTER M and WEILAND U 2011 *Applied urban ecology: A global framework* John Wiley& Sons 2011.

[6] PAN H, ZHANG L CONG C, Deal B and Wang Y 2019 A dynamic and spatially explicit modeling approach to identify the ecosystem service implications of complex urban systems interactions *Ecological Indicators* 102 426-436.

[7] CUMMINGS C and ROSENTHAL S 2018 Climate change and technology: examining opinion formation of geoengineering *Environment Systems and Decisions* 38 2 208-215.

[8] FINNEY D. L, MARSHAM J. H, WILKINSON J. M, FIELD P. R, BLYTH A. M, JACKSON L. S, KENDON E. J, TUCKER S. O and STRATTON R. A 2020 African Lightning and its Relation to Rainfall and Climate Change in a Convection Permitting Model *Geophysical Research Letters* 47 23 e2020GL088163.

[9] BHATTI A, FAROOQUE A, KROUGLICOF N, PETERS W, ACHARYA B, LI Q and AHSAN M. S 2021 Climate change impacts on precipitation and temperature in Prince Edward Island, Canada *World Water Policy*.

[10] MARTIN Y, VAN DYCK H, LEGENDRE P, SETTELE J, SCHWEIGER O, HARPKE A and TITEUX N 2020 A novel tool to assess the effect of intraspecific spatial niche variation on species distribution shifts under climate change *Global Ecology and Biogeography* 29 3 590-602.

[11] HUBER D G and GULLEDGE J 2011 Extreme weather and climate change: Understanding the link, managing the risk Arlington : Pew Center on Global Climate Change.

[12] CUOMO MT, CAPRINO RM, GENOVINO C and TORTORA D 2021 Climate change, environmental auditing and corporate/brand strategy *Sustainable Branding*. Routledge 115-127.

[13] WANG S and ZHANG Z 2011 Effects of climate change on water resources in China *Climate Research* 47 1-2 77-82.

[14] MBACH N, MEZA FJ, GILABERT H and MIRANDA M 2013 Impacts of climate change on the distribution of species and communities in the Chilean Mediterranean ecosystem *Regional Environmental Change* 13 6 1245-1257.

[15] GANG C, ZHOU W, CHEN Y, SUN Z, LI J and ODEH I 2014 Quantitative assessment of the contributions of climate change and human activities on global grassland degradation *Environmental Earth Sciences*72 11 4273-4282.

[16] MACARY F , OMBREDANE D and UNY D 2010 A multicriteria spatial analysis of erosion risk into small watersheds in the low Normandy bocage (France) by ELECTRE III method coupled with a GIS *International Journal of Multicriteria Decision Making* 1 1 25-48.

[17] THIERRY E 2019 Pressions exercées sur les espèces dulcicoles piscicoles d'intérêt communautaire.

[18] BARICHIVICH J, KEITH RB, RANGA B, TIMOTHY J , THOMAS M, PHILIPPE C , SHILONG P and COMPTON T 2013 Large-scale variations in the vegetation growing season and annual cycle of atmospheric CO2 at high northern latitudes from 1950 to 2011 *Global Change Biology* 19 10 3167-3183.

[19] Tian X, Gala U, Zhang Y, Shang W, Nagarkar Jaiswal S, di Ronza A, Jaiswal, M, Yamamoto S, Sandoval H, Duraine L, Sardiello M, Sillitoe R.V, Venkatachalam K, Fan H, Bellen H.J and Tong C 2015 A Voltage-Gated Calcium Channel Regulates Lysosomal Fusion with Endosomes and Autophagosomes and Is Required for Neuronal Homeostasis PLoS Biol. 13 3 e1002103.

[20] BROCK R and KEITH S.T 2020 Making claims about learning: a microgenetic multiple case study of temporal patterns of conceptual change in learners' activation of force conceptions *International Journal of Science Education* 42 8 1388- 1407.

[21] ROUSE J, HAAS R, DEERING D, SCHELL J and HARLAN JC 1974 Monitoring the Vernal Advancement and Retrogradation (GreenWave E_ect) of NaturalVegetation; NASA/GSFCType III Final Report National Aeronautics and Space Administration NASA Greenbelt, MD USA.

[22] MATSUSHITA B, YANG W, CHEN J, ONDA Y and QIU G, 2007 Sensitivity of the enhanced vegetation index (EVI) and normalized di_erence vegetation index (NDVI) to topographic A case study in high-density cypress forest *Sensors* 7 2636–2651.

[23] FAYECH D and TARHOUNI J 2021 Climate variability and its effect on normalized difference vegetation index (NDVI) using remote sensing in semi-arid area. *Modeling Earth Systems and Environment* 7 3 1667-1682.

[24] RODRIGUEZ M, CONCI E, BECKER A, GRUMELLI M, LEDESMA M. M, RANGONE A and BOUZA P 2021 Environmental Risk in the Fluvio-Aeolian Plain of the Region Center of the General San Martín Department, Córdoba, Argentina *Advances in Geomorphology and Quaternary Studies in Argentina* Springer 244-284.

[25] BAGHERZADEH A, HOSEINI A and TOTMAJ L 2020 The effects of climate change on normalized difference vegetation index (NDVI) in the Northeast of Iran *Modeling Earth Systems and Environment* 6 2 671-683.

[26] AMRI R, ZRIBI M, LILI-CHABAANE Z, DUCHEMIN B, GRUHIER C and CHEHBOUNI A 2011 Analysis of vegetation behavior in a North African semi-arid region, using SPOT-VEGETATION NDVI data. *Remotesensing*, 3 12 2568-2590.

[27] ZRIBI M, DRIDI G, AMRI Rim and LILI-CHABAANE Z 2016 Analysis of the effects of drought on vegetation cover in a Mediterranean region through the use of SPOT-VGT and TERRA-MODIS long time series *Remote Sensing* 8 12 992.

[28] SHISANYA C. A, RECHA C and ANYAMBA A 2011 Rainfall variability and its impact on normalized difference vegetation index in arid and semi-arid lands of Kenya *International Journal of Geosciences* 2 01 36.

[29] AL-BAKRI J, SULEIMAN A, ABDULLA F and AYAD J 2011 Potential impact of climate change on rainfed agriculture of a semi-arid basin in Jordan *Physics and Chemistry of the Earth Parts A/B/C* 36 5-6 125-134.

[30] ZHANG Q, WANG J and WU F 2018 Spatial heterogeneity of surface roughness on tilled loess slopes in erosion stages *Soil and Water Research* 13 2 90-97.

[31] WU D , ZHAO X, LIANG S, ZHOU T, HUANG K, TANG B and ZHAO W 2015 Time-lag effects of global vegetation responses to climate change *Global change biology* 21 9 3520-3531.

[32] LIU J, FOLBERTH C, YANG H, RÖCKSTRÖM J, ABBASPOUR K and ZEHNDER A. J 2013 A global and spatially explicit assessment of climate change impacts on crop production and consumptive water use *PLoS One* 8 2 e57750.

[33] BANIYA B, TANG Q, HUANG Z, SUN S, and TECHATO K. A 2018 Spatial and temporal variation of NDVI in response to climate change and the implication for carbon dynamics in Nepal *Forests* 9 6 329.

[34] ZHAO D and LI YR 2015 Climate change and sugarcane production: potential impact and mitigation strategies *International Journal of Agronomy* 2015.

[35] WANG S and ZHANG Z 2011 Effects of climate change on water resources in China *Climate Research* 47 1-2 77-82.

[36] HE S, WANG D, FANG Y and LAN H 2017 Guidelines for integrating ecological and biological engineering technologies for control of severe erosion in mountainous areas–A case study of the Xiaojiang River Basin, China *International Soil and Water Conservation Research* 5 4 335-344.

[37] CHARRIÈRE A 1990 Héritage hercynien et évolution géodynamique alpine d'une chaine intracontinentale : le Moyan Atlas au SE de Fes (Maroc) These Doc. Toulouse France 589.

[38] EL ARABI H 1987 Etude stratigraphique et sédimentologique du lias aux confins du causse moyen atlasique et du moyen atlas plissé (Maroc) These Toulouse 192.

[39] CHARROUD M 1990 Evolution géodynamique de la partie sud-ouest du moyen atlas durant le passage jurassique-crétacé, le crétacé supérieur et le paléogène : un exemple d'évolution intraplaque These Doc. Rabat Morocco 234

[40] HINAJE S 2004 Tectonique cassante et paléochamps de contraintes dans le Moyen Atlas et le Haut Atlas central (Midelt-Errachidia) depuis le Trias jusqu'à l'actuel These Doc. Rabat Morocco 363.

[41] AMRANI S, HINAJE S and GHARMANE Y 2019 Application des méthodes paramétriques (DRASTIC et SI) pour l'étude de la vulnérabilité à la pollution potentielle par les nitrates de la nappe d'eau superficielle de Timahdite-Almis Guigou (Moyen Atlas, Maroc) *Revue des Sciences de l'Eau Journal of Water Science* 32 3 237–252.

[42] MACARY F 2013 Évaluation des risques de contamination des eaux de surface sur des bassins versants agricoles. Approches multiscalaires par modélisation spatiale et analyse multicritère pour l'aide à la décision. Sciences de l'environnement These Doc. Toulouse INPT France .

[43] CHEN B, ZHANG X, TAO J, WU J, WANG J, SHI P and YU C 2014 The impact of climate change and anthropogenic activities on alpine grassland over the Qinghai-Tibet Plateau. *Agricultural and Forest Meteorology* 189 11-18.

[44] DUCROT D 2017 Télédétection et Photogrammétrie pour l'étude de la dynamique de l'occupation du sol dans le bassin versant de l'Oued Chiba (Cap-Bon, Tunisie). *Revue Française de Photogrammétrie et de Télédétection* 215.

[45] RULINDA C, BIJKER W and STEIN A 2011 The chlorophyll variability in Meteosat derived NDVI in a context of drought monitoring *Procedia environmental sciences* 3 32-37.

[46] YENGOH G, DENT D, OLSSON L, TENGBERG A. E and TUCKER C. J 2015 Applications of NDVI for land degradation assessment. In : Use of the normalized difference vegetation index (NDVI) to assess land degradation at multiple scales *Springer Cham* 17-25.

[47] STOUR L and AGOUMI A 2008 Sécheresse climatique au Maroc durant les dernières décennies *Hydroécologie appliquée* 16 215-232.

[48] BOLLIET V and BARDONNET A 2018 Impact of Embeddedness on Salmo trutta at Different Periods of their Early Ontogenesis *Brown Trout Biology Ecology and Management* 201.

[49] BARIŠIĆ J, MARIJIĆ V, MIJOŠEK T, COŽ-RAKOVAC R, DRAGUN Z, KRASNICI N and ERK M 2018 Evaluation of architectural and histopathological biomarkers in the intestine of brown trout (Salmo trutta Linnaeus, 1758) challenged with environmental pollution *Science of the total environment* 642 656-664.

[50] TAHIRI M , HAKDAOUI M, EMRAN A, ACHAB M, EL HADI H and TAHIRI A 2014 Contributions du SIG et de la Télédétection à la Corrélation Entre le NDVI et la Pluviométrie Pour L'étude de la Variabilité Climatique Dans le Bassin de Tahaddart (Rif, Maroc) *European Journal of Scientific Research* 122 3 253-274.

[51] THORSON J, SCHEUERELL M, OLDEN J, and SCHINDLER D. E 2018 Spatial heterogeneity contributes more to portfolio effects than species variability in bottom-associated marine fishes *Proceedings of the Royal Society B* 285 1888 20180915.
[52] JARDIM E, EERO M, SILVA A, ULRICH C, PAWLOWSKI L, HOLMES S. J and MOSQUEIRA I 2018 Testing spatial heterogeneity with stock assessment models *PloS one* 13 1 e0190791.
[53] MA Q, JIAO Y, REN Y and XUE Y 2020 Population dynamics modelling with spatial heterogeneity for yellow croaker (Larimichthyspolyactis) along the coast of China. *Acta Oceanologica Sinica* 39 10 107-119.
[54] ZHANG Q, WANG J and WU F 2018 Spatial heterogeneity of surface roughness on tilled loess slopes in erosion stages *Soil and Water Research* 13 2 90-97.
[55] STAENTZEL C, KONDOLF G, SCHMITT L, COMBROUX I, BARILLIER A and BEISEL J. N 2020 Restoring fluvial forms and processes by gravel augmentation or bank erosion below dams: A systematic review of ecological responses *Science of The Total Environment* 706 135743.
[56] CHAFIN T, DOUGLAS M, MARTIN B and DOUGLAS M. E 2019 Hybridization drives genetic erosion in sympatric desert fishes of western North America *Heredity* 123 6 759-773.

Impacts of climate change on the dysfunction of Mediterranean wetlands: Case of Aguelmam Afennourir - Ramsar Site - the Middle Atlas – Morocco.

Mhamed Khaffou [1*], El Hassan Abba [1] and Abdelkader Chahlaoui [2]

[1] The Higher School of Technology of Khénifra. Sultan Moulay Slimane Univesity - Beni Mellal. Morocco.
[2]Faculty of Sciences. Moulay Ismail University. Meknes. Morocco.

* Corresponding author:m.khaffou@usms.ma

Abstract: The water is a limiting factor for wetlands, especially for those with shallow deep. They are more vulnerable to precipitation disturbance. However, most of these wetlands are suffering from repetitive natural droughts. Example Aguelmam Afennourir wetland classified Ramsar site since 1980. After calculating the Normal Difference Water Index (NDWI), it has undergone severe and repeatedly drying and we have noticed four total dry-outs between 1981 and 2016. It is completely dried in July 1995, July 2000, June 2008, and in November 2016. While its largest extension, 100 % (122 ha) occurred in May 2009, the one before analyzing the historic precipitation for 1981 to 2016 and open water area evolution in the site we have brought that the drying of Aguelmam is closely linked to the irregularity of precipitation. Their drying up causes an ecological disequilibrium and has negative impacts on biodiversity since it has lost water which constitutes the defense of its habitats and biodiversity. Meanwhile, the fauna must leave the site and live the migration risks of searching the living habitat. The analysis of the hydrographic network of the sub-basin in the site reveals the possibility of diverting neighbouring streams toward the Afennourir wetland for heightening its water volume. As a preventive measure to protect the biodiversity of the site and to reduce the anthropogenic action, we propose strengthening the dyke with local materials, to increase the volume and the level of water in the wetland.

Keys Words: Wetland, Morocco, Aguelmam Afennourir, Climate change, MNDWI, dry-out.

1. Introduction

Wetland loss and associated degradation is a serious environmental problem and it is a vital issue to the environmentalist for the last 5 decades all over the World [1, 2 and 3]. This highly productive, diverse wetland (10% of the earth's surface and 15% out of which flood plain wetland) [4] ecosystem provides a number of goods and services which possess immense importance to society [5]. The global extent of wetland losses in the twentieth century were 64–71% and for some areas it is even higher and the loss of natural wetlands continues in the twenty-first century even faster rate [6].

Hydrology is the key to wetland persistence and conservation [7]. Wetlands are systems that depend basically on hydrology. Hydrology determines its occurrence through the landscape and its persistence over seasons and years. All wetlands are the result of physiographic settings and water balances that favor the accumulation or retention of soil water and/or surface water for a period of time [8]. Affected by the seasonal variations of water level and surface area, floodpath lakes and wetlands are extremely sensitive to natural and anthropogenic impacts, such as climate change, human – induced intervention on hydrological regimes, and land use and land cover change [9-10]. A strong relationship exists between lake area variation and climatic and hydrological variables such as temperature, precipitation, and evaporation. Thus, changes in lake surface area can be viewed as indicators of climate variability [11].

Mediterranean wetlands are part of a global biodiversity hotspot, hosting many endemic species [12], but unfortunately, the rate of their degradation is very excessive with serious consequences on the state of the aquatic fauna and flora [13-14]. The Middle Atlas is the water tower and the most important water reservoir of Morocco. Throughout the North African mountains, it is the richest in wetlands, mainly in natural lakes, cold rivers and springs. These aquatic ecosystems fulfill important hydrological, socio-economic and ecological functions. Also their biodiversity offer them international status according to the Ramsar convention. However, most of them suffer from several anthropogenic disturbances, which are amplified by repetitive natural droughts. [15]. Among these naturals' ecosystems, Aguelmam Afennourir wetland, which is the first continental Moroccan Ramsar site, since 1980. It is located in a high plateau with low depth; mostly does not exceed two meters.

In November 19, 2016, we visited the site and found it completely dried up. That's why we took the initiative to study the history of changes in the surface of the wetland.

The structure and function of the wetland is very important for the determination of causes and consequences of their degradation and the previsions of solution of their protection and restoration. Aerial photographs are the most commonly used form of remote sensing data for mapping aquatic habitats [16]. Remote sensing has been used for many years and has been prove to be an indispensable tool in mapping and assessing natural resources. GIS is a valuable tool for studying the nature wetland and the potential of their restoration [17]. In order to accurately analyze the dynamic geographical phenomena related to wetland it is necessary to take into account their change in space and time [18]. Studies which are based on the analysis of remote sensing images allow for evaluating different aspects of hydrology and water resources over extended areas [19-20], floods [21]. Also, lake area variation is studied by remote sensing [19, 22]. Particularly, some studies showed that changes in lake area extension are indicators of climate variability because they reflect changes in the water balance and heat conditions [23-24]. Availability of open-access satellite data, for example the Landsat [25], created an unprecedented opportunity of converting the efforts and assets from a multi-decadal, time-series Earth observation for the use of societal benefits [26].

The main objectives of our study are to understand the draining of the Afennourir wetland, its impacts on the biodiversity of the region and also to propose solutions to ensure the continuity of these values and its biological and ecological functions.

2. Material and methods

2.1. Study area

Aguelmam Afennourir wetland it situated in the Moroccan Middle Atlas Mountains (Prefecture of Ifrane, 33° 17'N - 5° 16'W) (Figure 1). The wetland is spread over in area of 612 ha, 122 the pond area that is seasonally flooded and 490 ha presented by a wet meadow (Figure 2). In this site, the climate is sub-humid characterized by cold winter and important precipitations. The wetland have a great importance at both the national (Site of Biological and Ecological Interest (SBEI) and they are occur in Ifrane National Park) and the international level (Ramsar site, Important Bird Area (IBA) and it is located inside the Atlas cedar biosphere reserve). The fact that Afennourir is a grazing place, the riparian have built a dyke made by soil and wool, at its outlet, to raise the water level for livestock watering in summer. Subsequently, the Ministry of Agriculture built a dike made of rock and cement placed on a basalt substrate, which gave way to cracking and loss of water. In this context, it is more necessary than ever to know the history of the floods and their relationships with climatic conditions and anthropogenic activities. These details will assist decision makers in understanding the Afennourir hydrological regime and in making adaptive management decisions in this context, it is more necessary than ever to know the history of the floods and their relationships with climatic conditions and anthropogenic activities. These details will assist decision makers in understanding the Afennourir hydrological regime and in making adaptive management decisions.

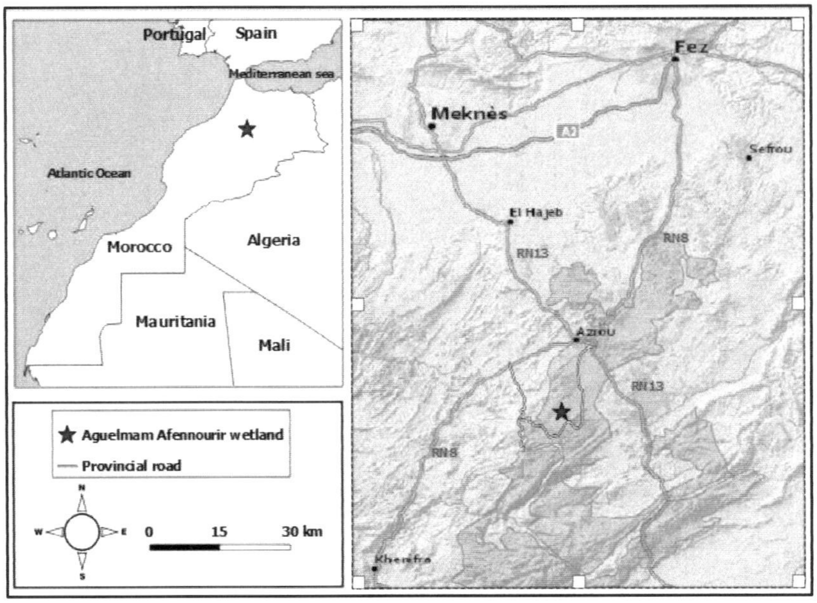

Figure 1. The geographical location of Aguelmam Afennourir wetland.

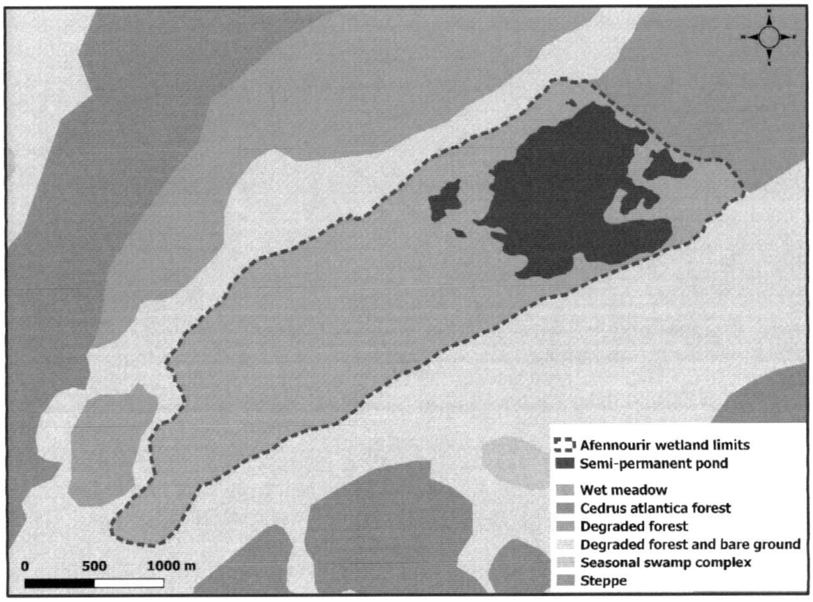

Figure 2. Landuse in Aguelmam Afennourir wetland

2.2. *Methods*

The general methodology followed to realizing this article is presented in table 1. The main tasks are: (1) analyze the pluviometry index diagram to determine the dry period. (2) Consultation of the Google Earth history for the site, (3) download of Landast images, (4) pre-processing of Landsat images, (5) Calculation of Normal Difference Water Index (NDWI) and Modified Normal Water Difference Index (MNDWI) and (6) extraction of the water open area for each scene.

To assess the impact of dewatering the study wetland on biodiversity, we based our assessment on habitat degradation and biodiversity loss.

Table 1: General methodology followed for this study

Pluviometry Index diagram	This diagram allows us to define the periods of pluviometry deficit
Google Earth historic	First idea about drying up of Aguelmam Afennourir wetland
Satellites data	1981, 1984, 1987, 1995, 1998, 2000, 2001, 2008, 2009, 2013, 2016
Pre-processing	QGIS / Semi-Automatic Classification Plugin
	Atmospheric correction
Spectral and Image processing	GRASS GIS and QGIS
	Satellite-derived index
	NDWI and MNDWI
	Water Feature Extraction
	Classification and Measurement
	Maps and Images

2.2.1. *Satellite data*

One scene of Landsat 3 MSS, seven scenes of Landsat-5 TM, one scene of Landsat 7 ETM+ and tow scene of Landsat-8 OLI were obtained from (U.S. Geological Survey website) in GeoTiff format. One satellite image per year was processed to analyze the interannual open water area fluctuations in the wetland. We analyzed 11 Landsat images for 1981–2016, using mostly summer images from Jun or July. The lack of precipitation in the summer will inform us about the continuity of the water in the wetland. Additionally, summer images contained less cloud. Table 2 presents the main characteristics of these satellite images.

2.2.2. *Data Pre-Processing*

Atmospheric and radiometric corrections were applied to all images to correct for distortions related to the sensor, the satellite and the Earth. The software QGIS [27] and GRASS GIS [28] were used to treat and view satellite data in the present study.

Table 2: Landsat data specifications

Satellite	Sensor	Acquisition date
Landsat 3	MSS	17 Jun 1981
Landsat 5	TM	10 July 1984
Landsat 5	TM	26 July 1987
Landsat 5	TM	19 July 1995
Landsat 5	TM	18 mars 1998
Landsat 7	LE	05 July 2000
Landsat 5	TM	28 July 2001
Landsat 5	TM	26 Jun 2008
Landsat 5	TM	28 May 2009
Landsat 8	OLI	18 August 2013
Landsat 8	OLI	14 November 2016

2.2.3. Spectral index methods

In order to detect the spatial-temporal changes of Aguelmam Afennourir wetland between 1981 and 2016, the open water area of this wetland in each temporal image of the Landsat MSS, TM, ETM and OLI (Table 2) was extracted individually. By doing so, the Normalized Difference Water Index (NDWI) [29] and the Modified Normalized Difference Water Index (MNDWI) [30] are calculating for the extraction surface water by choosing the threshold talking classification by GRASS GIS and Semi-automatic classification (SCP) plugin [31].

Satellite-derived indexes were used for water feature extraction for 1981. Since, there is no band MIR, we restrictedly obliged to calculate the NDWI index. NDWI = (Green – IR) / (Green + IR). And for other date, we calculate MNDWI = (Green – MIR) / (Green + MIR). After calculating the NDWI and MNDWI and extract the open water area. The numbers of pixels above the threshold value were together counted to represent the water outline recorded in white color. Finally, we calculate the open water area of each date.

2.2.4. Climatic variables

Since there is no meteorological station on the study site, we use the climatic data from the Sidi El Makhfi station located 13 km northeast of the site. The data is retrieved from Sebou Hydraulic Agency (ABH). We have limited to precipitation factor to explain the relationship between the drying up of the Afennourir wetland and the climatic conditions. In 1982 Lamp [32] defined a rainfall index which makes it possible to study climatic variability. This index reflects an excess or deficit in rainfall for the year under consideration compared to a reference period that is well established.

L = (Xi-X)/E

Xi: Rainfall of the year i

X : Average rainfall over the reference period

E : Standard deviation of the rainfall series over the reference period

This factor presented in figure 3 helps us to determine the periods of water deficit to determine the years to be evaluated. Also he tells us the wettest year which will represent a maximum area. To do this we choice: 1981, 1983, 1987, 1995, 2000, 2001, 2008, 2009 (reference year), 2013 and 2016.

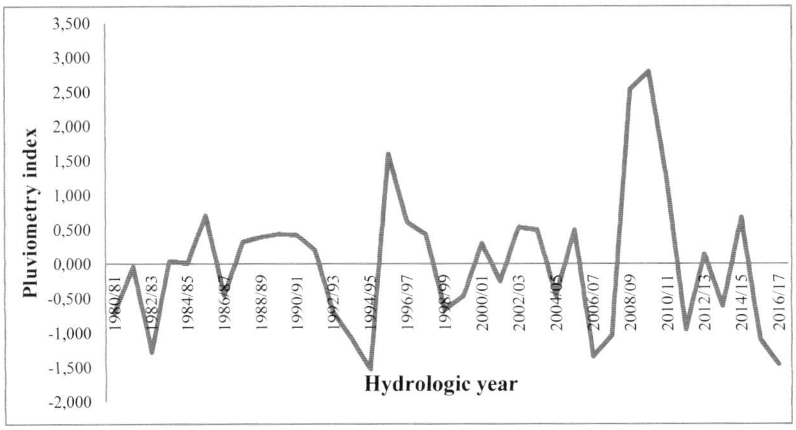

Figure 3: Evolution of pluviometry index in Sidi Elmakhfi, the nearest meteorological station for Afennourir.

2.2.5. Land-use changes in Aguelmam Afennourir wetland

Aguelmam Afennourir is a site exposed to limitless anthropogenic activities, overgrazing, livestock watering, water withdrawals, poaching, fishing and camping. These activities have caused changes and degradation of the main habitats on the site, also disturbances of the fauna. These activities have resulted in site changes. However, while the biological and ecological functions of the site require specific habitats mainly for the breeding and security of waterfowl chicks.

3. Results and discussions

3.1. Evolution of Aguelmam Afennourir open water area between 1981 and 2016

The area calculated by NDWI and MNDWI index from Landsat product were note accurate enough to analyze the open water area variation in the Aguelmam Afennourir wetland. That it related to the image resolution (30m) and the interference between open water and vegetation especially in the pond boundary where the water deep is low. The evolution of the open water area in the Afennourir wetland between 1981 and 2016 it presented in table 3 and figures 4 and 5. The analysis of reference sample (open water area calculated with Landsat image) shows that, during the study period, the open water area presented an inter-annual variation. The open water mean area was 44.28 ha with a standard deviation 43.07 ha. The Afennourir pond reached its lowest extension, 0 % (00 ha) and it is completely dried in July 1995, July 2000, June 2008 and in November 2016. While its largest extension, 100 % (122 ha) was occurred in May 2009. Four different periods of open water area changes, in Aguelmam Afennourir wetland could be defined:

- The 1980 decade: This period it characterized by defect precipitations (Figure 3), nevertheless there is no a totally drying up of the pond but it lose 62.81% in Jun 1981 and 56.62 % in July 1987 to it's totally area (Figure 4 and Figure 5).
- The 1990 decade: The beginning and the ending of this period it characterized by defect precipitation (Fig 3) which give a totally drying of the pond in July 1995 (Figure 4 and 5).
- The 2000 decade: This period it characterized by defect precipitation (Figure 3). Therefore, in 2000 the pond has totally drying up. While the open water area in the wetland it increasing by 68.50 % in 2001 to the reference area. In addition, we recorded a totally drying up and disappearance of open water area in Jun 2008 (figure 3 and 4). The ending of this decade, the site is characterized by the important precipitation and the wetland recorded its maximum area flooding in 2009.
- The period 2010 to 2016: The beginning of this period it characterized by flooding. But since 2012 the site has experienced a deficit in precipitation (Figure 3), which led to a total drying up in November 2016 where the pond loses 99.41% of its open water area (Figure 4 and 5).

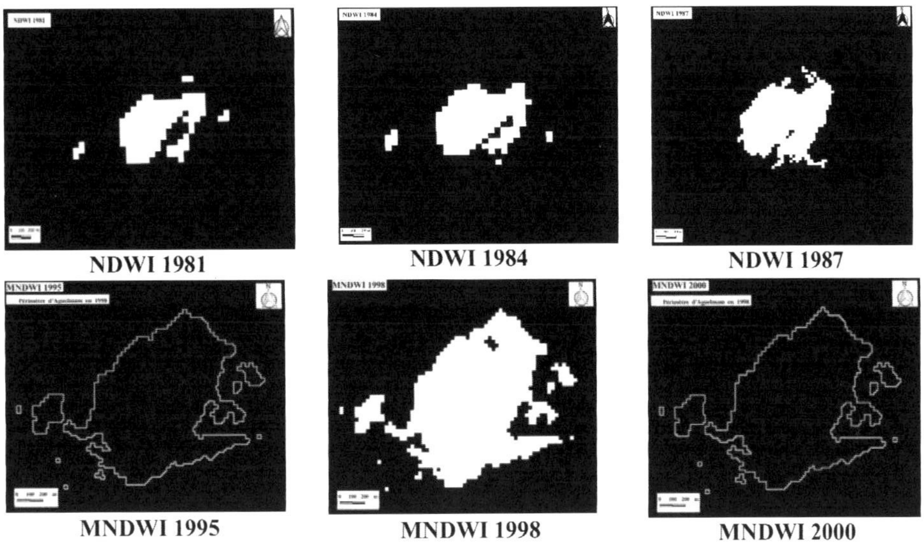

IOP Conf. Series: Earth and Environmental Science 1090 (2022) 012037 doi:10.1088/1755-1315/1090/1/012037

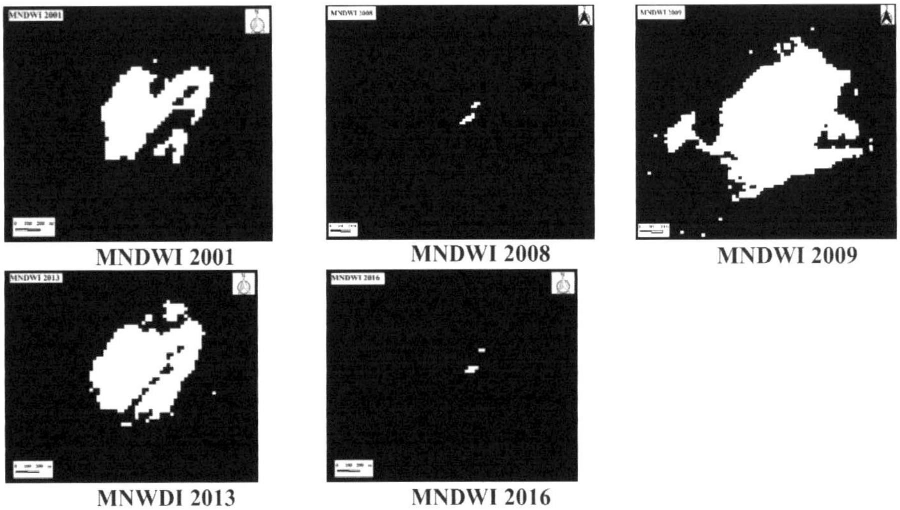

Figure 4. Resulting MNDWI images after applications of various thresholds for Aguelmam Afennourir between 1981 and 2016. White color refers to water and black color refers to no water.

Table 3: The area of Aguelmam extracted after calculate NDWI and MNDWI

Hydrologic year (September – August)	Area variation ha	Area variation %	Situation date	Pluviometry Index
1980-1981	45.36	-62.81%	06/17/1981	-0.70
1983-1984	52.92	-52.75%	07/10/1984	**0.038**
1986-1987	60.30	-56.62%	07/26/1987	**-0.45**
1994 - 1995	**00**	**-100 %**	07/19/1995	**-1.52**
1997 - 1998	110	-9.83 %	03/18/1998	0.34
1999 - 2000	**00**	**-100 %**	07/05/2000	**-0.51**
2000 - 2001	38.43	-68.50 %	07/28/2001	0.22
2007-2008	**1.08**	**-99.11%**	06/26/2008	**-1.039**
2008 - 2009	**122***	**100 %**	05/28/2009	2.53
2012 - 2013	56.34	-53.81 %	08/18/2013	0.07
2015 - 2016	**0.72**	**-99.40 %**	11/14/2016	**-1.11**

* Reference date where the site presented a maximum open water area (112 ha), ie 100%.

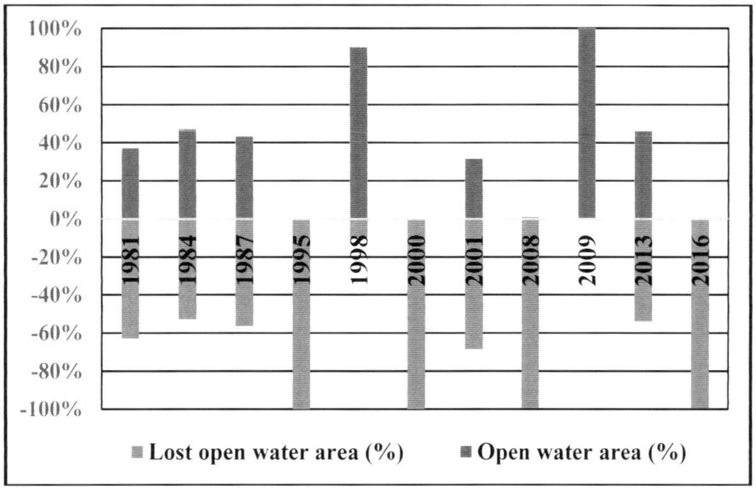

Figure 5. Deviation of the Aguelmam water surface between 1981 and 2016 with the 2009 state as a reference.

3.2. Relationship between open water area and rainfall

The water flight in the dam accelerates the drying up of the wetland. Likewise, [33] evaluated the water withdrawals to 180 m3 per day for livestock and domestic use. The water volume of Afennourir wetland it not estimated and we cannot do it for its logistics and cost. Therefore, the water flight in the dam and water withdrawals distorts the possible relationship between rainfall and open water area in the Afennourir wetland. To explain the drying up and evolution of open water area in Afennourir wetland, we based on figure 3, which presents the evolution of the pluviometry index and explaining the climatic variations in the site. During the decade of 1980, the area experienced low rainfall (Figure 3), but the site never dried, because there was a minimum of precipitation to maintain the water in the wetland and it is possible there was no water loss through the dike. During the decade of 1990, the site experienced a period of drying up between 1990 and 1995 which caused the total drying up of the site in July 1995 (Figure 4 and Figure 5). The total drying up of the pond in July 2000 is due to the absence of precipitation for 8 months less at the site between the hydrological year 1998/1999 and the hydrological year 1999/2000. In 2001, the site began to adapt to the total drying up of 2000 despite the deficit in precipitation recorded during this hydrological year. The totally drying up of Aguelmam Afennourir wetland in 2008 is caused by deficit precipitations since hydrological year 2005/2006 to hydrological year 2007/2008 (Figure 3). The boom of the wetland is the period 2009-2011, a period of high rainfall where the site was recorded a maximum flood in May 2009. After this period and since the hydrological year 2011/2012 the site experienced a deficit in precipitation which led to a total drying in November 2016. Our study showed that there is a relationship between the water area in the wetland and annual precipitation. This is approved by other study as example [34-35-36-37]. The climate type of the site characterized by precipitations abundant and according to the precipitation data, the duration of the total dryings of the pond is 5 months on 1995, 2 month on 2000 and 2008 and 1 month on 2016.

3.3. Causes of Aguelmam drying up

To reveal the potential causes of the drying up of Aguelmam Afennourir wetland, we establish the watersheds of the site (Figure 6). The site is divided between two watersheds. The biggest with an area of 14868 ha drained its water to the pond. While the other watershed (1505 ha) does not drain its water adjacent to the pond. The watershed alimenting the pond has a very weak slopes and presents like a plain, which slows down the flow speed. It characterized by three water retention factors, the forest, the seasonal swamps, and the volcanic craters (Figure 6). The volcanic craters absorb water and drain them directly to the aquifer. For these reasons, the pond receives small water quantity from its watershed. At the pond and for direct causes we cited:

(1) Evaporation, a phenomenon that cannot be assessed due to lack of means, (2) The water flight in the dam, (3) water withdrawals and (4) livestock watering on-site and off-site. The water quantity used for livestock drink and domestic use is estimated by [33] at nearly 180 m3 per day. Along the scenes of satellite image since 1981 to 2016, we do not mark any flooding on the pond shores, which means that the input of water is insufficient to fill the pond. The solution to ensure water continuation in the Afennourir wetland is strengthening the dyke with local materials to increase the water level.

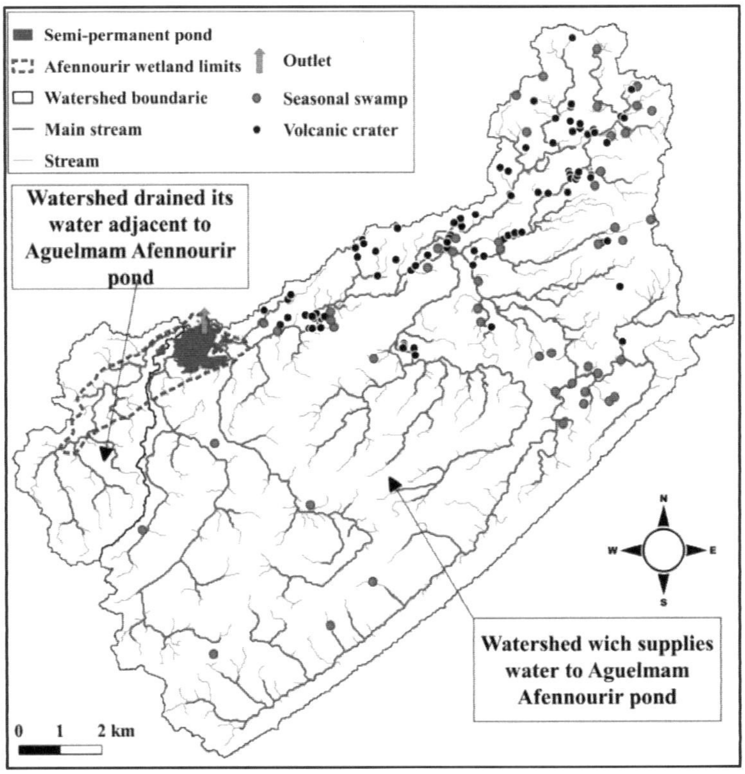

Figure 6. The watershed of Aguelmam Afennourir wetland

3.4. Drying up and impact on biodiversity

Each drying up of the pond causes habitat contraction and loss in ecological functions. Habitat loss is the major threat to biodiversity. The habitat protection is the most important means of conserving biodiversity [38]. The drying up of Aguelmam Afennourir has conducted the site in ecologic deficiency because its bio- capacity is unable to satisfy requirements for the biodiversity. In particular, to fill the directives of Ramsar convention to maintain a minimum water level in the wetland to protect species for theme the site has provided Ramsar statute. The plants species are *Persicaria lapathifolia*, *Damasonium alisma* and *Juncus inflexus,* while the waterfowl species are Ruddy Shelduck (Tadorna ferruginea), crested coot (Fulica cristata) and Black-necked Grebe (Podiceps nigricollis). The advanced reduction of open water area gives access to livestock to grazing the aquatic plants, which are the main refugees for waterfowl chicks, while the successful breeding in such areas often requires exceptional flooding events that last long enough for completion of the breeding cycle [39-40-41]. The drying up of Afennourir wetland can gives naissance for comportment for species not to back to the wetland. Therefore, this created habitude gives a decline of specific biodiversity in Aguelmam Afennourir wetland; Finlayson confirmed this in 2018: a long drought affects the dispersal and breeding of many migratory or nomadic species. Along the drying up of November 2016, we notice the transience of fish (Figure 7) and absence of waterfowl that frequently wintering in the site. We observed only three species (1 *Ardea cinerea*, 1 *Tringa ochropus* and 1 *Tringa nebularia*). For 1998, as the wet year, the wetland welcomed 6321 birds spread over 14 species [42] and without wintering bird in 1995 among the dry years [43]. The drying of Aguelmam favors the human access to the wetland inside and picks the aquatic plants (*Scirpus holosehoenus* and *Joncus bufonius), as* the only waterfowl chick's refuge. The Figure 8 and 9 present the disappearance of scripts and rush massifs in the Aguelmam. Bathymetry plays a key role in plant and animal community dynamics [44-45].

Figure 7. The death of fish after totally drying up of Aguelmam Afennourir (November 2016)

Figure 8. Aguelmam with aquatic plants (2001)

Figure 9. Aguelmam without aquatic plants (2016)

4. Conclusion

The present study brings useful information to elucidate the open water area evolution in Aguelmam Afennourir wetland. It presented the inter-annual evolution of the open water area in the wetland and that raised total drying up of the site in 1995, 2000, 2008 and 2016 and for other years studied, in the summer, the site loses more than 50% of its open water area. The total or partial drying up has negative impacts on the biodiversity of the site. For example, in 2016 we noted a highly catastrophic death of fishes, aquatics plants and macro-invertebrates and quasi-absence of water birds in 2016. To urgently overcome this drying up, we have established the watershed of the site and have seeked to increase the amount of water received by the deviation of the water of an under watershed to Aguelmam and reconstructed the dike throughout the rammed soil. Aguelmam Afennourir wetland is a Ramsar site and requires an utmost care to ensure the continuity of its ecological and biological functions to contribute to sustainable development goals. The main cause of this drying up and the decrease open water area in the site is the leaks at the dam. The other water supply activities from the pond are a human nature and we suggest alternatives that will take into consideration the health aspects of humans and livestock. We propose the installation of drinkers in different places and well distributed spatially. faced with this repetitive situation of ecological disturbance in Afennourir wetland, [46] noted an increase in breeding pairs of the Black-necked Grebe on the site, between 2009 and 200, hence the necessity to increase the water level of the pond.

5. References

[1]. J.B. Zedler, S. Kercher, *Wetland Resources: Status, Trends, Ecosystem Services, and Restorability.* Annual Review of Environmental Resources, **30**, 39-74 (2005). http://dx.doi.org/10.1146/annurev.energy.30.050504.144248
[2]. B. Gopal, *Future of wetlands in tropical and subtropical Asia, especially in the face of climate change.* Aquatic Sciences, **75(1)**, 39–61(2013)
[3]. C. B. V. Rees, J. M. Reed, *Wetland loss in Hawai'I since human settlement.* Wetlands, **34**, 335–350 (2014)
[4]. K. Tockner, S. E. Bunn, C. Gordon, R. J. Naiman, G. P. Quinn, J. Standord, *A Flood plains: Critically threatened ecosystems.* In N. V. C. Polunin (Ed.), Aquatic ecosystem (pp. 45–61). Cambridge: Cambridge University Press (2008).
[5]. Brander, L. M., Flora, R. J. G. M., and Vermaat, J. E. (2006). The empirics of wetland valuation: A comprehensive summary and a meta-analysis of the literature. Environmental and Resource Economics, 33(2), 223–250
[6]. Davidson, N. C. (2014). How much wetland has the world lost? Long-term and recent trends in global wetland area. Marine & Freshwater Research, 65(10), 934–941
[7]. Sánchez-Carrillo, S., and Angeler, D.G. (eds.) (2010). Ecology of Threatened Semi-Arid Wetlands: Long-Term Research in Las Tablas de Daimiel, DOI 10.1007/978-90-481-9181-9_3, © Springer Science + Business Media B.V
[8]. Winter, TC. (1988). A conceptual framework for assessing cumulative impacts on the hydrology of nontidal wetlands. Environ Manage 12:605–620
[9]. Dunne, T., Mertes, L.A., Meade, R.H., Richey, J.E., Forsberg, B.R. (1998). Exchanges of sediment between the flood plain and channel of the Amazon River in Brazil. Geol. Soc. Am. Bull. 110, 450–467
[10]. Mei, X., Dai, Z., Du, J., Chen, J. (2015). Linkage between three gorges dam impacts and the dramatic recessions inchina's largest freshwater lake, Poyang Lake. Sci. Rep. 5, 18197.
[11]. Yan, L., and Zheng, M. (2015). The response of lake variations to climate change in the past forty years: a case study of thenortheastern Tibetan Plateau and adjacent areas, China. Quat Int. 371:31–48
[12]. Darwall, W., Carrizo, S., Numa, C., Barrios, V., Freyhof, J., and Smith, K. (2014). Freshwater Key Biodiversity Areas in the Mediterranean Basin Hotspot: Informing Species Conservation and Development Planning in Freshwater Ecosystems. Cambridge, UK and Malaga, Spain, IUCN. https://portals.iucn.org/library/sites/library/files/documents/ SSC-OP-052.pdf
[13]. Finlayson, M., Hollis, T., and Davis, T. (1992). Managing Mediterranean Wetlands and their Birds. IWRB Spec Publ. No. 20. Slimbridge, 285 pp
[14]. Pearce, F., and Crivelli, A.J (1994). Characteristics of Mediterranean Wetlands. MedWet publications, No.1. Tour du Valat
[15]. Chillasse, L., and Dakki, M. (2004). Potentialités et statuts de conservation des zones humides du Moyen- Atlas (Maroc), avec référence aux influences de la sécheresse. Science et changements planétaires / Sécheresse > Volume 15, numéro 4, octobre-novembre-décembre 2004
[16]. Lehmann, A., Jaquet, JM., and Lachavanne, JB. (1997). A GIS approach of aquatic plant spatial heterogeneity in relation to sediment and depth gradients, Lake Geneva, Switzerland. Aquat Bot 58: 347-361
[17]. Gottgens, JF., Swartz, BP., Kroll, RW., and Eboch, M. (1998). Long term GIS based record of habitat changes in a lake Erie costal marsh. Wetland Ecol Manag 6 : 5-17
[18]. Jensen, JR., Rutchey, K., and Koch, MS. Narumalani S (1995). Inland wetland change detection in the Everglades Water Conservation Area 2A using a time series of normalized remotely sensed data. Photogram Eng Rem Sen 61: 199-209

[19]. Peng, D., Xiong, L., Guo, S., Shu, N. (2005). Study of Dongting Lake area variation and its influence on water level using MODIS data. Hydrol. Sci. J. 50, 31–44
[20]. Soti, V., Tran, A., Bailly, J.S., Puech, C., Lo Seen, D., and Bégué, A. (2009). Assessing optical earth observation systems for mapping and monitoring temporary ponds in arid areas. Int. J. Appl. Earth Obs. Geoinf. 11, 344–351
[21]. Sakamoto, T., Van Nguyen, N., Kotera, A., Ohno, H., Ishitsuka, N., Yokozawa, M. (2007). Detecting temporal changes in the extent of annual flooding within the Cambodia and the Vietnamese Mekong Delta from MODIS time-series imagery. Remote Sens. Environ. 109, 295–313
[22]. Bohn, V.Y., Delgado, A.L., Piccolo, M.C., Perillo, G.M.E. (2016). Assessment of climate variability and land use effect on shallow lakes. *Environ. Earth. Sci. 75*, 818
[23]. Shi, Y.F., and Ren, J.W. (1990). Glacier Recession and Lake Shrinkage Indicating a Climatic Warming and Drying Trend in Central Asia. Ann. Glaciol. 14, 261–265
[24]. Scordo, F., Perillo, G.M.E., Piccolo, M.C. (2018). Effect of southern climate modes and variations in river discharge on lake surface area in Patagonia. Inland Waters. https://doi.org/10.1080/20442041.2018.1487118
[25]. Woodcock, C.E., Allen, R., Anderson, M., Belward, A., Bindschadler, R., Cohen, W., Gao, F., Goward, S.N., Helder, D., Helmer, E., and Nemani, R. (2008). Free access to Landsat imagery. Science 320, 1011
[26]. Wang, Y., & Yésou, H. (2018). Remote Sensing of Floodpath Lakes and Wetlands: A Challenging Frontier in the Monitoring of Changing Environments. Remote Sensing, 10(12), 1955. doi:10.3390/rs10121955
[27]. QGIS Development Team. (2016). QGIS Geographic Information System. Open Source Geospatial Foundation Project. URL http://qgis.osgeo.org/
[28]. GRASS Development Team. (2016). Geographic Resources Analysis Support System (GRASS) Software, Version 7.0.3 Open Source Geospatial Foundation. Electronic document :. http://grass.osgeo.org
[29]. McFeeters, S.K. (1996). The use of the normalized difference water index (NDWI) in the delineation of open water features. Int. J. Remote Sens. 17, 1425–1432
[30]. Xu, H. (2006). Modification of normalized difference water index (NDWI) to enhance open water features in remotely sensed imagery. Int. J. Remote Sens. 2006, 27, 3025–3033
[31]. Congedo, L. (2016). Semi-Automatic Classification Plugin Documentation. DOI: http://dx.doi.org/10.13140/RG.2.2.29474.02242/1
[32]. Lamb, P. J. (1982). Persistence of Subsaharandrought, Nature (Internaional Journal of Science) n°299, p. 46 – 47
[33]. Aaouine, M., Aoussar, H., El Mataouat, S., Loudiyi, M. et Sghir, A. (2008). Aguelmam Afennourir. Rapport de stage de l'école nationale forestière d'ingénieurs, Salé, Maroc, 43 p
[34]. George, G. Hurley, M. Hewitt, D. (2007). The impact of climate change on the physical characteristics of the larger lakes in the English Lake District. Freshwater Biol. 52:1647–1666
[35]. Whitehead, PG. Wilby, RL. Battarbee, RW. Kernan, M., and Wade, AJ. (2009). A review of the potential impacts of climate change on surface water quality. Hydrolog Sci J. 54:101–123
[36]. Williamson, CE., Saros, JE., Vincent, WF., And Smol, JP. (2009). Lakes and reservoirs as sentinels, integrators, and regulators of climate change. Limnol Oceanogr. 54:2273–2282
[37]. Woolway, RI., Meinson, P., Nõges, P., Jones, ID., and Laas, A. (2017) Atmospheric stilling leads to prolonged thermal stratification in a large shallow polymictic lake. Clim Change. 141:759–773
[38]. Primack, R.B. (1995). A Primer of Conservation Biology. Sinauer Associates, Sunderland
[39]. Gehrt, S.D., Fritzell, E.K., (1998). Resource distribution, female home range dispersion and male spatial interactions: group structure in a solitary carnivore. Animal Behaviour 55, 1211–1227.
[40]. Hoover, J.P. (2003). Water depth influences nest predation for a wetland-dependent bird in fragmented bottomland forests. Biological Conservation 127 (2006) 37 – 45
[41]. Finlayson, M. (2018). Climate Change and Wetlands. The wetland book. DOI: 10.1007/978-90-481-9659-3_126
[42]. Dakki M., Qninba A., El Agbani M.A. et Benhoussa A. (2003). Recensement hivernal d'Oiseaux d'eau au Maroc : 1996-2000. Trav. Inst. Sci., Rabat, série Zool., 45, 28 pp

[43]. Dakki M., El Agbani M.A., Qninba A. & Benhoussa A. (1995). Recensement hivernal d'Oiseaux d'eau au Maroc : janvier 1995. Doc. Inst. Sci., Rabat, N° 18. 1- 32 pp
[44]. Van der Valk, AG. (1981). Succession in wetlands: a Gleasonian approach. Ecology 62:688–696
[45]. Ripley, BJ., Holtz, J., and Simovich, MA. (2004). Cyst bank life-history model for a fairy shrimp from ephemeral ponds. Freshw Biol 49:221–231
[46]. Cherkaoui, I., Bouajaja, A., Elbanak, A., Lahrouz, S., and Hanane, S. (2014). The Black-necked Grebe (Podiceps nigricollis): an expanding species in the Middle Atlas wetlands, Morocco. Wetlands Ecol Manage (2014) 22:93–98 DOI 10.1007/s11273-013-9321-7.

Effects of meteorological droughts on water resources in Northern Morocco: Case study of the Martil watershed

I El-Khantoury*, K Obda, H Achiban

Sidi Mohamed Ben Abdellah University, Fez, Morocco

*Corresponding author: ismail.elkhantoury@usmba.ac.ma

Abstract. Recent droughts in the Northern Morocco have highlighted the increasing vulnerability of water resources to this natural hazard, with reduced potential for drinking water supply in the city of Tetouan and its coastal area. Therefore, the main objectives of this work are: the identification of dry events that affected the Martil watershed on the one hand, and the study of the effects of meteorological droughts on water resources on the other hand. For this purpose, a meteorological-hydrological modelling framework was adopted, using standardized precipitation index (SPI) values and the volume of surface water resources recorded at the Torreta station. The results obtained show that the last decade of the 20th century and the beginning of the 21st century have seen the heaviest and most critical droughts in the study area, namely the drought of 1994-95, 1998-99, 2004-05, 2006-07, 2011-12 and 2013-14. Also, water resources have been reduced by 62% to 82% during critical dry years compared to the long-term average for the period 1970-71/2014-15. Finally, this study showed that the rainfall variability is the determining factor in the fluctuations of surface water resources at the Martil watershed because it explains 84% of their total variance.

1. Introduction

Due to its location on the margins of the oceanic disturbances of the middle latitudes on the one hand, and its proximity to a center of large-scale anticyclonic action on the other hand, Morocco has a climate characterized by high rainfall variability and recurrent droughts. While sometimes the South of the country suffers the most, other times the central region is the most affected, but it is often the whole of Morocco which is seriously affected by this phenomenon [1]. In this case, the droughts of 1944-1945, 1980-1984, 1998-2000, 2004-2005 and 2015-2016 had a significant negative impact on both surface and groundwater resources.

The effects of the drought periods on the water resources have significantly increased the deficit in water flow observed since 1970, when the deficit cycle began at the level of the whole country. In Morocco, the average surface water runoff, estimated at about 19 billion m^3, was reduced by 47% in 1980-85, 74% in 1992-1993, 72% in 1994-95 and by 68% in 1998-2000 [2]. Thus, one of the major current challenges in water management is related to the drought phenomena, from genesis to the control of the impacts [3].

Worse still with climate change, Moroccan climate does not appear to be so rainy. Thus, climate projections predict that Morocco's climate will become increasingly arid by 2100. Indeed, annual rainfall is expected to decrease by 10 to 40%, including 10 to 30% during the rainy season, from October to April, and 10 to 40% during the dry season, from May to September [4]. At the same time, taking into account a 30% reduction in water resources due to climate change, nearly 4 billion m^3 could be missing from the water supply and demand balance by 2030 [5].

Nowadays, the drinking water supply of the city of Tetouan is almost exclusively based on the surface water resources from the Martil watershed. But, recent droughts have shown how water resources are so vulnerable to climatic constraints with reduced potential for drinking water supply. This has proved to be problematic especially after the 2015-2016 drought. This drought, which came after three years of rainfall deficits, caused a strong decrease in runoff, resulting in a dramatic decline in dam reservoirs in the region. Thus, the quantities of water supplied by National Office of Electricity and Drinking Water to the company in charge of managing the water sector (Amendis) have fallen significantly.

This situation has led, among other things, to disruptions and temporary suspensions of drinking water supply in the Tetouan region. It should also be mentioned in this respect that restrictions and rationing by districts were introduced in the city of Tetouan during the drought of 1995, a year in which the deficit in drinking water was estimated at between 30 and 35% for daily needs averaging around 103 liters per person [6].

The water deficit situation will probably become very critical in a region where water needs are constantly increasing. Indeed, according to the study by the Loukkos Water Basin Agency, daily consumption in Tetouan city rose from 42.7 thousand m^3 to 50.5 thousand m^3 between 1994 and 2005 and should reach 100 thousand m^3 in 2030 [7]. In relation to this, it should be noted that in 1974, the water needs of the city of Tetouan were estimated at 7.5 million m^3/year (i.e. about 20 thousand m^3/day) [8].

This work is structured as follows: Section 1 is dedicated to the introduction. While section 2 presents the study area, and describes data and methods. The results obtained are presented and discussed in Section 3. The conclusions of the work are presented in Section 4.

2. Data and methods

2.1. Presentation of the study area

The study area is located in the extreme north-west of Morocco, on an area of 1131 Km^2. Its geographical coordinates are successively 35.25° and 35.75° North, and 5.26° and 5.65° West (figure 1). Its altitudes vary between 0 and 1791 m.

Geologically, the general litho-stratigraphic features of the Rifan Mountains are well represented [9]. Thus, the study area is characterized by a diversity of rock formations: there are essentially the Cretaceous marls and marl-schists of the Tangier unit, the sandstones and marls of the Beni Ider unit, the carbonate formations of the limestone ridge, and finally the Sebtides and Ghomarides of Paleozoic Era.

The climate of the region is Mediterranean, which is characterized by hot, dry summers and mild, wet winters. The quantities of rainfall reflect a significant spatial variability due to factors of altitude, exposure and continentality [10]. The average annual rainfall for the period 1970-71 and 2014-15, is 720 mm but maybe exceeds 1000 mm on the summits. Besides the spatial variability, the precipitation is characterized by its monthly and interannual irregularity (figure 2).

From a hydrological point of view, river flow show strong monthly and interannual variation. As an example, at the Torreta station on the Wadi Martil, the average annual streamflow vary between 2 m^3/s and 40.3 m^3/s during 1970-71/2014-15. During the same period the maximum instantaneous streamflow often exceeded 2000 m^3/s. Thus, Annual water resources in this region vary greatly depending on rainfall conditions.

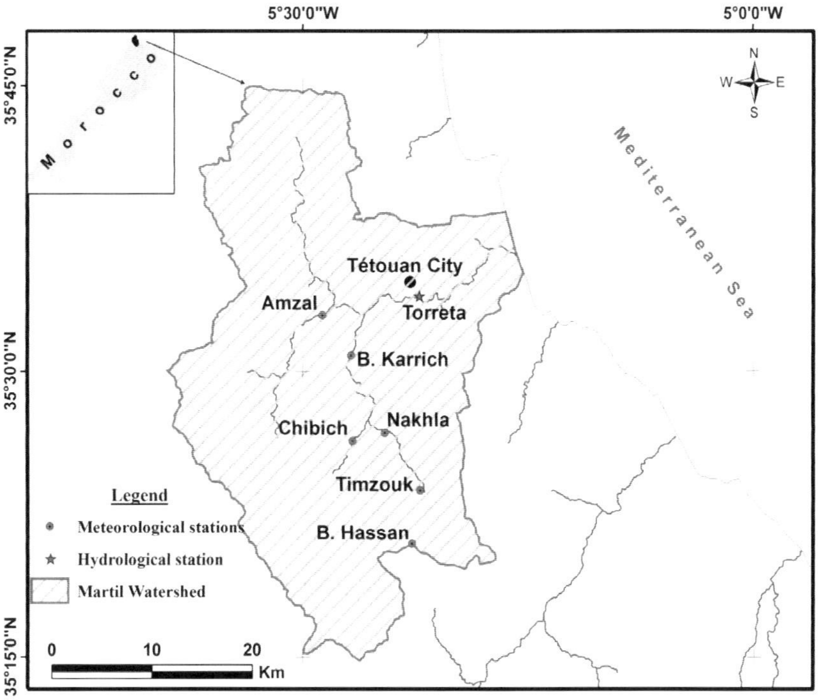

Figure 1. Location of the Martil Watershed

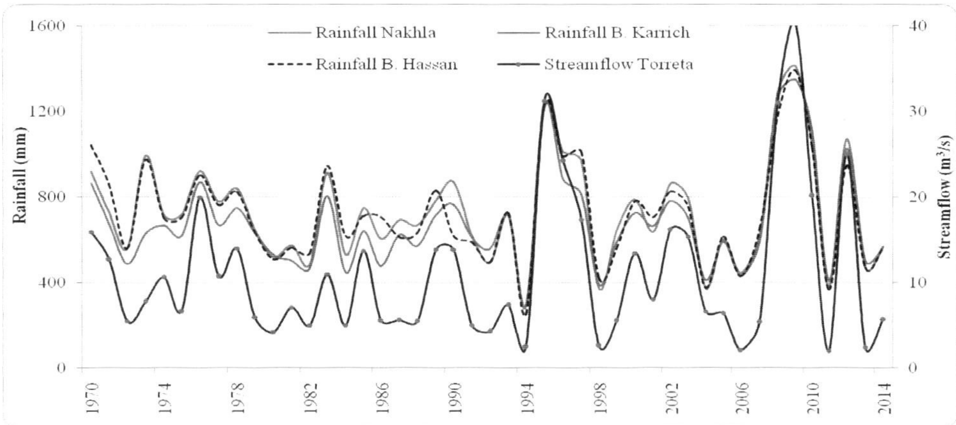

Figure 2. Rainfall and streamflow variability at the Martil watershed

2.2. Data

In order to investigate the effects of meteorological droughts on water resources at the Martil watershed, 45 years (1970-71/2014-15) of continuous data from 6 meteorological stations and the hydrological station of Torreta on Wadi Martil were selected. This period has the particularity of representing the extreme meteorological and hydrological events that Morocco has experienced since the 1970s.

2.3. Methods

The methodology adopted for this work consists of two steps:
- Calculation of the Standardized Precipitation Index (SPI) for 6 meteorological stations;

- Analysis of the effect of meteorological droughts on water resources through Standardized Precipitation Index (SPI)-water resources modeling.

2.3.1. Approach to calculating the drought index [SPI] Several drought indices have been developed for the study and analysis of drought, of which the Standardized Precipitation Index (SPI) is probably the most widely used.

According to [11] and [12], the process of calculating the SPI for a given location and over different time steps includes the following steps:

The first step is to find the probability density function (PDF) that best describes the distribution of precipitation relative to a given time scale. According to [13] the two-parameter (α and β) Gamma distribution is better adapted to rainfall time series. The Gamma distribution is defined by its probability density function:

$$g(x) = \frac{1}{\beta^{\alpha}\Gamma(\alpha)} x^{\alpha-1} e^{-x/\beta} \text{ for } x > 0 \qquad (1)$$

where $\alpha > 0$ and $\beta > 0$ are the shape and scale parameters respectively, x is the amount of precipitation, and Γ is the Gamma function which is defined as follows:

$$\Gamma(\alpha) = \int_0^{+\infty} y^{\alpha-1} e^{-y} dy \qquad (2)$$

The shape and scale parameters of the Gamma probability density function are estimated for each station, for each time scale of interest, and for each year. The maximum likelihood is used for optimal estimation α and β:

$$\alpha = \frac{1}{4A}\left(1 + \sqrt{1 + \frac{4A}{3}}\right) \text{and} \beta = \frac{\bar{x}}{\alpha} \qquad (3)$$

and

$$A = \ln(\bar{x}) - \frac{\sum \ln(x)}{n} \qquad (4)$$

where n for number of observations and $\bar{x}$ being the average of observed rainfall.

The resulting parameters are then used to find the cumulative probability of an observed precipitation event for a given time scale by integrating the probability density function g(x). The cumulative probability function/distribution function is defined by:

$$G(x) = \int_0^x g(t)dt = \frac{1}{\hat{\beta}^{\hat{\alpha}}\Gamma(\hat{\alpha})} \int_0^x t^{\hat{\alpha}-1} e^{-\frac{t}{\hat{\beta}}} dt \qquad (5)$$

G(x) represents the cumulative probability that a certain amount of rain has been observed over a given period of time for a given time scale.

Since the Gamma function is not defined in $x = 0$, and a precipitation series may contain zeros, the cumulative probability becomes so:

$$H(x) = q + (1 - q)G(x) \qquad (6)$$

where q is the probability of not having any precipitation estimated from the observed data. H(x) is the cumulative probability or distribution function of observed precipitation.

The method described in Edwards and McKee (1997) uses the technique of converting the cumulative probability into a standard normal variable Z of mean 0 and standard deviation 1 [14]. This is given by the following formula:

$$SPI = -\left(t - \frac{c_0 + c_1 t + c_2 t^2}{1 + d_1 t + d_2 t^2 + d_3 t^3}\right) for\ 0 < G(x) \leq 0.5 \qquad (7)$$

$$SPI = +\left(t - \frac{c_0 + c_1 t + c_2 t^2}{1 + d_1 t + d_2 t^2 + d_3 t^3}\right) for\ 0.5 < G(x) < 1 \qquad (8)$$

where $c_0 = 2.515517$; $c_1 = 0.802853$; $c_2 = 0.010328$; $d_1 = 1.432788$; $d_2 = 0.189269$; $d_3 = 0.001308$

and

$$t = \sqrt{\ln\left(\frac{1}{(G(x))^2}\right)} \text{ for } 0 < G(x) \leq 0.5 \qquad (9)$$

$$t = \sqrt{\ln\left(\frac{1}{(1 - G(x))^2}\right)} \text{ for } 0.5 < G(x) < 1 \qquad (10)$$

According to [15], drought begins at a value of -1 or less, but there is no general rule; some researchers set the threshold at less than 0 but not as low as -1, while others consider that a drought begins when the index is below -1 [16]. In order to determine the duration of a dry episode, the criterion used by the US Drought Monitor was adopted, i.e. a SPI≤-0.80 and thus count as a dry year, and the following year if SPI≤-0.80 and when the year has an SPI ≥0.79 the episode is terminated (table 1).

Table 1. Classification of drought categories for standardized precipitation index (SPI) according US Drought Monitor [17]

SPI values	Drought category
-0.5 to -0.7	Abnormally dry
-0.8 to -1.2	Moderate drought
-1.3 to -1.5	Severe drought
-1.6 to -1.9	Extreme drought
-2 or less	Exceptional drought

2.3.2. Modeling the relationship: meteorological drought - water resources During this step a coupled meteorological-hydrological modeling framework based on the SPI values and the water resources recorded at the hydrological station of Torreta was adopted. This modeling framework focuses on the physical impacts of the recurrence of dry years on water resources.

3. Results and discussion

3.1. Identification and characterization of meteorological drought

Annual time series with dry years are shown for a number of stations in figure 3. Thus, according to the 12-month standardized precipitation index (hydrological year September-August) calculated for all stations, it is found that, on average, normal classes represent 58% and 42% for abnormal conditions, of which 21% are dry. The most critical dry and generalized episodes took place during the years 1982-83, 1994-95, 1998-99, 2004-05, 2006-07 and 2011-12. However, a year of extreme/exceptional drought is never observed over the entire study area.

The frequency of occurrence of a dry year over the whole study area is on average 20%, i.e. one year in five. The maximum frequency (22%) is recorded at the Ben Karrich station, against a minimum of 18% at the Beni Hassan station.

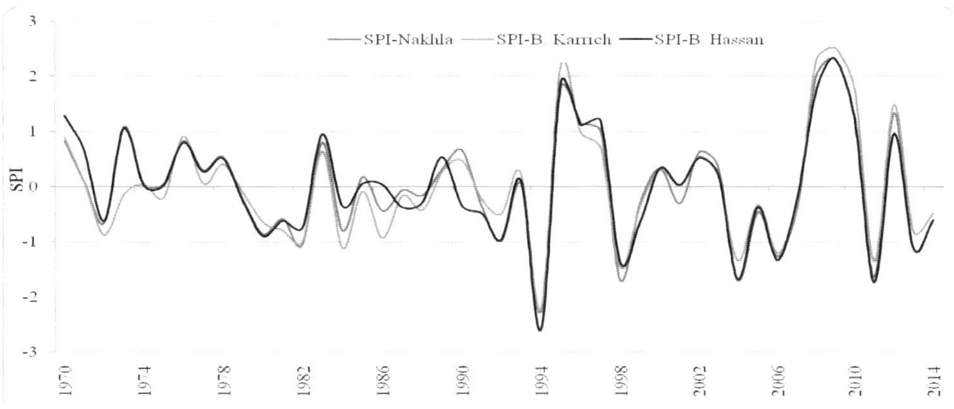

Figure 3. Change of the annual SPI index for the period 1970-71/2014-15

The recorded droughts have caused a significant rainfall deficit. Indeed, during 1994-95 the deficit reached 65%, 60% and 57% for the stations of Beni Hassan, Nakhla dam and Ben Karrich respectively. As for the drought of the year 2011-12, it was accompanied by a drop in rainfall equivalent to 46% on average. During this drought, the Nakhla dam station recorded only 382.5 mm compared to 731 mm in a normal year.

The most severe droughts recorded in the study area date back to the 1980s and 1990s, which corresponds to the period of the most unusual North Atlantic Oscillation (NAO) observed between the 1970s and 1990s, when the NAO index showed an increasing tendency towards positive phases [18]. The 2004-05 drought was caused by several factors favoring atmospheric circulation anomalies, including positive values of the NAO index between November 2004 and February 2005 [19]. While the 1994-95 drought was associated with a high frequency of anticyclonic synoptic situations, and a low percentage of wet weather types (cyclonic, west and southwest) [20].

Table 2 shows a strong correlation/relationship of 0.918 on average between all meteorological stations used in this study, which is an indication of the homogeneity of climatic conditions at the Martil watershed.

Figure 4 shows the spatial extension of the different drought classes during two dry years considered to be the most severe during the period 1970-71 / 2014-15. In 1994-95, the entire study area was affected by extreme to exceptional drought, of which 87% was affected by exceptional drought, while during 2011-12, 84% of the total area was dominated by extreme drought, 15% by severe drought and 1% by exceptional drought.

Table 2. Pearson correlation coefficient between rainfall stations (n=45); p $value$<0.0001

	Nakhla dam	Ben Karrich	Beni Hassan	Timzouk	Chibich
Ben Karrich	0.96				
Beni Hassan	0.97	0.92			
Timzouk	0.91	0.92	0.86		
Chibich	0.94	0.94	0.91	0.89	
Amzal	0.92	0.95	0.89	0.87	0.91

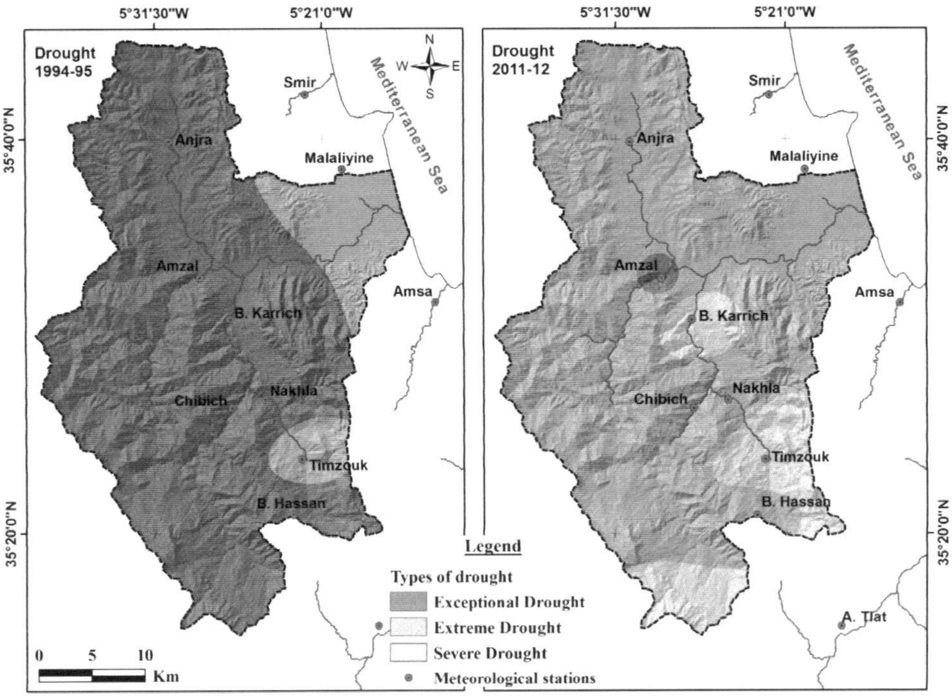

Figure 4. Spatial extension of the drought of 1994-95 and 2011-12 using inverse distance weighting

3.2. Impact of meteorological drought on water resources

The results obtained show that the droughts that occurred during the period 1970-71 / 2014-15, strongly affected the water resources observed at the Torreta station, especially during so-called critical years, such as 1994-95, 2006-07, 2011-12 and 2013-14 (table 3). In the drought year 2011-12, water resources were significantly reduced by up to 82% compared to the average, which represents the largest drop in water resources ever recorded. In 2006-07 and 2013-14, water resources were reduced by 81% and 79% respectively. While during the drought year 1994-95, water resources were decreased by 78%, compared to only 72% at the national level.

Table 3. Overview of years of critical water deficit

Dry year	Water deficit
1980-81	-63
1992-93	-62
1994-95	-78
1998-99	-77
2006-07	-81
2011-12	-82
2013-14	-79

The results also show that 15 of the 45 years had a critical water deficit exceeding 50%, especially after 1992-93. This confirms the trend towards increasingly extreme hydrological phenomena. It should also be noted that the water situation at the Martil watershed is marked by the succession of extremely dry and very wet years, such as during 1994-95 and 2011-12 (dry years) and 1995-96 and 2010-11 (wet years). Thus, if the deficit was 78% and 82% in the first two years, the second years recorded a surplus of 181% and 81% respectively.

In order to find the meteorological station that correctly reflects the evolution of water resources over time; the water resources at the Torreta station was correlated with the drought indices (SPI) calculated for six meteorological stations (table 4). According to the values of the Pearson correlation, it is evident that the water resources at the Torreta station are better correlated with the rainfall variability at the Ben Karrich meteorological station. The values of the Pearson correlation and determination coefficients (r and r^2) calculated are r = 0.916 and r^2 = 0.84. Thus, the meteorological drought (explanatory variable) explains 84% of the total variance of the dependent variable (water resources) compared to 87% and 73% at the Smir and Laou watersheds respectively [21]. That said rainfall is the determining factor in the fluctuations of surface water resources at the watershed of Martil.

Table 4. Pearson correlation coefficient between water resources at Torreta station and standardized precipitation index (SPI) for six meteorological stations (*n*=45); *p value*<0.0001

	Nakhla dam	Ben Karrich	Beni Hassan	Timzouk	Chibich	Amzal
Correlation coefficient (r)	0.878	0.916	0.846	0.857	0.892	0.884
Determination coefficient (r^2)	0.77	0.84	0.71	0.73	0.79	0.78

Figure 5 shows that a linear relationship between annual standardized precipitation index (SPI) at the Ben Karrich station and water resources at Torreta station is evident. The equation that describes the volume of water resources as a function of the value of the standardized precipitation index (SPI) is the following:

$$Y = 246.6X + 350.7 \qquad (11)$$

where Y for water resources and X being the standardized precipitation index (SPI).

Thus, low streamflow coincide with negative SPI values and high streamflow with positive SPI values. As was the case during 1994-95 and 1995-96, where the first year had a SPI value equal to -2.21 against +2.18 during the second year, i.e. the equivalent of 76.8 million m^3 compared to 984.4 million m^3 respectively.

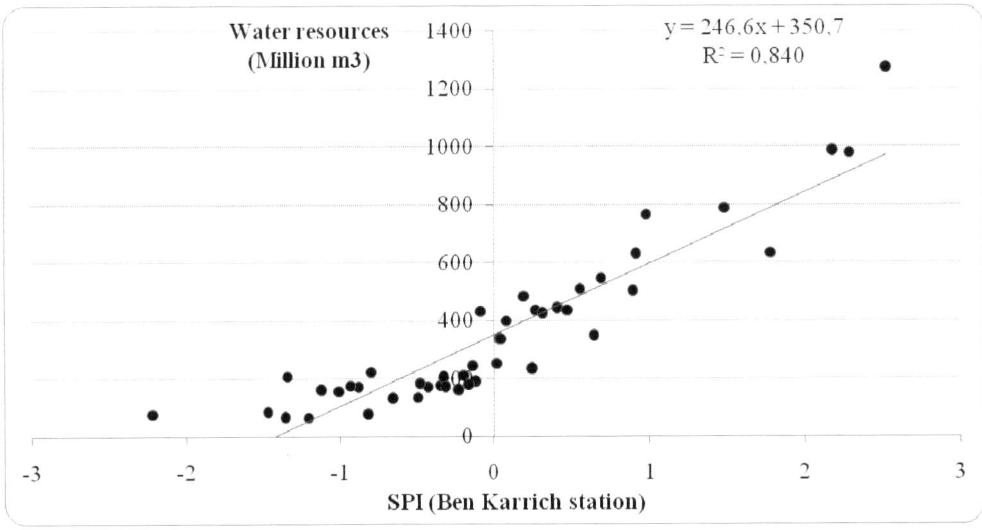

Figure 5. Modeling standardized precipitation index (SPI) - Annual water resources at Torreta station

4. Conclusion

According to this study, although the Martil watershed is located in an area of high rainfall at the national level, experiences periods of rainfall deficit. The results obtained show that the last decade of the 20th century and the beginning of the 21st century have seen the heaviest and most critical droughts in the study area, namely the droughts of 1982-83, 1994-95, 1998-99, 2004-05, 2006-07 and 2011-12.

The meteorological factor explains 84% of the total variance for water resources, whereas the rest 16% of the variance is likely due to other additional factors, such as watershed geology and topography, soil characteristics, and human impacts.

The scarcity of drinking water that the city of Tetouan experienced during the autumn of 2016 is an explanation of the negative impact that meteorological drought can have on water resources. This situation is expected to cause, among other things, disruptions and temporary suspensions of drinking water supply in this region.

The local water infrastructure and water management modes have developed under so-called normal conditions. Thus, any change in the frequency and duration of drought is likely to lead to conflicts between water's users. In addition, given the increasing use of water for agricultural, industrial and urban purposes, a situation of critical water stress can be expected.

In the future, in order to meet water needs, it will be necessary to multiply water supply sources and to implement water resources management plans during drought periods.

5. References

[1] Chbouki N 1994 Une synthèse dendroclimatique du Cèdre de l'Atlas *Annales de la recherche forestière au Maroc* **27** 33-59.

[2] Bzioui M 2004 *Rapport national 2004 sur les ressources en eau au Maroc* (Rabat: UN Water-Africa).

[3] Jellali M 1997 Développement des ressources en eau au Maroc *Options Méditerranéennes A* **31** 51-68.

[4] Verner D, Treguer DO, Redwood J, Christensen JH, Mcdonnell R, Elbert C, Konishi Y and

Belghazi S 2018 *Climate Variability, Drought, and Drought Management in Morocco's Agricultural Sector* (Washington DC: World Bank Group).

[5] Bzioui M 2012 Gestion des ressources en eau au Maroc et changement climatique, *Environnement et Changement Climatique au Maroc: Diagnostic et Perspectives* (Rabat: Konrad-Adenauer-Stiftung) 63-69.

[6] Ennabili A 2005 Problématique de la gestion des ressources en eau au Maroc méditerranéen *Revue AFN* **1** 87-106.

[7] A.B.H.L 2005 *Etude d'actualisation du plan directeur d'aménagement intégré des ressources en eau des bassins Loukkos, Tangérois et Côtiers Méditerranéens: évaluation et projection des besoins et des demandes en eau* (Tétouan, Morocco: Agence du bassin hydraulique du Loukkos).

[8] Province de Tétouan 1977 *Plan quinquennal 1978-1982: monographie de la province de Tétouan* (Tétouan, Morocco: Province de Tétouan).

[9] Karrouchi M, Ouazzani Touhami M, Oujidi M and Chourak M 2016 Cartographie des zones à risque d'inondation dans la région Tanger-Tétouan: Cas du bassin versant de Martil (Nord du Maroc) *International Journal of Innovation and Applied Studies* **14** 1019-35.

[10] El-Khantoury I, Obda K and Achiban H 2020 Wavelet Analysis: A Links between the North Atlantic Oscillation and Winter Drought in the Mediterranean Watersheds of the Western Rif (North Morocco) *European Scientific Journal* ESJ **16** 99-114. doi: 10.19044/esj.2020.v16n15p99

[11] Tsakiris G, Tigkas D, Vangelis H and Pangalou D 2007 Regional drought identification andassessment. Case study in Crete, *Methods and Tools for Drought Analysis and Management* (Dordrecht: Springer) 169-91.

[12] Şen Z 2015 *Applied Drought Modeling, Prediction, and Mitigation* (Amsterdam: Elsevier).

[13] Thom HCS 1958 A note on the gamma distribution *Monthly Weather Review* **86** 117-22.

[14] Zelen M, Severo NC 1964 Probability Functions, *Handbook of Mathematical Functions With Formulas, Graphs, and Mathematical Tables* (New York : Dover Publications) 925-64.

[15] Mckee TB, Doesken NJ and Kleist J 1993 The relationship of drought frequency and duration times scales, *8th Conference on Applied climatology*, American Meteorological Society 179-84.

[16] Svoboda M, Fuchs BA 2016 *Handbook of Drought Indicators and Indices* (Geneva: World Meteorological Organization WMO and Global Water Partnership GWP).

[17] Wilhite DA, Svoboda MD and Hayes MJ 2005 Monitoring drought in the United States: Status and trends, *Monitoring and Predicting Agricultural Drought* (New York: Oxford University Press) 121-131.

[18] Osborn TJ, Briffa KR, Tett SFB, Jones PD and Trigo RM 1999 Evaluation of the North Atlantic Oscillation as simulated by a climate model *Climate Dynamics* **15** 685-702. doi: 10.1007/s003820050310

[19] Trigo RM 2011 The Impacts of the NAO on Hydrological Resources of the Western Mediterranean, *Hydrological, Socioeconomic and Ecological Impacts of the North Atlantic Oscillation in the Mediterranean Region* (Dordrecht: Springer) 41-56.

[20] El-Khantoury I 2021*Caractéristiques et causes des sécheresses hydroclimatiques aux bassins versants méditerranéens du Rif Occidental.* Doctoral Thesis (Sidi Mohamed Ben Abdellah University, Fez-Morocco).

[21] El-Khantoury I, Obda K, Achiban H and Qadem Z 2019 Application de « ISP » pour la quantification de l'impact de la sècheresse météorologique sur les ressources en eau: cas des bassins versants méditerranéens du Rif occidental, *Les ressources naturelles au Maroc : Diagnostic, Exploitation et Outils de Protection* (Oujda: Edition Qurtuba) 55-70.

AUTHOR INDEX